SCIENTIFIC TECHNOLOGY
AND SOCIAL CHANGE

Readings from
**SCIENTIFIC
AMERICAN**

SCIENTIFIC TECHNOLOGY AND SOCIAL CHANGE

With Introductions by
Gene I. Rochlin
University of California, Berkeley

W. H. Freeman and Company
San Francisco

Some of the SCIENTIFIC AMERICAN articles in
SCIENTIFIC TECHNOLOGY AND SOCIAL CHANGE
are available as separate Offprints. For a complete list
of more than 975 articles now available as Offprints,
write to W. H. Freeman and Company, 660 Market
Street, San Francisco, California 94104.

Library of Congress Cataloging in Publication Data

Rochlin, Gene I   comp.
   Scientific technology and social change.

    1. Technology—Social aspects—Addresses, essays,
lectures. I. Scientific American. II. Title.
T14.5.R62   301.24′3   74–3282
ISBN 0–7167–0501–X
ISBN 0–7167–0500–1 (pbk.)

Printed in the United States of America.

9 8 7 6 5 4 3 2 1

PREFACE

The interaction between science, technology, and society has been the subject of many *Scientific American* articles over the years. This collection contains a selection from among these, treating a wide variety of topics from several different points of view. Each of the thirty-one essays by physical and biological scientists, historians, and social scientists explores, in its own way, some aspect of the relations of science and technology to one another and to their social and cultural milieu. I have chosen to restrict this collection to physics, and to the technology based on physics, largely to keep the discussion within reasonable bounds; my choice is meant to reflect no judgment on the relative influences of chemical and biological sciences and technologies. It is my hope that, even within these limits, the articles and introductions will convey some sense of the importance of science-based technology as a component of social change and of the futility of trying to determine a social future without bringing technology under social control.

The articles are arranged primarily by the historical order of their subject matter, with a secondary grouping by topic. Part I chronicles the era of empirical technology, when technics were nearly independent of science. Part II treats the coupling of science and technology in the nineteenth century, whereas the subsequent rise and ultimate triumph of scientific technology are examined in Part III. Parts IV and V deal with modern technology and its ultimate physical limits—energy as a resource and heat as a waste product. In choosing these articles from more than one hundred possibilities, I have tried to select in favor of general interest and comprehensiveness. This necessarily has resulted in some gaps in the presentation and the inclusion of some articles that may no longer reflect the current views of their authors. I hope they, and the reader, will nevertheless find this volume to be of value. John Heilbron gave the introductions a thorough scrutiny and made several valuable suggestions, some of which were followed; the responsibility for such errors or oversights as may remain is entirely mine. I also thank Anne Middleton for her patient ear and many helpful comments, and Linda Billard, whose skilled and rapid typing and constructive proofreading sped the completion of this book enormously.

January 1974   Gene I. Rochlin

CONTENTS

Note on cross-references: References to articles included in this book are noted by the title of the article and the page on which it begins; references to articles that are available as Offprints, but are not included here, are noted by the article's title and Offprint number; references to articles published by SCIENTIFIC AMERICAN, but which are not available as Offprints, are noted by the title of the article and the month and year of its publication.

SCIENTIFIC TECHNOLOGY
AND SOCIAL CHANGE

INTRODUCTORY ESSAY

Human knowledge and human power meet in one; for where the cause is not known, the effect cannot be produced. Nature to be commanded must be obeyed, and that which in contemplation is as the cause is in operation as the rule.

Francis Bacon
Novum Organum, Aphorism III

Science and technology have been important factors in shaping human society and culture from prehistoric times, but until recently they have served quite different purposes. In the past, technology sought to manipulate and control the physical world, whereas science attempted primarily to comprehend it. That these separate efforts could be conjoined to forge a tool of nearly irresistible force—a technology based on science—was a realization born of the scientific revolution and brought to fruition by nineteenth-century industrialization. With this new tool, the world could be molded into almost any desired shape—except back into its original one. For the cost of technological intervention has been very high, and each succeeding "technological fix" only creates a newer and more extensive disruption.

The present crises in energy supply, resource depletion, environmental pollution, and arms control are each in their way consequences of a lack of technical restraint, of the almost uncontrolled growth and careless use of modern science and technology. It seems at times as if our machines have become our masters against whom we must ultimately rebel. Yet, since our civilization is based largely on scientific technology, it is now as necessary for our survival as air or water. Human beings are part of their environment—this much, at least, has been rediscovered in recent years—and the machine has become part of the human environment as well. The alternative to a careful and critical analysis of both our present dependence on science and technology and the route by which we came to that dependence will very probably be destruction. And whether it be destruction of humanity by the machine, or destruction of the machine by humanity, our present civilization will not survive. We must learn to reassert control over the scientific-technological forces we have unleashed. But without understanding, there can be no control, and without knowledge, no understanding.

The main purpose of this collection of articles from *Scientific American* is to explore the development, growth, and impact of science-oriented technology, in the hope of providing some of the knowledge that can lead to an understanding of its importance to society. Although the thirty-one articles I have chosen are concerned primarily with technology based on physics, this choice is not meant to be a judgment on the relative importance or impact of other technologies. The long-range effects of intervention in the biosphere with chemically and biologically active substances is one of the most pressing problems of our time, and industries based on chemical and biological science and technology have been among the most influential in determining the character of modern society. Indeed, these areas have been treated extensively in the pages of *Scientific American* and in recent collections of *Scientific American* readings. Physical technology, however, has the longest, best documented, and most continuous history; and even a selection of articles taken from this single thread may serve to reveal the fabric of technological society.

The articles included in this volume discuss: (1) the growth of *ad hoc*, empirical technology, which proceeded without the guiding light of science; (2) the coupling of science and technology by the demands of the Industrial Revolution; (3) the growth of the new science-oriented technologies into a preeminent social and economic force in the nineteenth and twentieth centuries; (4) the ultimate physical limits of a highly technologized society—the consumption of energy; and (5) the production of heat, which is, of all our problems, the least susceptible to technological fixes. Over the years, *Scientific American* has carried a wide-ranging collection of articles on these and related topics. The articles I have selected are neither comprehensive nor wholly representative; they have been chosen to illustrate specific key features of the growth and character of modern, physics-oriented, scientific technology, and its influence on shaping our present culture.

In the 1950s, when physics and technology produced the H-bomb, many feared that there might not be a future. In the 1960s, there was widespread concern over the accelerating growth of world population due to advances in medical and agricultural science and technology, a concern over the quality of the future we could foresee. In some respects, our chances for a livable future now seem better. We are still precariously perched on the razor's edge of the nuclear arms race, but our balance appears to have improved. The population is still growing, but at least the rate of growth appears to have stopped increasing.

With the growing possibility of an intact, stabilized world in the next century, the issues of what that world will look like, and whether our supply and use of energy can be stabilized to maintain an adequate quality of life have become serious concerns. Nuclear disarmament and limits to growth will be of little value unless gross inequities in the distribution of energy and goods can be removed, and adequate living conditions can be created worldwide. We ourselves must learn to use less and share more of the earth's resources, for we can not afford to maintain permanently our present overconsumption, nor are we likely to be allowed to do so.

An improvement in the world situation will not discredit those who have been predicting disaster over the past decades. These prophets have lacked honor, not influence, in their own country. Without their warnings, neither the arms race, population growth, environmental pollution, nor ecological destruction could have been controlled even as minimally as they now are. But even a Jeremiah sometimes grows hoarse as "crises" crowd closer and closer under the accelerated pace of technical change. We must evolve a better way to avoid or resolve these issues until the acceleration itself can be halted, until technology can be brought under effective social control. The methods for achieving these goals do not exist; the means are not yet at hand; the will to act is still in the formative stage. But it is my hope that this volume will be of some use to the growing number of persons who are actively seeking to develop those methods, to strengthen that will.

Francis Bacon has often been considered the intellectual progenitor of scientific technology, but his aphorisms have frequently been inverted and misapplied. Power must be derived from knowledge; nature must first be obeyed and not merely commanded. Science and technology can and must be properly used to extricate us from the labyrinthine dilemmas of technical overdevelopment and runaway growth. But they must be used wisely and with a regard to the finite resources and fragile ecology of this very small and still beautiful planet.

I

TECHNICS AND EMPIRICAL TECHNOLOGY

The fossil records of the human race and its tools are so closely linked that, until recently, the archaeological discovery of tools or remnants of a controlled fire was considered sufficient evidence to attribute the site to Homo sapiens *or a closely related ancestral hominid. The history of our species, our technics, and our civilization are inextricably intertwined; and our ability to manipulate and control our environment differentiates us from all other life on earth. Yet, for all except the last ten of the five thousand or more human generations, the development of the tools and machines that facilitated this control proceeded on a purely empirical basis.*

EARLY TECHNICS   I

INTRODUCTION

*We recognize as cultural all activities and resources which are useful to
men for making the earth serviceable to them, for protecting them against
the violence of the forces of nature, and so on. As regards this side of
civilization, there can be scarcely any doubt. If we go back far enough,
we find that the first acts of civilization were the use of tools, the gaining
of control over fire, and the construction of dwellings.*

Sigmund Freud
Civilization and Its Discontents

The importance of technics, the control of nature through the creation
and use of tools and related crafts, has been recognized since the dawn
of civilization. As a subject for speculative thought, the how and why
of the dominion of the human race over its environment and other species
of animals has given precedence only to the question of the creation of the
world itself. One of the most enduring legends of technics is that of Prometheus, who not only stole fire from the gods but also taught the arts of domestication and invention. In *Prometheus Bound*, Aeschylus describes the condition
of humanity before the arrival of the Promethean gifts:

> For seeing they saw not, and hearing they understood not, but like as shapes
> in a dream they wrought all the days of their life in confusion. No houses of
> brick raised in the warmth of the sun they had, nor fabrics of wood, but like
> the little ants they dwelt underground in the sunless depth of caverns.

There is still a tendency, despite the work of many cultural anthropologists,
to consider primitive peoples subhuman, purely on the basis of their inferior
technical skill. We connect ourselves so firmly with our artifacts that it has
even been proposed that our correct species name should be *Homo faber*, the
maker of tools. It is most appropriate, then, that this collection should open
with "Tools and Human Evolution," by the noted anthropologist Sherwood
L. Washburn. Although it has long been recognized that toolmaking was one
of the most important factors in the evolution of culture, new archaeological
evidence indicates that it may also have been of primary importance in our
physical evolution. The remarkable and controversial interpretation of the
Olduvai Gorge excavations of L.S.B. and Mary Leakey in 1959 caused a reexamination of Charles Darwin's hypothesis that the use of tools was not
merely a consequence of an erect posture, but one of its causes. Washburn
argues persuasively that the physical evolution of genus *Homo* has been determined in large part by continuing selection pressure that favored the more
adept user of tools. The use of tools by early hominids became in effect a
self-determined environmental pressure that tended to lower the survival

value of sharp teeth, powerful jaws, an armored skull, and quadrupedal stability in comparison to the advantages of the opposable thumb, the extended cortex, and the freeing of the hands by an upright, bipedal posture.

Critics of this view have often questioned the importance attached to the notion of *Homo faber*, the toolmaking animal. They argue that our species, and perhaps our genus, is differentiated from others primarily by our capacity for introspection, by our search for self-discovery, self-mastery, and self-transformation. In their view, the preeminently human trait is not the understanding of such externalities as tools or the forces of nature, but the comprehension of self. The polarization of this argument was certainly not apparent to early writers. Aeschylus saw the two positions as being totally interdependent, as did most pre-Renaissance philosophers, authors, scientists, and poets. Accepting the importance of the process of growth of consciousness as a determining factor in evolution does not derogate the significance of the discovery of tools, which undoubtedly played a large part in raising that consciousness and nurturing it. The growth of individuation of self and control of nature were mutually reinforcing, and the evidence of Olduvai and other sites shows that the development of society and culture is intimately linked with technical development.

The history of technics, however, encompasses far more than just the study of tools and their use. A tool is only an extension of an innate animal capability. Civilization and culture began with the discovery of machines, in the general sense, of devices able to perform functions that a single human body cannot perform. Fire, being generally available, was not so much invented or discovered as it was controlled. The earliest inventions, such as housing, clothing, domestication, or the wheel, attest to a unique capability for abstraction that goes far beyond the purely mechanical and that truly characterizes *Homo sapiens* alone. Some machines, perhaps most, arise more from the propensity for luxury and comfort (relatively speaking) than from necessity. The invention of the wheel occupies a unique place in the history of technics because it is so early an example of such a luxury.

In his essay on "The Beginnings of Wheeled Transport," Stuart Piggott takes us on a tour of Mesopotamia and Eurasia in the pursuit of the origin of the wheel. The evidence of the New World cultures shows that the invention of wheeled transport did not arise as a result of direct survival pressure. It may therefore be taken to be one of the first examples of ingenious creation of a totally new device that was developed, not from necessity, but from a synthesis of the available materials and perceived needs of a culture by the action of creative human intelligence. The wheel is also an excellent case study for comparing the bases of the two leading theories regarding the spread of early technics. The diffusionist theory holds that inventions occur as isolated events and spread from their point of inception by cultural contact and export. The local-origins theory suggests multiple invention at different times and places according to local conditions. In subtly varying forms, these themes are also woven repeatedly into discussions of the origins of technology and science. The diffusionist theory, in its most extreme form, attributes human progress in such endeavors to the occurrence from time to time of singular geniuses whose ideas remake our vision of the world. The local-origins theory, taken to its extreme form, holds that scientific and technical change are largely the result of social development and cultural milieu; it views even such acknowledged luminaries as Isaac Newton as having merely seized the opportunity provided by their age.

In our own era, such questions have come to be of great importance in dealing with technological change. In the struggle to control modern technology, the battle has frequently been joined over such issues as whether the production and use of a new invention is inevitable and therefore unstoppable; whether we *must* do what we *can* do (the technological imperative); whether each of us must consistently strive to be the first to invent new methods of destruction because what can be invented will be. The story of the gift of Prometheus (forethought) is not the entire legend. Zeus sought revenge on the human race as well and fashioned Pandora (the all-gifted) as a punishment. Thus Pandora's box of evils from the gods and Prometheus' gift of technics have been connected since antiquity. There also remained one ameliorating gift when the sorrows and plagues had flown out—hope. Yet, hope alone is not enough—at times it even seems to have been one of the plagues. The problems raised by modern technology and science will not "hopefully" disappear, nor can we merely hope for solutions to nuclear war and environmental disaster. We must ask ourselves whether we are separable from our technology; whether we can limit technical and economic growth and still remain a viable and dynamic species; whether we can choose *not* to invent a new device for fear of its destructive capability. And only through an examination of our history, by a study of how our society has developed, can we begin to construct a rational basis for discussion, a way to formulate these questions.

Tools and Human Evolution

by Sherwood L. Washburn
September 1960

*It is now clear that tools antedate man, and that their
use by prehuman primates gave rise to Homo sapiens*

A series of recent discoveries has linked prehuman primates of half a million years ago with stone tools. For some years investigators had been uncovering tools of the simplest kind from ancient deposits in Africa. At first they assumed that these tools constituted evidence of the existence of large-brained, fully bipedal men. Now the tools have been found in association with much more primitive creatures, the not-fully bipedal, small-brained near-men, or man-apes. Prior to these finds the prevailing view held that man evolved nearly to his present structural state and then discovered tools and the new ways of life that they made possible. Now it appears that man-apes—creatures able to run but not yet walk on two legs, and with brains no larger than those of apes now living—had already learned to make and to use tools. It follows that the structure of modern man must be the result of the change in the terms of natural selection that came with the tool-using way of life.

The earliest stone tools are chips or simple pebbles, usually from river gravels. Many of them have not been shaped at all, and they can be identified as tools only because they appear in concentrations, along with a few worked pieces, in caves or other locations where no such stones naturally occur. The huge advantage that a stone tool gives to its user must be tried to be appreciated. Held in the hand, it can be used for pounding, digging or scraping. Flesh and bone can be cut with a flaked chip, and what would be a mild blow with the fist becomes lethal with a rock in the hand. Stone tools can be employed, moreover, to make tools of other materials. Naturally occurring sticks are nearly all rotten, too large, or of inconvenient shape; some tool for fabrication is essential for the efficient use of wood. The utility of a mere pebble seems so limited to the user of modern tools that it is not easy to comprehend the vast difference that separates the tool-user from the ape which relies on hands and teeth alone. Ground-living monkeys dig out roots for food, and if they could use a stone or a stick, they might easily double their food supply. It was the success of the simplest tools that started the whole trend of human evolution and led to the civilizations of today.

From the short-term point of view, human structure makes human behavior possible. From the evolutionary point of view, behavior and structure form an interacting complex, with each change in one affecting the other. Man began when populations of apes, about a mil-

lion years ago, started the bipedal, tool-using way of life that gave rise to the man-apes of the genus *Australopithecus*. Most of the obvious differences that distinguish man from ape came after the use of tools.

The primary evidence for the new view of human evolution is teeth, bones and tools. But our ancestors were not fossils; they were striving creatures, full of rage, dominance and the will to live. What evolved was the pattern of life of intelligent, exploratory, playful, vigorous primates; the evolving reality was a succession of social systems based upon the motor abilities, emotions and intelligence of their members. Selection produced new systems of child care, maturation and sex, just as it did alterations in the skull and the teeth. Tools, hunting, fire, complex social life, speech, the human way and the brain evolved together to produce ancient man of the genus *Homo* about half a million years ago. Then the brain evolved under the pressures of more complex social life until the species *Homo sapiens* appeared perhaps as recently as 50,000 years ago.

With the advent of *Homo sapiens* the tempo of technical-social evolution quickened. Some of the early types of tool had lasted for hundreds of thousands of years and were essentially the same throughout vast areas of the African and Eurasian land masses. Now the tool forms multiplied and became regionally diversified. Man invented the

STENCILED HANDS in the cave of Gargas in the Pyrenees date back to the Upper Paleolithic of perhaps 30,000 years ago. Aurignacian man made the images by placing hand against wall and spattering it with paint. Hands stenciled in black (*top*) are more distinct and apparently more recent than those done in other colors (*center*).

OLDUVAI GORGE in Tanganyika is the site where the skull of the largest known man-ape was discovered in 1959 by L. S. B. Leakey and his wife Mary. Stratigraphic evidence indicates that skull dates back to Lower Pleistocene, more than 500,000 years ago.

bow, boats, clothing; conquered the Arctic; invaded the New World; domesticated plants and animals; discovered metals, writing and civilization. Today, in the midst of the latest tool-making revolution, man has achieved the capacity to adapt his environment to his need and impulse, and his numbers have begun to crowd the planet.

The later events in the evolution of the human species are treated in other articles in the September 1959 issue of SCIENTIFIC AMERICAN. This article is concerned with the beginnings of the process by which, as Theodosius Dobzhansky says in the concluding article of the issue, biological evolution has transcended itself. From the rapidly accumulating evidence it is now possible to speculate with some confidence on the manner in which the way of life made possible by tools changed the pressures of natural selection and so changed the structure of man.

Tools have been found, along with the bones of their makers, at Sterkfontein, Swartkrans and Kromdraai in South Africa and at Olduvai in Tanganyika. Many of the tools from Sterkfontein are merely unworked river pebbles, but someone had to carry them from the gravels some miles away and bring them to the deposit in which they are found. Nothing like them occurs naturally in the local limestone caves. Of course the association of the stone tools with man-ape bones in one or two localities does not prove that these animals made the tools. It has been argued that a more advanced form of man, already present, was the toolmaker. This argument has a familiar ring to students of human evolution. Peking man was thought too primitive to be a toolmaker; when the first manlike pelvis was found with man-ape bones, some argued that it must have fallen into the deposit because it was too human to be associated with the skull. In every case, however, the repeated discovery of the same unanticipated association has ultimately settled the controversy.

This is why the discovery by L. S. B. and Mary Leakey in the summer of 1959 is so important. In Olduvai Gorge in Tanganyika they came upon traces of an old living site, and found stone tools in clear association with the largest man-ape skull known. With the stone tools were a hammer stone and waste flakes from the manufacture of the tools. The deposit also contained the bones of rats, mice, frogs and some bones of juvenile pig and antelope, showing that even the largest and latest of the

SKULL IS EXAMINED *in situ* by Mary Leakey, who first noticed fragments of it protruding from the cliff face at left. Pebble tools were found at the same level as the skull.

SKULL IS EXCAVATED from surrounding rock with dental picks. Although skull was badly fragmented, almost all of it was recovered. Fragment visible here is part of upper jaw.

14

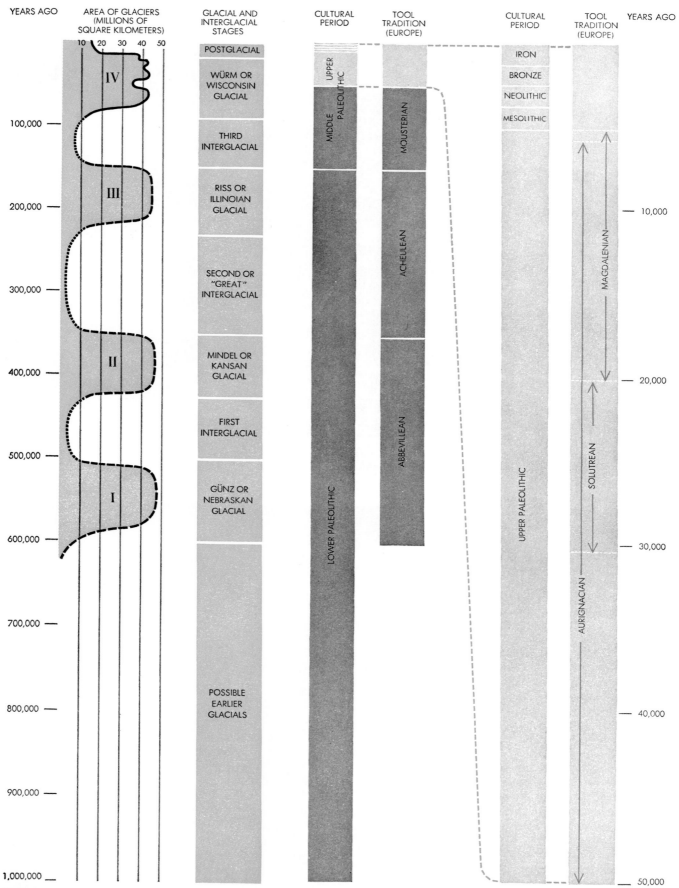

| YEARS AGO | AREA OF GLACIERS (MILLIONS OF SQUARE KILOMETERS) | GLACIAL AND INTERGLACIAL STAGES | CULTURAL PERIOD | TOOL TRADITION (EUROPE) | CULTURAL PERIOD | TOOL TRADITION (EUROPE) | YEARS AGO |

TIME-SCALE correlates cultural periods and tool traditions with the four great glaciations of the Pleistocene epoch. Glacial advances and retreats shown by solid black curve are accurately known; those shown by broken curve are less certain; those shown by dotted curve are uncertain. Light gray bars at far right show an expanded view of last 50,000 years on two darker bars at center. Scale was prepared with the assistance of William R. Farrand of the Lamont Geological Observatory of Columbia University.

man-apes could kill only the smallest animals and must have been largely vegetarian. The Leakeys' discovery confirms the association of the man-ape with pebble tools, and adds the evidence of manufacture to that of mere association. Moreover, the stratigraphic evidence at Olduvai now for the first time securely dates the man-apes, placing them in the lower Pleistocene, earlier than 500,000 years ago and earlier than the first skeletal and cultural evidence for the existence of the genus Homo [*see illustration on next two pages*]. Before the discovery at Olduvai these points had been in doubt.

The man-apes themselves are known from several skulls and a large number of teeth and jaws, but only fragments of the rest of the skeleton have been preserved. There were two kinds of man-ape, a small early one that may have weighed 50 or 60 pounds and a later and larger one that weighed at least twice as much. The differences in size and form between the two types are quite comparable to the differences between the contemporary pygmy chimpanzee and the common chimpanzee.

Pelvic remains from both forms of man-ape show that these animals were bipedal. From a comparison of the pelvis of ape, man-ape and man it can be seen that the upper part of the pelvis is much wider and shorter in man than in the ape, and that the pelvis of the man-ape corresponds closely, though not precisely, to that of modern man [*see top illustration on page 19*]. The long upper pelvis of the ape is characteristic of most mammals, and it is the highly specialized, short, wide bone in man that makes possible the human kind of bipedal locomotion. Although the man-ape pelvis is apelike in its lower part, it approaches that of man in just those features that distinguish man from all other animals. More work must be done before this combination of features is fully understood. My belief is that bipedal running, made possible by the changes in the upper pelvis, came before efficient bipedal walking, made possible by the changes in the lower pelvis. In the man-ape, therefore, the adaptation to bipedal locomotion is not yet complete. Here, then, is a phase of human evolution characterized by forms that are mostly bipedal, small-brained, plains-living, tool-making hunters of small animals.

The capacity for bipedal walking is primarily an adaptation for covering long distances. Even the arboreal chimpanzee can run faster than a man, and any monkey can easily outdistance him.

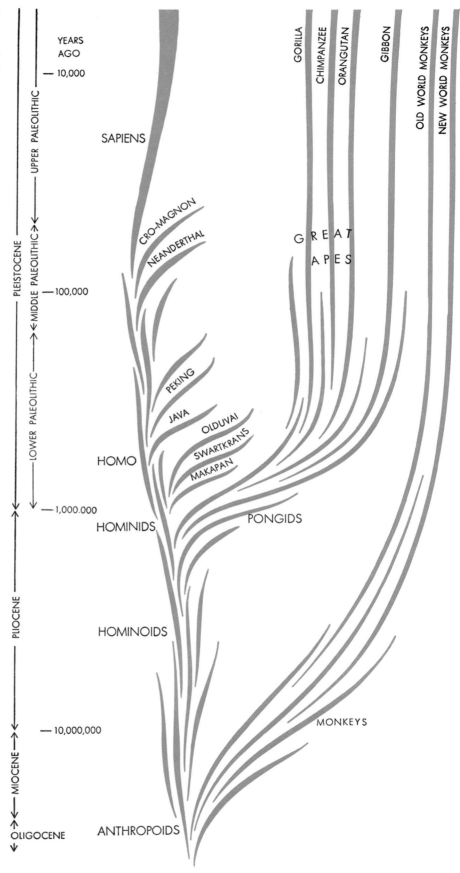

LINES OF DESCENT that lead to man and his closer living relatives are charted. The hominoid superfamily diverged from the anthropoid line in the Miocene period some 20 million years ago. From the hominoid line came the tool-using hominids at the beginning of the Pleistocene. The genus *Homo* appeared in the hominid line during the first interglacial (*see chart on opposite page*); the species *Homo sapiens*, around 50,000 years ago.

FOSSIL SKULLS of Pleistocene epoch reflect transition from man-apes (*below black line*) to *Homo sapiens* (*top*). Relative age of intermediate specimens is indicated schematically by their posi- tion on page. Java man (*middle left*) and Solo man (*upper center*) are members of the genus *Pithecanthropus,* and are related to Peking man (*middle right*). The Shanidar skull (*upper left*) be-

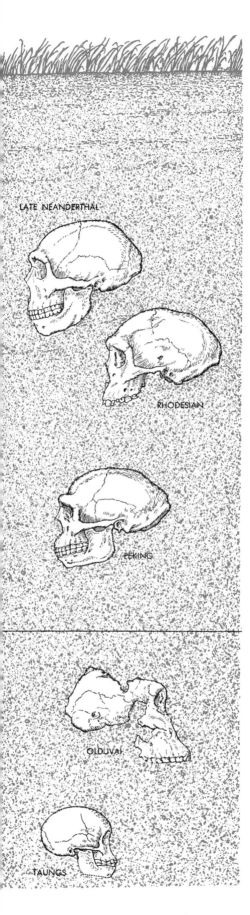

longs to the Neanderthal family, while Mount Carmel skull shows characteristics of Neanderthal and modern man.

A man, on the other hand, can walk for many miles, and this is essential for efficient hunting. According to skeletal evidence, fully developed walkers first appeared in the ancient men who inhabited the Old World from 500,000 years ago to the middle of the last glaciation. These men were competent hunters, as is shown by the bones of the large animals they killed. But they also used fire and made complicated tools according to clearly defined traditions. Along with the change in the structure of the pelvis, the brain had doubled in size since the time of the man-apes.

The fossil record thus substantiates the suggestion, first made by Charles Darwin, that tool use is both the cause and the effect of bipedal locomotion. Some very limited bipedalism left the hands sufficiently free from locomotor functions so that stones or sticks could be carried, played with and used. The advantage that these objects gave to their users led both to more bipedalism and to more efficient tool use. English lacks any neat expression for this sort of situation, forcing us to speak of cause and effect as if they were separated, whereas in natural selection cause and effect are interrelated. Selection is based on successful behavior, and in the man-apes the beginnings of the human way of life depended on both inherited locomotor capacity and on the learned skills of tool-using. The success of the new way of life based on the use of tools changed the selection pressures on many parts of the body, notably the teeth, hands and brain, as well as on the pelvis. But it must be remembered that selection was for the whole way of life.

In all the apes and monkeys the males have large canine teeth. The long upper canine cuts against the first lower premolar, and the lower canine passes in front of the upper canine. This is an efficient fighting mechanism, backed by very large jaw muscles. I have seen male baboons drive off cheetahs and dogs, and according to reliable reports male baboons have even put leopards to flight. The females have small canines, and they hurry away with the young under the very conditions in which the males turn to fight. All the evidence from living monkeys and apes suggests that the male's large canines are of the greatest importance to the survival of the group, and that they are particularly important in ground-living forms that may not be able to climb to safety in the trees. The small, early man-apes lived in open plains country, and yet none of them had large canine teeth. It would appear that the protection of the group must have shifted from teeth to tools early in the evolution of the man-apes, and long before the appearance of the forms that have been found in association with stone tools. The tools of Sterkfontein and Olduvai represent not the beginnings of tool use, but a choice of material and knowledge in manufacture which, as is shown by the small canines of the man-apes that deposited them there, derived from a long history of tool use.

Reduction in the canine teeth is not a simple matter, but involves changes in the muscles, face, jaws and other parts of the skull. Selection builds powerful neck muscles in animals that fight with their canines, and adapts the skull to the action of these muscles. Fighting is not a matter of teeth alone, but also of seizing, shaking and hurling an enemy's body with the jaws, head and neck. Reduction in the canines is therefore accompanied by a shortening in the jaws, reduction in the ridges of bone over the eyes and a decrease in the shelf of bone in the neck area [see illustration on page 20]. The reason that the skulls of the females and young of the apes look more like man-apes than those of adult males is that, along with small canines, they have smaller muscles and all the numerous structural features that go along with them. The skull of the man-ape is that of an ape that has lost the structure for effective fighting with its teeth. Moreover, the man-ape has transferred to its hands the functions of seizing and pulling, and this has been attended by reduction of its incisors. Small canines and incisors are biological symbols of a changed way of life; their primitive functions are replaced by hand and tool.

The history of the grinding teeth—the molars—is different from that of the seizing and fighting teeth. Large size in any anatomical structure must be maintained by positive selection; the selection pressure changed first on the canine teeth and, much later, on the molars. In the man-apes the molars were very large, larger than in either ape or man. They were heavily worn, possibly because food dug from the ground with the aid of tools was very abrasive. With the men of the Middle Pleistocene, molars of human size appear along with complicated tools, hunting and fire.

The disappearance of brow ridges and the refinement of the human face may involve still another factor. One of the essential conditions for the organi-

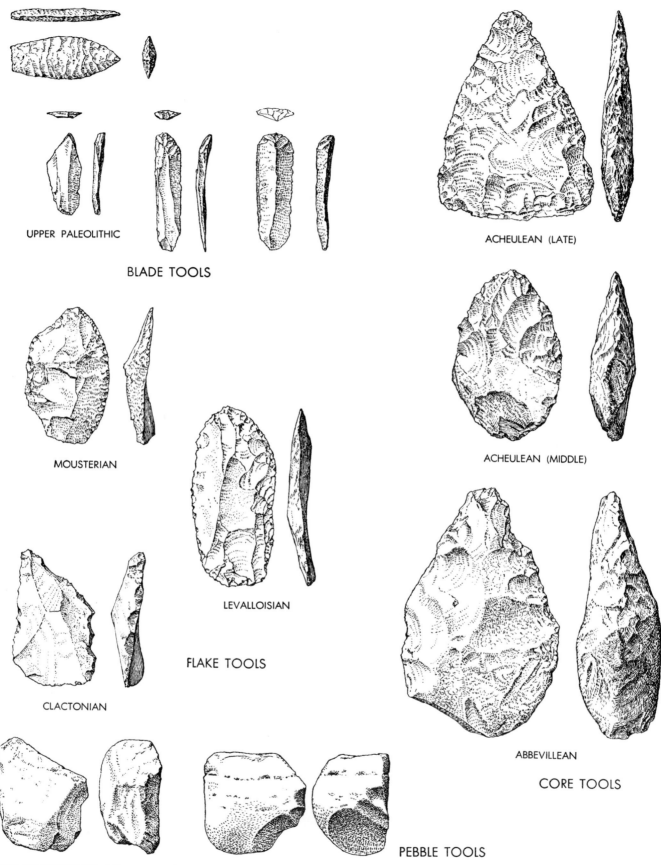

UPPER PALEOLITHIC

BLADE TOOLS

ACHEULEAN (LATE)

MOUSTERIAN

LEVALLOISIAN

FLAKE TOOLS

ACHEULEAN (MIDDLE)

CLACTONIAN

ABBEVILLEAN

CORE TOOLS

PEBBLE TOOLS

TOOL TRADITIONS of Europe are the main basis for classifying Paleolithic cultures. The earliest tools are shown at bottom of page; later ones, at top. The tools are shown from both the side and the edge, except for blade tools, which are shown in three views. Tools consisting of a piece of stone from which a few flakes have been chipped are called core tools (*right*). Other types of tool were made from flakes (*center and left*); blade tools were made from flakes with almost parallel sides. Tool traditions are named for site where tools of a given type were discovered; Acheulean tools, for example, are named for St. Acheul in France.

zation of men in co-operative societies was the suppression of rage and of the uncontrolled drive to first place in the hierarchy of dominance. Curt P. Richter of Johns Hopkins University has shown that domestic animals, chosen over the generations for willingness to adjust and for lack of rage, have relatively small adrenal glands. But the breeders who selected for this hormonal, physiological, temperamental type also picked, without realizing it, animals with small brow ridges and small faces. The skull structure of the wild rat bears the same relation to that of the tame rat as does the skull of Neanderthal man to that of *Homo sapiens*. The same is true for the cat, dog, pig, horse and cow; in each case the wild form has the larger face and muscular ridges. In the later stages of human evolution, it appears, the self-domestication of man has been exerting the same effects upon temperament, glands and skull that are seen in the domestic animals.

Of course from man-ape to man the brain-containing part of the skull has also increased greatly in size. This change is directly due to the increase in the size of the brain: as the brain grows, so grow the bones that cover it. Since there is this close correlation between brain size and bony brain-case, the brain size of the fossils can be estimated. On the scale of brain size the man-apes are scarcely distinguishable from the living apes, although their brains may have been larger with respect to body size. The brain seems to have evolved rapidly, doubling in size between man-ape and man. It then appears to have increased much more slowly; there is no substantial change in gross size during the last 100,000 years. One must remember, however, that size alone is a very crude indicator, and that brains of equal size may vary greatly in function. My belief is that although the brain of *Homo sapiens* is no larger than that of Neanderthal man, the indirect evidence strongly suggests that the first *Homo sapiens* was a much more intelligent creature.

The great increase in brain size is important because many functions of the brain seem to depend on the number of cells, and the number increases with volume. But certain parts of the brain have increased in size much more than others. As functional maps of the cortex of the brain show, the human sensory-motor cortex is not just an enlargement of that of an ape [*see illustrations on last three pages of this article*]. The areas

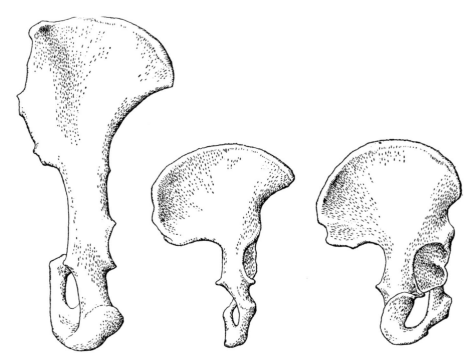

HIP BONES of ape (*left*), man-ape (*center*) and man (*right*) reflect differences between quadruped and biped. Upper part of human pelvis is wider and shorter than that of apes. Lower part of man-ape pelvis resembles that of ape; upper part resembles that of man.

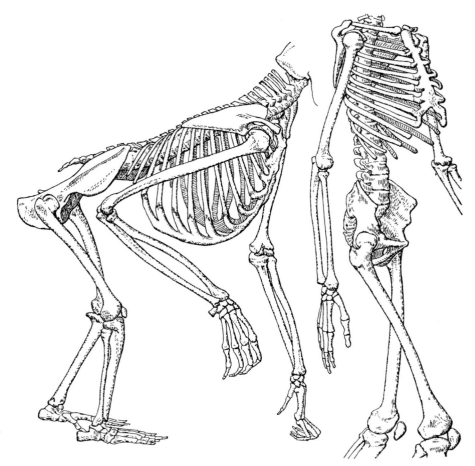

POSTURE of gorilla (*left*) and man (*right*) is related to size, shape and orientation of pelvis. Long, straight pelvis of ape provides support for quadrupedal locomotion; short, broad pelvis of man curves backward, carrying spine and torso in bipedal position.

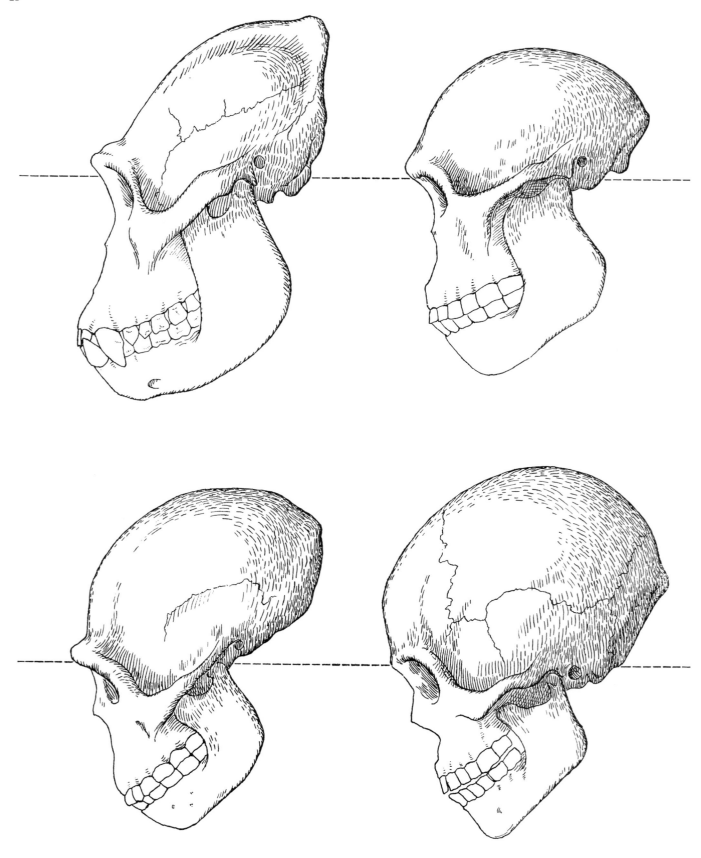

EVOLUTION OF SKULL from ape (*upper left*) to man-ape (*upper right*) to ancient man (*lower left*) to modern man (*lower right*) involves an increase in size of brain case (*part of skull above broken lines*) and a corresponding decrease in size of face (*part of skull below broken lines*). Apes also possess canine teeth that are much larger than those found in either man-apes or man.

for the hand, especially the thumb, in man are tremendously enlarged, and this is an integral part of the structural base that makes the skillful use of the hand possible. The selection pressures that favored a large thumb also favored a large cortical area to receive sensations from the thumb and to control its motor activity. Evolution favored the development of a sensitive, powerful, skillful thumb, and in all these ways —as well as in structure—a human thumb differs from that of an ape.

The same is true for other cortical areas. Much of the cortex in a monkey is still engaged in the motor and sensory functions. In man it is the areas adjacent to the primary centers that are most expanded. These areas are concerned with skills, memory, foresight and language; that is, with the mental faculties that make human social life possible. This is easiest to illustrate in the field of language. Many apes and monkeys can make a wide variety of sounds. These sounds do not, however, develop into language [see "The Origin of Speech," by Charles F. Hockett, Offprint 603]. Some workers have devoted great efforts, with minimum results, to trying to teach chimpanzees to talk. The reason is that there is little in the brain to teach. A human child learns to speak with the greatest ease, but the storage of thousands of words takes a great deal of cortex. Even the simplest language must have given great advantage to those first men who had it. One is tempted to think that language may have appeared together with the fine tools, fire and complex hunting of the large-brained men of the Middle Pleistocene, but there is no direct proof of this.

The main point is that the kind of animal that can learn to adjust to complex, human, technical society is a very different creature from a tree-living ape, and the differences between the two are rooted in the evolutionary process. The reason that the human brain makes the human way of life possible is that it is the result of that way of life. Great masses of the tissue in the human brain are devoted to memory, planning, language and skills, because these are the abilities favored by the human way of life.

The emergence of man's large brain occasioned a profound change in the plan of human reproduction. The human mother-child relationship is unique among the primates as is the use of tools. In all the apes and monkeys the baby clings to the mother; to be able to do so,

MOTOR CORTEX OF MONKEY controls the movements of the body parts outlined by the superimposed drawing of the animal (*color*). Gray lines trace the surface features of the left half of the brain (*bottom*) and part of the right half (*top*). Colored drawing is distorted in proportion to amount of cortex associated with functions of various parts of the body. Smaller animal in right half of brain indicates location of secondary motor cortex.

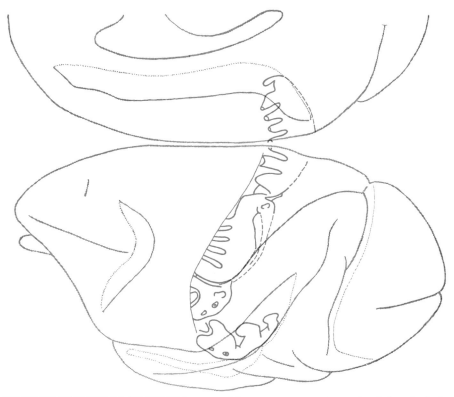

SENSORY CORTEX OF MONKEY is mapped in same way as motor cortex (*above*). As in motor cortex, a large area is associated with hands and feet. Smaller animal at bottom of left half of brain indicates location of secondary sensory cortex. Drawings are based on work of Clinton N. Woolsey and his colleagues at the University of Wisconsin Medical School.

the baby must be born with its central nervous system in an advanced state of development. But the brain of the fetus must be small enough so that birth may take place. In man adaptation to bipedal locomotion decreased the size of the bony birth-canal at the same time that the exigencies of tool use selected for larger brains. This obstetrical dilemma was solved by delivery of the fetus at a much earlier stage of development. But this was possible only because the mother, already bipedal and with hands free of locomotor necessities, could hold the helpless, immature in-

fant. The small-brained man-ape probably developed in the uterus as much as the ape does; the human type of mother-child relation must have evolved by the time of the large-brained, fully bipedal humans of the Middle Pleistocene. Bipedalism, tool use and selection for large brains thus slowed human development and invoked far greater maternal responsibility. The slow-moving mother, carrying the baby, could not hunt, and the combination of the woman's obligation to care for slow-developing babies and the man's occupation of hunting imposed a fundamental pat-

tern on the social organization of the human species.

As Marshall D. Sahlins suggests ["The Origin of Society," SCIENTIFIC AMERICAN Offprint 602], human society was heavily conditioned at the outset by still other significant aspects of man's sexual adaptation. In the monkeys and apes year-round sexual activity supplies the social bond that unites the primate horde. But sex in these species is still subject to physiological — especially glandular — controls. In man these controls are gone, and are replaced by a bewildering variety of social customs. In no other primate does

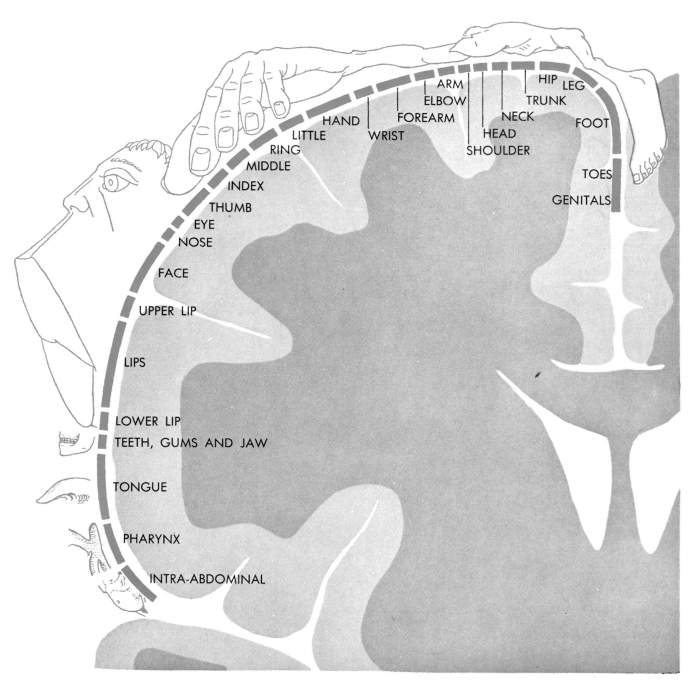

SENSORY HOMUNCULUS is a functional map of the sensory cortex of the human brain worked out by Wilder Penfield and his associates at the Montreal Neurological Institute. As in the map of the sensory cortex of the monkey that appears on the preceding page, the distorted anatomical drawing (*color*) indicates the areas of the sensory cortex associated with the various parts of the body.

a family exist that controls sexual activity by custom, that takes care of slow-growing young, and in which—as in the case of primitive human societies—the male and female provide different foods for the family members.

All these family functions are ultimately related to tools, hunting and the enlargement of the brain. Complex and technical society evolved from the sporadic tool-using of an ape, through the simple pebble tools of the man-ape and the complex toolmaking traditions of ancient men to the hugely complicated culture of modern man. Each behavioral

stage was both cause and effect of biological change in bones and brain. These concomitant changes can be seen in the scanty fossil record and can be inferred from the study of the living forms.

Surely as more fossils are found these ideas will be tested. New techniques of investigation, from planned experiments in the behavior of lower primates to more refined methods of dating, will extract wholly new information from the past. It is my belief that, as these events come to pass, tool use will be found to have been a major factor, beginning with

the initial differentiation of man and ape. In ourselves we see a structure, physiology and behavior that is the result of the fact that some populations of apes started to use tools a million years ago. The pebble tools constituted man's principal technical adaptation for a period at least 50 times as long as recorded history. As we contemplate man's present eminence, it is well to remember that, from the point of view of evolution, the events of the last 50,000 years occupy but a moment in time. Ancient man endured at least 10 times as long and the man-apes for an even longer time.

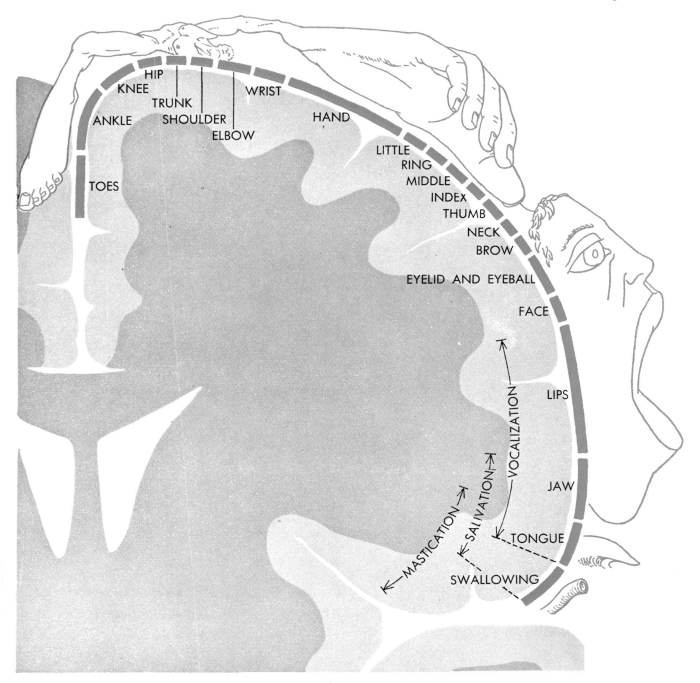

MOTOR HOMUNCULUS depicts parts of body and areas of motor cortex that control their functions. Human brain is shown here in coronal (ear-to-ear) cross section. Speech and hand areas of both motor and sensory cortex in man are proportionately much larger than corresponding areas in apes and monkeys, as can be seen by comparing homunculi with diagram of monkey cortex.

2

The Beginnings of Wheeled Transport

by Stuart Piggott
July 1968

Mankind has traveled on wheels for at least 5,000 years. The recent discovery of ancient wagons at sites in the U.S.S.R. casts doubt on the accepted hypothesis that vehicles were invented in Mesopotamia

Professor Marshall McLuhan, in one of his oracular pronouncements, defined the relationship of the automobile to modern man as that of the mechanical bride. Recent archaeological studies help to trace the earliest stages in man's romance with the wheel that ultimately led to this strange, if not unholy, consummation. Like all first courtships, it was inexpert and tentative in its beginnings, but more than 5,000 years ago the bride of wheeled transport had been won in Eurasia. For whatever reason, the early Americans failed to duplicate this invention.

It is not excessively determinist to suggest that certain prerequisites are needed for the development of wheeled vehicles. The vehicles will be invented only in societies that have a need to move heavy or bulky loads considerable distances over land that is fairly flat and fairly firm. A suitable raw material, such as timber, must be on hand for building the vehicle. And a prime mover stronger than a man must be available to make the wheels turn. In the Old World the power problem had been solved at least 7,000 or 8,000 years ago by the domestication of cattle. Once it was realized that castration produced a docile, heavy draft animal, oxen were available for traction; their strength and patience more than compensated for their slowness. Timber was available in quantity in the parts of the Near East that were neither desert nor steppe. These are the regions that saw the emergence of the earliest agricultural communities, beginning about 9000 B.C. The same communities were among the first to possess polished stone axes and adzes and, soon thereafter, copper and bronze tools suitable for elaborate carpentry.

The archaeological evidence shows that the first stages of wheeled transport depended on heavy vehicles with disk (as opposed to spoked) wheels. The wheels were either cut from a single massive plank or were made from three (and occasionally more) planks doweled and mortised together. Light vehicles with spoked wheels, harnessed to swift draft animals, were a later development that combined an advanced technology in bronze tools—and thus in the wheelwright's craft—with the domestication of the small wild horse of the steppe. Such vehicles first appear in the Near East in response to military needs during the first half of the second millennium B.C. Here, however, we are concerned mainly with developments earlier than the second millennium, when vehicles were usually drawn by oxen.

From the standpoint of the archaeologist wood is a miserable material; it is resistant to decay only in exceptional conditions of waterlogging or desiccation. Under normal circumstances to detect and recover traces of wood encountered in an excavation calls for a high degree of technical skill. It may therefore surprise the reader to learn that in Europe and Asia nearly 50 wheeled vehicles—or their wheels—have been recovered from sites that predate the second millennium B.C. This type of direct evidence concerning early vehicles is supported by discoveries of other kinds, such as models of vehicles or their wheels made from pottery, which of course is much less susceptible to disintegration than wood.

The earliest examples of wheeled vehicles have all been found within a region no more than 1,200 miles across centered between Lake Van in eastern Asia Minor and Lake Urmia in northern Iran. Presumably the first wheeled vehicle originated somewhere within this region. The oldest evidence dates back to the final centuries of the fourth millennium B.C., indicating that wheeled transport came into existence somewhat more than 5,000 years ago.

The region within which wheeled vehicles made their first appearance embraces desert and open steppe as well as forested slopes along the mountain belt that includes the ranges of the Taurus, the Caucasus and the Zagros. Deciduous timber does not grow below the 1,000-foot contour of these mountains and often not below 3,000 feet. A mosaic of communities, with economies based on mixed agriculture and copper or bronze metallurgy, flourished in the region from about 3000 B.C. onward. In Mesopotamia to the south the population was already literate and urban societies were beginning to form. To the north the zone of farming communities probably merged gradually into the area occupied by the pastoralists of the steppe beyond the Caucasus. All three societies were ones in which wheeled transport would constitute a valuable technological addition to the existing economy.

Our earliest evidence for vehicles with wheels, as opposed to simple sledges that could be dragged overland, is provided by symbols in the pictographic script of Uruk, a Sumerian city in southern Mesopotamia. The Uruk pictographs represent man's earliest known writing; they are believed to date from somewhat before 3000 B.C. Some Uruk signs depict a schematized profile view of a sledge; others show the sledge pictograph with two little disks added below it—an abbreviated symbol for a four-wheeled vehicle. Beginning about 2700 B.C. in Mesopotamia the evidence is no longer symbolic but concrete. By that time the Sumerians and their neighbors buried vehicles along with their dead; sometimes the vehicles even contained the dead. The vehicle remains often survive as nothing more than stains in the soil such as have been de-

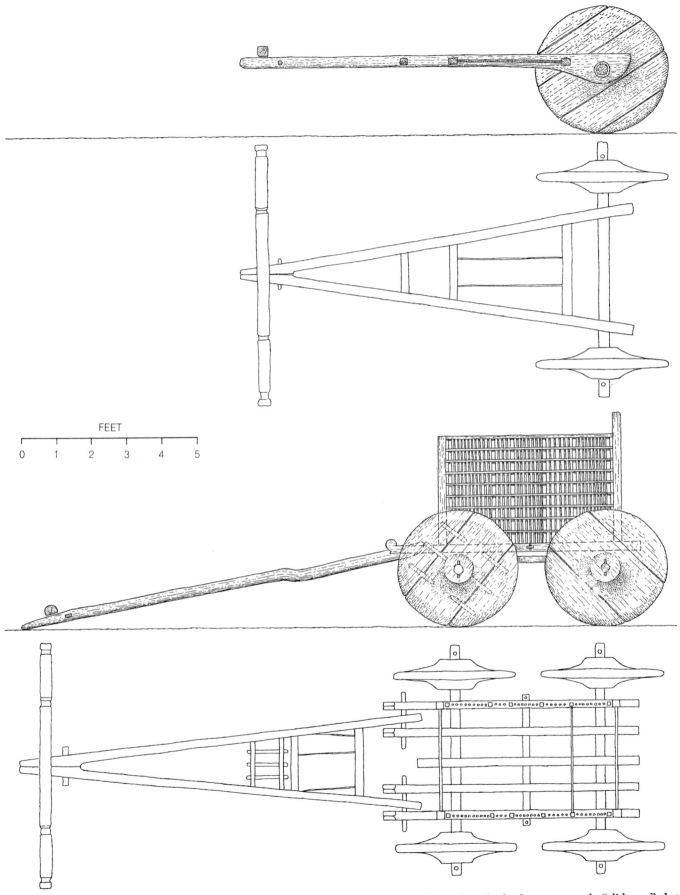

FEET

0 1 2 3 4 5

CART AND WAGON from the latter half of the second millennium B.C. were found by Soviet archaeologists in tombs at Lake Sevan in the Armenian S.S.R. They closely resemble the wheeled vehicles of much earlier times. The simple design of the A-frame cart (top) suggests that these vehicles came into being through the addition of an axle and wheels to a two-pole "slide car" that was formerly dragged along the ground by draft animals. The wagons at Lake Sevan (bottom) were complex and utilized mortise-and-tenon joining. Their draft poles, however, were apparently nothing more than cart A-frames, pegged to the wagon's chassis.

tected in the Royal Tombs at Kish and Ur and at Susa in Elam. They are of two kinds: vehicles with two wheels (carts) and vehicles with four wheels (wagons).

Carts and wagons alike were drawn by oxen or by Asiatic asses (*Equus onager*), cousins of the horse that the early Mesopotamians had managed to domesticate. The vehicles' wheels were light disks made by joining three planks. The representations in Mesopotamian art and the model vehicles that have survived from this period and the periods that follow it show that disk-wheeled vehicles were known both in Mesopotamia and among the nonliterate peoples along the Mesopotamian frontier, from Asia Minor on the west to Turkmenia on the east. The vehicles were present along most of the periphery before 2000 B.C.; soon thereafter they were common throughout it.

It had been assumed until recently that the adoption of wheeled transport among peoples to the north and west of the central zone outlined above, as well as the eventual adoption of vehicles by the peoples who inhabited Europe, were events that took place measurably later than adoption of vehicles within the central zone. Indeed, the spread of wheeled transport is often cited as a classic example of diffusion from a primary center. Since World War II, however, the picture has changed as archaeologists in southern Russia and in the Soviet republics of Georgia and Armenia have unearthed large quantities of new prehistoric material.

Among the discoveries are more than 25 burials in which vehicles were included; these apparently date from at least 2500 B.C. up to about 1200 B.C. Indirect evidence from several Soviet sites for even earlier knowledge of wheeled

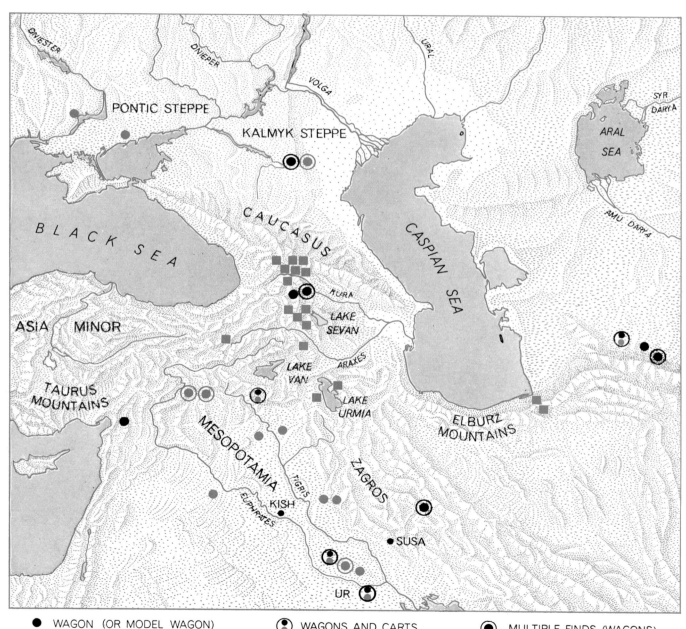

● WAGON (OR MODEL WAGON)
● CART   (OR MODEL CART)

◉ WAGONS AND CARTS
■ MODEL WHEEL

◉ MULTIPLE FINDS (WAGONS)
◎ MULTIPLE FINDS (CARTS)

VEHICLES built before 2000 B.C. in a zone between the Black Sea and central Asia are found in two main concentrations. One is Transcaucasia and the open steppe to the north. The other is Mesopotamia, including the headwaters of the Tigris and Euphrates. It was formerly believed that the first wheeled vehicles were made in Mesopotamia. The discovery that such vehicles were made in Soviet Georgia and Armenia before the second millennium B.C. diminishes the probability of a Mesopotamian origin of wheeled transport.

vehicles—in the form of model wheels made from pottery—pushes the starting date back perhaps as far as 3000 B.C. Nothing has been found at the new Soviet sites that is demonstrably as old as the pictographs from Uruk. Nonetheless, the Soviet evidence considerably weakens the case for absolute priority in the invention of wheeled transport previously conceded to Mesopotamia. The challenge is a serious one because the Mesopotamian claim rests on the pictograph of a modified sledge and on nothing else.

Because the recent evidence from the U.S.S.R. is little known outside that country it is worth describing in some detail. By way of preface I should explain that a number of excavations in Transcaucasia—the region between the Black Sea and the Caspian Sea lying south of the greater Caucasus—have made it evident that this region was once occupied by a single homogeneous culture. Marked by a complex of sedentary mixed farming, village settlements and some copper-working, the culture extended from the river valleys of the Kura and the Araxes in Georgia and Armenia, southward to Lake Urmia and westward well beyond Lake Van [see illustration on opposite page]. The period during which the Kura-Araxes culture flourished is known on the basis of carbon-14 determinations. It began about 3000 B.C., continued until sometime after 2500 B.C. and may even have lasted down to the end of the third millennium B.C. Pottery models of disk wheels with well-marked hubs are found at a number of Kura-Araxes sites. They are identical with the wheels of model vehicles unearthed in the Near East; evidently the existence of wheeled transport was at least known in Transcaucasia at the same time that actual vehicles were being entombed at Kish and Ur.

As a matter of fact the Kura-Araxes culture possesses vehicle burials of its own. One tomb at Zelenyy, in the Tsalka region of the Georgian S.S.R., was found to have contained a wagon. It had evidently been interred in working order, since the floor of the tomb bore long grooves made by the vehicle's wheels. The burial at Zelenyy, a pit grave covered by a round kurgan, or barrow mound, belongs to a style of burial that moved into the Caucasus from the southern Russian steppe. On the steppe the burials have given their name to the Pit Grave culture, which flourished during much of the third millennium B.C. Similar burials—including in two instances the remains of wagons—have been unearthed at Trialeti, another site in the Tsalka district.

The waterlogged soil of one of these tombs, excavated in 1958, contained a wagon with massive three-piece wheels. The wagon had an A-shaped draft pole, of which the stumps were preserved. It had apparently been equipped with an arched canopy to shelter its occupants. As we shall see, several more or less complete examples of similar "covered wagons" have been found elsewhere. The Trialeti burial probably took place sometime before 2000 B.C., although the date is not known precisely.

Some 350 miles north of the Tsalka district, beyond the passes of the greater Caucasus range and well into the southern Russian steppe, Soviet archaeologists have unearthed several other buried carts and wagons. The sites are located in the Elista region of the Kalmyk Steppe, no more than a month's ox-trek distant from Transcaucasia. The first Elista burials were found in 1947; others were located in 1962 and 1963. They belong to the final phase of the Pit Grave culture or to a culture that overlapped and succeeded it, and appear to be dated between 2400 and 2300 B.C.

The carts buried in the Elista graves are represented by pairs of wheels and by one pottery model of a cart with an arched canopy. The most interesting of the Elista burials, however, are those containing four-wheeled wagons. Like the model cart, the Elista wagons had arched canopies; in some cases remains of the wickerwork of which the canopies were made have survived. The Soviet excavators maintain that one of the Elista wagons had a pivoted front axle, a device that would have permitted steering the wagon. This is remarkable. If accepted, the Elista innovation antedates by many centuries the first previously known appearance of a most important advance in vehicle design. Heretofore no ancient vehicle was known to have a pivoted front axle until the time of the Celtic ritual wagons at Djebjerg in Denmark, in the first century B.C. Indeed, the very existence of the feature before medieval times has sometimes been called into question, and we are certain of pivoted axles only from the Middle Ages onward.

The resemblance between the Elista vehicle burials and vehicle burials in the Georgian S.S.R. is not the only evidence that implies contact between the steppe and Transcaucasia. Near the Black Sea and the Dnieper River in southern Russia, an area that also lies within the ancient boundaries of the Pit Grave culture, two more vehicle burials

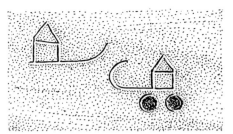

OLDEST PORTRAYAL of a vehicle with wheels is a Sumerian pictograph used shortly before 3000 B.C. It is derived from the sign for a sledge (left). The addition of two circles (right) turns it into the sign for a wagon.

CHINESE PICTOGRAPHS for a chariot or cart, apparently with spoked wheels, are first seen in inscriptions of the Shang dynasty, somewhat later than 1500 B.C. Two inverted "y" strokes (top right) represent the horses.

have been found. The graves were dug late in the third millennium B.C.; in both instances the vehicles are carts with one-piece disk wheels.

In 1956 the initiation of a hydroelectric project at Lake Sevan in the Armenian S.S.R. lowered the lake's level by many feet. Near Lchashen, as the level fell, a number of formerly submerged tombs were revealed; in them were found nearly a dozen carts and wagons. The vehicles are comparatively recent, having been buried over a period of some centuries, beginning about 1400 B.C. They are so well preserved, however, and have so many features in common with carts and wagons of much greater age that they merit special attention. Each burial was made in a huge boulder-lined pit, originally with a sloping ramp at one end. The vehicles were apparently maneuvered down the ramps and the pits were then covered with stone cairns. Soon afterward the level of Lake Sevan rose and immersion preserved the wood of the vehicles. In addition to vehicles with three-part disk wheels some of the tombs contained light carts—virtually chariots—that had spoked wheels.

From the viewpoint of technological development two-wheeled vehicles are more primitive than four-wheeled ones. In spite of their relatively late date the

MESOPOTAMIAN CHARIOT, pulled by Asiatic asses, was modeled in copper by an artisan at Tell Agrab around 2800 B.C. Although such vehicles were used for sport and war rather than for cartage, their three-piece disk wheels are identical with earlier cart wheels.

Lchashen carts reflect this. They are of the simplest kind and embody a design that is still found today among nonindustrialized peoples in parts of Europe and Asia from the Iberian peninsula and the Mediterranean coast to Asia Minor, the Crimea, the Kalmyk Steppe and the Caucasus. (The same simple carts are found even farther east, of course, but their distribution in the Orient need not concern us here.) The basic design is an A-frame. The design presumably evolved from a simple travois, or slide car, made by lashing the butts of two poles together and letting the tips of the poles trail along the ground behind the draft animals. The addition of an axle and a pair of wheels near the wide end of the A-frame turns such a travois into a cart.

Wagons, on the other hand, are comparatively complex structures. With their intricate frames and often elaborate ornamentation, the Lchashen wagons were plainly vehicles of prestige just as much as today's Cadillac. The tombs at Lchashen contained six wagons in all. Four of them had arched canopies and one had upright wickerwork sides and a decorated panel at the back. Their complicated carpentry testifies to the need for adequate coachbuilders' tools. One covered wagon was an assembly of 70 component parts; the parts were either pegged together or joined by a mortise-and-tenon system that required cutting no fewer than 12,000 mortises. (The frame

of the canopy alone required 600.) In spite of the excellence of their workmanship, the Lchashen wagons must have been slow and clumsy: the estimated unloaded weight of the wagon with the wickerwork sides is two-thirds of a ton.

Although the covered wagons of Lchashen are nearly 1,000 years younger than the steppe vehicles of the Elista region, they have counterparts among them. Moreover, the same form of wagon was common in the Near East during the third millennium B.C., as is indicated by pottery models unearthed in northern Iraq and Syria. The draft poles of the Lchasen wagons provide a lesson in vehicle evolution. They are plainly derived from cart A-frames; each wagon looks as if a cart A-frame had been pegged to the front of its chassis [see illustration on page 25]. This suggests continued to use the familiar A-frame cart was the earliest form of vehicle known and that, when four-wheeled wagons came to be built, the designers continued to use the familiar A-frame shape instead of devising a single central pole for the draft animals.

What do the various Soviet discoveries signify as far as the beginning of wheeled transport is concerned? One way of interpreting this evidence is to suggest that during the third millennium B.C. the wagon found its way to Transcaucasia from the Russian steppe to the north, along with a funeral rite that re-

quired the burial of the vehicles in pit graves. At the time of the wagons' emergence, however, wheeled transport must already have existed in Transcaucasia, perhaps in the form of A-frame carts (the evidence for this being the pottery models of wheels found in Kura-Araxes sites of earlier date). Another interpretation might suggest instead that the covered wagons came to Transcaucasia from the south and that their presence in the Pit Grave sites represents an exotic intrusion into the steppe that has its ultimate origins in the early civilizations of the Near East.

The problems presented by such alternative explanations will be discussed later. Meanwhile one should remember that the Caucasus do not in fact form an insuperable barrier to movement across them in either direction. There are good passes through the greater Caucasus, particularly the one through which the Georgian Military Highway runs from Tiflis to Ordzhonikidze. Whichever way the current of diffusion may have run between urban and barbarian zones during the third millennium B.C., carts and wagons with one-piece and three-piece wheels certainly were in use throughout the region well before 2000 B.C.

Let us now turn to the spread of wheeled transport into prehistoric Europe and see how, if you will, the West was won by the covered wagons of antiquity. Recently a number of large one-piece disk wheels have been discovered in the Netherlands. Carbon-14 determinations indicate that they were made a century or so before 2000 B.C. Two similar disk wheels, slightly earlier in date, have been found in Denmark. The wheels that most closely resemble the Dutch and Danish discoveries are ones found in the cart burials of the Pontic Steppe in southern Russia. This area is some 2,500 miles removed from the North Sea, even as the crow flies, and is considerably farther in terms of feasible overland routes of travel. What connections can be found between two such widely separated areas?

In all the land between the steppe and the North Sea the only direct evidence of the ancient use of wheeled vehicles consists of model wheels made of pottery and of a single model wagon. The pottery objects all appear to have been made before the end of the third millennium B.C., although precise dating is difficult. The model wagon, equipped with disk wheels, was found in a cemetery of the copper-using Baden culture, located at Budakalasz, on the outskirts of Budapest. It has a stunted one-piece draft pole. Above the chassis the wagon's sides rake outward; their concave upper edges suggest a body made of matting, supported by four corner poles [see *lower illustration on page 31*].

The Baden culture appears to have flourished in the period between 2700 and 2300 B.C. Chronologically this is not far removed from the era that saw wagons being buried at Ur, and it is contemporary with the vehicles found to the north and south of the Caucasus. The pottery models of single wheels are distributed at random. The one found deepest in central Europe was unearthed near Brno in Czechoslovakia; in general they are all contemporary with the later centuries of the Baden culture.

These few bits of clay constitute our only direct evidence of linkages between Europe and the steppe, but they are not the only evidence. In the Baden culture and in other roughly contemporaneous societies that flourished in what are now

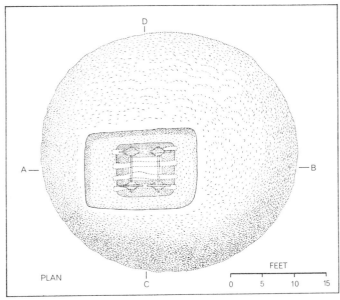

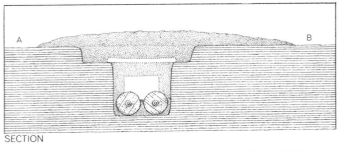

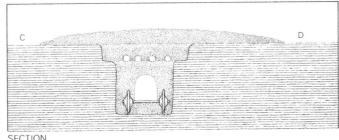

WAGON IN A TOMB was unearthed by Soviet archaeologists at Trialeti, in the Tsalka district of the Georgian S.S.R., in 1958. The wagon's wooden wheels and parts of its draft pole and chassis were preserved in the burial pit's waterlogged soil (*see illustration below*). Traces were found of an arched canopy that sheltered the occupants. The wagon probably was buried before 2000 B.C.

TRIALETI BURIAL PIT resembles a bog as the diggers probe for remnants of an entombed wagon. The vehicle's distinctive three-piece wheels have been almost wholly exposed. Trialeti is one of two Caucasus sites containing vehicles that predate 2000 B.C.

Poland and East Germany it was not uncommon to give ceremonial burial to animals as well as to men and women. Frequently the buried animals were pairs of cattle; more than 15 such burials belonging to the third millennium B.C. have been unearthed. One of them was found in the same Hungarian cemetery that yielded the model wagon. There a pair of oxen occupied one end of a long grave and human remains occupied the other end. In other cases paired oxen have been found lying at one end of a grave that was dug longer than necessary to accommodate the animals alone.

One inference to be drawn from the burials is that we are seeing pairs of draft animals; the discovery in Poland of two models of yoked pairs of oxen, of about the same age as the animal burials, lends weight to the inference. It is uncertain, however, whether we are seeing burials that originally included wheeled vehicles. The vehicles might have decayed without leaving a trace, or the traces could have gone unrecognized by the excavators. There also could have been no buried vehicles at all; the oxen may have been plow teams or animals that pulled a sledge. Nonetheless, if we take into account the burials of vehicles that have survived elsewhere, it seems most probable that burial of pairs of oxen is

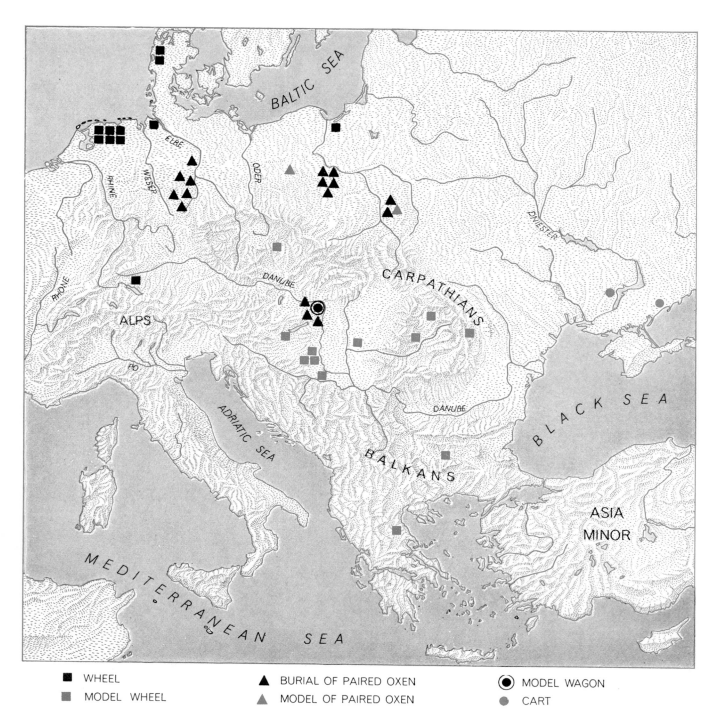

| ■ WHEEL | ▲ BURIAL OF PAIRED OXEN | ◉ MODEL WAGON |
| ■ MODEL WHEEL | ▲ MODEL OF PAIRED OXEN | ● CART |

PRESENCE OF WAGONS in Europe before 2000 B.C. is shown by the discovery of one-piece disk wheels in the Netherlands and in Denmark that yield carbon-14 dates earlier than the second millennium B.C. Direct evidence, in the form of pottery models of wheels and a single pottery model of a wagon, suggests that some vehicles entered central Europe via the Ukraine, the Romanian plain and Hungary. Indirect evidence, in the form of buried pairs of oxen and models of yoked oxen, suggests that another influx of wagons moved northwest from the Ukraine, skirting the Carpathian Mountains and arriving in the forested plain of northern Europe.

another variant of a general funeral rite in which at times a vehicle and its animals were buried together (as at Uruk), and at other times the animals were buried alone as a token representation. Accepting such an interpretation tentatively, we find that the indirect evidence of the animal burials fits in well with the direct evidence provided by the model wagon and model wheels. Both lines of evidence give added substance to a picture of Europe in which wheeled transport was used from the middle Danube to the Low Countries and Jutland at least by 2500 B.C. and probably earlier.

Thus, as in the Soviet excavations, we see that new or reassessed archaeological evidence, given the secure dating provided by carbon-14 analysis, is serving to narrow the ancient Near East's supposed margin of priority in the innovation of wheeled transport. What remains to be seen is whether valid inferences can be drawn with respect to two interlinked questions. The first question is whether or not the available evidence is sufficient to test the traditional diffusionist hypothesis about wheeled vehicles. This is the contention that the first such transport originated in a restricted region of western Asia where other technological innovations were under way among the precociously developing societies that immediately preceded the literate civilization of Sumer and Elam. The other question is more restricted in scope but is nevertheless important: Assuming that the area can be found in which wheeled vehicles were first used, by what routes and in what context of prehistory was the technology transferred from the point of innovation to Europe?

The answers to both questions depend heavily on the acceptability of the estimated ages of many archaeological finds and even of the actual dates of past events. In the context where the evidence is most needed, alas, it is not precise. For example, the date assigned to the earliest Uruk pictographs is derived from reasoned guesses and historical computation backward from the 24th century B.C., the point at which history of a sort begins in Mesopotamia, along with a glance at one or two relevant carbon-14 dates. Yet physicists have recently questioned whether "carbon-14 years" are exactly equivalent to calendar years during the period in question. The carbon-14 readings apparently give "true" dates that are several centuries earlier than the ones now in use, and correlation of the historical time scale with the carbon-14 time scale is fraught with difficulties. Even if carbon-14 dates themselves are no more than expressions

of statistical probabilities, comparison of one with another should still provide a good relative scale. In this way, for example, one could validly equate part of the Kura-Araxes culture with part of the Baden culture. The scarcity of carbon-14 determinations for the Soviet vehicles is particularly regrettable when one considers the wealth of wood available for analysis. In spite of these handicaps, however, it is hard to escape the conclusion that the closely spaced dates of early wheeled vehicles unearthed from the Caucasus to the Netherlands must reflect a basic reality that indicates a rapid transmission of ideas over great distances.

Where was the first wheeled vehicle made? Let us return briefly to some of the factors considered at the outset. Timber would be needed both for the

HEAVY WHEEL, fashioned from a single massive plank of wood, was found by Dutch archaeologists in Overijssel in 1960. Carbon-14 analysis dates it earlier than 2000 B.C.

MODEL OF A WAGON was found in a Hungarian cemetery that contains remains of the Baden culture. The Baden culture flourished in the middle of the third millennium B.C.

wheels and for the chassis; thus one is inclined to look toward regions adjacent to natural woodlands. In the case of Mesopotamia, timber would have had to come from the Zagros Mountains or from the Kurdish highlands. In support of Mesopotamia as the scene of the invention we should bear in mind that the Sumerians themselves seem to have come down from hill country, perhaps as early as the sixth millennium B.C.

Alternatively, the first wheeled vehicles could have been made within easy reach of the timber of the Caucasus. From there the invention could have been transmitted on the one hand to Mesopotamia (in the context of long-standing Sumerian ancestral relationships with the mountain peoples) and on the other into the treeless steppe to the north. Without drawing on resources beyond its bounds, however, the steppe itself certainly could not have been the birthplace.

In the light of our limited present knowledge it seems prudent to assume that the invention of wheeled vehicles took place in a wide area rather than a narrow one. The area should include Transcaucasia. Furthermore, the inventors should have not only access to raw materials but also suitable draft animals and adequate metal tools. Finally, the possibility of multiple invention is not beyond imagining. A vehicle of Sumerian design, with small one-piece wheels, could have provided a starting point. Later such developments as the covered wagon, with its heavy three-piece wheels, may well have taken place elsewhere and then been introduced into Mesopotamia. Mesopotamian wagon wheels are of noticeably lighter construction than those of the steppe; they

can hardly have been the prototypes of the massive, doweled and mortised wheels of Transcaucasia. They could, however, represent a timber-saving version of a Transcaucasian original.

As for tracing the routes by which a knowledge of vehicles, or for that matter the vehicles themselves, moved westward into Europe, we must depend largely on inference. The evidence for cultural connections between the southern Russian steppe and the areas to the north and west is general rather than specific. Moreover, a new technological addition to a culture, such as the use of wheeled transport, does not necessarily carry any other traits of the parent culture with it. In spite of these caveats it has long been recognized that the cultures of the Hungarian plain in the late third millennium B.C. possess features that are difficult to explain on a basis of evolution from indigenous antecedents alone. A likely source for at least some of the obviously intrusive elements, such as new types of copper implements that resemble Caucasus copperwork, is southern Russia. The route of the intrusion could have been by way of the Ukraine and the plains of Romania, and thence into Hungary either over the mountains or by way of the Danube's Iron Gate. Along this route too could have come knowledge of the first wheeled vehicles.

The evidence of buried oxen in Poland and buried wheels in the Netherlands and Denmark may be related to an entirely separate intrusion. The pattern of cultures in the northern European plain around the middle of the third millennium B.C. includes a culture complex characterized by cord-ornamented pottery, stone battle axes and the custom of burying the dead in single graves cov-

ered by earth barrows. It has long been held that the complex is an intrusive one and is ultimately to be derived from sources in southern Russia related to the Pit Grave culture, although this interpretation has recently been disputed, and a case has been made instead for indigenous evolution.

The championing of local origins arises perhaps in part as a healthy reaction to earlier, overworked models of European prehistory that too strongly emphasized "invasions" and "folk movements." Nevertheless, even if some features of early northern European cultures can better be explained in terms of local growth, there still remains ample evidence of contact between southern Russia and northern Europe during the period. Dutch archaeologists have sought the origin of their disk-wheeled vehicles in Russia with good reason. A practicable route for the contact would also involve the Ukraine steppe, but from there it would cross the forest steppe and run beside the Dniester River, skirting the northern slopes of the Carpathian Mountains until it reached northern Europe's forested plain.

As in all questions of prehistory, we can at most advance working hypotheses that seem, in accordance with Occam's law, to account most economically for the archaeological facts. Indeed, what we call the facts are themselves only inferences derived from the surviving material culture of extinct communities. The investigation is nonetheless worthwhile, since it was during prehistoric times that the foundations of all our technology were laid down. No innovation was more fraught with ambiguous consequences than the invention and development of wheeled transport.

EMPIRICAL TECHNOLOGY I

INTRODUCTION

Anthropos apteros for days
Walked whistling round and round the Maze,
Relying happily upon
His temperament for getting on.

W. H. Auden
The Labyrinth

We now jump forward in time to Medieval Europe, to the beginnings of a society recognizably like our own and from there trace the continuous thread of technical evolution linking us with our past. Proper justice will not be done to Egypt, Mesopotamia, China, Greece, Alexandria, Rome, Byzantium, or Islamic civilizations. To study even such a limited aspect of these cultures as their technics would require an investigation of the political, social, cultural, and economic forces that helped to shape them. The discussion must now include technology, the systematic development of technics. Empirical technology, being largely *ad hoc* and therefore culturally relative, is best understood by examining the society in which a given technique arises.

The next three articles span the period of growth of empirical technology in Europe from the ninth century through the nineteenth, an era that can be divided into two sharply delineated periods—the period before and the one after the scientific revolution and the mechanization of the world picture. In the first period, extending from early medieval times to about A.D. 1550, science remained aloof from technology for the most part, and the subjects were rarely thought to have a causal relationship with one another. Technical progress, proceeding empirically and without scientific foundation, largely focused on the acquisition of power—power to make war, power to till fields, power to mill and grind, power to pump and haul. The search for newer and better means of producing power was particularly vigorous in Europe, where the development of labor-saving devices helped the newly emergent Christian societies to accommodate themselves to a gradual renunciation of human slavery. This was not solely a moral decision, since slaves were one of the greatest sources of power to be used since antiquity for the performance of massive, tedious, and killing work.

The supply of human power dropped abruptly when the Black Death swept over Europe after 1348, wiping out a quarter of the populace almost immediately. By 1400, the plague had claimed nearly a third of the population. The great cities were hit worst, losing as much as two-thirds of their inhabitants. The population was not to regain the level of the 1340s for nearly two centuries, and during that period the survivors struggled mightily to retain a civilization geared for twice their numbers. The search for new sources of power and new means to save labor had been important in 1340. By 1400, it might well be called urgent. The development of new techniques and the improvement of old devices responded to that urgency with a slow but marked acceleration.

A most appropriate question is posed by the famed historian of medieval technology, Lynn White, Jr. in his article "Medieval Uses of Air": "How did technicians operate before they learned to get innovative ideas from science—indeed, before science had reached a state that could provide many such ideas?" That such a question probably could not have been asked before the middle of the nineteenth century is as accurate a method as any for gauging the progress of the coalescence of science and technology. Some appreciation of the difficulties of these empirical technologists may be acquired by noting the progress made in the uses of air, despite the fact that it was not until the seventeenth century that Galileo Galilei and his followers put on a scientific basis the idea that the atmosphere was like an ocean, and the air a fluid similar to water. Nevertheless, craftsmen and builders made empirical use of the pressure of air without waiting for the analysis, much in the fashion of Moliere's would-be gentleman who discovers that he has been talking prose for forty years without knowing it. White examines for us the wide range of medieval technology concerned with the uses of air, from gliders to gas turbines, with a variety of applications ranging from entertainment (the pipe-organ) to the generation of power (the windmill). The windmill, in particular, was not only a technical device of great cleverness but also a major factor for social change. With windmills standing the length and breadth of the countryside, every town and castle had its own private source of mechanical energy, which freed it from the stringent siting requirements of the water wheel and the attendant dependence on controlling the source of water during a war or a siege. Since continuous availability of power was not necessary at that time, the replacement of ten or fifteen horses by each windmill was a positive social benefit of the first magnitude.

Otto Mayr's essay "The Origins of Feedback Control" discusses an even more extreme case of a technology that preceded its science, and in this case, was even largely responsible for the existence of the related science. Feedback, the recirculation of part of the output back to the input in order to provide a mechanism for automatic self-regulation, is a fairly clear example of reinvention. Known in ancient Greece and Alexandria, discussed by the Islamic scholars of the early Middle Ages, the idea then disappeared until its apparently independent rediscovery in England at the beginning of the Industrial Revolution. Considering the importance of feedback control for regulation of power-generating devices such as the steam engine, we are surprised to discover that the analytic theory of dynamic feedback control is one of the most recent developments of twentieth-century science. Both in industry and in electronics, the use of feedback control to stabilize systems is indispensable; it distinguishes a sophisticated, efficient technology from one based purely on the massive application of force. Yet, in the curious history of feedback control, the theory developed not as a result of the needs of engineering and technology, which were well advanced in the empirical use of feedback, but rather in response to the needs of social and biological scientists who were adapting the engineering model to their own fields.

Continuing our progression from the general to the particular, we conclude this section with a specific and well-documented, if somewhat anachronistic, case study in empirical technology—the bicycle. In "Bicycle Technology," S. S. Wilson analyzes the development of the bicycle itself, such technical offshoots as the roller chain, and the impact of this apparently trivial toy on society. Having been developed as a machine specifically adapted to the human form, the bicycle quickly evolved by trial and error to the near perfection of design that it retains to this day. Recent examinations of the dynamics of bicycle stability have shown that the fork angle and offset of the modern bicycle constitute a near-optimal tradeoff of stability versus sensitivity. Only the brakes have been found to be in need of improvement. It is as nearly efficient a machine and as well adapted to the proper muscles of its motive force as could be desired. Yet, the indirect social and economic consequences of this seemingly humane and efficient machine have been devastating. The wide acceptance of the bicycle as a personal mode of transportation was largely responsible for creating the demand for a mode of mechanized transport that did not depend on the care and feeding of horses and that was not limited to running on tracks. In this sense, the bicycle was to transportation what the windmill was to energy. The windmill accustomed society to the possession of independent, decentralized sources of power; the bicycle raised the possibility of independent, self-determined, long-distance travel. It was a natural step to add an engine, and the combination led directly to the development of the automobile. Thus, from this apparently innocuous example of empirical technology sprang the demand for personal, mechanized transportation, which has ultimately led to traffic jams, freeways, smog, the death of the passenger railroads, and the decline of mass transportation.

There is a double lesson to be learned here. The problems of our modern society in dealing with our machines do not stem solely from science or technology but also from the unplanned nature of technical development and the unconcern with which new technologies are foisted upon us. The effects of such new technologies, on the other hand, are often wide ranging, indirect, and therefore extremely difficult to foresee. In fact, the uses and impacts of such *ad hoc* technologies may be even more difficult to plan and evaluate than those of the scientific technologies to be discussed in the next section. Scientific technology is at least the result of a determined policy to explore a specified phenomenon. Except in rare cases of unusually divergent serendipity, its results are preplanned. *Ad hoc* technology remains a weed; it tends to pop up and grow in totally unexpected places and in largely unforeseen directions. The orientation of the twentieth century, however, is largely toward scientific technology, and it is to this subject that we shall turn in the next section.

3

Medieval Uses of Air

by Lynn White, Jr.
August 1970

*Technology came before science in the invention of the
blast furnace, the windmill and the suction pump.
Medieval workers also constructed a manned glider
and a useful gas turbine and conceived the parachute*

Most of us who live in the latter decades of the 20th century take it for granted that advances in technology represent practical applications of earlier scientific discoveries. In our vernacular the term "science and technology" is subconsciously hyphenated; once when I tried to refer in print to technology and science the typesetter "corrected" the sequence. In actuality the marriage of technology and science is not more than a century and a quarter old. Architects since antiquity have used mathematics to achieve aesthetic effects, but mathematics played no significant part in their engineering. Although astronomers, from the 15th century on, worked at improving the art of navigation, the vast majority of sea captains, even on great oceanic voyages, long preferred to rely on their own empirical methods of getting where they wanted to go. During the 18th century chemical science began to stimulate industrial chemistry, particularly in France, but it was not until the middle of the 19th century that our present assumption—that a better way of doing something is almost always based on a new scientific discovery—became widespread.

Before that time science and technology lived apart. Science had been a largely speculative effort to understand nature, whereas technology was an exclusively practical attempt to use nature for human purposes. The two seldom interacted; Francis Bacon had said that "knowledge is power," but he did nothing to validate his dictum and not many were listening. It was only in the decades of the younger Queen Victoria, Napoleon III and Bismarck that this idea became normal in Europe. Its general recognition in America took a bit longer (even though when Benjamin Franklin founded the American Philosophical Society in the middle of the preceding century, its purpose was specified as "the promotion of *useful* knowledge").

How did technicians operate before they learned to get innovative ideas from science—indeed, before science had reached a state that could provide many such ideas? The Greek and Arabic mathematics and physics that western Europe absorbed almost completely when they appeared in new Latin translations during the 12th and 13th centuries were not of the slightest assistance to the engineers of that age. Yet the same engineers, with unprecedented daring and ingenuity, raised the vast Gothic cathedrals that still grip us by the throat when we see them. We underestimate today what keen empirical perception can produce by way of technological novelty quite unrelated to scientific thinking.

In retrospect basic inventions often seem incredibly simple. The long delays between their appearance in the course of human history, however, make one wonder what we mean by "simple." The Greeks and Romans were sophisticated in many ways: in drama, law, sculpture, philosophy, geometry and much else. But no Greek or Roman ever saw the common, twice-bent lever we call a crank. To be sure, the Chinese who were contemporary with imperial Rome knew about cranks, but this most ordinary means of connecting reciprocating motion with continuous rotary motion, which in machine design is second in importance only to the wheel, does not show itself in the West until it is pictured in the Utrecht Psalter, a work produced near Reims between A.D. 816 and 834. Did it diffuse from China? A crank

MEDIEVAL PIPE ORGAN, with a crew of four busy at the bellows, is shown in one of the illuminations of the Utrecht Psalter, a work produced between A.D. 816 and 834. Larger and larger bellows, worked by water power rather than manpower (see *illustration on opposite page*), led eventually to the first European blast furnaces and the casting of iron.

TWO PAIRS OF BELLOWS, driven by crankshaft and connecting rod attached to a waterwheel, provide the air current needed to keep up high furnace temperatures in a 16th-century smithy. The plate is in *Le diverse et artificiose machine,* published in 1588.

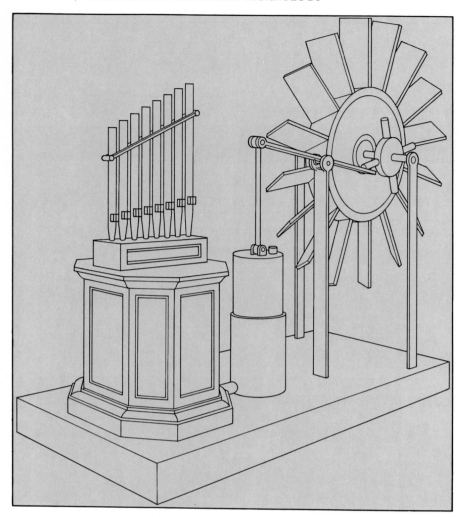

WINDMILL with a horizontal axle bearing cogs was proposed by the mechanical genius of
the first century after Christ, Hero of Alexandria, to raise the piston of the gravity-powered
air pump for a pipe organ. Hero's suggestion for a new source of power was evidently
ignored and the windmill was independently reinvented twice in the following centuries.

is scarcely borrowable apart from some more complex apparatus; the picture in the Utrecht Psalter shows it turning the first rotary grindstone known anywhere. The second use of the crank in Europe was with the hurdy-gurdy in the 10th century. Since neither rotary grindstones nor hurdy-gurdies were known in China, it would seem that the crank was invented independently by two artisans widely separated in time and space: first by a Chinese of the Han Dynasty and then by a Frank under the Carolingians. To bend a lever twice is indeed "simple." To see that something new and useful can be done with this distortion demands empirical genius. Elaboration of something already known, it appears, is a far easier achievement than a genuine new insight, however simple. It is my purpose here to examine a few insights that occurred during the later medieval period.

Outside the laboratory we experience the physical world in terms of somewhat naïve categories. The ancients named four: earth, air, fire and water. Of these air is the most puzzling in our normal contacts with it because, although we are enveloped by it, we neither see nor grasp it. Yet air can be exploited for technological purposes, and in medieval times a number of empirical efforts were made to put air to work. All these efforts were simple. Some were of great immediate importance and others were of great seminal significance in the growth of Western culture.

Of course, scientists as well as artisans were interested in air in the Middle Ages. Indeed, the greatest controversy in medieval physics—the challenge of Aristotle's theory of motion—had much to say about air. Aristotle believed a moving object could continue in motion only as long as something moved it. In the case of a projectile, air displaced by compression in front of the object came around behind it and pushed it along. The new medieval theory of "impetus" insisted that a moving object continues to move until it is stopped by resistance: air has nothing to do with motion except insofar as it is a cause of friction. The anti-Aristotelian scientists of the 14th century used the rotary grindstone (a device that, as we have seen, was unknown to Aristotle) as a favorite nonspeculative proof of their position. The grindstone still turned for a time after the grinder's hand left the crank, but its motion, unlike the motion of a traveling projectile, displaced no air. Therefore the grindstone moved not by pressure of air but by impetus until the resistance of friction at the axle stopped it. The new physics of Jean Buridan and others laid the foundations for the work of Galileo and Newton. Still, we have no indication that at the time there was any relation between the science of air and the technology of air.

Since at least Neolithic times men have made moving air do work by using sails to capture the force of the wind and propel boats. In the first century B.C. the fore-and-aft rig appeared in the Mediterranean, enabling small boats to run into the wind at a fairly sharp angle. Big merchant ships were not equipped with triangular lateen sails, however, until the sixth century. The delay was probably due to the slow development of keels that bit deep into the water and thus reduced the sideways drift involved in tacking. Because keels are not discussed in any contemporary document, and because all pictures extant show ships afloat with the keel hidden, how the deep keel actually evolved is a question that must be left to underwater archaeologists. In any case, mariners continued to experiment with sails and rigging throughout the Middle Ages, adding masts and subdividing sails to achieve greater power and flexibility of control. Their cumulative successes by the end of the 15th century were an element in making possible the great voyages of Columbus and Vasco da Gama.

How to use compressed air in pipe organs was a discovery of the Hellenistic age. The "boxes of whistles" were costly, and their machinery was so complicated that maintenance was difficult. Indeed, the pipe organ is the most intricate apparatus known before the invention of mechanical clocks in the 14th century. It is therefore not surprising that organs vanished from the West after the barbarian invasions. The Byzantine world, however, enjoyed perfect continuity with the Roman Empire, and organs continued to be used there for secular ceremonies and celebrations, notably in the imperial palace at Constantinople.

In A.D. 757 the Greek Emperor sent a pipe organ as a gift to Pepin the Short, King of the Franks. Pepin used it not in his chapel but in his palace; its fate is unknown. In 826 the Western Emperor, Louis the Pious, commissioned a Venetian priest named George, who had doubtless learned the art of building organs in Constantinople, to construct one for his palace at Aachen in the Rhineland. George's construction may well have looked like an organ shown in the Utrecht Psalter; clearly the Benedictine monk who illuminated that manuscript was interested in machines.

George later became an abbot, and organs began to appear in Western monasteries. Toward the end of the 10th century Elfeg, bishop of Winchester, installed the first monster organ known to us in his cathedral there. It took 70 men pumping 26 bellows to supply wind for the Winchester organ's 400 pipes. The Greeks never allowed organs in their churches: mechanized music was considered alien to religion. The Latin church, however, took the box of whistles to its bosom and made it the typical instrument of Western religious music. With the later invention of the mechanical clock the same contrast of attitudes showed itself again. Clocks proliferated as exterior and interior displays in Western churches, but the Eastern church banned them from its shrines on the ground that eternity must not be contaminated by time's measurement. Until recently the West has felt no such antagonism between mechanization and spirituality.

The organs that reappeared in the West were kept in repair by local craftsmen. Working with the big bellows of these machines would have given such men ideas. It is no accident that at about the time of the building of the Winchester organ we find the first evidence for the use of water power at a smithy. The first documents do not tell us whether the waterwheel drove the smithy hammers or pumped the bellows for the forge fire. Water-powered pounding mills for fulling, a step in the manufacture of cloth, also first appear at this time, so that we are reasonably sure that water-powered hammers or stamps were used in the metallurgical industries by A.D. 1000; the probability is strong that power bellows were also in use by then. Power bellows gradually increased in size and efficiency until, by 1384 at Liège in Flanders, they were producing temperatures high enough to make cast iron in the first known blast furnaces.

This production of cast iron, the greatest metallurgical development since the discovery of iron itself, qualifies as one of the medieval uses of air that was of immediate importance to Western culture. Although cast iron had been known in China much earlier than the 14th century, most Chinese iron ores contain phosphorus. This inclusion lowers the melting point of the iron so that it can be cast at lower temperatures, but the cast iron that results is a metal of limited versatility. The European blast furnace of the late 14th century could

FIRST WINDMILL, its vanes attached to a vertical axle, was invented somewhere in the Middle East during the 10th century. This is a later model, developed in Europe during the 16th century.

SECOND WINDMILL, with a horizontal axle, was invented independently in the 12th century in the North Sea region of Europe. The one shown is a "tower mill." Only its top is turned to the wind.

FIRST AIRSCREW was drawn in plan view (*left*) in the notebook of Mariano Taccola, an engineer in Siena in the 15th century. The words under the launcher for the airscrew, sketched in perspective at right, are *puerorum ludus est*, Latin for "[it] is a boys' toy."

smelt superior ores, and its development vastly expanded both production and consumption of iron in the West.

A bellows produces wind. Anyone who watches a hawk soaring can see that its flight involves resistance to the wind. Although at least two Muslims are known to have attempted to fly earlier,

it is perhaps not a coincidence that the first European to try it was a younger Anglo-Saxon contemporary of the organ-building Bishop Elfeg who was likewise a Benedictine. Sometime between A.D. 1000 and 1010, Eilmer, a monk of Malmesbury in Wiltshire, built a glider, took off from the abbey's tower and flew

600 feet before crashing and breaking his legs. His own diagnosis of his failure was entirely matter-of-fact: he said that it was because he forgot to put a tail on the rear end of the glider—"*caudam in posteriore parte*." Eilmer lived to a ripe and respected old age, and his feat was never forgotten, although he had few imitators.

More than two centuries later, to be sure, Friar Roger Bacon wrote that "flying machines can be constructed in which a man...may beat the air with wings like a bird" and even claimed to know the designer of one. But the fact that no one after Eilmer actually tried to fly (before an instance, badly documented, at Perugia in the 15th century) may be more an indication of common sense than of timidity. The human musculature is such that it cannot power birdlike flight. Gliding, moreover, is very nearly suicide in the absence of fairly detailed knowledge of how air currents move. The great development of gliding in the 19th century rests on observations of kites in flight, but the only news of kites received in medieval Europe, where otherwise they were unknown, was tucked away in the Latin version of Marco Polo's memoirs preserved in the cathedral of Toledo. European kites are a 17th-century development and were still something of a novelty in Benjamin Franklin's time.

Medieval Europeans were destined, however, to become increasingly conscious of the properties of air and its role in motion. The great mechanical pioneer of the first century after Christ, Hero of Alexandria, had sketched a little windmill that, he proposed, might operate the pump of a pipe organ [*see illustration on page 38*]. The mechanism is impractical; it had no subsequent influence, and the best modern opinion calls it the armchair invention of a genius. In the 10th century a quite different kind of windmill was produced in eastern Iran or Afghanistan. A vertical axle, fitted with a series of vanes, was set in a millstone; the wind, which in that region blows constantly from one direction, entered the mill through a door placed off-center to expose only the vanes on one side of the axle, causing the millstone to rotate. The Chinese vertical-axle windmills of the 13th century were derived from this source. There is no firm evidence, however, that knowledge of any kind of windmill spread from the Middle East to the other parts of Islam. This is surprising; the dearth of water power in the largely arid Islamic lands might seem to make windmills particularly useful.

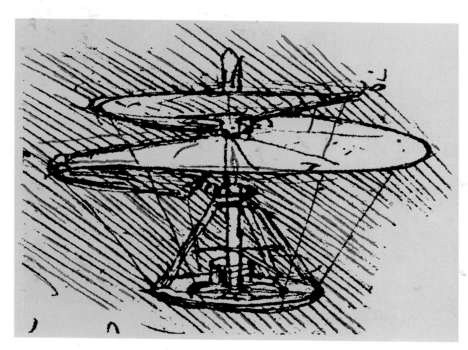

LARGER AIRSCREW, designed to lift a man, is shown in a drawing by Leonardo da Vinci, who proposed making it out of starched linen. The lack of an adequate power source barred practical use of airscrews, however, until the internal-combustion engine was developed.

The windmill as we know it, with vanes on a horizontal axle, is seen in the illustration at the right on page 39. It was invented independently in the North Sea region of Europe by A.D. 1185, probably by applying the principle of Vitruvius' water mill (an invention of classical times that was still in common use and had a horizontal axle) to the harnessing of wind power. The spread of the new machine was explosive: within seven years of its first appearance in Europe the windmill was known in Palestine, where it had been imported by German Crusaders. By the 1190's the Pope was trying to tax windmills—a sure sign of their success.

In flat country, where good sites for water mills are scarce, the windmill offered an invaluable new source of mechanical power. By the 13th century, for example, at least 120 windmills stood around Ypres in Flanders. Because their location was not restricted to the banks of streams, windmills came into use to some extent in all parts of Europe. Castles were seldom built directly beside running water; windmills were often installed in castles to produce flour during sieges. Early in the 14th century an English chronicler lamented that the search for long wooden beams suitable for the vanes of windmills was a major cause of deforestation. Dante, in his *Inferno*, was sure that all his readers had seen windmills when he described Satan flailing his arms like *"un molin che il vento gira."* Don Quixote's astonishment at windmills (and at fulling mills too, as is often forgotten) is one of Cervantes' ways of emphasizing that La Mancha was the backwoods of Spain.

Water mills had appeared in the last years of the Roman Republic. They spread steadily throughout the early Middle Ages until by the 11th century every European community had one. Now, with the diffusion of the windmill, a second big power machine was blanketing Europe. In no other society had the constant presence of such engines been a stimulus to further engineering adventures. The windmill was not only a major new source of power; it was also a prime element in building the technological mentality of our culture.

If moving air can turn axles, then the friction of static air against moving vanes can decelerate them. About 1250, at the court of St. Louis, King of France, an immense illustrated version of the Bible was produced. One of the pictures, illustrating King Hezekiah's dream, shows a large water clock, presumably modeled after a clock in the king's palace in Paris. Every hour on the hour the water clock triggered a mechanical chime driven by a falling weight. To slow the fall of the weight a brake has been provided in the form of whirling vanes. Similar "fan escapements" are found in the striking mechanisms of weight-driven clocks of the 14th century and later.

Early in the next century a variant appears. Someone saw that, if fan vanes on a vertical axle are tilted slightly and the fan is spun with the uptilted edges of the vanes leading, the rotating fan will tend to rise. Such a device is first found, labeled *puerorum ludus* ("boys' toy"), in a manuscript of the 1430's produced by a citizen of Siena, Mariano di Jacopo Taccola, who was the most creative engineer of his generation. It has been stated that this little protohelicopter came to Europe from China, but the evidence for its earlier presence there is not yet firm. We have several other pictures of such aerial tops from later in the 15th century; the fact that technicians other than Taccola were aware of their potential is shown by Leonardo's famous sketch of an airscrew. As in the case of winged flight, however, nothing practical could be achieved until the development of the internal-combustion engine (which, incidentally, traces its ancestry back through the experiments of Denis Papin in 1690 and Christiaan Huygens in 1673 to the invention of the cannon early in the 14th century).

The helicopter toy, boring its way upward through the air, took the initiative.

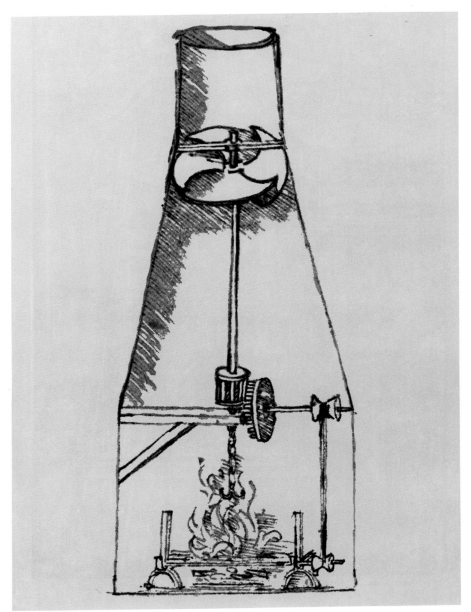

GAS TURBINE applied the principle of the airscrew to a static situation. As another sketch by Leonardo indicates, the flow of hot air up the chimney makes a set of blades rotate, and their motion, transmitted by gears and a belt drive, turns the spit near the fire.

Could anything be done if instead the air passed swiftly upward through a fixed rosette of slightly tilted vanes? In chimneys the hot air rises with some force; late in the 15th century Leonardo gives us our first picture of a roasting spit activated by an air turbine placed in a chimney [*see illustration on preceding page*]. Similar machines are known from other sources shortly thereafter, so that

in this case Leonardo was probably recording rather than inventing. These powered spits were elegant in their automatic control; the hotter the fire, the more rapidly the roast rotated.

The uses of the air in the Hellenistic world did not extend to pumps; the Hellenistic pumps were force pumps, which work without the aid of atmospheric pressure. The first evidence of a

suction pump, which utilizes the weight of air to raise water, appears by a curious coincidence in the same manuscript by Taccola that contains the first airscrew. The device is still primitive: the piston is raised by a cord and evidently falls simply by gravity; moreover, the compound crank is defective. These faults

COMPRESSED AIR is serving as a propellant, forcing wine through a tube from one barrel to another, in a painting made in Nuremberg in A.D. 1474. Large-scale movements of materials with compressed air had to await the development of better blowers.

ADVANCED SUCTION PUMP appears in 1556 treatise on mining by Georgius Agricola. A crank, driven by a waterwheel, drives the pistons of a three-stage pump. Water is lifted from one stage to the next until it pours into a trough that leads the water out of the mine.

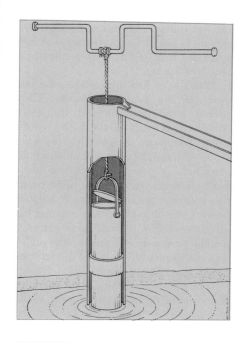

PRIMITIVE SUCTION PUMP appears in the same notebook of Mariano Taccola's that shows the first airscrew. This modern drawing is based on his sketch: the piston is raised by a cord and gravity makes it fall.

were soon remedied, and the use of suction pumps spread swiftly, first within Italy and then to the mining districts north of the Alps. Now at last we see an input from the technology of air to the science of air. At this time a vacuum could not be observed in nature. By providing a "laboratory situation," suction pumps did much to stimulate men such as Galileo (and Otto von Guericke, whose demonstration of the vacuum-sealed "Magdeburg hemispheres" astonished the imperial court in 1654) to investigate the phenomenon of variable air pressure.

Another new gadget led to further application of air pressure. The flow of spices to Europe from the Indies, which had been established in classical times, was never seriously interrupted. It is therefore not astonishing that around A.D. 1425 the Malay blowgun reached Italy, accompanied by its native name, *sumpitan* (which, as the Arabic *zaba-tānah*, became successively the Italian *cerbottána* and the English *sarbacane*). As a French manuscript painting of about 1475 shows, the blowgun was used in Europe chiefly for hunting birds. If one could propel a dart with compressed air, why not other things? A Nuremberg picture of 1474 shows the earliest use of compressed air to convey materials: a winehandler is using a special bellows to force wine through a tube from one barrel to another.

To us who see the third millennium approaching and who take moon landings almost for granted, our ancestors' practical applications of the force and substance of air during the five centuries that separate Eilmer's glider in Malmesbury from Leonardo's automated spit may seem rudimentary. Such a judgment commits the intellectual sin of anachronism. To evaluate what those generations of technicians accomplished we must take our historical stance not in 1970 but in 970. Looked at from there, the engineering accomplishments of the later Middle Ages not only are spectacular but also provide the essential basis for what the West has achieved since then. The earlier growth of technology had been steady, but it was incredibly slow by modern standards. Between the seventh and the 10th century the medieval West had taken the initiative in developing a new and more productive agriculture and a superior military technology. Around A.D. 1000 these impulses carried over into engineering. Before that time the Chinese had been the most inventive of the great cultures. By about A.D. 1350, however, Europe began to surpass China in technological innovation. Today's mechanical technology, which aims at the use of natural power and automatic control to save human labor, is the extrapolation of a concerted and unbroken cultural movement that is now some 10 centuries old. The inventory of medieval explorations of the uses of air reviewed here shows one aspect of that concerted thrust in its earlier phases.

Of the items reviewed here four—the improvement in sails, the development of the blast furnace and the invention of the windmill and the suction pump—were of immediate importance. The elaboration of some of the other innovations took more time. The obstacles in the way of achieving flight are obvious. The large-scale use of compressed air for materials transport had to await more powerful blowers than were then feasible. The failure of medieval engineers to turn all their dreams into realities is less significant than the vigor with

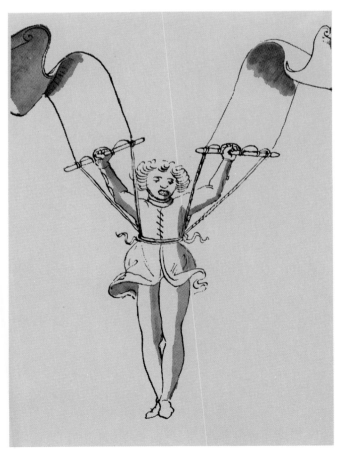 

PARACHUTER'S PROGRESS is evident in successive sketches made by an anonymous Italian engineer late in the 15th century. At left the jumper's fall is decelerated only by friction of the air against a pair of cloth streamers. At right the engineer has improved on his original scheme by substituting a conical canopy of cloth for the streamers. The first recorded jump was made in 1783.

which they conceived new technologies. The parachute, a means of using air friction for deceleration, is a case in point.

Preserved in the British Museum is the notebook of an anonymous Italian engineer, probably a citizen of Siena, that covers a period during the late 1470's or the early 1480's. On one of its pages a man is pictured jumping through empty air [see illustration at left on page 43]. At first glance the expression of his mouth is odd; the reason is that he is gripping a sponge between his teeth to protect his jaws from the shock of landing. All that is braking his fall is a pair of cloth streamers. He looks scared. He should be.

The next few pages of the manuscript show fairly routine devices: derricks, military machines and the like. Our engineer-sketcher, however, is also worried about that jumper. Something better must be done for him. After 21 pages the jumper reappears [see illustration at right on page 43]. The sponge is now secured in his teeth by a strap that runs around his head, so that if he cries out in terror he will not drop it. Above him appears a much more efficient, although still far too small, decelerator: a conical parachute. It is the earliest depiction of the device so far discovered.

Slightly later Leonardo drew a pyramidal parachute; by the 1480's the idea, if not the thing, was in the air. It reached actual publication in 1615–1616 in a famous book, New Machines, by Fausto Veranzio, a bishop in Hungary. Thenceforth everyone concerned with mechanics knew its possibilities. It was not until some 300 years after the parachute's conceptual invention by the anonymous Italian engineer, however, that anybody actually jumped in one. Not until the brothers Montgolfier began ballooning did parachuting become functional; the first actual jump was made in 1783. The notion of the parachute had been available for 10 generations. When it was needed, the notion was there to be applied.

The virtue of early engineers lies not only in what they accomplished but also in what they imagined. In our own time, when the parachute is a familiar adjunct to airplanes and space vehicles, it is a symbol of why we may well respect the creativity of the technicians of centuries long gone. Quite without the aid of science they greatly expanded not only the repertory of engineering but also the Western concept of the scope of technology.

4

The Origins of Feedback Control

by Otto Mayr
October 1970

The evolution of the concept of feedback can be traced through three separate ancestral lines: the water clock, the thermostat and mechanisms for controlling windmills

Every animal is a self-regulating system owing its existence, its stability and most of its behavior to feedback controls. Considering the universality of this process and the fact that the operation of feedback can be seen in a great variety of phenomena, from the population cycles of predatory animals to the ups and downs of the stock market, it seems curious that theoretical study of the concept of feedback control came so late in the development of science and technology. The term "feedback" itself is a recent invention, coined by pioneers in radio around the beginning of this century. And the exploration of the implications of this principle is still younger: it received its main impetus from the work of the late Norbert Wiener and his colleagues in the 1940's.

Feedback control is an instance of technology giving birth to science. Application of the feedback principle had its beginnings in simple machines and instruments, some of them going back 2,000 years or more. The thermostat and the flyball governor are well-known modern examples. Although the simple early inventions have been developed to a high order of sophistication, feedback control as an abstract concept did not receive much attention until the 1930's, when biologists and economists began to note striking parallels between their own objects of study and the feedback control devices of engineers. Certain regulatory processes in living organisms and in economic behavior showed the same cyclic structure of cause and effect and apparently obeyed the same laws. It became evident that the concept of feedback control could be a versatile and powerful tool for investigating many forms of dynamic behavior. Today the feedback control principle is not only widely embodied in hardware but also recognized as an important unifying concept in science.

The subject of this article is the historical growth of the concept. Its career can be traced with some assurance because feedback control can be rigorously defined. Wiener described it as "a method of controlling a system by reinserting into it the results of its past performance." A more formal definition, offered in 1951 by the American Institute of Electrical Engineers, states: "A Feedback Control System is a control system which tends to maintain a prescribed relationship of one system variable to another by comparing functions of these variables and using the difference as a means of control." The purpose of such a system is to carry out a command automatically, and it functions by maintaining the *controlled variable* (the output signal) at the same level as the *command variable* (the input) in spite of interference by any unpredictable disturbance. The command signal may be either constant, as it is in the case of the temperature setting on a thermostat, or continuously variable, as it is in the case of the steering wheel position in the power-steering system of an automobile. In all cases, if the feedback control system is to function effectively, it must be so designed that the controlled variable follows the command signal with the utmost fidelity.

The main characteristic of a feedback control system is its closed-loop structure. The state of the output signal is

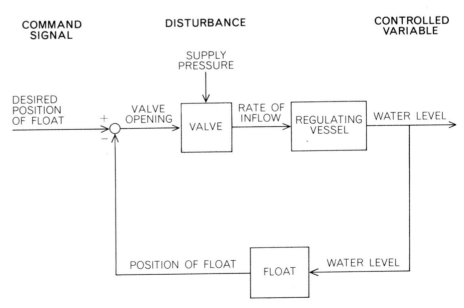

TYPICAL CLOSED FEEDBACK LOOP is evident in this simplified block diagram depicting the operation of Ktesibios' water-clock flow regulator. The arrows represent signals and the blocks represent the physical components on which the signals operate. By expressing the signals as mathematical variables and the blocks as functions, the diagram can be reduced to a differential equation describing the dynamic behavior of the system.

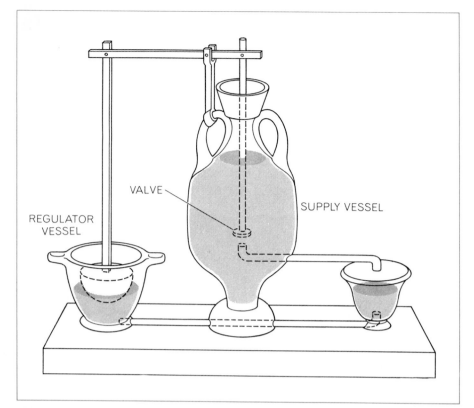

"WINE DISPENSER" designed by Hero of Alexandria around A.D. 50 incorporated an improved float regulator in which the valve (the control element) was not directly attached to the float (the sensing element). The level in the communicating vessels was maintained by the float in one of the vessels (*left*) acting on the valve in the supply vessel (*center*).

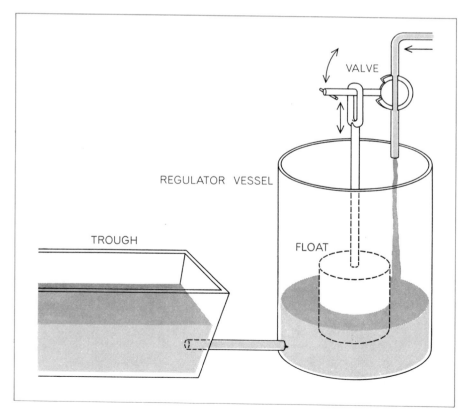

FLOAT REGULATOR for an animal drinking trough was described in a ninth-century book titled *Kitāb al-Ḥiyal (On Ingenious Mechanisms)* by three brothers from Baghdad named Banū Mūsā. Water was drawn from a river through a pipe into two communicating vessels. The float in the regulating vessel controlled a stopcock valve in the intake pipe.

monitored by some sensing device that feeds the signal back to the input side. There it is subtracted from the command signal; if the result is not zero, the system responds with a corrective action whose size and direction depend on the magnitude of the deviation, or "error signal." In the case of a home thermostat, for example, if the room temperature has dropped below the desired temperature, the system responds with an increased supply of heat, that is, a negative change in the output signal evokes a positive corrective action. In general a signal that has traveled around the loop of a feedback system returns with a reversed sign. The change of sign is essential for the stability of the system; if the signal were not changed in sign, it would create a vicious circle, building up the deviation of the output from the desired level.

The origin and main lines of development of the feedback concept are illustrated by three devices: the ancient water clock, the thermostat and mechanisms for controlling windmills. Let us trace the history of each of these applications and see where they led.

The earliest known construction of a device for feedback control was a water clock invented in the third century B.C. by a Greek mechanician named Ktesibios, working in the service of the Egyptian King Ptolemy II in Alexandria. He was probably associated with the illustrious museum that was then the principal cultural center of the Mediterranean world and attracted Greece's foremost scholars. Ktesibios' own descriptions of his inventions (which in addition to the water clock included a force pump, a water organ and several catapults) are now lost, but fortunately an account of them is preserved in *De Architectura,* the great work of the Roman architect and engineer Vitruvius.

Vitruvius' description of Ktesibios' water clock is not clear; however, the German classicist Hermann Diels translated its obscurities into a plausible reconstruction of the device. The water clock measures the passage of time by means of a slow trickle of water, flowing at a constant rate into a tank where an indicator riding on the water tells the time as the water level rises [*see illustration on page 47*]. Ktesibios solved the problem of maintaining the trickle at a constant rate by inventing a device resembling the modern automobile carburetor. Interposed between the source of the water supply and the receiving tank, this structure regulates the water

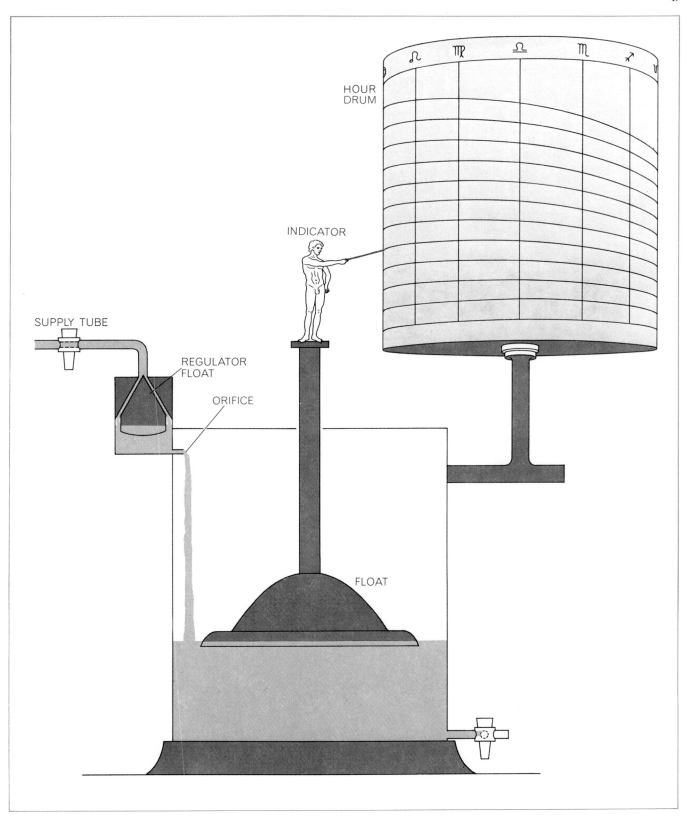

ANCIENT WATER CLOCK is the earliest known device for feed-back control. It was invented in the third century B.C. by a Greek mechanician named Ktesibios, working in Alexandria. This drawing is based on a reconstruction by the German classicist Hermann Diels. The indicator figure is mounted on a large float (*bottom*), which rises inside a tank as a result of a slow trickle of water into the tank. The 12 hours, which vary in length with the seasons of the year, are indicated on the drum at top right. The change in the length of the hours can be represented by simply turning the drum to the proper month. The float regulator at top left controls the rate of water flowing into the main tank by maintaining a constant water level in the adjacent regulator vessel. If the level rises (as a result, say, of an increase of static pressure in the external supply line), the regulator float will rise, throttling the inflow into the regulator vessel. The device is remarkably similar in operation to the carburetor of a modern automobile.

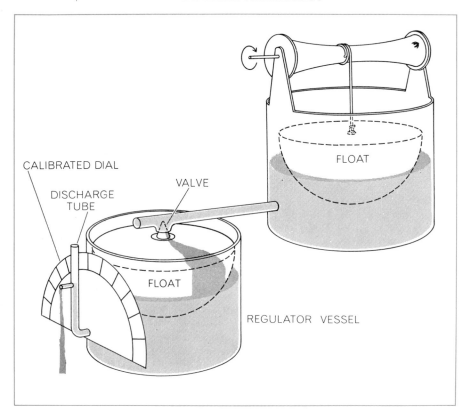

NINTH-CENTURY WATER CLOCK was described by the Islamic author known as "Pseudo-Archimedes." The time-indicating mechanism (*not shown*) was driven by the constantly falling water level in the main float chamber. A constant discharge from this chamber was maintained by means of a float valve in the regulating vessel. The outflow from this vessel could in turn be calibrated by turning the discharge tube around its axis, thereby making it possible to adjust the clock for seasonal variations in the length of the hour.

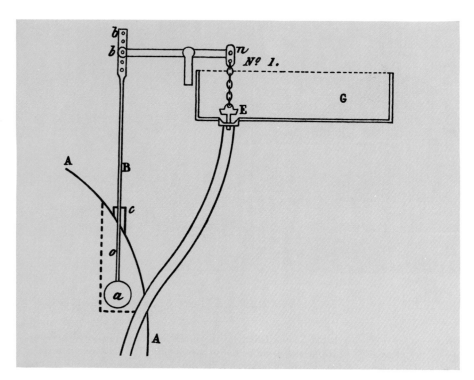

FLOAT REGULATOR WAS REINVENTED in 18th-century England, apparently without knowledge of its earlier career. This drawing, showing a water-level regulator for steam boilers, is from a British patent awarded to Sutton Thomas Wood in 1784. The level in the boiler (*A*) was sensed by a float (*a*) that controlled the water supply through a valve (*E*).

flow by means of a float valve. When the float is at a certain level, the valve attached to it is open just far enough to feed water into the timekeeping tank at the desired rate. If for some reason the water in the regulator falls below or rises above that level, the float responds by opening or closing off the water supply until the float returns to the specified level.

About three centuries after Ktesibios we again encounter automatic regulators of the float type in the *Pneumatica* of Hero of Alexandria. Hero was a prolific author of books on mathematics, surveying, optics, mechanics, pneumatics, automatons and military engineering, and his *Pneumatica* contains a number of amazing anticipations of modern inventions. Among other things, it describes several float regulators considerably more refined than that of Ktesibios. In one, called a "wine dispenser," the valve is not directly attached to the float, thus demonstrating that in a feedback control device the sensing element and the control element can be widely separated from each other [*see top illustration on page 46*]. The playful applications Hero suggested for his devices are frowned on by some scholars; it is clear, however, that he was a serious scientist who was primarily interested in describing principles and used trivial but readily understandable examples the better to make his points.

In the ninth century, some 800 years after Hero, we find the float regulator cropping up again, this time in Arabic. In a book called *Kitāb al-Ḥiyal* (*On Ingenious Mechanisms*), evidently inspired by Hero's *Pneumatica*, a trio of authors in Baghdad presented eight applications of the float valve for feedback control. The authors were three brothers, Banū Mūsā, who were high officials at the court of the Abbaside caliphs. Their devices added a few refinements to the float-valve system. One was the use of a proper stopcock as the regulating valve instead of the primitive contrivance of a plate held against the end of a pipe [*see bottom illustration on page 46*].

The float valve inspired some of the proudest achievements of Islamic technology in the period preceding the Middle Ages. Its artisans built monumental water clocks in which the time was told by elaborate theatrical displays performed by automatons. These are described in detail in three surviving books on water clocks. The first, probably written in the ninth century, is by an anonymous author usually called "Pseudo-

Archimedes"; the other two, based on that work, are by 13th-century writers named Ibn al-Sāʿātī and al-Jazarī. The clocks described in these three books employ the float-level regulator of Ktesibios, but it now regulates the *outflow* from the main float chamber instead of the flow into the chamber. Hence time is measured by the sinking, rather than the rise, of the water level [*see top illustration on opposite page*].

After these accounts in the early 13th century the float valve drops out of sight. No references to the employment of the device for water-level regulation have been found in the technological literature of the Middle Ages or the Renaissance or Baroque periods. Even a beautifully illustrated Latin translation of Hero's *Pneumatica,* which was published in 1575 and had a powerful impact on the development of technology, failed to induce engineers to take up the float regulator as a method of feedback control.

In the middle of the 18th century the device was reinvented in England, apparently without knowledge of its earlier career. The float regulator's rebirth was first mentioned in a 1746 building manual, *The Country Builder's Estimator,* by William Salmon, as a device for regulating the water level in domestic cisterns. In 1758 the British bridge and canal builder James Brindley obtained a patent for a steam engine that incorporated a float valve to regulate the water level in the steam boilers. A few years later I. I. Polzunov, a Russian pioneer in the development of the steam engine, designed such a device for the same purpose. In 1784 Sutton Thomas Wood in England patented the same invention once more in a design strikingly similar to Hero's 17-centuries-old system [*see bottom illustration on opposite page*]. The float regulator soon won general acceptance as a method of feeding water to boilers. Today it is widely used for many purposes.

The thermostat does not have so ancient a history. Its first prototype was invented early in the 17th century by Cornelis Drebbel, a Dutch engineer who had migrated to England and worked in the service of James I and Charles I. A highly original inventor, Drebbel would be much better known today if he had committed his inventions to writing. According to an account by Francis Bacon, Drebbel devised his temperature regulator only incidentally, as an instrument to serve another purpose: alchemy. He believed he could transmute base metals to gold if he could keep the temperature of the process constant for a long time.

Drebbel's apparatus consisted basically of a box with a fire at the bottom and above this an inner compartment containing air or alcohol with a *U*-shaped neck topped by mercury [*see illustration at left*]. As the temperature in the box rose, the increased pressure of the heated air or alcohol vapor pushed up the mercury, which in turn pushed up a rod; this mechanical force was applied to close a damper and throttle down the fire. Conversely, if the temperature in the box fell below the desired level, the gas pressure was reduced, the mercury dropped and the mechanical linkage opened the damper.

Drebbel used his contrivance not only for smelting experiments but also to maintain an even temperature in incubators. His regulator seems to have worked with some success; members of the Royal Society of London, including Robert Boyle, Christopher Wren and in the following generation Robert Hooke, showed interest in it. Detailed descriptions of the device were given in a laboratory book by Drebbel's grandson (whose manuscript is preserved in the University of Cambridge library) and in the journals of a French devotee of science, Balthasar de Monconys, who investigated Drebbel's furnaces. Over the following century there were occasional reports of similar furnaces, evidently inspired by Drebbel's, that were built in Germany, France and America. None of these reports gave credit to Drebbel. The French natural philosopher and inventor René-Antoine de Réaumur described such a furnace for the artificial hatching of chickens and attributed its invention to a member of the French royal family, the Prince de Conti.

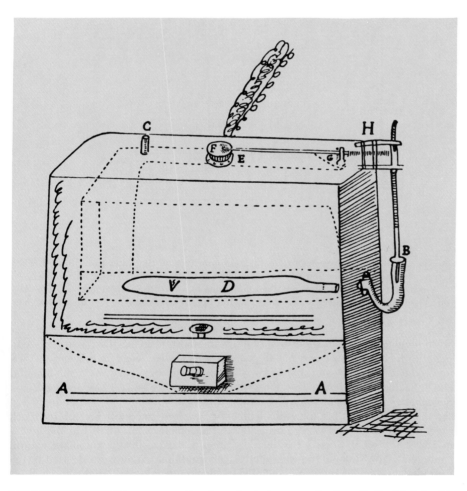

FIRST THERMOSTAT was invented early in the 17th century by Cornelis Drebbel. In this drawing, made by Drebbel's grandson, the device is shown adapted as a temperature regulator for an incubator. Smoke rising from the fire (*A–A*) passes by the water-jacketed incubator box (*dotted lines*) and escapes at the top through an opening (*E*). A glass vessel (*D*) containing alcohol is inserted into the water jacket and is sealed by mercury contained in a *U*-shaped portion of the vessel (*right*). As the temperature rises the increasing volume of the evaporating alcohol forces the mercury to rise in the right leg of the vessel, raising a float (*B*) and, through a linkage (*H*) pivoted at a point (*G*), closing a damper (*F*).

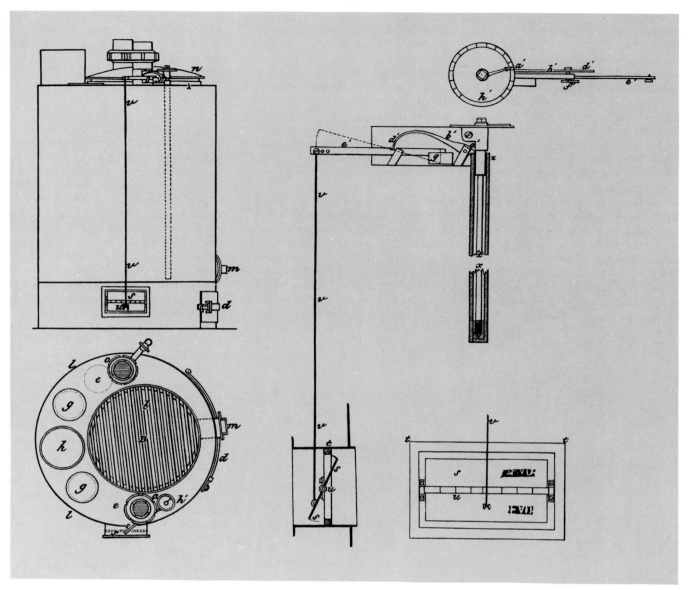

IMPROVED THERMOSTAT, shown here as it was applied to regulate the temperature in a hot-water furnace, was designed in 1783 by a Parisian inventor named Bonnemain. A sensitive two-metal temperature feeler, consisting of an iron rod (x) surrounded by a lead tube (z), was immersed in the water to be heated. The motion of the upper rim of the lead tube caused by thermal expansion was then employed to adjust the air damper (s). The desired temperature inside the furnace could be set on a dial (h).

For two centuries Drebbel's idea of temperature regulation by feedback drew little notice apart from these few sporadic reports. Then the idea suddenly aroused the interest of the entire engineering community. Credit for this achievement belongs to a Parisian inventor named Bonnemain. In 1783 Bonnemain, presumably having got the idea from reading of Réaumur's success in hatching chickens with an artificial incubator, built a *régulateur de feu* himself and obtained a French patent for it. He proceeded to employ his self-regulating incubator with success in a large farm supplying chickens to the royal court and the Paris markets. Bonnemain's apparatus was far superior to the earlier temper-

ature regulators: it had a sensitive temperature feeler made of two metals (an iron rod encased in a lead tube) and several refinements in design [*see illustration above*]. Bonnemain refrained from sharing the details of his apparatus with the world at large until he was over 80; in 1824 the French Society for the Encouragement of National Industry finally prevailed on him to publish a detailed description of his system of temperature regulation. The leading technical journals in Britain and Germany promptly published translations of this account, and Bonnemain's temperature regulator soon found its way into encyclopedias. The author of one of these, the Scottish chemist Andrew Ure, coined the term

"thermostat" in his *Dictionary of Arts, Manufactures, and Mines,* which in 1839 described Bonnemain's regulator and some that Ure himself had designed.

The third ancestral line of feedback mechanisms originated in the invention of devices for the automatic control of windmills. They were devised in the 18th century by millwrights in England and Scotland, a resourceful group who combined craft skills with the beginning of a scientific attitude. Many of the famous British mechanical engineers of the 18th and 19th centuries began their careers as millwrights.

The first of the millwrights' feedback devices, patented in 1745 by Edmund

Lee, was a fantail designed to keep the windmill facing the wind [see top illustration at right]. The fantail is a small windwheel mounted at right angles to the main wheel. It is attached to the rear side of the movable cap that turns the big wheel into the wind. Through a train of gears the fantail controls the turning of the cap, so that any rotation of the fantail will cause the cap to turn. When the main wheel squarely faces the wind, the fantail, at right angles, is aligned parallel to the wind direction and does not rotate. Whenever the wind shifts so that the main wheel no longer faces it squarely, the wind will strike the tail wheel, causing it to rotate and slowly turn the mill cap until the fantail again is parallel to the wind and the main wheel faces it. In short, the system forms a closed loop. Under actual conditions, with the wind direction constantly changing, the fantail can be considered a rudimentary servo system.

Lee's windmill also contained an invention that was designed to control the speed of the mill in spite of changes in the wind velocity. Regulation of the speed of rotation was needed to protect the millstones from excessive wear and to produce flour of uniformly fine quality. Lee attacked this problem by allowing the windmill sails to pivot around the arms that held them. The sails were connected to a counterweight that pitched their leading edge forward in moderate winds. When the wind rose to excessive velocities, so that its force on the sails was greater than that of the counterweight, the tilt of the sails was reversed and the wheel's rotation velocity was checked.

This system was not a case of feedback control, because it does not try to sense the controlled variable: speed. For genuine feedback control of a windmill's speed a method of measuring the speed with some sensitivity had to be found.

An approach to meeting that need was discovered in a mechanism known as the "lift-tenter." This device was designed to counteract the tendency of millstones to move apart as their speed of rotation increased. The lift-tenter operated to press the millstones together with a force proportional to the rotation speed [see bottom illustration at right]. In 1787 Thomas Mead, an English millwright and inventor, combined the lift-tenter idea with the use of a centrifugal pendulum to produce a speed-control system that genuinely embodied the feedback principle. The whirling pendulum measured the speed of the millstones' rotation, and through appropriate mechani-

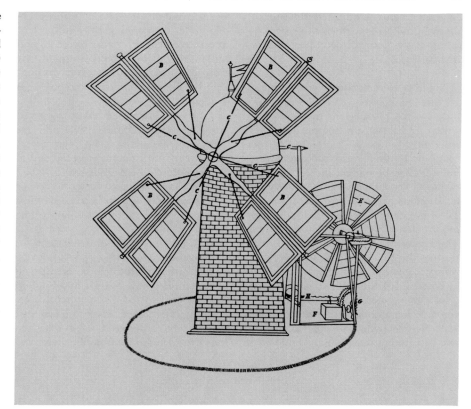

EARLY WINDMILL CONTROLS are shown in this drawing, from a British patent awarded to Edmund Lee in 1745. The regulatory devices consisted of a fantail designed to keep the windmill facing the wind and a mechanism to control the speed of the mill in spite of changes in the wind velocity. The tail wheel (E) attached to the movable cap of the mill drove a chain of gears that engaged a circular rack on the ground. If the mill was not facing the wind, the fantail would rotate, turning the main wheel into the wind. The main sails (B) of the mill were pivoted along the crossbeams and were held forward by means of a counterweight (F), to which they were attached by chains (C) running through the hollow main shaft.

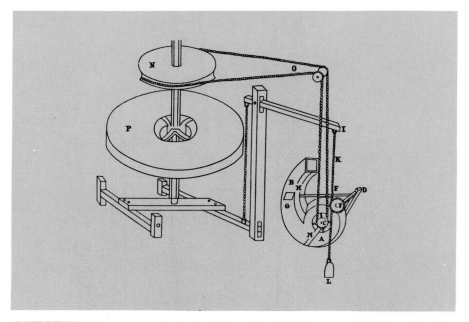

"LIFT-TENTER" was a control device designed by the 18th-century British millwrights to counteract the tendency of millstones to move apart as their speed of rotation increased. In this drawing of a lift-tenter invented by Robert Hilton in 1785 the "runner" millstone (P) was lowered in proportion to the speed of the mill. The speed was measured by means of the displacement of a baffle (B) in the discharge shroud of a centrifugal fan (A).

cal connections it adjusted the area of the windmill sails to keep the wheel rotating at the desired speed [*see illustration below*].

The idea of the centrifugal pendulum was immediately greeted with gratitude by the pioneers in the new technology of the steam engine, just then emerging. James Watt and his partner Matthew Boulton were building a large mill (later to be named the Albion Mill) where the capabilities of Watt's new rotary engine were to be demonstrated. The new engine presented totally new requirements for its regulatory system. There was no way to adapt the existing devices to the continuously operating rotary engine.

Watt and Boulton hired John Rennie, then a young man of 23, to supervise the construction and operation of the Albion Mill. Rennie (who was later to become one of Britain's most famous builders of

bridges) had just finished his apprenticeship under the noted Scottish millwright Andrew Meikle. In a visit to the Albion Mill in May of 1788 Boulton found that a lift-tenter had been installed, presumably by Rennie. Boulton promptly sent a detailed and enthusiastic description of it to Watt. The idea fell on prepared ground. By November, Watt and his colleagues had designed a "centrifugal speed regulator," and around the end of the year the first governor was installed on the "Lap" engine. The picture of Watt's governor was to become perhaps the most familiar one in the entire history of technology.

Watt did not take out a patent for the governor. He considered the device merely an adaptation of the centrifugal pendulum to a new use. He and Boulton tried to protect it from competitors by keeping its existence secret; the first customers who ordered it were asked to hide the governor from public view. The

device soon became known, however. Within a few years after its invention it was recognized everywhere as a symbol of the steam engine. Rotating dramatically at the top of every steam engine, it demonstrated the action of feedback control more widely and more forcefully than words could have done. The governor soon entered the textbooks and handbooks of engineering, and inventors began to develop feedback devices in other areas of technology.

It is curious that all the inventions of feedback devices that came in with the beginning of the Industrial Revolution originated in Britain. Even those inventors who were not British-born, notably Drebbel and Denis Papin (the Frenchman who invented the safety valve, a rudimentary feedback device), presented their inventions while working in England. Why was the Continent so backward? Why was it, for instance,

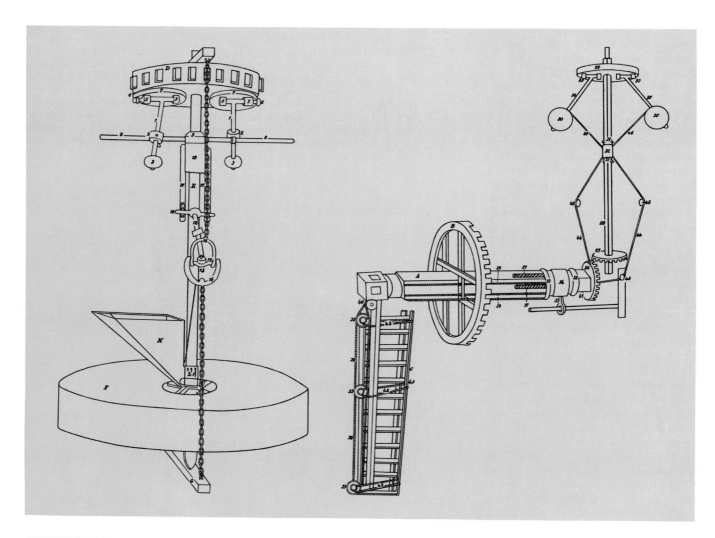

CENTRIFUGAL PENDULUMS were employed as feedback control devices by the English millwright Thomas Mead in his 1787 windmill patent. The speed of rotation of the mill sensed by one set of centrifugal pendulums drove the mill's lift-tenter mechanism (*left*). The motion of another set of pendulums in turn regulated the speed of the mill by reducing the area of the sails (*right*).

that engineers and inventors on the Continent ignored the float valve presented in the widely read translations of Hero's *Pneumatica* in 1575 and took serious notice of it only after the device was rediscovered in Britain two centuries later?

One can speculate plausibly that in the 16th to 18th centuries the Continental mind rejected feedback control because it was preoccupied with a different conception of control: control by rigidly predetermined program. In technology this was evidenced by almost countless inventions of automatons, monumental clocks, music boxes and clock-driven planetariums. The fascination with ordered programs was reflected in the Continent's prevailing attitude toward the state (absolute government) and in the economic system (mercantilism). In Britain, on the other hand, scientists, inventors and philosophers early in the 18th century began to turn to a different concept of control, one in which the system was truly autonomous, containing inherent mechanisms that maintained its equilibrium and viability. In technology such thinking led to the creation of feedback devices, in economics to the free-market system of Adam Smith and in political science to the division of powers and constitutional government.

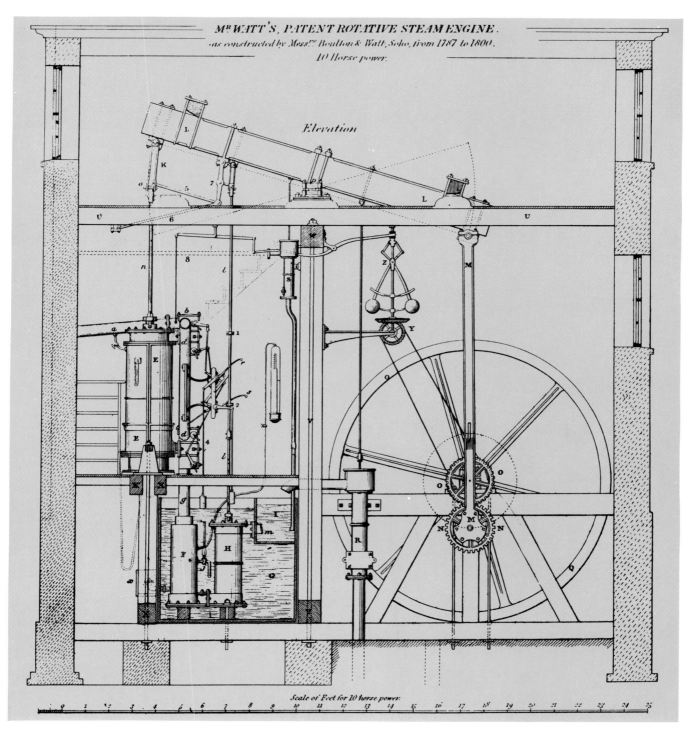

FLYBALL GOVERNOR incorporated into the first continuously operating rotary steam engines in the 1790's by James Watt and Matthew Boulton was based directly on the windmill lift-tenter. In this 1826 drawing the centrifugal pendulum (*top center*), driven by a pulley, is linked to a throttle valve in the engine's steam line, enabling it to throttle the steam supply with increasing speed.

Bicycle Technology

by S. S. Wilson
March 1973

*This humane and efficient machine played a central
role in the evolution of the ball bearing, the pneumatic
tire, tubular construction and the automobile and
the airplane*

We tend to take the bicycle too much for granted, forgetting the important role it played in the evolution of modern technology. The first machine to be mass-produced for personal transportation, the bicycle figured prominently in the early development of the automobile. Thus in addition to its own considerable direct impact on society the bicycle was indirectly responsible for substantial social and economic changes. A remarkably efficient machine both structurally and mechanically, the bicycle continues to offer distinct advantages as a means of personal transportation in both developed and underdeveloped countries.

When one considers how long the wheel has served in transportation (more than 5,000 years), it seems odd that the first really effective self-propelled wheeled vehicle was developed only about 100 years ago. As with most epoch-making inventions, many men and many nations can claim a share in its development. The earliest legitimate claimant would be Baron von Drais de

ROVER SAFETY BICYCLE, introduced in 1885 by J. K. Starley of England, is widely regarded as marking the final development of the bicycle form. The Rover had most of the major features of the modern bicycle: rear-wheel chain-and-sprocket drive with a "geared up" transmission, ball bearings in the wheel hubs, tangentially mounted wire spokes, lightweight tubular-steel construction and a diamond-shaped frame. Unlike most modern bicycles, the Rover incorporated two curved tubes without the extra diagonal tube from the saddle to the bottom bracket; also the front forks, although sloping, were straight instead of curved. Unless otherwise noted, the old vehicles shown in the photographs used to illustrate this article are now in the Science Museum in London.

DRAISIENNE, a two-wheeled "pedestrian hobby-horse" devised between 1816 and 1818 by Baron von Drais de Sauerbrun of Baden-Württemberg, is considered the earliest fore-runner of the bicycle. The vehicle, which was propelled by the feet pushing directly on the ground, was not, however, regarded as a serious means of transportation. This par-ticular model, dating from 1817, was photographed at the city museum in Heidelberg.

Sauerbrun of Baden-Württemberg, who between 1816 and 1818 devised the Draisienne, a two-wheeled "pedestrian hobby-horse" propelled by the feet push-ing directly on the ground [see illustra-tion above]. The vehicle had a brief vogue, but it was not taken seriously as a means of transportation. In 1839 Kirk-patric Macmillan, a Scottish blacksmith, succeeded in making a treadle-driven two-wheeled machine, which was copied but was never a popular success [see il-lustration below].

The first commercially important ma-chine in this lineage was the French velocipede, developed by Pierre and Ernest Michaux in Paris in 1863 [see top illustration on opposite page]. This vehi-cle, sometimes called "the bone-shaker,"

had cranks fixed directly to the hub of the front wheel, like the simplest child's tricycle. As a result it suffered from the limitation of having too low a "gear ra-tio," to use the modern term. This meant that one turn of the pedals advanced the machine a distance equal to the cir-cumference of the front wheel, perhaps only 10 feet. (In a modern bicycle one turn of the pedals, by means of a chain drive from a large sprocket to a small one, advances the machine 16 feet or more.) The only way to overcome this limitation while retaining the simplicity of direct drive was to use a very large front wheel. Thus the next stage in the evolution of the bicycle, the famous "high-wheeler," was characterized by a front wheel as much as 60 inches in

diameter accompanied by a back wheel with a diameter of 20 inches or less.

The high-wheeler, also known as the "penny farthing" or "ordinary" machine, evolved primarily at Coventry in En-gland, now the center of the British automobile industry. It was largely the work of one family, the Starleys. This sequence of events ensued from the ac-cident that the Coventry Sewing Ma-chine Company had a representative in Paris, one Rowley Turner, who brought a Michaux velocipede back to Coven-try in 1868. James Starley, a self-taught engineer and inventor, was works man-ager at the time, and he immediately saw the possibilities of bicycle manu-facture. The firm promptly became the Coventry Machinists' Company, Limit-ed, and began to take orders for the manufacture of several hundred ma-chines of the Michaux type for the Paris market. As it happened, the Franco-Prussian War of 1870 intervened, so that the machines were sold mostly in England. This episode led to a period of intense technical and commercial devel-opment that resulted not only in the bi-cycle's achieving its definitive form but also in the emergence of the motorcycle, the motor tricycle and the automobile.

Before considering these developments in detail it is worth asking why such an apparently simple device as the bi-cycle should have had such a major ef-fect on the acceleration of technology. The answer surely lies in the sheer hu-manity of the machine. Its purpose is to make it easier for an individual to move about, and this the bicycle achieves in a way that quite outdoes natural evolu-tion. When one compares the energy consumed in moving a certain distance as a function of body weight for a va-riety of animals and machines, one finds that an unaided walking man does fairly well (consuming about .75 calorie per gram per kilometer), but he is not as efficient as a horse, a salmon or a jet transport [see illustration on page 64]. With the aid of a bicycle, however, the man's energy consumption for a given distance is reduced to about a fifth (roughly .15 calorie per gram per kilo-meter). Therefore, apart from increasing his unaided speed by a factor of three or four, the cyclist improves his efficiency rating to No. 1 among moving creatures and machines.

In order to make this excellent per-formance possible the bicycle has evolved so that it is the optimum design ergonomically. It uses the right muscles (those of the thighs, the most powerful in the body) in the right motion (a

TREADLE-DRIVEN TWO-WHEELER was built in 1839 by Kirkpatric Macmillan, a Scot-tish blacksmith, for his own use. Although copied, the machine was never a popular success.

smooth rotary action of the feet) at the right speed (60 to 80 revolutions per minute). Such a design must transmit power efficiently (by means of ball bearings and the bush-roller chain); it must minimize rolling resistance (by means of the pneumatic tire), and it must be the minimum weight in order to reduce the effort of pedaling uphill.

The reason for the high energy efficiency of cycling compared with walking appears to lie mainly in the mode of action of the muscles. Whereas a machine only performs mechanical work when a force moves through a distance, muscles consume energy when they are in tension but not moving (doing what is sometimes called "isometric" work). A man standing still maintains his upright posture by means of a complicated system of bones in compression and muscles in tension. Hence merely standing consumes energy. Similarly, in performing movements with no external forces, as in shadowboxing, muscular energy is consumed because of the alternate acceleration and deceleration of the hands and arms, although no mechanical work is done against any outside agency.

In walking the leg muscles must not only support the rest of the body in an erect posture but also raise and lower the entire body as well as accelerate and decelerate the lower limbs. All these actions consume energy without doing any useful external work. Walking uphill requires that additional work be done against gravity. Apart from these ways of consuming energy, every time the foot strikes the ground some energy is lost, as evidenced by the wear of footpaths, shoes and socks. The swinging of the arms and legs also causes wear and loss of energy by chafing.

Contrast this with the cyclist, who first of all saves energy by sitting, thus relieving his leg muscles of their supporting function and accompanying energy consumption. The only reciprocating parts of his body are his knees and thighs; his feet rotate smoothly at a constant speed and the rest of his body is still. Even the acceleration and deceleration of his legs are achieved efficiently, since the strongest muscles are used almost exclusively; the rising leg does not have to be lifted but is raised by the downward thrust of the other leg. The back muscles must be used to support the trunk, but the arms can also help to do this, resulting (in the normal cycling attitude) in a little residual strain on the hands and arms.

A less comfortable attitude is adopted by a racing cyclist in order to lessen wind resistance, perhaps the worst feature of

FRENCH VELOCIPEDE, produced by Pierre and Ernest Michaux in Paris in 1863, was the first commercially important machine on the way to the modern bicycle. The "bone-shaker," as the vehicle was sometimes called, had cranks with pedals fixed directly to the hub of the front wheel and hence suffered from the limitation of having too low a gear ratio.

HIGH-WHEELER, developed primarily by the Starley family, was designed to overcome the low gear ratio of the Michaux-type velocipede while retaining direct drive. The high-wheeler, also known as the "penny farthing" or "ordinary" machine, had a front wheel as much as 60 inches in diameter and a back wheel only 20 inches or less. This particular all-metal design, called the Ariel, was produced in 1870 by James Starley and William Hillman. Its spokes, which were radial, were not well adapted to resist the large torque exerted by the pedals on the hub. Hence the two extra rigid bars, each with its own adjustable spoke, were added to help transmit the torque from the hub to the rim. The definitive solution to this problem, the tangent-spoked wheel, was patented by Starley four years later.

COVENTRY LEVER TRICYCLE was designed by James Starley in 1876 as a way of circumventing the difficulties encountered in mounting the high-wheeler and then staying aloft.

ROYAL SALVO TRICYCLE, also designed by James Starley, attracted the attention of Queen Victoria, who ordered two of the machines, thereby establishing the respectability of the new fad of cycling. This tricycle, which incorporated one of the earliest differential gears, invented by Starley in 1873, was considered particularly well suited for female riders.

the bicycle for energy loss. Wind resistance varies as the square of the velocity of the wind with respect to the cyclist. Hence if one were to cycle at 12 miles per hour into a wind blowing at six miles per hour, the wind resistance would be nine times greater than if one were to maintain the same road speed with a following wind of six miles per hour. In practice, as every cyclist knows, one's speed can be adjusted to suit the wind conditions with a change of gear ratio to maintain an optimum pedaling speed. Apart from wind resistance the only significant form of energy loss is due to rolling resistance, which with normal-size wheels and properly inflated tires is very small on a smooth surface and is almost independent of speed.

It is because every part of the design must be related to the human frame that the entire bicycle must always be on a human scale. The lightness of construction, achieved mainly through the development of the wire-spoke wheel and the tubular frame, was dictated not only by the fact that the machine has to be pedaled uphill but also by the desirability of making it easy to lift. Since the bicycle makes little demand on material or energy resources, contributes little to pollution, makes a positive contribution to health and causes little death or injury, it can be regarded as the most benevolent of machines.

To return to the story of the bicycle's evolution, in 1870 James Starley and William Hillman (who later founded the automobile firm named after him) designed the Ariel machine: an elegant all-metal high-wheeler with wire-spoke wheels [*see bottom illustration on preceding page*]. The spokes, which were radial and could be tightened as desired, were not well adapted to resist the large torque exerted by the pedals on the hub. Four years later Starley patented the definitive solution to the problem: the tangent-spoked wheel. In this design, now universal, the spokes are placed so as to be tangential to the hub in both the forward and the backward direction, thus forming a series of triangles that brace the wheel against torque during either acceleration or braking. The usual number of spokes in a modern bicycle is 32 in the front wheel and 40 in the back; they are of uniform thickness or else butted (thickened toward both ends) for greater strength with lightness.

The major flaws of the high-wheeler bicycle were the difficulty in mounting the machine and then staying aloft. To overcome these dangers tricycles were developed during the 1870's. Again

Starley took the lead with his Coventry lever tricycle of 1876 [*see top illustration on opposite page*]. Another notable attempt to make the high-wheeler safer was the Star bicycle made by the Smith Machine Company of New Jersey in 1881 [*see illustration at right*]. That machine had the small wheel in front and used a system of levers to drive the large wheel at the rear. The Star bicycle had some success, but the tricycle was much more suitable for women riders, one of whom attracted the attention of Queen Victoria; she ordered two of Starley's Royal Salvo tricycles, met the designer and presented him with an inscribed watch. Nothing could have been more effective in establishing the respectability of the new craze of cycling. It made it possible for well-brought-up young ladies to get out and away from the stuffiness of their Victorian homes and led to such new freedoms as "rational dress," a trend led by Amelia Jenks Bloomer. It is not too farfetched to suggest that the coincidence of cycling with the gradual spread of education for women played a significant part in the early stages of women's movement toward political and economic equality.

The Starleys were also responsible at about this time for a technical innovation in one of their machines that was to be of major importance for the automobile. That development arose from a difficulty encountered with a side-by-side two-seater machine in which James Starley and one of his sons, William, each pedaled a driving wheel. The result of young William's superior strength was a spill into a bed of nettles for his father. While recovering, James thought up the idea of the differential gear (actually a reinvention), which spreads the effort equally between the wheels on each side and yet allows the wheels to rotate at slightly different speeds when turning a corner. William Starley himself went on to be a prolific inventor, having 138 patents to his name, many for bicycles, when he died in 1937.

Two major developments of 1877 were the introduction of the tubular frame and ball bearings. In neither case was the concept altogether new, but it was its widespread adoption for the bicycle that brought each technique to fruition and universal use. This pattern was to be repeated later with the pneumatic tire and other innovations.

The thin-walled tube of circular cross section is a most efficient structural member; it can resist tension or compression, bending, torsion or the combination of stresses that are exerted on the frame of a vehicle. Although for bending in a par-

STAR BICYCLE, built by the Smith Machine Company of Smithville, N.J., in 1881, was another notable attempt to make the high-wheeler safer. The machine had a small wheel in front and used a system of levers, drums and straps to drive the large wheel at the rear.

ticular plane an *I*-section joist may be more efficient, if the bending load can be applied in any plane, then the thin tube is to be preferred. It is for this reason that tubes are used as a strut or compression member in which failure could occur by elastic instability, or bowing. For torsion there is no better section, hence the tube in the typical main transmission shaft of an automobile. The stem of the bamboo plant is an excellent example of the properties of a hollow tube, as is attested by its use in the Far East for buildings, bridges, scaffolding and so on. Indeed, bicycles have been made of bamboo.

For the smallest stresses the design of a structure should be such as to result in tension or compression, not bending or torsion. That is the principle of the "space frame," which is used in bridge trusses, tower cranes and racing cars. Such construction is not practical in a bicycle, and so a compromise emerged in the classic diamond frame. In such a frame the main stresses are taken directly, even though there are bending stresses in the front fork and torsional stresses in the entire frame as the rider exerts pressure first on one pedal and then on the other. On a high-wheeler the rider feels these forces through the handlebars.

An alternative to the diamond frame

that first appeared in 1886 and has recently reappeared is the cross frame. It consists of a main tube, extending from the steering head above the front wheel directly to the rear axle, crossed by a second tube from the saddle to the bottom bearing carrying the pedals. This is a simple arrangement, but it relies entirely on the strength and stiffness of the main tube, unless further members are added to obtain a partial triangulation of the frame.

Here an improvement is to increase the cross-sectional area of the main tube, thereby obtaining the full benefit of thin-tube construction. This principle, used in most airplane fuselages since the 1930's, is described as monocoque or stressed-skin construction. Some recent motorized bicycles ("mopeds") have gone further and incorporate an enlarged main tube that forms the gasoline tank, a principle applied also in the construction of certain modern racing cars, which have a tubular fuel tank on each side of the driver. One advantage of the cross-frame bicycle is that it is equally suitable for men and skirt-wearing women.

At least one early example of a fully triangulated frame achieved some success: the Dursley-Pedersen, which used small-diameter tubes in pairs and weighed only 23 pounds [*see illustra-*

FULLY TRIANGULATED FRAME was the most distinctive feature of the Dursley-Pedersen bicycle, which achieved some success in England in the 1890's. This machine used small-diameter tubes in pairs and was remarkably light, weighing in at only 23 pounds.

tion above]. Some remarkably low weights were achieved at a very early date, mainly for racing machines. For example, a Rudge "ordinary" of 1884 weighed only 21.5 pounds. For purposes of comparison, a modern racing bicycle weighs about 20 pounds and a typical modern tourer weighs about 30 pounds.

Steel tubes are still the normal choice, but light alloys, titanium and even plastic reinforced with carbon fibers have been used. The steel tubes, usually of a chrome-molybdenum or a manganese-molybdenum alloy, are butted and then brazed into steel sockets to form a complete frame. The result is a structure able

to carry about 10 times its own weight, a figure not approached by any bridge, automobile or aircraft.

The use of a roller to reduce friction finds its ultimate development in ball bearings and roller bearings. The main bearings of a modern bicycle have at least 12 rows of balls, all rolling between an inner cone and an outer cup. All the bearing surfaces are hardened steel for resistance to deformation and wear. If such bearings are lubricated and properly adjusted, their life can be surprisingly long. Even if they are neglected, they continue to function for a long time, and all their parts can easily be renewed if necessary.

Another important advance, the adoption of chain-and-sprocket drive to the rear wheel, was made by Harry J. Lawson in 1879. The following year Hans Renold produced the definitive form of the bicycle chain, the bush-roller chain, which combines the virtues of long life, efficiency and low weight. At first sight the design seems to have little subtlety but a closer look reveals just how significant its various features are [*see illustration on page 62*]. The progenitor was the pin chain, or stud chain, in which the pins bear directly on the sprocket teeth and the link plates swivel on the studs at each end of the pins. In such a chain there is undue wear and friction both at the teeth and at the holes in the plates. An improvement devised by James Slater in 1864 was the bowl chain, or roller chain, in which friction and wear of the sprocket teeth was reduced by rollers on the pins, but wear of the plates on the studs was still too great. Renold's design, by the addition of hollow bushes to spread the load over the entire length of the pin, overcame this final shortcoming and led to the foundation of the precision-chain industry. The bush-roller chain spread from bicycles to textile machinery and other power-transmission applications, in competition with the Morse "silent chain." Bush-roller chains replaced belt drives for motorcycles, and they served as the main drive for the rear wheels of automobiles until they were replaced by shaft drive. Today the bush-roller chain drive remains virtually the universal choice for automobile camshafts, although it is threatened now by the toothed-belt drive. The demands of increased power, greater speed and longer life for chain drives meant that the makers were pioneers in metallurgy, heat treatment, lubrication and production.

The satisfactory solution to the problem of an efficient drive giving a "step up" ratio of any desired value made possible the final evolution of the bicycle to

STEAM-DRIVEN BICYCLE was built in 1869 by Pierre Michaux using a Perreaux steam engine. Although earlier steam-driven road vehicles had proved too heavy and cumbersome, the technology of the lightweight bicycle seemed at this time to offer new possibilities. The later evolution of the internal-combustion engine ended this line of development.

its modern form. This last step was achieved mainly by J. K. Starley, a nephew of James Starley. James, honored in Coventry as the "father of the bicycle industry," had formed a partnership with William Sutton in 1878 to produce tricycles. In 1885, however, J. K. independently brought out his famous Rover safety bicycle [see illustration on page 55]. This machine can be regarded as the final development of the bicycle form. From that form the bicycle has not departed, in spite of a recent attempt to use a spring frame in conjunction with small wheels. The Rover bicycle had a diamond frame, incorporating two curved tubes without the extra diagonal tube from the saddle to the bottom bracket. The front forks were straight, although sloping. The slope was used to give a self-centering action to the steering. The later development of curved front forks such that the line of pivoting of the steering head meets the ground at the point of contact with the tire results in a reduction of steering effort because side forces do not tend to turn the handlebars.

The appearance of the Rover safety bicycle started a boom in bicycles that quickly established them as an everyday means of transport, as a sport vehicle and as a means of long-distance touring. For Coventry and other Midlands cities there was a trade boom that lasted until 1898. Then there was a disastrous slump, largely because of the financial manipulations of one Terah Hooley. The bicycle spread all over the world, and it was adapted to local needs in such machines as the bicycle rickshaw of the Far East. Among the early manufacturers in the U.S. were the Duryea brothers, builders of the first American automobile in 1893. By 1899 there were 312 factories in the U.S. producing a million bicycles a year.

Before this time, however, there was one more signal development: the pneumatic tire. This feature had actually been patented as far back as 1845 by R. W. Thomson, a Scottish civil engineer, to reduce the effort needed to pull horse-drawn carriages. It failed to establish itself until it was reinvented in 1888 by John Boyd Dunlop, a Scottish veterinary surgeon practicing in Belfast. This time success was rapid, owing to the enormous popularity of the bicycle and to the obvious superiority in comfort and efficiency of the pneumatic tire over the solid-rubber tire. Further developments came quickly. Charles Kingston Welch of London produced the wire-edged tire in 1890, and at almost the same time William Erskine Bartlett in the U.S. pro-

DAIMLER MOTOR BICYCLE, the forerunner of the modern motorcycle, was designed and built by Gottlieb Daimler in 1885. The vehicle was equipped with a single-cylinder internal-combustion engine. The two smaller jockey wheels retracted when the machine was under way. The original Daimler machine, shown in this photograph from the Bettmann Archive, is now in the restored Daimler workshop at Bad Cannstatt near Stuttgart.

BENZ MOTOR TRICYCLE, the forerunner of the modern automobile, also made its appearance in 1885. This first attempt of Carl Benz incorporated such features as electric ignition, effective throttle control, mechanical valves, horizontal flywheel and even a comfortably upholstered seat. In tests through 1886 the vehicle was developed to deliver a reliable nine miles per hour. The original Benz car is now in the Deutsches Museum in Munich; this photograph shows a replica of the original in the Daimler-Benz Museum in Stuttgart.

duced the bead-edged tire. Then David Mosely of Manchester patented the cord-construction tire, consisting of layers of parallel cords rather than a woven fabric, which gave rise to undue internal friction. Even the tubeless tire, which was not used for automobiles until the 1940's, was invented for bicycles in the late 1890's. E. S. Tompkins, a student of the pneumatic tire, has written: "This richness of invention in the very earliest years arises from the fact that the cycling enthusiasts were young folk and they had a young and enquiring approach to the design of tyres for their machines."

A further result of the popularity of

bicycles was the demand for better roads. In Britain the Cyclists' Touring Club was founded in 1878 and the Roads Improvement Association in 1886, leading to successful pressure for better roads. A similar movement in the U.S., founded by Colonel Albert A. Pope, a pioneer bicycle manufacturer of Boston, was also influential. The bicycle quite literally paved the way for the automobile.

How the bicycle led directly to the automobile is described by the inventor Hiram P. Maxim, Jr., in *Horseless Carriage Days* (published in 1937): "The reason why we did not build mechanical road vehicles before this [1890] in my opinion was because the bicycle had not yet come in numbers and had not directed men's minds to the possibilities of independent long-distance travel on the ordinary highway. We thought the railway was good enough. The bicycle created a new demand which was beyond the capacity of the railroad to supply. Then it came about that the bicycle could not satisfy the demand it had created. A mechanically propelled vehicle was wanted instead of a foot-propelled

one, and we now know that the automobile was the answer."

Looking backward after another 36 years, one is tempted to ask whether the automobile is really as good an answer as it once appeared to be. Steam-propelled road vehicles had been tried earlier, but they had failed to establish themselves because they were heavy and cumbersome. The technology of the lightweight bicycle seemed to offer new possibilities, and a steam-driven motorcycle was actually built in France in 1869 by Pierre Michaux, using a Perreaux enplane designed by Christopher Roper [*see illustration on page 65*]. This airwas successfully applied in 1885 by Gottlieb Daimler to a velocipede of the Michaux type [*see top illustration on preceding page*] and by Carl Benz to a lightweight tricycle [*see bottom illustration on preceding page*]. From then on development was rapid, particularly in France and later in the U.S.

Many of the pioneer automobile manufacturers started as bicycle makers, including Hillman, Colonel Pope, R. E. Olds, Henry M. Leland and William Morris (later Lord Nuffield). Meanwhile in Coventry the firm of Starley and Sut-

ton became the famous Rover Company, producing an electrically driven tricycle in 1888, a motorcycle in 1902 and an automobile in 1904. Both Morris and Rover are now part of the British Leyland Motor Corporation, the largest British automobile company. Henry Ford's first car used bicycle wheels and chains, as did other early vehicles of the "motorized buggy" type. The Wright brothers were bicycle makers, and the early flying machines benefited considerably from the lightweight and efficient design features evolved so successfully for bicycles.

Perhaps the most interesting modern application of the technology of the lightweight bicycle to flying is in man-powered aircraft. There were attempts to develop such aircraft in Germany and Italy before World War II, and recently much interest has been aroused in England by the offer of a prize of £ 10,000 by Henry Kramer for the first man-powered flight over a figure-eight course around two poles half a mile apart. Several designs have succeeded in flying for distances of up to some 1,000 yards in a straight line; the best flight so far is one made by Flight Lieutenant John Potter of the Royal Air Force in a monoplane designed by Christopher Roper [*see illustration on page 65*]. This aircraft, with a wingspan of 80 feet and a weight of only 146 pounds, clearly shows the debt owed to the bicycle in the design of efficient and lightweight machinery for the production and transmission of power. The Kramer prize has not yet been won, but it represents a goal that will no doubt eventually be achieved.

Production engineering also owes much to the sudden demand created by the bicycle for precision parts in quantity. The average bicycle has well over 1,000 individual parts. Admittedly nearly half of these parts are in the chain, but the rest of them call for high standards of pressing and machining, and the methods worked out for producing them represented a big step forward. A comparable demand for automobile parts had to await the Ford assembly line. Perhaps the most ingenious process used in producing bicycle parts is the procedure for forming the complicated bottom bracket of the frame, which has four tube sockets and a threaded barrel to receive the outer races for carrying the pedals. The bottom bracket is made in a series of press operations, a method evolved by the Raleigh Cycle Company of Nottingham in 1900. This company, which is now by far the largest bicycle

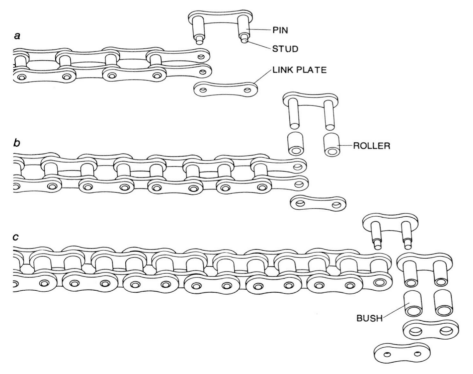

DEVELOPMENT OF BICYCLE CHAIN went through several stages before arriving at the definitive form, the bush-roller chain invented by Hans Renold in 1880. The progenitor was the pin chain, or stud chain (*a*), in which the pins bore directly on the sprocket teeth and the link plates swiveled on the studs at each end of the pins. In such a chain there is undue wear and friction both at the teeth and at the holes in the plates. A subsequent improvement was the bowl chain, or roller chain (*b*), in which friction and wear of the sprocket teeth was reduced by rollers on the pins, but wear of the plates on the studs was still too great. Renold's invention of the bush-roller chain (*c*), by the addition of hollow bushes to spread the load over the entire length of the pin, overcame this final shortcoming.

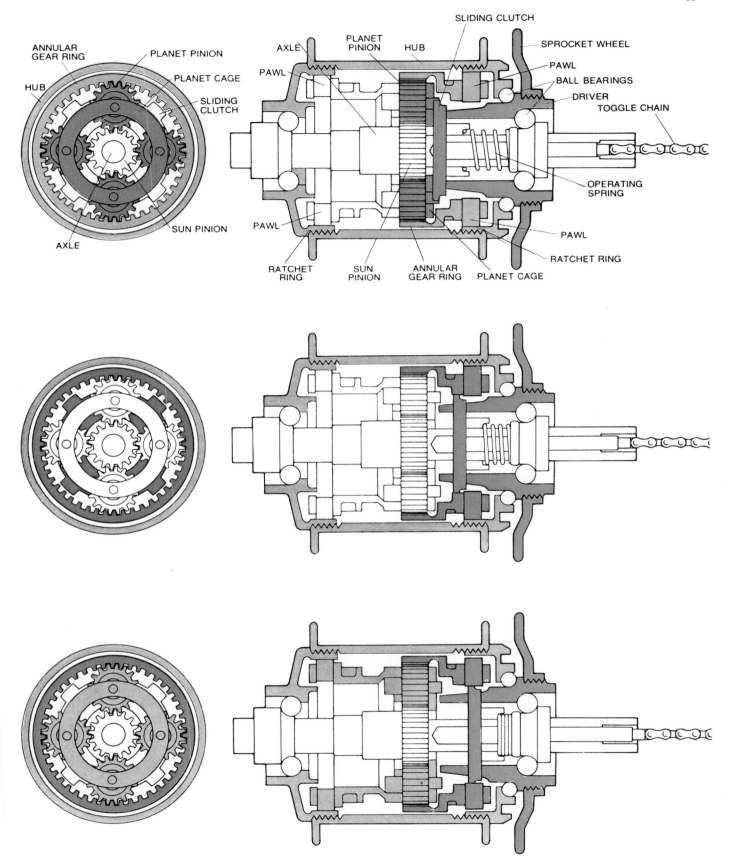

STURMEY-ARCHER HUB GEAR commonly used in British touring bicycles is shown here in both transverse section (*left*) and longitudinal section (*right*). In both cases the driving elements are indicated in dark color and the driven elements in light color. In this three-speed system a single epicyclic gear train is used in such a way that when the bicycle is in high gear (*a*), the cage is driven by the sprocket and the wheel is driven by the annulus. When the bicycle is in middle gear (*b*), the drive is direct to the wheel. When it is in low gear (*c*), the sprocket drives the annulus and the cage drives the wheel at a reduced speed. If the "sun" wheel (which is fixed to the stationary axle) has the same number of teeth as the "planet" wheels, and the annulus gear has three times this number, then high gear will have a step-up ratio of 4/3 compared with direct drive and low gear a step-down ratio of 3/4.

manufacturer in Britain, was founded by Sir Frank Bowden in 1887.

Other accessories that made their appearance during the 1880's and 1890's include the freewheel mechanism, originally introduced as an aid to mounting, since with the old "fixed" drive some form of mounting step was needed. The abandonment of a fixed drive created the need for better brakes. Originally bicycle brakes were of the simple "spoon" type, pressing on the front tire. Rim brakes came in later, first the stirrup brake, which acts on the underside of the rim, then the caliper brake, which presses on the flat outer sides of the narrow rim used in racing bicycles and lightweight touring machines. The original purpose of the caliper brake was to make it easier to change the wheel without disturbing the brake. Such a brake is operated by a cable in which an inner flexible wire in tension is contained within a flexible outer tube in compression. This system provides a most effective means for the remote operation of a

mechanism such as a brake, for which it was first developed. It has since been widely used for such purposes as operating the clutch and throttle mechanisms of motorcycles and the control surfaces of airplanes.

Two other types of brake made their appearance later. One is the coaster brake, or back-pedaling brake, which is particularly popular in the U.S. The other is the hub brake, or drum brake, of the type used in automobiles and motorcycles. Both types have the advantage of remaining effective in wet conditions.

The desirability of being able to change the gear ratio of a bicycle to provide for pedaling uphill or against a head wind is fairly obvious. This gear ratio is still defined by the diameter of the equivalent wheel of the original high-wheelers. Thus a wheel 27 inches in diameter driven by sprockets of 44 and 18 teeth has a "gear" of $(27 \times 44)/18$, or 66 inches. The idea of having different sizes of sprockets on the rear

wheel led to the *dérailleur* change-speed gear of 1899, in which the rider can transfer the chain from one sprocket to another while pedaling. The necessary variation in the length of the chain on the tight side of the drive is accommodated by the use of a spring-loaded jockey pulley on the slack side of the chain. This type of gear is light and efficient but needs proper adjustment and lubrication if it is to last.

The alternative form of change-speed gear is the Sturmey-Archer hub gear [see illustration on preceding page]. In this system a single epicyclic gear train is used in such a way that when the bicycle is in high gear, the cage is driven by the sprocket and the wheel is driven by the annulus. When the bicycle is in middle gear, the drive is direct to the wheel, and when it is in low gear, the sprocket drives the annulus and the cage drives the wheel at a reduced speed. If, for example, the "sun" wheel (which is fixed to the stationary axle) has the same number of teeth as the "planet" wheels, and the annulus gear has three times this number, then high gear will have a step-up ratio of 4/3 compared with direct drive and low gear a step-down ratio of 3/4. A five-speed version is available incorporating two epicyclic gear trains of different ratios. The advantage of hub gears is compactness and the fact that the mechanism is well protected against dust and damage.

A large number of firms grew up to supply bicycle parts or accessories. One firm in particular, Joseph Lucas Limited, owes its early prominence to the manufacture of the first successful bicycle lamp, an oil-burning wick lamp rejoicing in the name of the King of the Road Cycle Hub Lamp. Later this company made acetylene lamps and then electric lamps, so that they were in a good position to develop the market for automobile lights and other electrical equipment for automobiles. An early form of an electric generator for bicycles was invented by Richard Weber of Leipzig in 1886. The modern hub generator, which requires a minimum of additional effort on the part of the rider, was first produced by Raleigh in 1936.

Bicycle manufacture is still a big business, accounting for a worldwide production of between 35 and 40 million vehicles per year. The leading manufacturing country is still the U.S., with about six million bicycles per year, followed closely by China, with about five million. By any standards these figures demonstrate the importance of the bicycle. If one examines the extent to

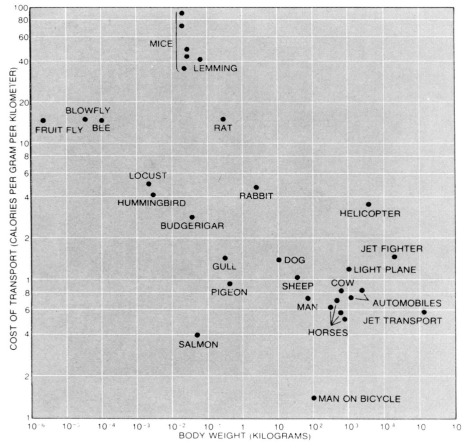

MAN ON A BICYCLE ranks first in efficiency among traveling animals and machines in terms of energy consumed in moving a certain distance as a function of body weight. The rate of energy consumption for a bicyclist (about .15 calorie per gram per kilometer) is approximately a fifth of that for an unaided walking man (about .75 calorie per gram per kilometer). With the exception of the black point representing the bicyclist (*lower right*), this graph is based on data originally compiled by Vance A. Tucker of Duke University.

which bicycles are in use today, one finds that in most of the world they play a role far more significant than that of the automobile. China with its 800 million inhabitants relies heavily on the bicycle for the transport of people and goods. So do the countries of Southeast Asia and Africa. Even the U.S.S.R., with only about 1.5 million automobiles, has an annual production of 4.5 million bicycles. Europe and North America are therefore in a minority in relying so heavily on the automobile. The true cost of doing so is becoming increasingly evident, not only in the consumption of resources but also in pollution and other undesirable effects on urban life.

For those of us in the overdeveloped world the bicycle offers a real alternative to the automobile, if we are prepared to recognize and grasp the opportunities by planning our living and working environment in such a way as to induce the use of these humane machines. The possible inducements are many: cycleways to reduce the danger to cyclists of automobile traffic, bicycle parking stations, facilities for the transportation of bicycles by rail and bus, and public bicycles for "park and pedal" service. Already bicycling is often the best way to get around quickly in city centers.

Two important factors must gradually force a reappraisal of the hypertrophic role the automobile plays in Western life. The first is the undoubted diminution of fossil-fuel resources and the accompanying increase in fuel prices. The second is the sheer inequity in per capita energy consumption between automobile-using and non-automobile-using countries. In these days of universal communication such a situation will appear more and more inequitable and a source of resentment. It is inconceivable that 800 million Chinese will ever become consumers of energy on the per capita scale of 200 million Americans, and the end result must be a gradual reduction of energy consumption in the U.S. To this end the bicycle can play a significant part and thereby become a great leveler.

For the developing countries the bicycle offers a different set of opportunities. With the continuing spread of bicycles from cities to towns and villages go the accompanying mechanical skills and essential spare parts. Thus bicycle technology serves the purpose of technical education on which the peoples of these countries can build in the same way that we in the developed countries

RECENT APPLICATION of bicycle technology to flying is in the design of man-powered aircraft, a goal that has attracted a great deal of interest in England lately following the offer of a prize of £10,000 by Henry Kramer for the first man-powered flight over a figure-eight course around two poles half a mile apart. The best flight so far is one made by Flight Lieutenant John Potter of the Royal Air Force in a monoplane designed by Christopher Roper. The Roper aircraft, shown in the photograph with Potter in the driver's seat, has a wingspan of 80 feet and a weight (without Potter) of only 146 pounds. In actual flight an aerodynamic nose canopy completely encloses the rider and drive mechanism.

did only 70 to 100 years ago. There is evidence of such a process at work. The Chinese are replacing the wooden wheel on their traditional wheelbarrow with a bicycle wheel, thereby making it both easier to push and kinder to road surfaces. Threshing and winnowing machines have been designed that incorporate bicycle bearings and chain drives. It is this kind of do-it-yourself, village-level technology that offers the best route to self-improvement, a route far more plausible than any form of large-scale aid from outside. The power of the bicycle with respect to the power of the most sophisticated modern technology is

perhaps best shown in Indochina, where the North Vietnamese have used it as a major means of transport. The Japanese who captured Singapore in World War II also traveled largely by bicycle. Nonetheless, the bicycle remains essentially a peaceful device, and we do not need to include it in strategic-arms-limitation negotiations. We might do well, however, to go all out in encouraging its use. If one were to give a short prescription for dealing rationally with the world's problems of development, transportation, health and the efficient use of resources, one could do worse than the simple formula: Cycle and recycle.

THE RISE OF SCIENTIFIC TECHNOLOGY

The distinction between science—the study of how and of what the world is made—and technology—the practical art of manipulating the world—did not become blurred until the middle of the Industrial Revolution. The acceleration of technical progress during this period exposed vast areas of ignorance about the working of the world. The reductionist scientific method and mechanistic world view that developed during the scientific revolution were admirably suited to exploring these uncharted territories. The marriage of science and technology in the nineteenth century gave birth to the first scientifically oriented technologists and opened the way to the systematic, planned, and controlled exploitation of nature and natural resources in the twentieth century.

II

THE INDUSTRIAL REVOLUTION

INTRODUCTION

Just as the computer who wants his calculations to deal with sugar, silk, and wool must discount the boxes, bales, and other packings, so the mathematical scientist, when he wants to recognize in the concrete the effects which he has proved in the abstract, must discard the material hindrances, and if he is able to do so, I assure you that things are in no less agreement than arithmetical computations.

Galileo Galilei
Dialogue Concerning the Two Chief World Systems

The beginning of the seventeenth century was the fountainhead of scientific technology. Between 1350 and 1630, Europe had passed through the decline of agriculture-based feudalism, the rise of mercantile capitalism and the consequent emergence of the middle class, the Renaissance, the great global navigations and discoveries, and the politico-religious wars of the Reformation. Medieval Europe had been a domain of lord and serf, town and field, with one great universal Church serving as a central, coherent force. Reformation Europe saw the emergence of nation-states possessed of preeminently national interests, state religions, and primarily mercantile economies based largely on technology. The cultural divisiveness of the Reformation had been paralleled by the revolution that transformed science— the replacement of the holistic, organismic world view with a reductionist, mechanistic one. Where nature had previously been treated with awe and respect as something to be investigated but perhaps never fully understood, it now came to be perceived as a giant clockwork, a machine whose workings were open to discovery. This perception was particularly influential upon the world view of the Protestant countries, where economics, politics, and religion were being blended into a moral philosophy of salvation through good works on earth, and the progress of industry and technology were becoming as much a moral as a technical value.

The scientific and mathematical foundation of this new edifice is found in the work of Galileo Galilei, the socio-economic and philosophical foundation in the writings of Francis Bacon. Galileo's technique of reducing a problem to a form susceptible to mathematical analysis, of stripping away those attributes that are irrelevant to the problem at hand, is the rock upon which the church of modern science and engineering has been built. The importance of the Galilean method lies in (1) the ability it confers to analyze a device by recognizing the underlying physical principles and (2) the confidence it instills that a set of geometrical figures on paper can be converted to a machine once the key structural factors are identified. Francis Bacon, though apparently failing to appreciate the full significance of the mathematical science of Copernicus, Kepler, and Galileo, was quick to promote the purposeful support of science, seeing in it the key to a rapid advance of technology and a new method of invention. In his proposal of a program to improve the human condition by the advancement of society through technical discovery, Bacon stated that there was a class of inventions, still unknown, which could be made only when their underlying scientific bases were understood. Thus he argued that one should, in all cases, encourage the advancement and acquisition of scientific knowledge, since through an understanding of nature, new and radical inventions could be made. The arguments of Bacon, René Descartes, and others in favor of generous support of science (which resulted, among other things, in the founding of the Royal Society and the *Académie des Sciences*) created a social ambiance that had much to do with the rise of industry. The Industrial Revolution became possible when the essence of the Galilean analytical and mathematical method had been thoroughly absorbed.

No machine could better exemplify these converging influences than the steam engine, the machine that supplied the power that made the Industrial Revolution a revolution rather than an evolution and whose origins are discussed here by Eugene S. Ferguson in "The Origins of the Steam Engine." The engines of Thomas Newcomen and Thomas Savery were based on the establishment (by Galileo, Evangelista Torricelli, and Blaise Pascal) that the atmosphere was an ocean of fluid less dense than, but resembling, water, and on the subsequent demonstration by Otto von Guericke and others that the pressure of air could be made to do work. Newcomen and Savery were scientific technologists in the original sense. They adapted and used the new scientific discoveries of their age and turned them into machines, inventions, and devices. James Watt, on the other hand, was the prototype of an entirely new genre of inventor. Combining scientific theory and experimentation with technical brilliance, he designed his condenser steam engine only after a careful and thorough study of the thermal efficiency of existing steam engines led him to conclude that too much of the machine's energy was being lost in heating and cooling the piston chamber on each stroke.

The half-century following Watt's steam engine was a period of interaction between science and technology, which culminated in Nicolas-Léonard-Sadi Carnot's precocious analysis, in 1824, of the maximum possible efficiency of heat engines. It took another quarter-century for pure science to absorb and dissect the events of this period, and for James Prescott Joule, Rudolph Clausius, Lord Kelvin, and others to create the analytical science of thermodynamics, upon which the next and most expansive phase of the Industrial Revolution was to be partially based. The century following Watt's engine was an era of revolution and political reformation. In the earlier phase, Britain, spared from the nineteenth-century revolutionary upheavals that kept most of Europe in turmoil, and spurred by the economic pressure of the Napoleonic wars, acquired an enormous industrial lead that promoted her to the position of the leading world power. But by 1871 the century of revolution was over. America, emerging from the Civil War with industry ascendant over agriculture; Germany, newly born as a unified nation-state; and Japan, breaking her old isolation, all turned to the new discoveries of science and created new methods of industrialization for their rapidly developing industries. With the political structure of Europe stabilized for the next half-century, the marriage of science and technology became a dominant influence on the culture of the new Golden Age.

The history of electricity is similar to that of the steam engine, although in this case the gap, though contemporary, was reversed. In thermodynamics there was a theory lag, which delayed the further development of heat engines. In electromagnetism there was a fifty-year engineering lag between the discovery of the scientific basis of electricity and magnetism and the first working generators. This period is the subject of Harold I. Sharlin's essay "From Faraday to the Dynamo." In 1796, Alessandro Volta had discovered how to construct a battery, a continuous source of electricity, from dissimilar metals. The availability of such a source led to a series of investigations of the behavior of electricity. Following Hans Christian Oersted's discovery that magnetism could be produced from electricity (when he showed that an electrical current could deflect a compass needle), Michael Faraday became intrigued with the idea that electricity could also be produced from magnetism. During an incredible ten-day period in 1831, Faraday discovered magneto-electric induction, analyzed it, and built the first working dynamo. At first, Faraday's discovery was of importance primarily to the development of pure science. The continuing exploration of the relationship of electricity and magnetism was to lead to one of the crowning achievements in the history of science: the electromagnetic theory of James Clerk Maxwell, which unified the theories of electricity, magnetism, and light. As Sharlin indicates, the history of the half-century between Faraday and the dynamo also illustrates the modern coalescence of science and technology. Early devices based on simple theory may be of some practical importance. The industrialist and the investor, seeking further improvement, encourage engineers and technicians to turn to science. When they do, the theory may prove to be inadequate or obscure, and the progress of the technology itself begins to contribute toward further advances in theory; these in turn serve to further refine the technology. By the time the cycle ends, both the devices and the theory have profited greatly, as have the industrialist and the investor. Thus, the gap in thermo-dynamic theory from Carnot in 1824 to the internal combustion engine in 1876, and the gap in electrodynamics from Faraday in 1831 to the dynamo in 1880 are opposite sides of the same coin.

As Asa Briggs points out in his essay "Technology and Economic Development," it is important to separate two distinct phases of the Industrial Revolution, not only for historical and historiographical purposes, but also for examining the problems attendant to the industrial development of the so-called "underdeveloped" nations. The early British phase of the Industrial Revolution was one of small-scale, laissez-faire capitalism based on free trade, local industry, and mercantile-oriented production. It was characterized by the spread of factories built to manufacture goods for trade, particularly textiles. That monument to early British industry, the Crystal Palace (built for the Great Exposition of 1851), was as much the mausoleum as the cathedral of the petit-industrialist era. The social values that created the achievements of this period were to have less influence in our century than those of its analysts and critics (such as John Stuart Mill, William Morris, Friedrich Engels, and Karl Marx). Once the factories had spread, the technical thrust shifted from the production of goods to be traded and sold by merchants to the improvement and expansion of the *means* of production in themselves. Industrialization itself became an industry. Mercantile capitalism, which had nurtured the bourgeoisie, was superseded by industrial capitalism, which created a new working class.

In the mid-nineteenth century, the great triumph of industrial technology was the development of precision machine tools and the coal, iron, and railroad industries that fed the ever-growing factories. The rapid qualitative shift in the pattern of industrialization in the last third of the century is now recognized as a second phase of the Industrial Revolution. It was based not only on the formation of new industries linked to the new scientific technology, but also on direct manipulation by the state of a social and political climate congenial to industrial growth and the tolerance (or even encouragement) of giant industrial cartels and monopolies. An ironic side-effect of the success of state-encouraged monopoly capitalism is that the socialist movement (whose founders had once been the foremost critics of megalithic industrialization) has become its foremost promoter, particularly in the less-developed nations.

The key question raised in Briggs' article is whether the European experience has any relevance to the problem of closing the gap between the "overdeveloped" nations and the rest of the world. The second phase of the Industrial Revolution began with an explosive industrial expansion, based on new forms of industry and entirely new products. The resultant commitment to rapid growth created an economic gap between the West and the rest, a gap which still continues to widen at an increasing rate. It is difficult to envision how any country can now retrace or even parallel that particular path. What is more, the growth of these industries was, in Briggs' words, based on a division between the black and the green—the manufacturing countries and the suppliers of raw materials. The newly emergent nations rightly aspire to the benefits conferred upon the great industrial powers by their economic strength and abundant manufacturing capacity. Whether they can attain these benefits without the dislocations, social inequities, and adverse environmental and economic side effects attendant to western development is a question no less valid today than it was a decade ago. Whether our path can be retraced at all without the colonization, imperialism, and exploitation that provided ready sources of cheap materials and captive markets for manufactured goods is also problematical. The poor countries are poor largely because the industrial countries are rich; their underdevelopment was part of the price of our progress. The world in which they seek to prosper cannot offer them an easy road to riches. However, in the ten years since Briggs wrote his article, there has been a gradual change in economic rhetoric, and the emphasis has shifted from the question of their "underdevelopment" to the problem of our "overdevelopment." Since the finite resources of the earth simply cannot supply a world in which every country consumes energy and resources at the rate our country does, we must either rein ourselves in or resign ourselves to the maintenance of a permanent division between rich and poor with all the political tension that this entails. Nevertheless, the mere fact that we are now beginning to pose these questions may, at last, represent the beginnings of an "imagination powerful enough to bridge the gulf between the different worlds of our own making."

6

The Origins of
the Steam Engine

by Eugene S. Ferguson
January 1964

*Fifty years before James Watt came on the scene
Thomas Newcomen built practical steam engines to
pump water out of mines. What is known of these
engines and how did they influence later ones?*

If one had a handbook of human history with a synoptic chart that opened out at the back, one might expect the chart to reduce the industrialization of England in the 18th century to the words "James Watt," "steam engine" and "textile mills." This familiar view is misleading on two counts. Watt did not invent the machine that supplied power to the looms of the textile mills. Steam engines had been put to work 50 years before Watt appeared on the scene, and the industry that created the demand for them was not weaving but mining. At the beginning of the 18th century two Englishmen from Devonshire, Thomas Savery of Shilston and Thomas Newcomen of Dartmouth, built steam-powered pumping machinery for the drainage of mines. The need for a way to remove water from mines had become more and more pressing as the mineral resources of England were exploited during the 17th century. The operators of tin mines in Cornwall and lead mines in Derbyshire were waging a losing battle against water seepage as their mines were dug deeper, and many coal mines around Birmingham and Newcastle were threatened with flooding as the inflow of water overcame the pumps then available. Savery's engine never succeeded as a mine pump, although it was useful for other purposes; Newcomen's did provide the power to lift water from mines.

The Newcomen engine also succeeded, two generations later, in stimulating the curiosity and imagination of James Watt. In 1769 Watt patented an engine soon brought to commercial status by industrialists who realized that its superior thermal efficiency would enable them to make greater practical use of steam power. The Newcomen engine should not, however, be considered a mere taking-off point for the genius of Watt. Its impact on the technology and economics of mining is today symbolized by the monotonous bobbing of the pivoted "walking beams" of oil-well pumping rigs, a familiar sight in the southwestern U.S. The vertical pump shaft is guided at its upper end by an odd protuberance on the beam called, in the graphic language of the industry, a horsehead. Whether or not the builder of the first such oil-pumping rig was aware of his source, he had borrowed these elements from a Newcomen engine originally used to pump water [*see illustration on opposite page*]. In 250 years the power unit has evolved from a steam cylinder to an electric motor, but its function—to pull one end of the beam down—has not changed at all.

In Newcomen's design one end of the beam was secured to the pump shaft and the other end was chained to a piston that fitted into a vertical steam cylinder. When steam supplied to the cylinder from a boiler directly below it was condensed by the injection of water, the resulting vacuum enabled the pressure of the atmosphere to force the piston down, thus drawing the beam down on one side and the pump rod up on the other. Long after the steam engine was being used for purposes other than pumping it retained the overhead beam of Newcomen's design. Watt himself experimented in 1770 with turning the cylinder upside down in order to eliminate the beam, but he quickly and permanently abandoned the idea.

Little is known about Thomas Newcomen, the man whose innovations were so original, influential and enduring. He was born in Dartmouth in 1663, made a living as a seller and perhaps small-scale manufacturer of iron products and died in London in 1729. The recent tricentennial of his birth gave impetus to the study of his life and work; such study, including this critical review of the Newcomen engine, would scarcely have been possible were it not for the dedicated men who in 1920 in London organized the Newcomen Society for the Study of the History of Engineering and Technology. The members of this group, the first to take a serious interest in Newcomen as an individual, combed the records for the origins and later history of the steam engine, publishing their findings in the *Transactions* of the society.

The source materials they uncovered tell us more about the state of technology at the time than about the events of Newcomen's life; nothing is revealed of his formal and informal education, the actual sources of his ideas and the steps by which his major innovations were thought out. It is unlikely that we shall ever learn the details of the steps taken by Newcomen during the 10 years he spent developing his engine. He was not prominent during his lifetime, and although his engine won immediate acceptance, it was seldom linked with his name, being known merely as the "fire engine" or "atmospheric engine." In this article I shall review the antecedents of the engine he designed and, as far as I am able to reconstruct it, the period of development immediately preceding its appearance.

A glance at the dozens of well-illustrated books devoted to machines that were published in Italy, France and Germany from the time of Georgius Agri-

NEWCOMEN ENGINE at a mine at Dannemora in Sweden was illustrated in a book of 1734 on Newcomen's principles; the illustration is reproduced on opposite page. The author was the Swedish engineer Mårten Triewald. Illustration describes the engine as the "Dannemora fire and air machine."

Dannemora Eld och Luft Machin,

Kongl. Maj:ts och Rikſens Höglofliga Bergs Collegio

Underdån-ödmiukaſt Dedicerad af Marten Triewald.

STEAM DEVICES of 16th century raised small amounts of water. At left is an apparatus of Giambattista della Porta; water poured into chamber *B* through funnel *A* is raised through pipe *C* by steam generated in flask *D*. At right is a device of Salomon de Caus adapting same principle to produce a fountain through the generation of steam in a copper sphere (*A*).

cola's *De Re Metallica* (1556) onward indicates that the problem of water-raising was one that occupied many mechanics and mechanical philosophers in the advanced countries of Europe. Except for Agricola's treatise on mining, which gave details of 14 kinds of pump for removing water from mines, the books were concerned less with mine drainage than with pumping water for town and castle water supplies and for the operation of fountains. Nevertheless, the techniques of pumping were well known and widely discussed. Some of the devices employed were an endless chain of buckets, the Archimedean screw and the rag-and-chain pump, in which a series of rag-wrapped balls, spaced a foot or two apart on a continuous chain, were drawn vertically upward through a wooden pipe, each forcing some water ahead of it. There were many alternative machines using manpower or horse-power for the hoisting of ordinary tubs of water. During the 17th century the possibility of using steam or gunpowder as a motive power was also being explored.

It has been said that science owes more to the steam engine than the steam engine owes to science. Such a generalization seems particularly inappropriate with respect to a machine that exemplifies the overlap between the empirical and the theoretical stages of the Indus-

trial Revolution. Although it is true that a clear understanding of the thermodynamic phenomena in the steam engine was not attained until around 1860, it is equally true that the sequence of ideas apparent in the work of Galileo, Torricelli and Pascal in establishing the fact of atmospheric pressure, and of von Guericke, Huygens and Papin in devising ways to make atmospheric pressure do work, was an indispensable prerequisite of the Newcomen engine.

Close to the Newcomen engine chronologically but not conceptually was the steam-powered machine patented by Thomas Savery in 1698. This engine, which promised to solve the problem of mine flooding, incorporated elements and principles not shared by the Newcomen engine and can be traced to a wholly different line of development. Savery, a gentleman of leisure and Fellow of the Royal Society of London, exhibited a model before the society in 1699. His engine consisted of a vessel in which steam was condensed to produce a vacuum, whereupon the vessel was filled by water rising through a suction pipe [*see illustration on opposite page*]. Steam at high pressure was then admitted to the same vessel, forcing the water to a higher elevation. The machine was a combination of steam pumping devices built or suggested earlier by Salomon de Caus and R. d'Acres and probably well known in Savery's circle.

In 1702 Savery expanded his patent application in a small book entitled *The Miner's Friend.* Here he addressed himself to the "Gentlemen Adventurers in the Mines of England: I am very sensible a great many among you do as yet look on my invention of raising water by the impellent force of fire a useless sort of a project that never can answer my designs or pretensions; and that it is altogether impossible that such an engine as this can be wrought underground and succeed in the raising of water, and dreining your mines.... The use of the engine will sufficiently recommend itself in raising water so easie and cheap, and I do not doubt but that in a few years it will be a means of making our mining trade, which is no small part of the wealth of this kingdome, double if not treble to what it now is."

In spite of Savery's optimism, the metalworking techniques at his command were inadequate to solve the problem of containing steam at several atmospheres of pressure. Hence the Savery engine was practical only in situations other than the one for which it was originally intended. The most successful application of the engine was in pumping water into building or fountain reservoirs that were no higher than about 30 feet, which called for only moderate steam pressure.

The Newcomen engine soon preempted the role of draining the mines, but the Savery engine was the first to be employed (around 1750) to turn machinery. For this purpose the engine pumped water into a reservoir some 15 or 20 feet above that supplied a conventional water wheel. Throughout the latter part of the 18th century the Savery engine was built in considerable numbers and used by manufacturers who could not or would not afford the larger, more efficient but initially more expensive Newcomen and Watt engines. As late as 1833 at least five Savery engines were at work in France; the engine was reinvented about 1870 in Germany, perhaps also in England. Now known as the pulsometer, it went on to a new career of pumping water containing solids in such applications as the drainage of shallow excavations.

The problem of following the sequence of events in the development of the Newcomen engine points up the meagerness of available source materials. What little information Newcomen's contemporaries have left us requires careful interpretation. One popular scientific lecturer of the early 18th century, John Theophilus Desaguliers, described him as an "ironmonger" and "Anabaptist." This .is the way he has been described by modern writers oblivious to the fact that "ironmonger" has come to imply "peddler," or perhaps "junkman," and that "Anabaptist" suggests the outlandish. Thus Newcomen is likely to be thought of as a ragged, gaunt pusher of a handcart, waiting for Dickens to be born so that he could get into one of his books.

The background of his assistant, John Cawley, is even less distinct. Desaguliers called him a glazier; another man who could have known him said he was a plumber. Elsewhere he is referred to as a brazier or coppersmith. This description seems proper because an ironmonger was a dealer in hardware and industrial supplies, sometimes manufacturing what he sold. He might have had an iron foundry as part of his establishment; he usually employed braziers and tinsmiths; he was likely to have a lathe and a smithy. It has been suggested, I think reasonably, that the ironmonger rather than the millwright (who generally built in wood) was the predecessor of the mechanical engineer. So we can forget the picture of Newcomen the indigent peddler and accept the more plausible likeness of a man well skilled in the machinery trade.

As for the significance of the work Newcomen did in developing the steam engine, Desaguliers states: "If the Reader is not acquainted with the History of the several Improvements of the Fire-Engine since Mr. *Newcomen* and Mr. *Cawley* first made it go with a Piston, he will imagine that it must be owing to great Sagacity, and a thorough Knowledge of Philosophy, that such proper Remedies for the Inconveniences and difficult Cases mention'd were thought of: But here has been no such thing: almost every Improvement has been owing to Chance."

Further detraction—or inverted praise —came from Mårten Triewald, a Swedish engineer who took plans for a Newcomen engine back to Sweden with him in 1726, attributing the design to the Almighty, who "presented mankind with one of the most wonderful inventions that has ever been brought into the light of day, and this by means of ignorant folk who had never acquired a certificate at any University or Academy." Triewald did mention, however, that Newcomen worked on his machine "for ten consecutive years."

Since Desaguliers and Triewald, our principal sources, were contemporaries of Newcomen's, it is perhaps presumptuous to question their judgment. But contemporaneousness does not ensure accuracy, and Desaguliers is known as a kind of press agent of science and the arts. He was the first to publish the absurd story about Humphry Potter, the boy who, while attending a manually controlled Newcomen engine, invented the automatic valve gear in order to keep the engine running when he went fishing. The work of both authors shows them to be vain and opinionated, and it is natural to wonder on what occasion Newcomen had pricked their pompous balloons.

The scant biographical information does not tell us unequivocally that Newcomen's design was complete when the engine was set to work near Birmingham in 1712. L. T. C. Rolt has recently assembled evidence that suggests the exist-

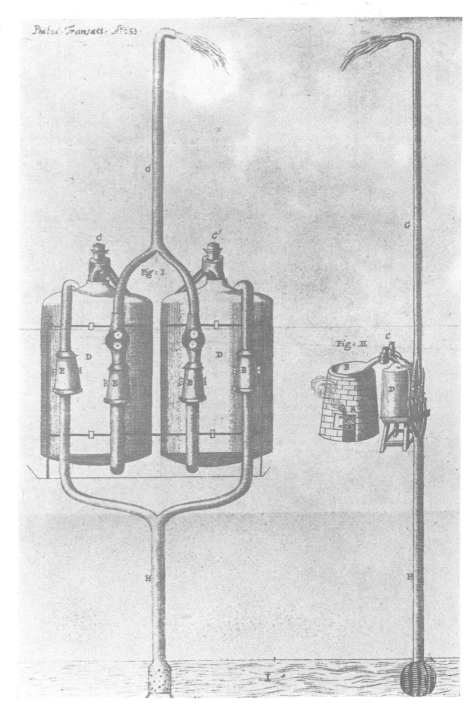

MINE PUMP designed by Thomas Savery in the late 17th century envisioned condensation of steam in vessels *D* to produce a vacuum, which would create suction to draw water up from the mine to fill the same vessels. High-pressure steam would then be introduced to dispose of the water through pipe *G*. The pump did not work in mines because metalworking techniques to contain high-pressure steam were not available. It did serve to pump water into low reservoirs for use in buildings and fountains; steam pressures for that were moderate.

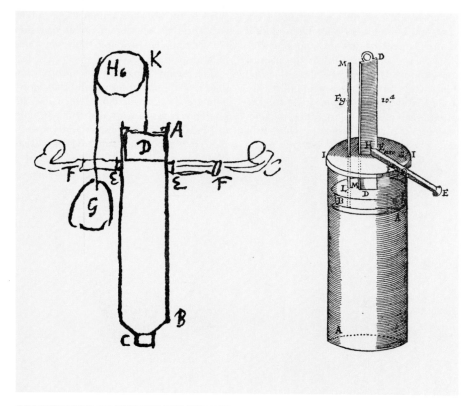

PRECURSORS OF NEWCOMEN ENGINE included (*left*) an idea sketched in a letter by Christian Huygens for using gunpowder to force air out of a cylinder; as remaining air cooled, piston *D* would drop. At right is design by Denis Papin using steam to make vacuum.

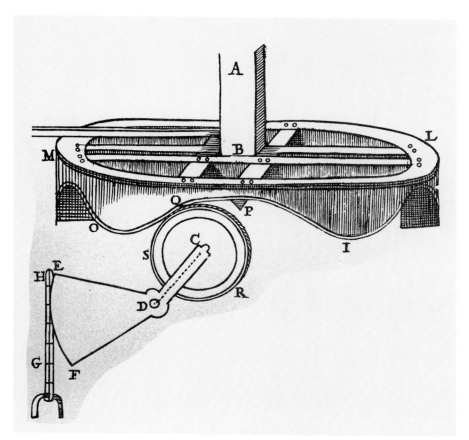

CAM ARRANGEMENT of a pump in 1696 was a precursor to the arch head devised by Newcomen. Horsepower turned scalloped cam, which raised and lowered wheel *C* around fixed pivot *D*, producing a pumping action by blade-shaped device attached to a lift pump.

ence a few years before 1712 of one or more unsuccessful Newcomen engines in Cornwall, near the inventor's home in Dartmouth. This would certainly make more credible the appearance of a definitive machine in 1712. In any case, a virtually anonymous ironmonger working in Dartmouth would hardly travel 175 miles to Birmingham, as Newcomen apparently did, to erect an engine unless he had connections farther afield than his home city. Although I cannot be certain, it seems probable to me that Newcomen was no stranger to London and that he quite possibly had traveled to the Continent, where he might have seen some of the great water-driven pumping engines around Paris. Just as Americans in the early 1800's went to England to learn the latest techniques in engineering, so in the 1700's Englishmen went to the Continent.

The design of the engine built by Newcomen in Birmingham in 1712 was, if not definitive, remarkably near completion. Certainly by 1717 it had been given its final form; we have an engraving made of the engine in that year. Fifty years later John Smeaton was to improve Newcomen's machine by determining after methodical empirical investigation the optimum operating conditions and proportions of parts of the engine, but Smeaton did not tamper with the inventor's essential design.

Even in its earliest manifestations the Newcomen engine was simple enough so that observers could understand its operating principle and cyclical sequence of events as soon as an explanation was provided. A vertical steam cylinder, fitted with a piston, was located under one end of the large, pivoted working beam; the piston rod was hung on a flat chain secured to the top of the arch-shaped head of the beam. Steam was supplied to the cylinder by the boiler directly below it. A vertical lift pump was located under the other end of the beam and the pump rod hung on a flat chain secured to the arch head just above it. Thus both the piston rod and the pump rod moved vertically, always tangent to a circle whose center was at the pivot of the beam.

A working stroke began after the steam cylinder had been filled with steam, at a pressure just slightly above atmospheric, from the boiler. The pump end of the working beam was held down by the weight of the reciprocating pump parts, which extended down into the mine. The steam-admission cock was closed, and water was then injected into the cylinder in order to condense the steam and produce a vacuum. The atmosphere, acting on the top of the piston, pushed the piston down into the evacuated cylinder, which caused the pump rod to be lifted by the other end of the beam. The cycle of operation was completed by again admitting steam to the cylinder in order to allow the pump end of the working beam to go down. As soon as the cylinder pressure reached atmospheric, the spent injection water was discharged into a sump.

The cylinder was large. The first engine cylinder was 21 inches in diameter and had a working stroke of more than six feet. The effective vacuum was about half an atmosphere, enabling a 21-inch piston to lift unbalanced pump parts and water weighing one and a quarter tons. Operating at 14 working strokes a minute, the engine would develop about six horsepower. Later engines increased in size to a cylinder diameter of seven feet and a stroke of 10 feet and developed well over 100 horsepower.

The late Henry W. Dickinson, author of the current standard history of the steam engine and a principal founder of the Newcomen Society, recognized that Newcomen's contribution was the "first and greatest step" in the development of the modern steam engine, but he diluted the effect of this judgment by writing: "When we look into the matter closely, the extraordinary fact emerges that the new engine was little more than a combination of known parts."

This statement brings to mind a remark made in 1853 by a correspondent of *Silliman's Journal*: "It appears that the human mind cannot arrive at simplicity except by passing through the complex; it is like a mountain more or less elevated, whose heights must be overcome before the plain at the opposite base can be reached; and when reached, the level seems to be that of the plain left behind. So when a simple solution of a problem is arrived at, we think it an easy natural thought and almost self-evident."

This, it seems to me, describes the problem we have in looking at the innovations of Thomas Newcomen from a 20th-century vantage point. In retrospect the idea of the steam engine is a natural thought, modified only by our occasional impatience with Newcomen's inability to see some obvious further development, such as the addition of a crank and flywheel, which came two generations later (shortly after having been rejected as impractical by so capable and forward-looking an engineer as John Smeaton). It is not easy for the

human mind to put what is now obvious back into the box labeled "Unknown."

In discussing Newcomen's achievement with reference to the "known parts" of the engine it should be noted that he was not simply a clever compiler of mechanical elements. He did not employ many devices, including the crank and flywheel, that were vastly better known than some he made use of in his "combination of known parts," and most of those he did use he modified in such a way as to make the distinction between adaptation and invention seem artificial.

Consider Newcomen's use of the steam cylinder and piston. The line of development leads straight from von Guericke through Huygens and Papin to Newcomen. The cylinder fitted with a piston and evacuated by the condensation of steam was clearly present in Papin's design published in 1690 and republished in 1695, and we ought to assume that Newcomen knew at least as much about Papin's work as had been published. The

steam in Papin's cylinder, however, was to be condensed by cold water dashed on the outside wall. Newcomen's essential improvement was to inject water directly into the cylinder, which sped the condensation and enabled the engine to operate at 12 or 14 strokes a minute instead of three or four.

One of Newcomen's experimental engines had employed a water jacket around the steam cylinder for cooling, and it may be, as Triewald reported, that the change from external to internal cooling resulted from the accidental leakage of jacket water into the cylinder, which "immediately condensed the steam, creating such a vacuum that . . . the air, which pressed with a tremendous power on the piston, caused its chain to break and the piston to crush the bottom of the cylinder as well as the lid of the small boiler." Even if this report is accurate, Newcomen was still faced with the nice diagnostic problem of determining from the wreckage what had caused the

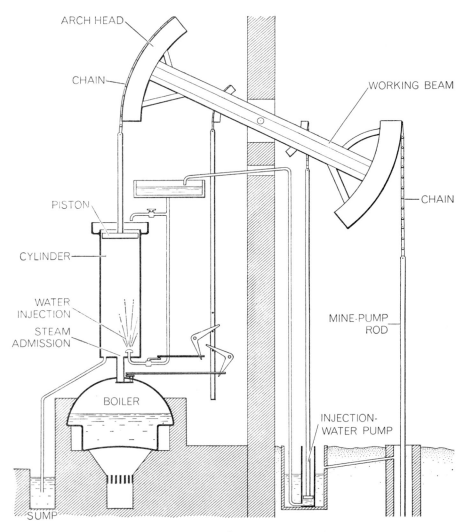

SUCCESSFUL ENGINE by Newcomen introduced steam into a cylinder that was then cooled with injection of water, creating partial vacuum. Atmospheric pressure forced piston down to achieve pumping; weight of mine-pump rod and equipment then raised piston for new cycle.

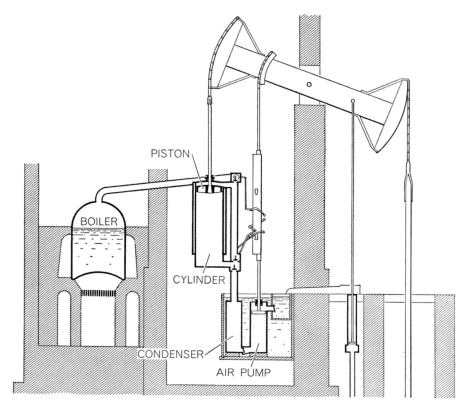

CONTRIBUTIONS BY WATT to the steam engine included the development of a separate condenser, as depicted here. Newcomen had effected the condensation in the main cylinder.

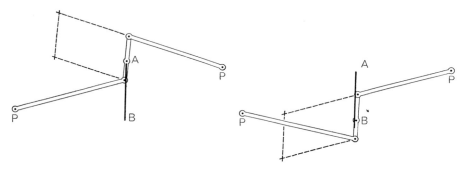

PARALLEL MOTION was another major contribution by Watt; it kept piston rod vertical as beam end moved in an arc. Three key elements (*solid lines*) worked from fixed pivots *P* so that rod end, at center of vertical element, moved along line *A-B*. Watt's final version made use of a parallelogram linkage (*broken lines*) that made whole apparatus more compact.

sudden smash. Serendipity in no way diminishes Newcomen's role in the innovation of injection condensation.

In his use of the boiler Newcomen was adopting a thoroughly developed "known part." Made of copper, the boiler probably was derived directly from the brewer's kettle. Since the steam pressure was low—Newcomen set his safety valve to open at about 1.5 pounds per square inch above atmospheric pressure—the difficulties of design and construction were few. Indeed, the boiler was similar to the one built by Savery.

The full synthetic ability of Newcomen, and his judicious critical sense, are revealed in his treatment of the working beam, the pump and the valve gear. The working beam and pump can be examined together, because their appearance is that of a greatly enlarged pump handle or well sweep attached to a common reciprocating lift pump. Before Newcomen's day few, if any, mines in England were drained by lift pumps attached to beams, pump handles or sweeps. Where the topography of the mining district permitted, long drainage tunnels called adits were dug from the lowest mine level to a lower open valley in the vicinity. Although the adits were small in cross section, some of them extended for two miles or more. Even after a mine was deepened beyond its

adit level, water had to be pumped only as high as the adit. When surface water was available, an underground water wheel, receiving its water from ground level and discharging into the adit, would operate a lifting device of some kind, usually a chain of buckets.

In some larger works, where horses could be used, the water was lifted in great tubs by a whim, or horse gin. The hoisting rope was wound on a horizontal drum geared to a vertical shaft. The vertical shaft, fitted with a hub with radiating arms, was dragged around by horses hitched to the ends of the arms. In smaller mines, where only manpower was available, a horizontal drum turned by hand cranks was used to hoist buckets or drive a rag-and-chain pump.

If Newcomen had seen a copy of Agricola's mining book, he would have found reciprocating lift pumps in profusion, but he would have come away from the treatise with the distinct impression that the proper way to move the rod of the lift pump up and down was to hang it on a crank arm, that is, to employ a crank and connecting rod. There is one simple beam pump in Agricola, but it is a small one operated by the power of a single man.

Among the actual devices that Newcomen might have seen was a large overhead pivoted beam, without arch heads, in the horse-driven water pump at York House in London. The London Bridge waterworks, although they employed cranks and connecting rods, had the lower third of a large pulley cut away in a manner that faintly suggests the arch heads at the ends of the Newcomen engine beam. The almost complete absence of the arch head from pump beams in the illustrations that Newcomen might have seen is most striking. A sketch in Leonardo da Vinci's notebooks could hardly have been known to Newcomen because the notebooks were effectively buried until the 19th century. Only one illustration remains as the possible—or probable—source of Newcomen's arch heads. In a book by Venturus Mandey and James Moxon—*Mechanick-Powers: or, the Mistery of Nature and Art Unvail'd*, published in London in 1696—there is a cam-operated pumping device that clearly shows the sector-and-flatchain arrangement adapted by Newcomen, who changed the shape of the beam from curved to straight. The drawing in the Mandey and Moxon book was copied directly from an earlier work edited by Philippe de la Hire, a French mathematician and member of the Académie des Sciences, who had directed the building of such a pump to supply water to a castle near Paris.

Thus the working beam of the Newcomen engine appears to be an elegant adaptation, not a copy, of ideas that existed before he designed his engine. I have labored this point in order to emphasize the fact that Newcomen was not merely adapting the steam cylinder to a widely used system of water-raising. His engine was a new and original system in itself.

The origin of the valve gear, which enabled the engine to operate automatically—opening and closing valves as required for the sequence of operations—is similarly obscure. The idea may have been suggested to Newcomen by a control mechanism of the automata—knights, maidens and animals—that performed at an appointed hour in the great medieval

clocks. The Newcomen engine valve gear was a sequential control; it remained for Watt to supply a regulatory feedback control system. Newcomen's system, however, was much more involved than, for example, the control of the rate of a common clock.

Even after all the elements of the steam pumping engine had been settled on, however, there was still the problem of the physical arrangement of the elements. Pictures of the gaunt and unsymmetrical profile of the Newcomen engine set against the English landscape, with its awkwardly tall stone enginehouse and its outlandish protruding beam threatening to topple the whole assemblage, make it difficult to believe that there was anything about the arrangement that could not have been built differently if the "right way" had not been shown bold-

ly by Newcomen. As an assemblage of elements, some adopted but most adapted, the engine was a clear statement of the builder's personal style of invention.

The genius of James Watt was of a different kind, and to discuss the difference in terms of superiority smacks of useless historical partisanship. Newcomen selected the components of a steam engine and gave to each its proper place and function. Watt, on the other hand, originated at least two new major components, and in making a brilliant adaptation of a third he introduced the world to the notion of feedback for automatic control.

Watt began his work on steam engines in 1763, when, as an instrument maker at the University of Glasgow, he undertook the repair of a teaching model of

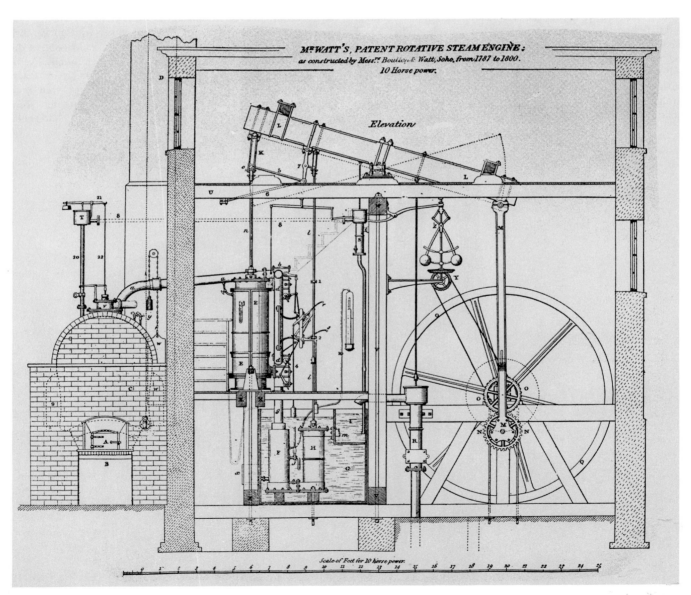

WATT ENGINE was depicted in this 1826 illustration. The engine transformed the vertical action of the piston rod *n* into rotary motion through the flywheel *Q*. Watt's special contributions included condenser *F*, parallel-motion linkage at left end of beam, centrifugal governor *Z* and "sun and planet" gear mechanism at center of flywheel, causing wheel to turn at double the speed of the engine.

a Newcomen engine. His careful and sustained study led him in 1765 to recognize that he might increase the thermal efficiency of the engine, as well as its capacity and operating speed, by condensing the steam in a chamber attached to, but separate from, the main steam cylinder. This was the first of his most important innovations.

His earliest patent, which included the separate condenser, was granted in 1769, but his first successful full-sized engine was not completed until 1775, the year in which Matthew Boulton became his partner. Parliament granted a patent extension to Watt that year, providing a virtual monopoly on the condensing steam engine for 25 years.

After Watt had devised a double-acting engine, in which steam moved the piston first in one direction and then in the other as it was admitted alternately to each end of the cylinder, the arch head and flat chain no longer sufficed to guide the upper end of the piston rod, because the chain transmitted force in tension only. Accordingly in 1783 Watt brought forth his second major innovation: the straight-line linkage that bears his name. Refining further his first ideas, Watt combined the straight-line linkage with a pantograph, a linkage system in parallelogram form, to produce the so-called "parallel motion" [see bottom illustration on page 78].

In these two inventions we find a measure of Watt's capacity: the separate condenser was neither anticipated nor invented independently by anyone else, and the parallel motion solved a problem whose existence was not even suspected until Watt overcame it. For the next 100 years mechanics and mathematicians occupied themselves in a search for alternative solutions.

Finally, in 1788 Watt adapted the centrifugal "flyball" governor to control the speed of his engine by linking the governor to the steam-inlet valve. The flyball governor had been used in grain mills to increase the distance between the flat grinding stones as their speed increased. Watt's use of the governor, however, added the far-reaching principle of feedback that made possible self-regulating, rather than merely automatic, machines. The ordinary steam engine and the Watt engine were built, in the words of Boulton, "with as great a difference of accuracy as there is between the blacksmith and the mathematical instrument maker." Thus the few astonishingly sophisticated Boulton and Watt engines in service toward the end of the century hurried a generation of machine builders to a higher order of accuracy, which in turn called for a whole new array of large, rugged and precise machine tools. The influence of the new tools on mechanization was profound and can be traced directly to the present. The effect of the separate condenser and self-regulating speed control on the direction of industrial technology can be appreciated if we recognize that their invention was an essential step toward the modern steam turbine. Undeniably Watt opened doors whose very existence might have gone unnoticed for 100 years after his time.

The Watt steam engine was twice as efficient, from the standpoint of fuel consumption, as even the best Newcomen engine. A recent study by two English economic historians, A. E. Musson of the University of Manchester and E. A. G. Robinson of the University of Cambridge, has shown, however, that both Savery and Newcomen engines were being built long after they had been

WORKING PUMP of Newcomen design is shown at a colliery in England. The pump was in use from 1791 until 1918. Subsequently it was taken down and re-erected at the Science Museum in London. Part of the working beam is visible below catwalk at top center.

made obsolete by Watt's improvements, and that Boulton and Watt supplied only about a third of all steam engines built during the 25-year period of the patent monopoly (1775–1800). It is also clear that a two-cylinder Newcomen engine capable of turning machinery was in existence, and that the high-pressure engines operating without condensers of any sort were soon to be built by Richard Trevithick in England and Oliver Evans in the U.S.

Since hindsight is one of our best-developed faculties, it has been possible for writers for more than 200 years to dismiss the appearance of the Newcomen engine of 1712 as well as the Watt engine of 1775–1788 as being merely normal responses to industrial demands. The well-established axiom of simultaneous but independent discovery, which can be interpreted to mean that a particular invention is inevitable, has been applied to suggest that if Thomas Newcomen had not built his engine, somebody else would have done so at about the same time.

This seems no more accurate in the case of Newcomen than in the case of Watt. In looking carefully at the Newcomen engine, it has become increasingly evident to me that it represents a unique solution to the problem the inventor set out to solve. There was no anticipation of the completed engine, and nobody came forward to contest Newcomen's priority of invention. The first radical modification occurred no sooner than 50 years later, when Watt conceived the separate condenser.

Newcomen was not the first man to "discover" the correct way to build a steam engine; there is no correct way. It is conceivable, for example, that he might have made the cylinder horizontal rather than vertical, that he might have supplied steam above atmospheric pressure (only eight pounds per square inch would have sufficed to do the work), or that he might have used a crank, connecting rod and flywheel. Any of these variations would have been possible if he had approached the problem differently. But by producing a machine that was a pumping engine, not easily adapted to the turning of wheels, Newcomen limited the options that lesser engineers could exploit in the future. He did the job his way, and he gave the world such a convincing statement of rightness in the machine he put together that he exerted an enormous influence on the direction in which English technology would proceed for the next several generations.

7

From Faraday
to the Dynamo

by Harold I. Sharlin
May 1961

Between the discovery of electromagnetic induction and the development of the electric generator 50 years elapsed. Why did it take so long for Faraday's basic work to be applied?

In 10 inspired days during the fall of 1831 Michael Faraday discovered electromagnetic induction, found essentially all the laws that govern it and built a working model of an electric dynamo. Then he moved on to other research. "I have rather," he wrote, "been desirous of discovering new facts and new relations dependent on magneto-electric induction, than of exalting the force of those already obtained; being assured that the latter would find full development hereafter."

Full development took a long time. Not until the 1880's were Faraday's theories, and the technical clues he provided, embodied in really efficient electric generators. The 50 years in between constitute an engineering gap: the period that separates a piece of basic research from its practical application.

PIONEERS OF ELECTRICITY assembled by Bernarda Bryson for this imaginary seminar are (*left to right*) André Marie Ampère, Alessandro Volta (*standing*), Hans Christian Oersted, Michael Faraday, James Clerk Maxwell and Dominique François Arago.

KEY EXPERIMENT is reported by Faraday on this page from his diary, a day-by-day laboratory notebook. The passage reads: "Aug. 29, 1831. Expts. [Experiments] on the production of Electricity from Magnetism, etc. etc. Have had an iron ring made (soft iron), iron round and 7/8 inches thick and ring 6 inches in external diameter. Wound many coils of copper wire round one half, the coils being separated by twine and calico—there were 3 lengths of wire each about 24 feet long and they could be connected as one length or used as separate lengths. By trial with a trough [battery cell] each was insulated from the other. Will call this side of the ring A. On the other side but separated by an interval was wound wire in two pieces together amounting to about 60 feet in length, the direction being as with the former coils; this side call B. Charged a battery of 10 pr. plates 4 inches square. Made the coil on B side one coil and connected its extremities by a copper wire passing to a distance and just over a magnetic needle (3 feet from iron ring). Then connected the ends of one of the pieces on A side with battery; immediately a sensible effect on needle. It oscillated and settled at last in original position. On *breaking* connection of A side with Battery again a disturbance of the needle. Made all the wires on A side one coil and sent current from battery through the whole. Effect on needle much stronger than before. The effect on the needle then but a very small part of that which the wire communicating directly with the battery could produce." This experiment showed Faraday that it was making or breaking the magnetic circuit—in other words, changing the field—that induced a current in the second coil. This was the key to the understanding of electromagnetic induction.

The history of the 10 days and the half-century furnishes an excellent example of the process by which science passes into technology.

The process is typically divided into two phases. During the first, a fundamental discovery has been made but no one sees a possibility of using it. The new field attracts only the pure scientist and the dabbler in curiosities, neither one aiming at a practical goal. This is the time when theory is ahead of application.

Then there appears a technological niche for the discovery, usually as a result of advances in some collateral area. The niche may open up in months or in years—perhaps never. When it does, the engineer, the inventor and the businessman enter the arena. But now they are likely to find that theory is inadequate to their purposes or that they do not understand it well enough to use it. In this second phase application suffers from inadequate theoretical support. Catalyzed by economic incentive, the pace quickens. A growing technology begins to contribute to science, as well as the other way around, and theory is extended and deepened; devices based on what man thinks he knows about nature are mirrors of truth that bring him closer to understanding the material world. Eventually the interaction closes the engineering gap.

When Faraday began, he knew just what he was looking for. He undertook his experiments with the explicit "hope of obtaining electricity from ordinary magnetism." That hope was prompted by Hans Christian Oersted's demonstration that magnetism could be obtained from electricity. Oersted had been trying to find out if electric current, made available by the recent invention of the chemical battery, exhibited the same attractive power as static electric charge. In 1820 he found that while a current flowing through a wire does not attract objects, it does cause a magnetic needle to line up perpendicularly to the wire.

As early as 1822 Faraday wrote in his notebook: "Convert magnetism into electricity." The same idea occurred to the two great French physicists André Marie Ampère and Dominique François Arago, but both soon decided there was nothing in it. There was no way to arrive at electromagnetic induction by reasoning from the scanty theory of the time. The effect would have to be discovered by experiment, and Faraday was the supreme experimenter.

Four times between 1822 and 1831 he

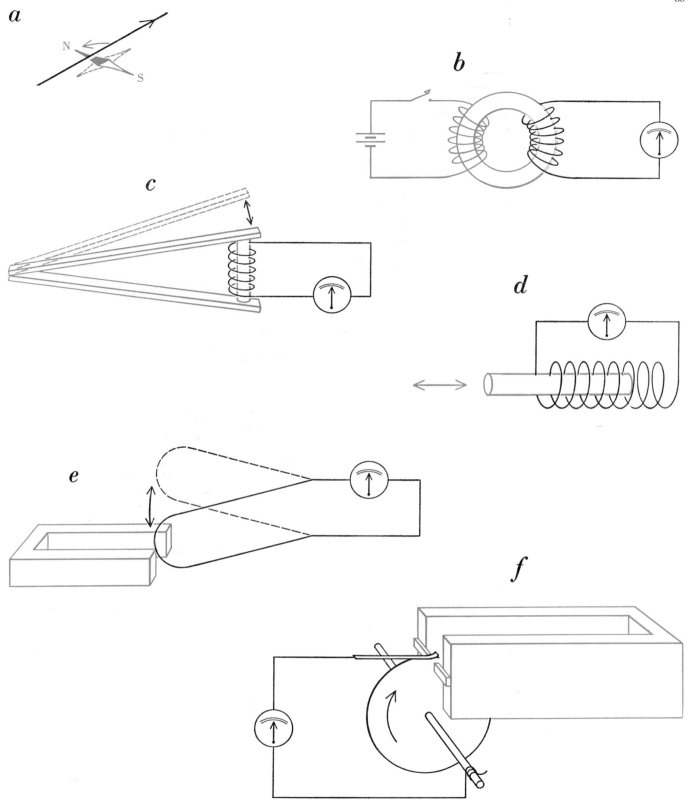

FARADAY'S EXPERIMENTS were prompted by Oersted's discovery that a current-carrying wire made a compass needle near it (*a*) swerve at right angles to the wire, showing that electricity produced magnetism. Faraday sought to show that magnetism could produce electricity. He found first that when a coil on one side of an iron ring (*b*) was connected to or disconnected from a battery, a surge of current was sent through a coil on the opposite side of the ring. Then he found that the same effect could be obtained by making or breaking the magnetic contact between two bar magnets and a coil wound on an iron core (*c*), by thrusting a magnet into a coil of wire or withdrawing it (*d*), or simply by moving a loop of wire up and down in a magnetic field (*e*). Finally he rotated a copper disk between the poles of a powerful magnet and found that a steady electric current was induced across the disk (*f*).

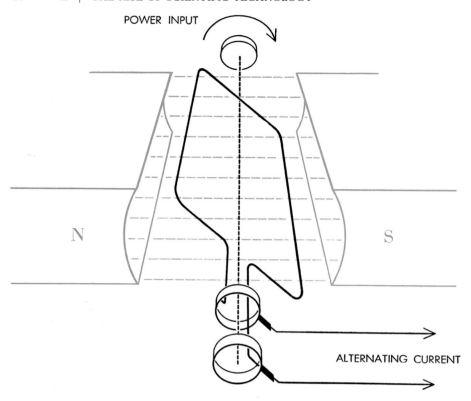

POWER INPUT

N

S

ALTERNATING CURRENT

DYNAMO PRINCIPLE is illustrated. A loop of wire (the armature) is rotated so as to cut the lines of force between magnetic poles. A current—clockwise in this case—is induced in the loop, which is connected to brass slip rings, and the current is led to the external circuit by two brushes. The current alternates because it reverses directions as the two sides of the loop cut the magnetic field first in one direction and then in the other.

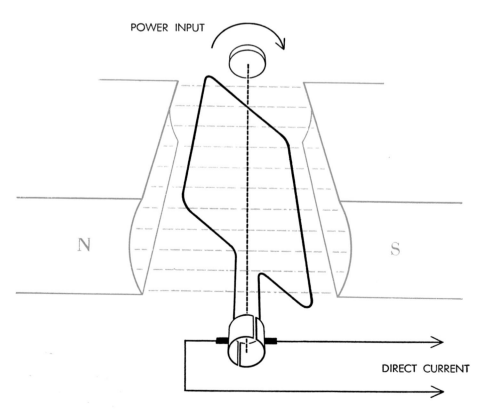

POWER INPUT

N

S

DIRECT CURRENT

DIRECT-CURRENT DYNAMO is made by substituting a commutator for slip rings. The commutator, a ring divided into two segments, switches the sides of the loop from brush to brush so that the current flowing through each brush always goes in the same direction.

tried and failed. On August 29, 1831, he began his fifth attempt and was rewarded almost at once by the happy accident that every experimenter hopes for. He had wound two coils of wire on opposite sides of an iron ring, insulated from each other and from the ring. A battery sent current through the first coil, magnetizing it, and a galvanometer was connected to the end of the second. As in all the previous trials, no current was detected in the second coil. But then Faraday noticed that whenever the battery was connected to or disconnected from the first coil, the galvanometer indicated a momentary current. He had at last found the key: a *change* in the magnetic field created by the first coil produced a current in the second.

Faraday immediately set out to investigate all possible types of this "transient effect," as he called it. He wound a coil of wire around a straight core and placed the core between two bar magnets arranged to form a V. When he pulled the magnets away, a current flowed through the coil. "Hence," he noted in his diary, "distinct conversion of Magnetism into Electricity." Similarly, he induced a current by thrusting a bar magnet into a coil of wire and obtained a current in the opposite direction by withdrawing the magnet. And he reduced his apparatus to its fundamentals when he induced a current in a simple loop of wire merely by passing it through a magnetic field.

All these experiments produced intermittent surges of current, lasting only as long as the relative motion of conductor and magnetic field. Faraday now arranged for continuous motion by rotating a copper disk between the poles of a permanent magnet. A wire around the axle of the disk ran to a galvanometer, and another wire led back from the meter to a metallic conductor held against the rim of the disk. As long as the disk was turned, the galvanometer indicated a continuous current. "Here therefore," Faraday wrote, "was demonstrated the production of a permanent current of electricity by ordinary magnets." He called the device a "new electrical machine" and suggested that its power could be increased by using several disks. Then he dropped the matter.

Faraday's experiments and his observations on them actually contained a number of clues to effective generator design. Had they been recognized, much of the trial, and more of the error, of the next half-century would have been avoided.

The most important clues were contained in his general statement: "If a terminated wire [*i.e.*, one forming part of a complete circuit] move so as to cut a magnetic curve, a power is called into action which tends to urge an electric current through it." This revolutionary idea of magnetic curves or lines of force was not accepted by most of the physicists of the time. It was not until James Clerk Maxwell published his mathematical interpretation of Faraday's model in 1864 that the idea took hold.

But Faraday had already shown in 1831 that, in each of his methods of producing electricity from magnetism, the cutting of lines of magnetic flux by a conductor is the crucial factor. This was true whether he changed the field (by connecting or disconnecting the battery), moved the magnet or moved the conductor. He had discovered the principle that came to be called Faraday's law, which states that the voltage induced in a conductor is directly proportional to the rate at which the conductor cuts lines of magnetic flux. To maximize the rate, the conductors in an ideal generator should pass through the field at right angles to its lines of force. This is perfectly obvious, but only to someone who visualizes the magnetic field as being made up of lines of force. Those who followed Faraday did not, and as a result an efficient armature did not appear for many years.

Another clue lay in the fact, duly recorded by Faraday, that coils wound on an iron ring gave an induced current "far beyond" that obtained from coils on a wooden core. The current was stronger because the iron provided a better magnetic circuit, concentrating the flux so that more lines of force passed through and cut the second coil. Neither Faraday nor his successors realized this, and in the early development of the dynamo the question of the magnetic circuit was ignored. It was simply adapted to fit each change of shape in the armature, sometimes by chance increasing the flux cut by the conductor, but just as often decreasing it.

Having used both permanent magnets and electromagnets in his experiments, Faraday remarked on the "similarity of action, almost amounting to identity, between common magnets and either electro-magnets or volta-electric currents." Yet he continued to distinguish between the two sources of magnetism. And the early builders of generators for some reason used clumsy permanent magnets exclusively, although electromagnets are lighter and more powerful. It was not until the 1860's that electromagnets were generally adopted.

With the conclusion of Faraday's 1831 experiments the first phase in the development of dynamo technology opened. The basic discovery had been made; there was theory, but no immediate interest in applying it. Electromagnetic induction seemed a far less powerful source of current than the chemical battery. More important, there was no apparent use for large currents of electricity and no incentive to develop machines to generate them.

As always, there were tinkerers. In 1832 Hippolyte Pixii exhibited a machine based on Faraday's principles in Paris. Producing very little power, it was in effect no more than a model of a generator. The device had stationary coils and a hand-driven rotating horseshoe magnet; its output was limited by the weakness of the magnet and the energy available in the operator's arm. Even so, Pixii could have increased its power substantially if he had understood the importance of the relation between magnet and conductors. He wound his conductor coils on two bobbins, a neat way of getting a long length of wire into a small space. At best, however, only a small proportion of the wire can ever be perpendicular to the field in this arrangement.

Pixii's first model produced alternating current as the rotating field cut the conductors first in one direction and then in the other. Alternating current seemed altogether pointless, and at Ampère's suggestion Pixii equipped his second version with a commutator so that it would deliver direct current, as a battery does. (A commutator is a rotating switch that reverses the connection between the armature winding and the outside circuit each time the current changes direction in the winding,

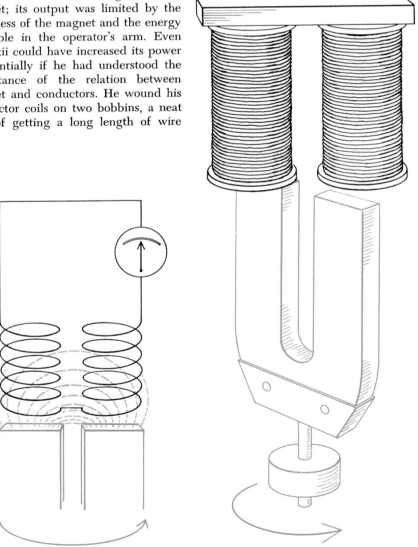

PIXII'S GENERATOR of 1832 had a permanent horseshoe magnet rotated by hand beneath two stationary coils wound on bobbins. Its shortcoming (*illustrated by the drawing at left*) was that, because of the manner of winding, only a small part of each coil was cut perpendicularly by the rotating magnetic field. This was a fault of all bobbin-wound generators.

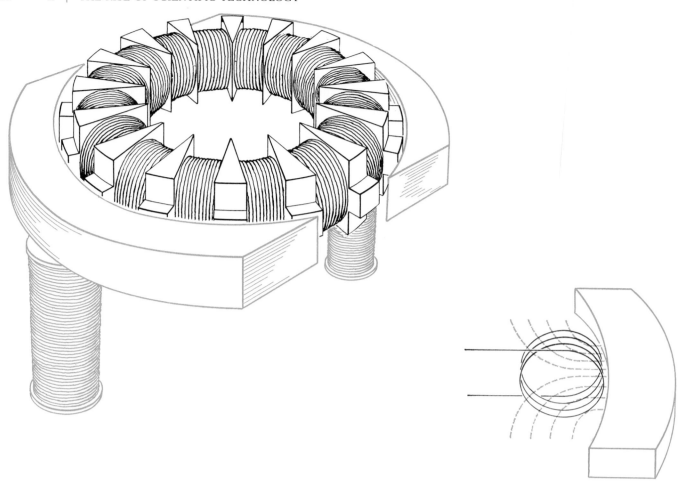

PACINOTTI'S GENERATOR of 1860 introduced the ring-wound armature, improved the magnetic circuit and utilized electromagnets. The advantage of his design (*illustrated at right*) was that the iron ring provided a good path for the magnetic flux and more of each coil was in position to cut lines of force at right angles. But the inner portion of each turn of wire was still ineffective.

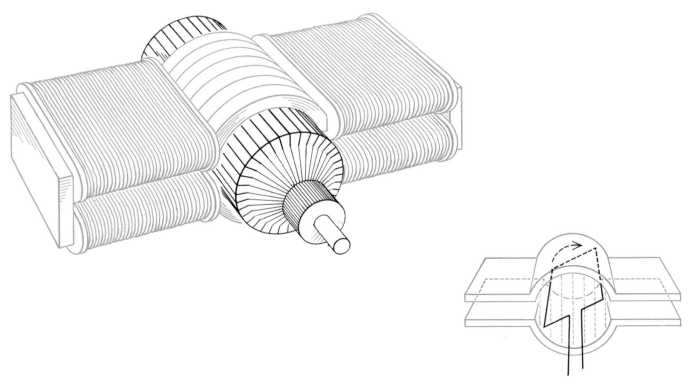

HEFNER-ALTENECK'S GENERATOR of 1872 brought the final basic step in armature design: the drum winding. In this method (*as shown at right*) almost all the armature wire was arranged so as to cut perpendicularly through the lines of magnetic force.

so that the external current always flows in the same direction.) Actually alternating current is better suited to electric power transmission, but it was many years before engineers broke away from the tradition of direct current that was established at this time.

Another experimental generator was built in 1833 by an American, Joseph Saxton. In his design a pair of bobbin-wound coils rotated just beyond the poles of a stationary horseshoe magnet. In the next few years similar machines, still very small, were developed in response to a prevalent medical fad. It was widely believed that a weak electric current sent through the body had a therapeutic effect. This was the first "commercial" application of Faraday's discovery.

The first real economic impetus, however, appeared only in 1839 with the invention of electrotyping. This process, in which copper is deposited by electrolysis on a mold of an engraving, shortly led to the general development of electroplating. Now there was a need for more current than batteries could conveniently supply.

An English engineer, John Stephen Woolrich, saw the possibilities here for an electric generator, and in 1842 he patented a modification of the spinning-bobbin machine. He increased the strength of the magnetic field by stacking several flat horseshoe magnets together and improved the design of the commutator to produce the more constant current required for electroplating. Woolrich's generator was driven by steam instead of by hand, and it delivered useful amounts of current for an industrial process, but it was essentially a beefed-up version of the simple permanent-magnet machine, or magnetoelectric generator.

About a decade later a much broader field of application began to open—electric lighting, originally for lighthouses. A Frenchman, F. Nollet, seems to have been the first to think of the dynamo in this connection, but he did so in a rather indirect way. His source of illumination was to be a "limelight," a block of lime heated to incandescence by an oxyhydrogen flame. He proposed to get the necessary oxygen and hydrogen from electrolysis of water, and for this purpose constructed a generator in which a number of coils rotated past the poles of horseshoe magnets. The system was unsuccessful but it suggested to Frederick Hale Holmes, an English engineer, that similar generators might better supply power for the new carbon-

arc lamps then being developed. In 1857 he rigged up a machine in which 36 permanent magnets rotated past stationary coils. It weighed 4,000 pounds and produced less than 1,500 watts. But the carbon arcs it powered did provide a brilliant light, and it was the precursor of several practical, if inefficient, generators.

With Holmes's big lighthouse installations the magnetoelectric design had gone about as far as it could. The next step was to switch from permanent magnets to the far more efficient electromagnet. Suggested by Søren Hjorth of Denmark in 1855, the idea was patented in 1863 by Henry Wilde of England. In the first models the electromagnets were supplied with current by batteries or by small magnetoelectric generators. Soon a number of workers recognized that the auxiliary source is not necessary—that the generator itself can supply the current needed to excite the magnets. The small amount of residual magnetism that always remains in the iron core of the electromagnet provides the initial field, and thereafter the strength of the field increases as the output of the generator builds up.

The self-excited "dynamoelectric machines" represented a considerable advance over the earlier magnetoelectric machines. In their armature design and magnetic circuits, however, they still reflected serious gaps in the theoretical understanding of induction. In 1860 an Italian physicist, Antonio Pacinotti, built a machine that incorporated large improvements in both. First of all he wound his coils on a ring that revolved in the plane of the lines of force between two electromagnets. This arrangement put a larger proportion of the winding in position to cut the lines of force perpendicularly than could any type of bobbin armature. Secondly, he made the ring out of iron, which, as has already been mentioned, increases the magnetic flux that threads the coil. Pacinotti's description of his apparatus in an Italian scientific journal attracted little attention. The electroplaters, who were still the major consumers of electric current, probably never saw the report, and the physicists who did read it presumably were not interested in application. In this case a gap between discovery and application resulted from lack of communication between scientists and engineers.

The ring winding was rediscovered by Zénobe Théophile Gramme of France in 1870. In principle his machine did not differ at all from Pacinotti's, but Gramme was associated with capable business-

men, who saw to it that the invention did not go unnoticed. Very soon the Gramme machine and modifications of it became standard equipment both for electroplating and for arc lamps in lighthouses and factories.

The final improvement in armature design came in 1872 with the invention, by F. von Hefner-Alteneck in Germany, of the drum winding. On a ring only the outer portion of each turn of wire produces a useful voltage; the voltage in the inner portion actually works in the wrong direction. The drum eliminates the inner portion completely and puts a much greater length of wire in a position to cut the field perpendicularly.

Hefner-Alteneck had started with a wooden drum but he shifted to iron. At that point the generator had almost reached its present form. Subsequent builders discovered the importance of minimizing the air gap in the magnetic circuit. This they did by such measures as curving the pole pieces of the magnets to fit around the armature and countersinking the windings in slots in the drum so that its iron surface could be brought closer to the magnets. In 1886 the British engineers John and Edward Hopkinson showed how to predict the performance of magnetic circuits, thereby finally taking generator design out of the trial-and-error stage.

By 1890 a flourishing electroplating industry, as well as the mushrooming lighting companies, could obtain direct-current generators about as efficient as those available today. One more giant step in electrical technology remained: the generator had yet to be teamed with the electric motor.

When Faraday sought to produce electricity from magnetism he was looking for the reverse effect of the motor principle—the force exerted by a magnet on a wire carrying a current—that he had demonstrated in 1825. But the electric motor was developed along different lines, and for the most part by different inventors, from the generator. Only gradually did it appear that the motor is the simple converse of the generator and the generator's natural complement in industry and transportation. The importance of reversibility was overlooked until central-station plants for lighting demonstrated that electricity is above all an efficient means of transmitting energy over long distances. By the end of the 19th century centrally generated electricity was beginning to replace steam as the motive power in railroading and, during the opening years of the 20th century, in industry generally.

8 Technology and Economic Development

by Asa Briggs
September 1963

How can nations attain a state of self-sustaining growth? This article outlines the history of development and of the division of nations into "rich" and "poor"

A circumstance new to history stretches the tensions of contemporary world politics. This is the widespread awareness of the division of nations into two classes: "developed" and "underdeveloped" in the parlance of the day or, in plainer words, rich and poor. The contour lines of international economic inequality are easily drawn. To the class of the rich belong the nations of northwestern Europe and those elsewhere in the Temperate Zones that were settled and organized by people of the same stock: the U.S., Canada, Australia and New Zealand. One non-European nation—Japan—should also be counted in the group, and recently another European nation, the U.S.S.R., has joined it. These nations, constituting less than a third of the human population, produce and consume more than two-thirds of the world's goods. Their output is increasing more rapidly than their population, and they boast rising incomes per capita.

Income per capita hardly serves as a measure of the position of the nations of the poor. The overwhelming majority of their populations are occupied in subsistence agriculture and live almost entirely outside the monetary systems of their meager economies. For what the economic indices are worth, they show that between the poorest 1.5 billion people—the bottom half of the human population—and the average standard of living prevailing in the rich countries the disparity is on the order of one to 10. More significantly, the indices show that the disparity between the two classes of nations is widening.

Poverty is not, of course, a new condition in human affairs. Some of the poor nations were once world powers and were held to be rich as well as powerful. But even though they have been placed at a disadvantage in recent years by the unfavorable terms of their relations with the rich nations, the situation in which their peoples live is not much worse than before. The rich nation is the novelty, and the development that makes entire nations rich is itself the pivotal development of modern history. To understand the increasing economic inequality of nations one must look outside the boundaries of economic theory. In the search for the causes, antecedents and "preconditions" of development it is necessary to turn to history, and the historian has his choice of starting points.

In the summer of 1454, one year after the fall of Constantinople to the Ottoman Turks, Enea Sylvio Piccolomini (later Pope Pius II), who has been described as one of the best-informed men in Europe, wrote gloomily that he could not see "anything good" in prospect. Christendom was weak and divided, and internal conflicts as well as external challenges foretold likely destruction. He did not add, as he might have done, that there had also been a downsweep in the medieval economy. This was not the language of the age. His modest humanist hope was that he would be proved entirely wrong and that posterity would call him a liar rather than a prophet.

Within less than 50 years Europeans had pushed out adventurously far beyond the confines of Europe around the coast of Africa, toward India and Southeast Asia and across the Atlantic. An Indian historian has described everything that has happened between then and our own times as the "Vasco da Gama epoch" in world history. The search for wealth outside Europe's boundaries preceded the full mobilization of wealth within. Long before our own times Adam Smith, writing on the eve of the great industrial changes that transformed both society and men's ways of thinking about it, declared that the discovery of America and of a passage to the East Indies via the Cape of Good Hope were the two most important events in the history of mankind.

Within less than 100 years after 1454 the great movements of thought and feeling to which historians long ago attached the labels of "Renaissance" and "Reformation" had further extended and disturbed the horizons of many Eu-

DEVELOPMENT IN NIGERIA is exemplified by the construction of a bridge across the Niger River, the first piers of which are seen in the aerial photograph on the opposite page. The bridge, at Onitsha, will greatly improve communication between Nigeria's Eastern and Western regions.

CHILD APPRENTICE is instructed in the workings of the spinning machine she will tend. This early industrial photograph was made in a U.S. textile mill. Child labor was an important source of the human capital invested in 19th-century industrialization.

ropeans. It is just as easy for 20th-century writers to place the beginnings of "modern times" in 15th- and 16th-century breaks with tradition as it was for Adam Smith. Those breaks now figure, however, less as spectacular events than as phases of processes, "preconditions" of what was to happen later. The invention of the steam engine or the French Revolution, the one carrying with it a universal technology, the other a universal ideology, may today look like even bigger breaks. It is part of the task of the historian to scrutinize old labels carefully, to qualify large-scale generalizations and to expose contradictory tendencies. Much that seems "modern" has origins more remote than the 18th century. Much that was old in the 15th and 16th centuries has survived on a massive scale.

The least modern element in the first predatory phases of discovery was that the underdeveloped countries of today then seemed to be the great centers of wealth: the "gorgeous East" and the South American El Dorado. The 17th-century English writer Thomas Mun, exaggerating and oversimplifying, maintained that the world commerce of his day consisted in the exchange of the mineral wealth of the new Indies in the West for the luxuries and refinements of the old Indies in the East. Francis Bacon referred to South America as "the money-breeder of Europe."

Between the beginnings of the age of world commerce, when new resources and markets were opened up, and the great industrial changes of the 18th and 19th centuries, when new methods of production were introduced, the wealth of nations was determined in large part by the struggle for empire and power. That struggle, which led to the eclipse of Spain and Portugal, the rise of the Netherlands and the protracted contest between England and France, was world-wide in scale. American independence was one aspect of it. Concurrently, within Europe, no less significant but less dramatic changes in economic life were under way, later to culminate in industrial revolutions and the postindustrial division of nations into developed and underdeveloped.

By the early 18th century there were present in parts of Europe many of those economic and social ingredients whose absence is taken today as a sign or a cause of "backwardness." Among them were transport and credit facilities, many deriving from international trade; supplies of relatively skilled labor, some of it employed in industries with scattered and potentially expanding markets, and—not least—well-trained acquisitive attitudes, congenial to both enterprise and capital accumulation. R. H. Tawney is not the only historian of capitalism to go back for his basic evidence not to the age of industrialism but to the shifts of values in the three centuries that preceded it.

It was during the last of these centuries that the "scientific revolution" created new climates of opinion. "The stream of English scientific thought, issuing from the teaching of Francis Bacon and enlarged by the genius of Boyle and Newton," T. S. Ashton has written, "was one of the main tributaries of the industrial revolution." The statement cannot be disputed, even though many of the first inventors who transformed ways of production were men of little science. Practical and empirical, they were more interested in solving an immediate problem than in speculating about nature. The technical ascendancy of science belongs to the 19th century, not the 18th.

Britain was the center of the first industrial revolution. Throughout the first decades of the 19th century more than half of the world's industrial output was concentrated in an island with only about 2 per cent of the world's population. The British industrial revolution, the first in a sequence, became a classic model, even if it was a misleading model. From it Karl Marx deduced that "the industrially more developed country presents to the less developed country a picture of the latter's future." The forecast, involving as it did both the premise of economic growth and the threat of social conflict, contrasts sharply with Adam Smith's preindustrial forecast of a "stationary state" in which the existing methods of production would have been "improved" as far as they could possibly be improved and economic growth would have ceased.

The "causes" of the British industrial revolution, therefore, have more than local interest. Historians are still arguing about the weighting of the various factors that contributed to the upsurge of growth, particularly in the 1780's. What seems clear is that, in addition to the cumulative build-up of economic power on the high seas and overseas and the social development of a community that encouraged innovation and thrift, there were urgent challenges that had to be overcome before there could be an immense spurt in invention, investment, production and trade. The slowing down of a previous rate of agricultural expansion and the peculiar exigencies of unprecedented population growth may explain difficult questions relating to timing. There are also long-term technical questions, however, in relation to the exploitation of iron and the development of steam power. It has even been argued that Britain had to leap ahead if it was not to lumber back.

Businessmen of the time often gave simple answers. "We want as many spotted Muslins and fancy Muslins as you can make," a Northern cotton spinner was informed by his London agents in 1786. "You have many competitors, we hear, coming forward.... You must give a look to Invention. Industry you have in abundance.... We expect to hear from you as soon as possible, and as the Sun shines let us make the Hay." There were no ways to increase output to meet rising demand without new processes and new forms of organization. As a knowledgeable Manchester man put it in 1783: "No exertion of the manufacturers or workmen could have answered the demands of trade without the introduction of spinning jennies."

Thus at the very time American independence was ratified Britain was finding new sources of economic strength. Between 1781 and 1800 the imports of raw cotton quintupled, pig iron production quadrupled, foreign trade (whether measured in shipping tonnage cleared from the ports or in export and import values) nearly tripled and total industrial production doubled.

A contemporary writer with a precocious statistical sense drew rhetorical conclusions. "An era has arrived in the affairs of the British Empire," wrote Patrick Colquhoun in 1814, just before the last of the great wars between Britain and France came to an end, "when resources have been discovered which have excited the wonder, the astonishment and perhaps the envy of the civilized world." He moved from rhetoric to social generalization. "It is with nations as it is with individuals who are in train of acquiring property. At first, progress is slow until a certain amount is obtained, after which, as wealth has a creative power under skillful and judicious management, the accumulation becomes more and more rapid, increasing often beyond a geometrical ratio, expanding in all directions, diffusing its influence wherever talents and industry prevail, and thereby extending the resources by which riches are obtained by communicating the power of acquiring it to thou-

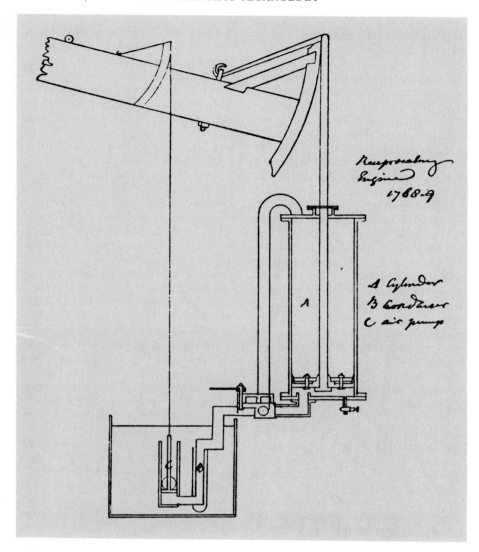

STEAM ENGINE as modified by James Watt was the basic source of power for the British industrial revolution. In this drawing prepared by Watt for a 1769 patent specification the cylinder is at right and the separate condenser, one of his important contributions, is at the bottom. The downstroke of the piston exerted a downward force on the rocker arm (*top*).

sands who have remained without wealth in countries less opulent."

The term "industrial revolution" seems to have been invented by a French economist, Jérôme Adolphe Blanqui, in 1827. Before this, however, James Watt and Richard Arkwright had already been compared with Mirabeau and Robespierre, and smoke with propaganda. Something more had happened than mere acceleration of existing economic trends. Man's position had changed in relation to nature. Poets and prophets were as fascinated by steam power as millowners and ironmasters. Erasmus Darwin, the grandfather of Charles, wrote in 1792:

*Soon shall thy arm, UNCONQUER'D
    STEAM, afar
Drag the slow barge, or drive
    the rapid car;*

*Or on wide-waving wings expanded
    bear
The flying chariot through the fields
    of air.
Fair crews triumphant, leaning
    from above
Shall wave their fluttering kerchiefs
    as they move;
Or warrior bands alarm the gaping
    crowd,
And armies shrink beneath the shadowy
    cloud.*

In the first flash of enthusiasm there was immense imaginative appeal in technical discovery, just as there had been in the discovery of America. It was the recognition that nature could be tamed and the environment controlled that distinguished the industrial revolutions of the 18th and 19th centuries from the only comparable revolution in

human productivity, that of the neolithic world, when settled agriculture took the place of hunting and food-gathering and a new division of labor transformed social and cultural processes.

The extent of the change can be measured not only in statistics of material progress but also in 19th-century social comment from Jean Charles Sismondi and Claude Henri Saint-Simon to John Stuart Mill and Marx. Saint-Simon wanted to change the words of the *Marseillaise* from "*enfants de la patrie*" to "*enfants de l'industrie.*" Politics for him was "the science of production." In 1848, the year of revolution, Mill wrote: "All the nations which we are accustomed to call civilized, increase gradually in production and in population; and there is no reason to doubt that not only these nations will for some time continue so to increase but that most of the other nations of the world, including some not yet founded, will successively enter upon the same career."

Only the word "gradually" is misleading. Industrialism was to establish itself in sharp bursts, and once established it was to develop unevenly through boom and slump. It was also to create new social conflicts. Later writers emphasized, as the elder Arnold Toynbee did in his pioneer study of the 1880's, *Lectures on the Industrial Revolution of the 18th Century in England,* the social consequence of steam power—its effects on men's relations not with nature but with one another.

Not only were owners of capital often pitted against owners of land—town versus country, competition versus monopoly, progress versus tradition were some of the battle cries—but also there was a new division between "capital" and "labor." It was to this division that Marx turned his attention, maintaining that in the very processes of industrial expansion "classes" were being formed that were inexorably antagonistic. The rich would become richer and the poor poorer. Unlike traditionalist writers who bemoaned the decay of an old social order, Marx welcomed the transformation and the social revolution he thought it would ultimately entail. The melancholy conservative reaction was well expressed by Henri Frédéric Amiel in 1851: "The statistician will register a growing progress and the novelist a gradual decline.... The useful will take the place of the beautiful, industry of art, political economy of religion, and arithmetic of poetry."

These contrasting pictures of the future were painted at a time when Europe was still primarily an agricultural continent with no more than patches of industry. Even these patches were frequently to be found among forests and beside streams rather than in concentrated industrial areas. Marx and the early British socialists before him might talk of a "working class," but in Europe craftsmen far outnumbered factory workers and even in Britain there were far more domestic servants than textile workers. When the Great Exhibition of 1851 was held in the specially built Crystal Palace to illustrate "the progress of mankind," there was no doubt that

Britain was a workshop of the world.

During the early stages of industrialization the two master commodities were coal and iron. They took the place of wood, wind and water at the center of the new technology. The two materials were associated both geographically and economically; their close geographical proximity often created "black country." Their economic interdependence was expressed most strikingly in the great symbol of early industrialization: the steam locomotive puffing its way over "iron roads." It is not surprising

that contemporaries saw the building of railroads as the beginning of a new world. "We who lived before railways," wrote William Makepeace Thackeray, "and survive out of the ancient world, are like Father Noah and his family out of the Ark. The children will gather round and say to us patriarchs, 'Tell us, grandpa, about the old world.'"

Others saw railroads as sinews of the economy and vitalizing influences on society. Heinrich von Treitschke, the ideologue of German nationalism, believed that railways "dragged the German nation from its stagnation." Count

INDUSTRIAL LANDSCAPE photographed in the Midlands of England about 1855 shows a mine with two shaft hoists and (center) a steam-driven water pump. In the foreground is a steam locomotive pulling a string of flatcars carrying loaded mine carts.

Sergei Yulievich Witte, the Russian engineer and exponent of industrialization, set out to make railroads the foundation of a new economy in Russia in the 1890's; economic historians have been unanimous in pointing to Russian railroad building as "the fulcrum round which the industrial level of the country was being rapidly lifted" during that decade.

The identification of industrialization with "carboniferous capitalism," which was simply one phase of industrialization, has had lasting results. Along with a waste of economic resources there was a marked deterioration in the human environment in the new industrial areas. At the very time that science was suggesting that "fate" was really amenable to social control, a new framework of social necessity was being constructed. The mill chimney and the slag heap dominated the horizon and set the scene for the social conflicts that also came to be identified with capitalism—conflicts centering not only on wages but also on status and authority, the length of the working day and the right to security.

Whatever the technology that activates industrial revolutions, the disturbance of old traditions and institutions and the imposition of unfamiliar rhythms of work and leisure are bound to bring social upheavals. But the coal-and-iron technology of the first industrial revolutions accentuated all the human difficulties. It is not surprising that in 19th-century Britain aesthetic and social protest converged in the writings of John Ruskin and William Morris. The same tradition of protest continued to exert a powerful influence, however, even after iron had given way to steel as a master material and electricity had provided a new source of power.

In this next age of technology Britain pioneered new schemes of social welfare but failed to hold its own economically. It was not only that other nations possessed greater physical resources they could develop at lower cost; there was also a withering of enterprise in Britain. The rate of expansion slowed and British industry failed to participate fully in the new developments in steel—even though some of the basic inventions were British—in machine tools, electrical engineering and chemicals.

Germany and the U.S. were the countries that took the lead in these industries of the future. By 1886 the U.S. had replaced Britain as the world's largest steel producer; Germany too was ahead by 1900. In machine tools the U.S. set a new pattern of standardization, the precondition of mass production. In the building of its electrical and chemical industries Germany, with the most advanced European system of scientific and technological education, became a "new model" of industrialization. Output of sulfuric acid and alkalies rose eight times between 1870 and 1900; that of dye-stuffs, in which Germany held a near monopoly, rose four times during the same period.

In both the U.S. and Germany the big industrial concern came to dominate manufacturing industry. The resulting concentration of economic power contrasted sharply with the diffusion of economic power during the early stages of the British industrial revolution. There was talk in the U.S. of "titans" and in Germany of "industrial Bismarcks." Many of the great corporations of the 20th century had their origins in the last decade of the 19th.

This was not the only difference between the British industrial revolution and the industrial revolutions that followed. In Britain little reliance had been placed on the state; everything had depended on a partnership of inventors and businessmen. The theory of the revolution—if there was a theory—was self-help in the industrial sphere and free trade between nations. In Germany the power of the state was harnessed to assist industrialization. In the U.S. the emerging industrial power enlisted the benevolent patronage of the Federal and state governments in the allocation of the continent's rich resources and the maintenance of a social and political climate congenial to its growth.

There was a traditional sanction for the German reliance on the state, but it was given eloquent new expression in the late 19th century. As Gustav Schmoller, an influential professor of economics, put it in 1884: "It was clearly those governments which understood best how to place the might of their fleets and the apparatus of their customs and navigation laws at the service of the economic interests of the nation with speed, boldness and clear purpose which thereby obtained the lead in the struggle and the riches of industrial prosperity." At the same time the riches of prosperity added to the power of the state. The newly unified German state grew stronger as the industrial economy grew stronger. Dependence on tariffs, the direct intervention by bankers in the structure and control of industry and the encouragement by the courts and the government of cartels and large-scale industrial organizations were all parts of the pattern.

The idea of using political power to hasten industrialization has become a commonplace in the 20th century. Protests against industrialism have been less vociferous than demands for more industrialization. Not only has nationalism come to be closely associated with industrial strength and economic independence, but also socialism, which began with industrial discontent, has dwelt increasingly on economic "planning." Whereas in 19th-century Britain transfers of political power followed industrialization as railroads followed factories and furnaces, in many 20th-century countries the existence of a "progressive" and "dynamic" directing political power has come to be considered a precondition to industrialization, much as Witte held railroads to be. Economic and social historians have spoken increasingly of innovating elites instead of entrepreneurs, the protagonists of classical economics. The more backward the economy, they argue, the more directly the state has had to intervene in the encouragement of industrialization, the greater has been the pressure for large-scale plant and the most up-to-date technology, and the more necessary has it been to proclaim a gospel of industrialism. In the "neo-classic" industrial revolutions the incentive to "get rich quick" has seldom proved a sufficient motive; "development" has had to be advocated in more general terms.

Even before 1914, when industrialization was widening the income gap between the countries now classified as "developed" and "underdeveloped," Japan had joined Germany and the U.S. as a new center of industrial revolution. In Japan, as in Germany, economic and political processes ran together in close harness. As a result of political and social revolutions in the 1860's Japan was able to break with enough of its tradition to carry out a deliberate industrial revolution. Samurai bureaucrats embarked on a sweeping "westernization," although they took care not to destroy existing social structures. The state itself initiated strategic enterprises, facilitated the borrowing of advanced technology from abroad and pursued a fiscal policy that encouraged the businessman and fixed the burden of forced savings on the farmer.

Between 1907 and 1914 Japan achieved an annual growth rate of more than 8 per cent. "By 1914," wrote William W. Lockwood, the historian of

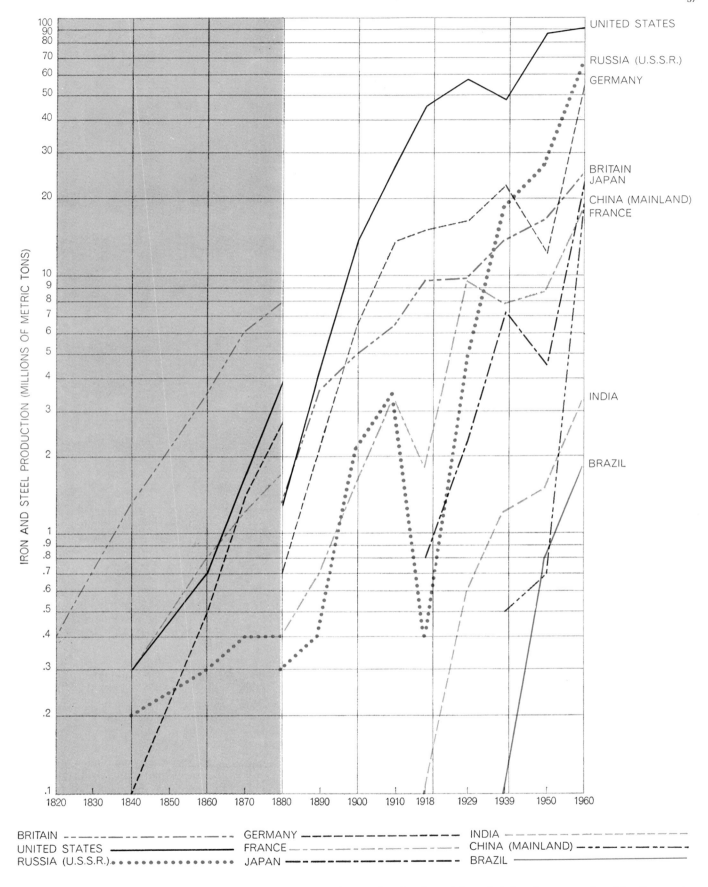

IRON AND STEEL PRODUCTION (MILLIONS OF METRIC TONS)

UNITED STATES
RUSSIA (U.S.S.R.)
GERMANY
BRITAIN
JAPAN
CHINA (MAINLAND)
FRANCE
INDIA
BRAZIL

BRITAIN	GERMANY	INDIA
UNITED STATES	FRANCE	CHINA (MAINLAND)
RUSSIA (U.S.S.R.)	JAPAN	BRAZIL

IRON AND STEEL PRODUCTION, charted here, tell the story of industrialization. In the era of iron (*colored part of chart*) Britain led the world. When steel technology supplanted iron, the U.S. and Germany began to move ahead. Over the years other nations developed substantial steel industries, some (notably Communist China) only in the past decade. The logarithmic scale makes it possible to compare rates of growth in production in different countries: parallel segments of the curves show growth at similar rates.

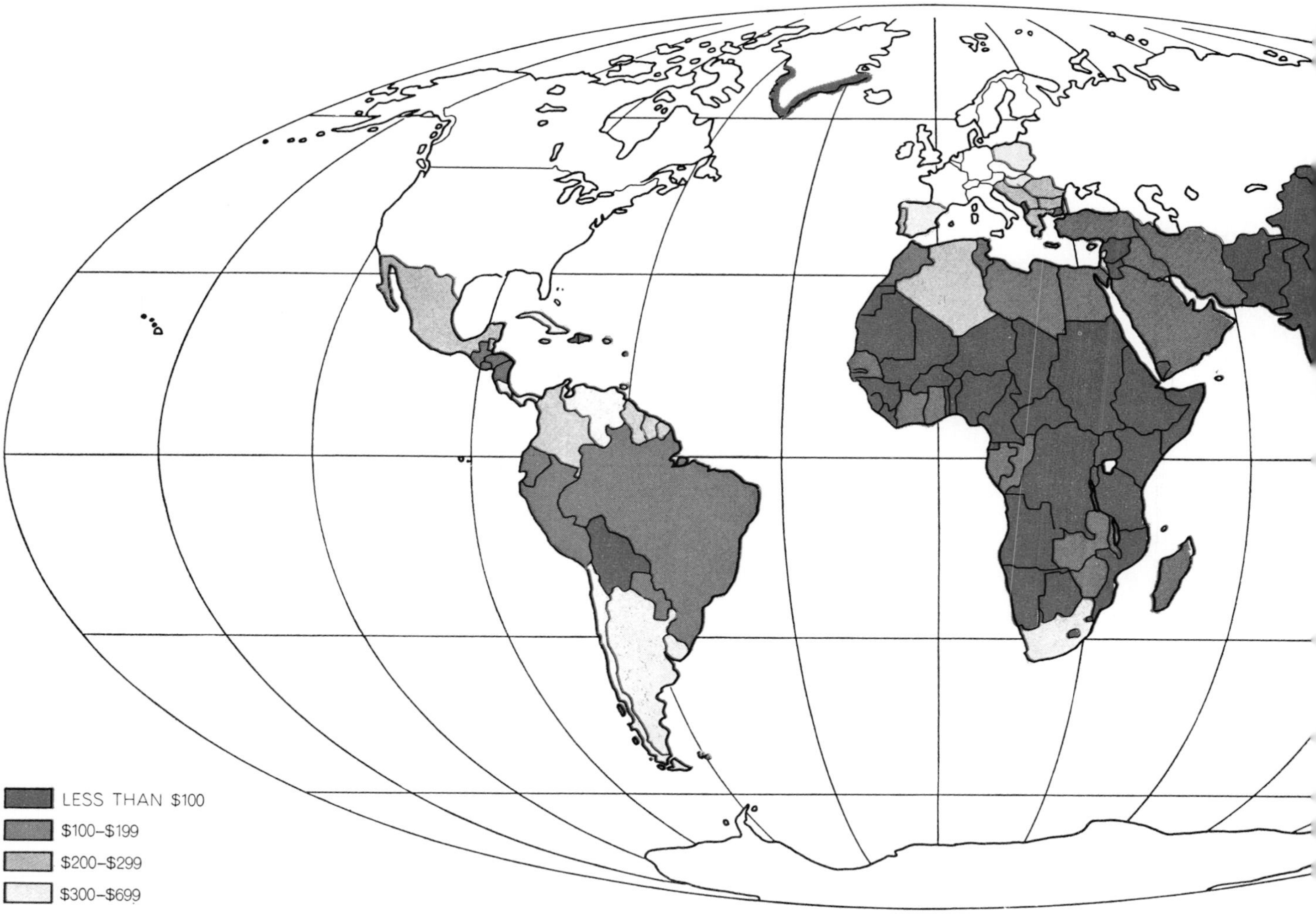

LESS THAN $100
$100–$199
$200–$299
$300–$699

UNDERDEVELOPED COUNTRIES are grouped here according to their average per capita incomes for 1957–1959, based on avail- able UN statistics and some estimates. Income, commonly accepted as a rough gauge of development, does not tell the whole story:

Japan's spectacular development, "Japanese industrial capitalism was still weak and rudimentary by comparison with the advanced economies of the West. But it had now emerged from its formative stage." During and after World War I it was to profit from its industrial lead in Asia as Britain had profited from its lead in Europe 150 years before. It was also to be in the vanguard of the third generation of industrial technology based on plastics, new metals and electronics.

Russia, which also increased its industrial output by 8 per cent per year during its great forward leap of the 1890's, relied on somewhat similar devices, notably fiscal pressure on the peasants and the acquisition of technology from abroad. After the revolution of 1917 and the economic vicissitudes leading up to the promulgation of the first and second Five Year Plans, there evolved a militant ideology of industrialization. In the 18th century Brit-

ain and France had offered different, if complementary, revolutions to the world; the first was economic, the second political. The U.S.S.R. sought to offer both in one package. The ideology of industrial change would appeal to "poor nations," it was felt, at least as strongly as socialist ideology had appealed to poor individuals or classes in the stormy years of iron and coal industrialization. Speed and scale were both emphasized. So too was sacrifice—the deliberate concentration on heavy industry and capital investment and the forcible limitation of consumption.

This model, reinforced as it has been with the factual evidence of an exceptionally high growth rate, has probably had more appeal than the ideology of communism itself. The appeal has influenced quite different kinds of society, although the universal applicability of the model is just as much open to question as the universal applicability

of the British model was open to question in the 19th century. More recently economists have directed increasing attention to the place in economic growth not of heavy industry but of agriculture. They have also stressed the subtleties of development. Industrial revolutions require more than political enthusiasm. "Economic development is a process," John K. Galbraith, then U.S. Ambassador to India, told an Indian audience in 1961, "that extends in range from new nations of Africa only slightly removed from their tribal structure to the elaborate economic and social apparatus of Western nations. At each stage along this continuum there is an appropriate policy for further advance. What is appropriate at one stage is wrong at another."

In the changing context of argument and action the industrial experience of the richest industrial country in the world—the U.S.—has general relevance. Industrialization in the U.S. proceeded in a number of clearly defined spurts:

Japan, for example, is considered developed although its per capita income is not high.

the first between 1837, a year of depression, and the Civil War; the second in the decade and a half following the end of the war; and the third in the 1890's. Although there was a marked trend in each of these three periods to increase the relative share in production of producers' goods, equipment and machines, emphasis in the 20th century has been increasingly placed on the great expansion of consumers' goods. The U.S., with its huge domestic market, has used the new technology to transform not only the standard of life of all its inhabitants but also their whole pattern of daily living. The consumer was deliberately placed at the center of the industrial complex. The resulting dazzle of affluence contrasts sharply with the grim facts of poverty in underdeveloped countries and with the compelling "puritan" philosophy of sacrifice based on investment for investment's sake.

Beyond doubt the problems centering on development and underdevelopment furnish the principal preoccupation of contemporary political, economic and social theory. Before World War I it was almost taken for granted that there was a natural division in the world between manufacturing countries and primary producing countries—the black and the green. This assumption did not suppose either that relative power within the group of manufacturing countries was fixed for all time or that the gap between the rich countries and the poor countries was destined to go on widening forever. Yet it did close minds to a number of problems that are now felt to be fascinating as well as important.

Within the developed countries the facts of inequality between societies have lately begun to shock. The body of world statistics, gathered for the first time by the technical agencies of the League of Nations and now powerfully amplified by the international civil service of the United Nations, has exposed not only inequality but also the ugly mechanism of the "Malthusian trap" in which two-thirds of mankind is imprisoned. Kingsley Davis shows in his article "Population" [SCIENTIFIC AMERICAN Offprint 645] how the 20 or so developed nations have made their escape and prays that increase in the rate of production may exceed the rate of population growth elsewhere and make it possible for other nations to follow. As for food, water, energy and minerals, the authors find that supply is a function of the dynamic international variable: technology.

Yet although technology is international, political facts and attitudes remain stubbornly national. The spell of nationalism is strongest in the ex-colonies of "developed" nations with empires. Political independence, it is felt, must be ratified by economic liberation. Any economic gains colonies may have secured through their place in the decaying empires are brushed aside, and the complicity of the rich in the poverty of the poor nations supplies the negative mold that shapes the plans of development. It is the "distortions" of colonialism that count, not its untapping of world resources. As Wassily Leontief shows in "The Structure of Development," [SCIENTIFIC AMERICAN, September 1963], many underdeveloped economies present a "mirror image" of the developed. Nigeria, for example, exports a few "single crop" products of plantation agriculture and imports the diversity of goods produced by the advanced industrial economies. Other underdeveloped countries are engaged in shipping out their "lifeblood" in the form of irreplaceable mineral resources. Each of these countries in turn, ex-colony or not, now seeks to import the technology necessary to the building of an indigenous, diversified, self-sustaining industrial economy.

Yet in spite of all the brave planning and even the beginnings of industrialization in countries such as India, the inequalities between nations are increasing. More than 90 per cent of the world's industrial output is still concentrated in areas inhabited by people of European origin. Even if underdeveloped countries were to increase their average incomes 10 times faster than the economically advanced countries, the gap would still widen. Given both population pressure and political pressure, is it possible to live peaceably in a world where such inequalities are being aggravated rather than attenuated and where dreams of development are sometimes frustrated?

Development is intimately bound up with 20th-century shifts in power and conflicts of power, with mainland China now taking the stage as well as the U.S.S.R. The dynamic forces of industrialization are as much an element in contemporary international politics as the quest for gold and spices was in the era of mercantile expansion. The countries of the Temperate Zones once turned to the Tropics for riches. Now the Tropics—Brazil, for example, whose gold found its way through Portugal and Spain to finance the industrial revolution of Britain—turn to the temperate countries for know-how that cannot be transported in ships or by formula.

Whether, as Abba Eban of Israel has proposed, the "new nations do not have to tread long and tormented paths . . . and can skip the turbulent phases through which Western industrial revolutions had to pass," depends in decisive part on the politics and social disposition of the developed nations. As Edward S. Mason shows in "The Planning of Development," [SCIENTIFIC AMERICAN, September 1963], the West is called on not only for "aid" but also for tolerance of the new modes in which the developing nations will assert their liberation from poverty. Attitudes deriving from inequality can help or hinder such tolerance. Consciences can be stirred, but there are also built-in feelings of "inferiority" and "superiority." In balance, a 20th-century Piccolomini might see some good in prospect, provided, that is, that we have imaginations powerful enough to bridge the gulf between the different worlds of our own making.

INVENTION IN THE ERA
OF MATERIALISM

INTRODUCTION

There is therefore much ground for hoping that there are still laid up in the womb of nature many secrets of excellent use, having no affinity or parallelism with anything that is now known, but lying entirely out of the beat of the imagination, which have not yet been found out. They too no doubt will some time or other, in the course and revolution of many ages, come to light of themselves, just as the others did; only by the method of which we are now treating can they be speedily and suddenly and simultaneously presented and anticipated.

Francis Bacon
Novum Organum, Aphorism CIX

At Versailles, on January 18, 1871, Wilhelm, king of Prussia, was proclaimed Wilhelm I, German emperor, and the nationalist wars of the nineteenth century came to an end. During the first two-thirds of the century, the balance of power had tilted overwhelmingly in favor of technology and organization. The newly industrialized nation-states, armed physically with industry and technology and morally with Social Darwinism or its racial equivalent, turned from dueling over each other's crowded territory toward capturing the vast and underdeveloped resources of the rest of the world. The next four decades were, on a historical scale, an era of peace. The few brief, though intense, clashes that occurred influenced the future primarily through their role of delineating the boundaries of the most immense conquests in history—the great imperial expansions. Between 1870 and 1900, Great Britain alone was to acquire nearly five million square miles of territory with an indigenous population of about ninety million and fulfill, in fact, the ancient boast of a global empire upon which the sun never set. The imperialist doctrine proved contagious and spread among the industrial powers at a rate damped only by the possession by some of internal frontiers that absorbed the initial urge for expansion. The industrialized nations, wielding rifled cannon and repeating rifles, armored fleets and Gatling guns, divided the world into merchants and markets, manufacturers and miners, colonizers and colonized, imperial powers and their empires. Africa, South Asia, Oceania, South and Central America were conquered rapidly. The initial cost of this immense conquest was ridiculously small at the time, but the immediate price has proved to be hardly a down payment; the former imperial powers have repeatedly been called upon to meet further, and far more costly, installments on a debt that has yet to be fully discharged. During these conquests, the political unity of the industrial nations was firmed by the strengthening of internal communication, particularly in the newly stabilized states; roads, railways, and telegraph lines linked the various regions and provinces, drawing them together into a vast web. The mobility and rapid communication thus acquired was to play a surprisingly decisive role in the opening battles of August 1914, when, having all too quickly run out of new worlds to conquer, the empires, moving inexorably on a course set forty years before, collided and shattered, leaving their fragments, their rivalries, and their hatreds strewn like unexploded bombs through the course of the twentieth century.

A solid, respectable, middle-class citizen of an imperial power, looking back over the end of the nineteenth century from the opening days of the twentieth, could well believe it was the dawn of a new golden age. The bicycle, the horseless carriage, and the railroad were replacing the exhaustible and mortal horse. The scale of time was compressed by rapid telegraph communication. Iron bridges spanned once formidable river barriers and chasms; iron ships plowed the oceans; iron rails crossed forbidding mountains and barren deserts. Standing at the apex of an economic pyramid, from whose strife-ridden base issued revolution, anarchy, socialism, and trade unionism (troubles that could be safely hidden behind class barriers and supremacist doctrines), the bourgeoisie possessed an abundance of food, clothing, and other necessities in addition to luxuries that exceeded the wildest Utopian dreams of preceding centuries. It was the Age of Invention, the Era of Materialism; in just thirty years, the supremacy of scientific technology had enabled industrial capitalism to dominate the world virtually unopposed.

The marriage of science and technology that contributed so much to these great changes was not entirely one of love nor even of convenience. It was also a union of necessity, with Industrial Capital holding the shotgun and the State performing the ceremony. The following four articles describe some of the issue of that union. The new devices based on science-oriented technology not only revitalized the industry of their age but also created whole new industries. Some of these were devoted entirely to their production, others were wholly dependent on their use. As Alfred North Whitehead has said in *Science and the Modern World* (Free Press, 1967): "The greatest invention of the nineteenth century was the invention of the method of invention. A new method entered into life." The four men described here, Nicolaus Augustus Otto, Rudolph Diesel, Thomas Alva Edison, and John Ambrose Fleming, present us with quite different facets of this new method. There can hardly be a better way to characterize the materialism of their time than to recall that, even during their lifetimes, men such as these were accorded a status and adulation that had never before been given by any society to creators of purely material, let alone mechanical, objects. Their stature has more closely resembled that of mythic folk heroes than that of even the most superb of craftsmen. As is the general case with mythic heroes of any era, their careers are well worth examining to investigate the history underlying the story, the motive behind the myth.

None of the inventions discussed in these essays can be said to have been socially or economically necessary, even in the limited sense of the engines of Newcomen or Watt. For example, there was no extant industry whose progress depended on the development of the internal combustion engine. Quite the opposite; it was the availability of a relatively inexpensive, efficient, and ultimately mobile engine with adequate reliability, which, together with the increasing desire for transportation, created the automobile industry. At the dawn of the twentieth century, it could no longer be argued that industries grew in order to satisfy human needs. The "needs" were now being created by the industries in order to promote new growth. Furthermore, in an era already corrupted by state-encouraged monopoly capitalism, the new industries had no history of coalescence from cottage industries and were centralized and monopolistic from the outset. They arose in and helped to perpetuate a social climate where the possession of material goods was becoming a sign of status, where desires were replacing needs as a reason for the acquisition of products. The search for market expansion in the automobile industry was to lead Henry Ford first to adopt mass production (previously used in the stockyards and at Sears Roebuck), then mass advertising, and then credit buying to broaden mass consumption. The auto industry is the archetype of the twentieth-century consumer industries; it was among the first to create a necessity out of a luxury and subsequently to alter the very fabric of society to perpetuate its own market.

"The Origin of the Automobile Engine," the first of the two essays by Lynwood Bryant, focuses on the man who made Ford possible, Nicolaus Augustus Otto. Of the four men whose inventions we are to examine, Otto is the closest to the early empirical technicians. His main advantage over them was attributable to the development of the science of thermodynamics in the mid-nineteenth century. Although, as Bryant points out, practical engineers may have known little thermodynamics in the 1870s, they had enough scientific knowledge to know that the Watt engine was appallingly inefficient, having a thermal efficiency below 10 percent. In 1860, Étienne Lenoir had unsuccessfully sought to improve on this low efficiency by building an engine that operated by internal, rather than external, combustion of the fuel. Otto, among others, was seeking to improve the thermal efficiency of the Lenoir engine by compressing the charge before igniting it. As this inevitably led to an extremely rapid combustion (an explosion), the engines tended to run in a series of violent and quite destructive impulsive thrusts. While seeking to decrease the shock by cushioning the explosion, Otto developed the valved four-cycle engine because he believed that it provided a stratified charge in which the exhaust gases would serve to absorb the impulse. As it turned out, he had developed the right engine on the basis of the wrong theory. It was not the stratification of charge but the critical adjustment of fuel-air mixture and ignition timing that was responsible for the smooth running of the "Silent Otto"; the valved, four-stroke Otto cycle had made such adjustment possible.

The necessity for such fine adjustment has become an even greater problem in recent years owing to the strong dependence of the type and quantity of atmospheric emissions on the state of engine tune. It is not simple to design and tune an Otto engine to maximize power and smooth running while minimizing air pollution. However, in all fairness to Otto, his engine was not only silent, but also relatively clean running, compared to the coal- and oil-burning boilers of his day. The advantages of perfecting the external combustion engine were unforeseen; in particular, it is readily converted to new or better fuels, and its emissions are more amenable to close control. The Otto engine, tied to high combustion temperatures and high-grade fuel, is far less adaptable; and, like many another overspecialized species, it is on a more or less rapid path to extinction as fossil-fuel supplies dwindle and smog increases.

In "Rudolph Diesel and His Rational Engine," Bryant presents us with the nicely complementary history of a man who, working in the same period as Otto, came up with a quite similar design for an internal combustion engine. But Rudolph Diesel's methods were not those of the early nineteenth-century practical engineer as were Otto's. They were closer to those of the modern scientifically-oriented engineer-enterpreneurs of the early twentieth century. Diesel was thoroughly educated in science and mathematics. He was not only familiar with the newly developed science of thermodynamics but also determined to apply its results directly to technology. In 1862, the French engineer Alphonse Beau de Rochas had outlined a program to duplicate the four steps of the Carnot cycle—which is the most efficient possible cycle for any engine—by means of a four-stroke internal combustion process. Diesel, having the advantage of familiarity with Otto's engine and with Beau de Rochas's pamphlet (of which Otto apparently knew nothing), set out from the first to build an ideal Carnot engine with the explicit aim of revolutionizing the production of power.

In order to duplicate the Carnot cycle, the combustion in a four-stroke engine should proceed at constant temperature and over an extended period of time. This condition was far removed from the rapid, high-temperature combustion in the engines of Otto and Beau de Rochas. Diesel's solution was to inject the fuel slowly after the air was compressed and thus force a relatively slow and cool fuel burn. In his article, Bryant describes the development of Diesel's thought and invention, from the original idea of fuel injection (patented in 1892) through two decades of struggle to construct a reliable engine. Diesel was not a modest man, as the title of his 1893 article describing his engine attests: "Theory and Construction of a Rational Heat Engine to Take the Place of the Steam Engine and of All Presently Known Combustion Engines"! In this, Diesel was, with Edison, one of the new technological entrepreneurs, promoting his ideas vigorously and advertising their social and economic benefits and consequent profitability in order to attract financial backers.

Of the four possible sources of power for early automobiles (steam, electricity, and the Otto and Diesel cycle engines), the Otto cycle was assuredly the worst choice in the long run, from the point of view of efficiency and pollution. It was, however, the simplest, cheapest, and most reliable of the early power sources and received the most promotional backing from automotive pioneers such as Benz and Ford. The industry, once committed, had little interest in further research on new types of engines, preferring not to risk its massive investment in tools, dies, and development. A moment's thought should convince any technically competent person that the basic operation of the Otto and Diesel engines is at best irrational. In essence, rotary motion is created by firing a series of small cannon whose captive projectiles are connected to a crooked shaft, the crank. Even if, as Lynn White, Jr. suggests in *Dynamo and Virgin Reconsidered* (M.I.T. Press, 1968), there are anthropological reasons for believing that reciprocating motion comes more naturally to humans than rotary motion, the absurdity of the reciprocating engine was clearly apparent to the new scientifically trained engineers of the 1890s. One of these was Charles Parsons who, in 1896, invented the powerful and efficient steam turbine (see Article 13, "Steam Turbines," by Walter Hossli). The steam turbine quickly supplanted the internal-combustion engine for large stationary installations such as ships and dynamos, and most of the remaining heavy work ultimately fell to the more efficient Diesel. Only in recent years has the development of the Wankel engine° even slightly shaken this pattern, which has thus far dominated our century.

The development of the internal combustion engine was only a part of the scientific-technological burst of the late nineteenth century and possibly not even one of the most significant parts. The development of the dynamo and the steam turbine made possible the efficient generation of an enormous quantity of electric power. The nineteenth century's crowning achievement in physics, the electromagnetic theory of James Clerk Maxwell, supplied the corresponding analytic method for the invention and construction of new devices. The man most responsible for turning electricity to the uses of society was also the most renowned inventor of the age of invention, Thomas Alva Edison, "The Wizard of Menlo Park."

In "The Invention of the Electric Light," Matthew Josephson disposes of the myth that Edison was a tinkerer, an empirical technologist who had no

°See "The Wankel Engine," by David E. Cole, *Scientific American*, August 1972.

use for mathematics or science. There has been, and still is to some degree, a powerful current of anti-intellectualism in America. The highly promoted cult of individualism, the elevation of the lone and taciturn gunslinger to the quintessential American folk hero, the identification of certain types of culture and intellectual activity with "foreigners" (that is, Europeans), and a remnant of the Puritan ethic that equates simplicity and just plain plainness with honesty and morality have, among other things, resulted in a mythic ideal of the inventor as part Tom Swift, part hermit—a lonely tinkerer who fools around with spare parts in his garage until one day he emerges and presents society with Hoover Dam or the SST. Only recently has the very apparent dynamics of invention in the space program begun to counteract this attitude, with not a little help from the post-Sputnik governmental promotion of science and technology. Edison was quite aware of these classic American values, and he was not at all reluctant to cultivate a public image that fit them. But as Josephson shows, the electric light was no simple, isolated idea. It was the key component in Edison's grand conception of a whole system of electrical generation and distribution—a conception that would do credit to the entire staff of any modern corporation. Edison set out not just to invent a new and better light, but to develop a high-resistance lamp, which, unlike the carbon arc, was powered by a high voltage and a low current. Since electrical losses in wires increase as the square of the current, such a light would make it economically possible to construct an elaborate low-current distribution system that would bring electricity into every household. Edison's invention was not just the electric light, but the electric home. Without a system for distribution, the dynamo was hardly more than an interesting device. Edison, like Diesel, neither ignored nor merely forecast the impact of his work, but actively sought to make inventions with the largest social and economic consequences.

The ready availability of electrical power was one of the prime factors in the rapid acceleration of the rate of technological change. This acceleration, which has continued unabated since the close of the nineteenth century, is well illustrated by George Shiers in his article "The First Electron Tube." The existence of electromagnetic radiation, as predicted by Maxwell, was experimentally proved by Heinrich Hertz in 1886, six years after Edison had first noted that current could flow through a vacuum. The bridge between the electric light and Hertzian waves was provided by John Ambrose Fleming, who had spent a decade investigating the Edison effect as a consultant to the Edison Electric Light Company before becoming technical adviser to the Marconi Wireless Telegraph Company in 1899, thirteen years after Hertz's discovery. Within five years, Fleming's search for a better detector of electromagnetic waves led to the discovery of the first high-frequency rectifier, the thermionic diode. There has been some argument on the subsequent history of the electron tube; but, as Shiers points out, it was Lee De Forest who actually applied for the first patent on a controllable electric device—the grid triode—in 1907.

By 1913, triodes were being widely used for amplifiers and oscillators, and the radio was becoming a commonplace article in the homes Edison had wired for electricity in the 1880s. The compression of the time scale for research and development is remarkable. In 1913, it had been only five years since the discovery of the triode, eight years since the first thermionic tube, and twenty-seven years since the first detection of electromagnetic waves. More significant, perhaps, is the statement in the last paragraph of Shier's article: "if [Fleming] had not devised the thermionic diode, someone else would have done so soon afterward." Progress in scientific technology was becoming independent of the existence of individual inventors; incremental change was becoming continuous. The rapid pace of further development is at least partly attributable to this new feature. The current stage of technological development now leads to the next in logical sequence, and, not just one, but many, of the growing number of science-trained engineers can follow the path. Technological change has developed its own *primum mobile* and, having fulfilled the dream of Francis Bacon, has passed into a new phase he could not foresee.

Individual inventors still exist; singular inventions are still made; but the process of invention has become too important to industry to be left to a handful of inventors alone. The industries and corporations finance research laboratories, subsidize "think tanks," encourage discovery, and generally support all activities that encourage better ways of creating new devices and techniques, with their enormous potential markets. The Age of Invention is not over, it has become incorporated. Inventions now come so thick and fast, change itself has become so permanent, that genus *Homo* (now apparently more *faber* than *sapiens*), has come close to losing control of its own technology. New devices no longer seem to be sought by us so much as they are imposed upon us. The surrender of the power of choice, the abdication of the right to determine and control the process of technological change, is in itself an adverse side effect of the fantastic success of *fin-de-siecle* technology.

9

The Origin of the Automobile Engine

by Lynwood Bryant
March 1967

The first internal-combustion engine to operate successfully on the four-stroke cycle was built in 1876 by Nicolaus August Otto. His "Silent Otto" was a good machine with a poor theory

The modern automobile is driven by a heat engine whose basic principle was first demonstrated 91 years ago. The principle is the Otto cycle, named after Nicolaus August Otto, a self-taught German engineer who stumbled on a way to burn a compressed mixture of gas and air in the cylinder of an engine without producing destructive explosions. The "Silent Otto," as the engine was somewhat extravagantly called, employed a scheme in which the piston required four strokes to complete one cycle: an inward (toward the crankshaft) fuel-and-air-intake stroke, an outward (away from the crankshaft) compression stroke, an inward power stroke and an outward exhaust stroke.

The term "Otto cycle" is sometimes used loosely to denote this four-stroke mode of operation, but actually an engine of the Otto type can be either two-stroke or four-stroke (or for that matter six-stroke, as some of the early ones were). The four functions of intake, compression, expansion and exhaust must be performed in any Otto engine, but they do not have to be performed in four distinct strokes. Strictly speaking, what distinguishes the Otto cycle from other cycles that can be used in piston engines is that an engine of the Otto type takes in a controlled mixture of fuel and air, compresses it to a moderate pressure and ignites it by some kind of ignition device, nowadays a spark plug.

The original Otto engine achieved an efficiency three or four times greater than the steam engines of the day, with the result that Otto's factory near Cologne became a world-famous source of stationary power plants. Two of his associates in this enterprise were Gottlieb Daimler, who later became a pioneer in the automobile business, and Wilhelm Maybach, who designed most of the early Daimler automobiles and went on to make excellent engines for aircraft.

Otto's engine would be important if only because it was the ancestor of the automobile power plant. It has a special appeal for the historian of technology because it was a good machine built on a bad theory. In the early days of the gas engine (and it was gas, not gasoline) the central problem was how to get a smooth flow of power out of a series of explosions. Otto solved the problem—or thought he did—by mixing fuel and air in a special way that yielded what he described as a stratified charge. This was supposed to cushion the shock of the explosions. Otto was wrong in attributing the success of his engine to the distribution of the gases in the cylinder, but his error led him to a mode of operation that is still employed in more than 10 million new engines a year.

The Silent Otto looks more like a

NICOLAUS AUGUST OTTO (1832–1891) was a traveling salesman for a wholesale grocer in the Rhineland in the early 1860's when he began his experiments with internal-combustion engines. He later was a partner in the firm of Gasmotorenfabrik Deutz near Cologne, for many years the largest manufacturer of internal-combustion engines in the world.

Versuchsmotor 1876   3 PS 180 U/min

EXPERIMENTAL MODEL of the original Silent Otto gas engine was built by Otto at the Gasmotorenfabrik Deutz in 1876. The single cylinder (*left*) is a converted steam-engine cylinder. The large flywheel is needed because only one out of four strokes is a power stroke. The Silent Otto developed three horsepower (in German *Pferdstarke*, abbreviated *PS*) at 180 revolutions per minute (*Umdrehungen in der Minute*, or *U/min*.). It was photographed at the museum of Klöckner-Humboldt-Deutz near Cologne.

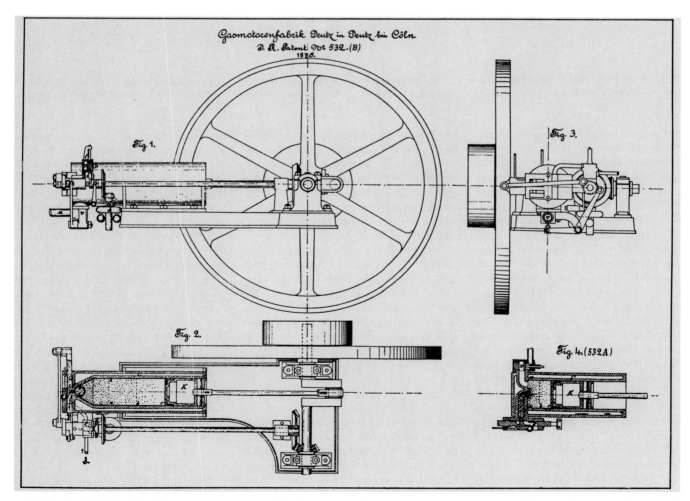

PATENT DRAWINGS of the Silent Otto accompanied Otto's original application for a German patent. The stippling in the cutaway view of the combustion chamber (*Fig. 2*) represents what Otto mistakenly believed was the reason for the smooth operation of his engine: a special mixture of illuminating gas and air that he described as a stratified charge (*see illustration on page 112*). A modified combustion chamber is shown in Fig. 4. An enlarged illustration of the slide-valve mechanism appears on page 114.

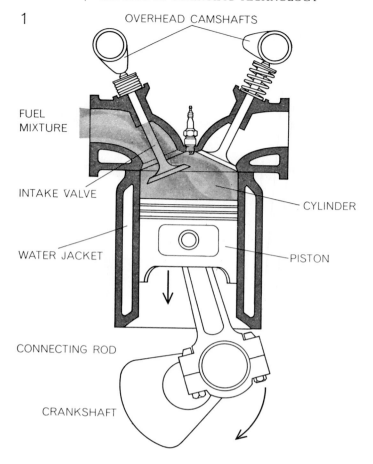

1

OVERHEAD CAMSHAFTS

FUEL
MIXTURE

INTAKE VALVE

CYLINDER

WATER JACKET

PISTON

CONNECTING ROD

CRANKSHAFT

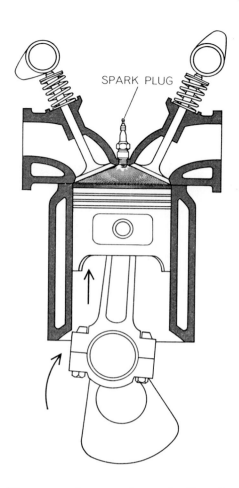

2

SPARK PLUG

MODERN AUTOMOBILE ENGINE utilizes the four-stroke cycle first demonstrated by Otto, which consists of a downward fuel-and- air-intake stroke (1), an upward compression stroke (2), a downward power stroke (3) and an upward exhaust stroke (4). Ignition

steam engine than an automobile engine [*see illustrations on preceding page*]. This is not surprising, because it was the steam engine that provided the theory, the experience and even the hardware used by early workers on gas engines. The cylinder in Otto's single-cylinder engine was in fact a converted steam-engine cylinder. It incorporated a slide valve like the valve of a steam engine, except that it had a more complicated system of passages because it had to control three fluids—fuel, air and the ignition flame—rather than steam alone. The Silent Otto developed about three horsepower at 180 revolutions per minute.

The Otto engine, demonstrated in 1876, was the first to use the four-stroke cycle. To observers brought up on the steam engine this must have seemed a wasteful way to run an engine because it yielded only one power stroke for every two revolutions of the crankshaft. The steam engine not only had a two-stroke cycle but also was usually double-acting; this meant the piston was pushed by steam in both directions, so that there were two power strokes for each revolution of the crankshaft—four times as many as in the single-acting four-stroke

gas engine. Otto seems to have adopted this mode of operation reluctantly and temporarily for lack of a better way to compress the charge. Such a drastic reduction in the frequency of power strokes must have seemed a high price to pay for the advantages of compression. Otto claimed the four-stroke cycle in his patent, but only incidentally. He promptly set to work to improve his engine by developing a two-stroke process, and so did a dozen other inventors. He was never able to improve on the four-stroke process, nor has anyone else been able to for engines of a size appropriate to automobiles.

Otto's fuel was illuminating gas. Although gasoline was known, it was regarded as being extremely dangerous. The problem of mixing gasoline vapor and air in the exact proportions required by an engine proved to be an intractable one that was not solved until the carburetor was devised in its present form in the 1890's. Gas, on the other hand, was a convenient and reliable fuel that was already in wide use for lighting. Usually produced by heating coal in the absence of air, it consisted chiefly of

hydrogen, methane and carbon monoxide. Gas technology had been evolving for some 50 years, and whenever a city installed a gas system someone was likely to get the idea that this new source of energy could be used for power as well as light.

Early inventors therefore envisioned a small gas engine the user could turn on whenever he needed power (as we now plug in an electric motor) and turn off when he was through. They hoped that such an engine might compete with the steam engine, particularly in small sizes, because of its convenience and adaptability to intermittent use. The more enthusiastic promoters dreamed of supplanting the steam engine entirely, but a more realistic goal was to provide stationary power plants for small enterprises—pumping stations, breweries, printing shops and the like—that could not afford a steam engine or did not use power continuously.

Otto and others naturally considered using a gas engine to drive a vehicle, but that was not a practical objective in the 1870's. The Silent Otto and the first generation of its rivals were much too heavy (they weighed about a ton per

3   4

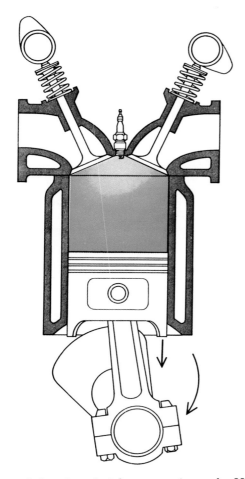

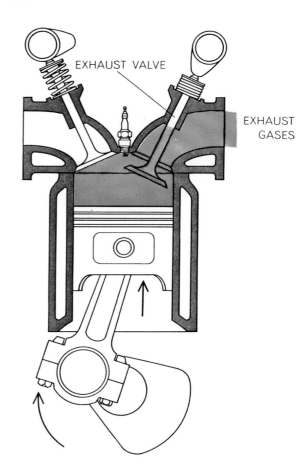

EXHAUST VALVE

EXHAUST GASES

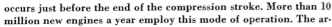

occurs just before the end of the compression stroke. More than 10 million new engines a year employ this mode of operation. The ar-

rangement of the valves and camshafts varies from one engine to the next; dual overhead cams are shown here for the sake of clarity.

horsepower) and they were tied to a stationary gas system. Nonetheless, Otto's engine of 1876 embodied the essential concepts that later made the automobile engine possible, after much refinement of detail, chiefly reduction of weight and adaptation to liquid fuel.

The concept of internal combustion—that is, the notion of burning the fuel inside the working cylinder of an engine and dispensing with firebox and boiler—was an attractive one in Otto's time, and scores of inventors were working on it. Practical engineers knew little thermodynamics in those days, but some knew enough to measure the thermal efficiency of the steam engine. They found it scandalously low, usually well under 5 percent. The internal-combustion approach seemed a promising one because it offered an opportunity to avoid the heat losses, not to mention the weight and expense, associated with firebox, smokestack and boiler. The heat would be generated at the face of the piston, so that it could immediately be converted into work without losses in transmission and storage.

At the time engineers also talked about another kind of economy: the sav-

ing of latent heat. The trouble with the steam engine, they said, is that much energy has to be spent in converting water into steam before any work is done, and that this investment in "latent" heat is not recovered if the exhaust steam is discharged into the atmosphere, as it usually is. The key advantage they saw in the internal-combustion process was that it utilized the products of combustion to drive the piston directly, without wasting energy in generating an intermediate working fluid such as steam. Actually the essential advantage, as more sophisticated engineers eventually learned, was that the internal-combustion engine was able to operate through a wider range of temperatures than the steam engine could; the nature of the working fluid made no essential difference.

Otto was not the first to try internal combustion. His most famous predecessor was Étienne Lenoir of France, who like Otto had no technical training. In 1860 Lenoir built a two-stroke engine much like a steam engine. It drew in a mixture of gas and air for the first half of each intake stroke and then ignited it with an electric spark. The second half

of the stroke was used for expansion, and there was no compression of charge. The engine was double-acting: it used both sides of the piston like a steam engine.

The Lenoir engine created a flurry of excitement in the early 1860's. *Scientific American* quoted French journals as saying that it marked the end of the age of steam. "Watt and Fulton will soon be forgotten," the article said. "This is the way they do such things in France." Then professional engineers who knew some thermodynamics ran tests on the engine and published discouraging reports. It used large quantities of oil and cooling water, it overheated badly and it was hardly better than a steam engine in efficiency. Several hundred Lenoir engines were made, but within a few years they were mostly scrapped or converted to steam.

One of the suggestions offered for improving the efficiency of the Lenoir engine was to try compressing the charge. The idea of compression is a simple and obvious one that came up frequently in the 15 years before the Silent Otto. A commonsense engineer

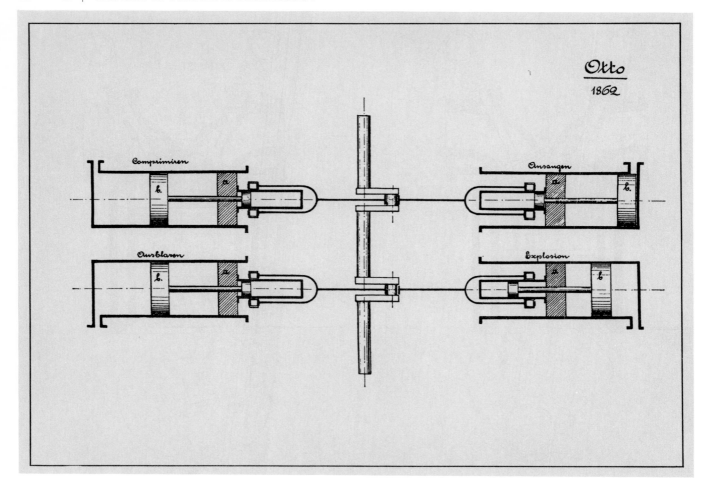

OTTO'S FIRST SOLUTION to the shock problem involved an extra free piston (*b*) that would act as a spring to absorb the shock of the explosion. Otto said that he had incorporated such an arrangement in a four-cylinder engine built in 1862. This drawing was prepared from memory in 1885 in the course of a patent litigation to support Otto's doubtful claim that the engine used the four-stroke cycle. The big defect of such an engine is that the instant of ignition cannot be fixed in relation to the instant of maximum compression, which is determined by the position of the free piston. Otto reported that the engine was ruined by destructive shocks.

would see compression as a way of increasing the power of an engine by packing in more fuel, or as a way of reducing the size of the engine by simply packing the same amount of fuel into a smaller volume. One of the advantages of gas as an engine fuel is that the designer can choose a convenient density.

Otto discovered the value of compression as he was experimenting with a model of the Lenoir engine. While working as a traveling salesman for a wholesale grocer in the Rhineland he read about Lenoir in a newspaper and had a mechanic build the model engine for him according to the newspaper description so that he could try it out. Once in the course of testing various mixtures and sizes of charge and times of ignition he drew in a full cylinder of gas and air and then compressed it on the next stroke of the piston. (Indeed, one could scarcely do anything else with a full charge.) When he set off the compressed charge, he got a surprisingly violent explosion that drove the flywheel through

several revolutions. He later said that this was the starting point of the four-stroke cycle.

Otto probably did not follow up this discovery at the time because a violent explosion was the last thing he wanted. His problem was to reduce the violence of the explosions. He was not looking for ways to increase the power of his engine but rather for ways of controlling the power he had. A worker on gas engines in the 1860's had two models to use in his thinking. One was the steam engine, in which the steam delivered an easy, steady pressure to the cylinder that could be turned on and off at will and also could be easily transferred to the other side of the piston. This was the ideal to be achieved in a gas engine.

The other model was the gun, which provided experience with the behavior of explosions in cylinders. The gun was the first internal-combustion engine, and it delivered energy in spectacular bursts. The problem of developing a practical gas engine was like the problem of get-

ting useful work out of a gun. One objective is to generate a series of explosions—to make it a machine gun. The other, the critical objective for Otto, was somehow to convert the series of explosive impulses into a smooth flow of power. A flywheel would obviously smooth out the impulses, but Otto still had the problem, or thought he did, of moderating or cushioning the explosions so that they would not destroy the engine.

Otto tried three different solutions to this problem. The first was to use an extra free piston in the cylinder that would act as a spring to absorb the shock of the explosion. Otto later said he had tried such an arrangement in 1862 [*see illustration above*]. It was incorporated in a four-cylinder engine (very surprising for 1862) that compressed the charge and was said to employ the four-stroke cycle. The drawing was prepared from memory by a mechanic in 1885 in the course of patent litigation to support Otto's doubtful contention that he had used the four-stroke cycle in 1862; its purpose here is only to illustrate Otto's first solution to the shock problem. Each cylinder had a working piston connected to the crankshaft in the usual way, and between this main piston and the end of the cylinder was a free piston mounted on a plunger that was free to move back and forth in a hollow passage in the connecting rod of the main piston. The free piston would act as a shock absorber; when the explosion came, it would be driven back against the main piston, and the air trapped between the two pistons and the air in the hollow passage of the connecting rod would cushion the shock. The main piston would therefore be driven back more smoothly and gradually than the free piston.

This double-piston arrangement had another function: it helped to drive out the exhaust gases. In an ordinary four-stroke engine the piston does not reach the end of the cylinder because it has to leave room for the compressed charge to be burned. Thus the space occupied by the charge at the end of the compression stroke is still there at the end of the exhaust stroke, and it contains unexpelled gases that remain to contaminate the next charge. At first engineers were concerned about this contamination. Nowadays they do not have to worry so much about it because they work with much higher compression ratios, say 10 to 1 (a tenfold increase in pressure). This means that the space not swept out by the piston is much smaller than it was, for example, in the Silent Otto, which had a compression ratio of 2.5 to 1. The free piston shown in Otto's 1862 arrangement would have been gradually forced back against the main piston by the increasing pressure of the charge being compressed, thereby leaving room for the charge at the end of the stroke. On the exhaust stroke the compressed air that had cushioned the explosion would have driven the free piston to the end wall of the cylinder and so have pushed out the exhaust completely.

It is doubtful that Otto actually ran such an engine, but he did try the idea of a shock-absorbing device. So did a number of others, including Lenoir. In 1893 Frank Duryea used a free piston like Otto's in the first American horseless-carriage engine. These engines were all failures. At that time engineers worried too much about expelling exhaust and not enough about properly timing the ignition of the charge. In a free-piston engine of this kind the instant of ignition could not be precisely fixed in relation to the instant of maximum compression. One could never tell from the

ATMOSPHERIC ENGINE, patented by Otto and his business partner Eugen Langen in 1866, was the first genuinely successful internal-combustion engine. Some 5,000 Otto & Langen engines were sold. An explosion drove a heavy free piston up the vertical cylinder as far as it would go. The gases then cooled and contracted to form a partial vacuum; atmospheric pressure, assisted by the weight of the piston, did the work on the way down. A clutch allowed the piston to rise freely but caused it to engage the output shaft on the downward power stroke. Atmospheric pressure placed a limit on the engine's performance, and it turned out to be a dead-end branch in the evolution of the internal-combustion engine.

outside where the free piston was at any given time, and it was the free piston that determined the instant of maximum compression. Premature explosions must have been common. Otto said that this engine was ruined by destructive shocks and that he despaired of ever getting a gas engine to work on the direct-acting principle, that is, with the expanding gases driving the output shaft directly.

Otto now turned to a second kind of solution to the shock problem. He gave up trying to moderate or absorb the shock of the explosion and instead disconnected the piston from the output shaft during the explosion. He had found in his early work on the Lenoir engine that the hot gases in a cylinder would cool quickly and form a partial vacuum, so that atmospheric pressure would drive the piston back. In the "atmospheric engine" he developed over the next five years the explosion drove a heavy free piston up a vertical cylinder as far as it would go [see illustration on preceding page]. The gases then cooled and contracted to form a partial vacuum, and atmospheric pressure, assisted by the weight of the piston, did the work on the way down. A clutch allowed the piston to rise freely but caused it to engage the output shaft on the way down.

The atmospheric engine was an ingenious and perfectly workable solution to the problem of getting useful power from a series of explosions. To someone accustomed to steam engines it was noisy and queer-looking, but it was more efficient than Lenoir's imitation steam engine (chiefly because it allowed more expansion of the hot gases). The principle of the atmospheric engine was not new, although Otto said he had got the idea from his own experience. The Newcomen steam engine, known for 150 years, created a vacuum by causing steam to condense and letting atmospheric pressure do the useful work, and the same principle had been applied to the gas engine in the 1850's. Otto's engine, however, went beyond earlier atmospheric engines in efficiency and reliability; it was the first genuinely successful internal-combustion engine. It was also the foundation of a successful business in which the guiding spirit was Eugen Langen, whose name appears with Otto's on the nameplate of the engine. Langen made the same kind of contribution to Otto's progress that Matthew Boulton made to James Watt's.

About 5,000 of these engines were built in various countries, but the Otto & Langen turned out to be a dead-end branch in the evolution of the internal-combustion engine. The reason was that atmospheric pressure placed a ceiling on its performance. The Otto & Langen came at a time of increasing demand for more powerful engines, and it seemed impractical to make an engine of more than about three horsepower. Since the working pressure could not be increased, the only way to increase the power was to enlarge the cylinder, and it was already so large that in the three-horsepower size it needed about 12 feet of headroom. Otto and his associates tried all kinds of ideas for improving the atmospheric engine, but the basic design proved to be inflexible and not easily adapted to further development. Nevertheless, they learned much from the engine. It was a good experimental tool for Otto's continuing work on combustion because it gave him an easy way to estimate the explosive force of different fuels and mixtures: he could see how far the piston was driven up by different charges. Otto continued to study combustion in cylinders, to test competing engines and to try various ideas for new types of engines, including at least one involving compression.

In 1876, with sales of the Otto & Langen dropping, the need for some kind of radical improvement became urgent. At this time Otto suddenly thought of a third solution to his key

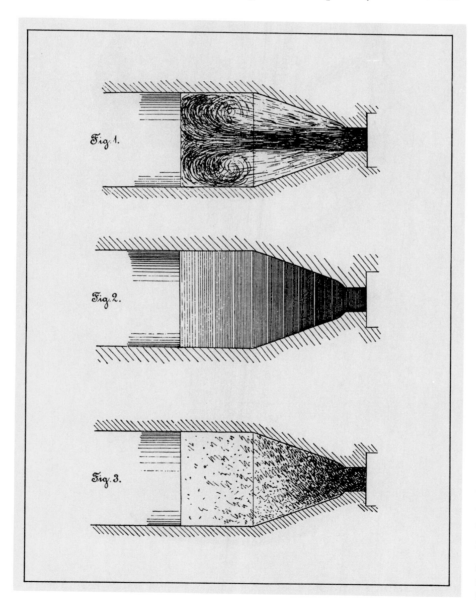

STRATIFIED CHARGE was Otto's third solution to the shock problem. He believed that at the moment of ignition the combustion chamber of the Silent Otto contained a layer of old exhaust gases nearest the piston (*left*), then a layer of air, then layers of fuel-air mixture of increasing richness, with the richest mixture nearest the point of ignition (*right*). It was not necessary to Otto's theory that there be distinct layers; he conceived of several possible charge structures, three of which he portrayed in these sketches, published in 1886.

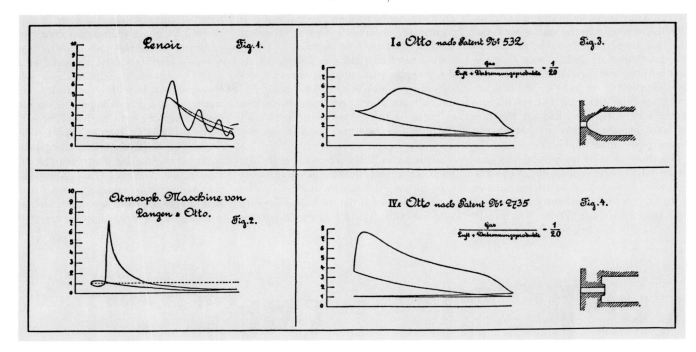

INDICATOR DIAGRAMS give the relation between the volume (*horizontal scales*) and the pressure (*vertical scales*) inside the cylinders of various engines Otto tested. The oscillations in the expansion curve of the Lenoir engine (*Fig. 1*) reflect the kind of shock waves Otto professed to find in engines operating without a stratified charge. In contrast the expansion curves in the diagrams of the three Otto engines decline smoothly. The sketches attached to the diagrams of two versions of the Silent Otto (*Fig. 3 and Fig. 4*) show a change in the shape of the combustion chamber that was supposed to improve the combustion of lean mixtures.

problem: cushioning the shock of the gas explosions. It sent him back to the direct-acting gas engine he had abandoned as being hopeless in 1862, and persuaded him to try a compressed charge without the shock-absorbing piston. Guided by this idea, he quickly built the Silent Otto of 1876 and worked out the details of the four-stroke cycle as a way of compressing the charge in the working cylinder. The engine was a great success, and it proved to be a flexible design that could be used for many purposes and was susceptible to evolution into a wide variety of forms.

The idea that led Otto to his third solution was the concept of stratified charge—the concept that the gases inside the cylinder could themselves perform the shock-absorbing function if they were mixed properly. The idea came to him, Otto said, as he was idly watching smoke rise from a factory smokestack and thinking about combustion problems. The smoke billowed and swirled into the air and then gradually disappeared. A combustible mixture might enter a cylinder in the same fashion, he thought, and then it occurred to him that if a combustible mixture spread itself out into an inert medium like smoke rising in air, the inert medium might somehow slow down the burning or moderate the explosion.

Otto had been studying combustion for 15 years. Back in 1860 his work on the Lenoir engine had taught him that the ratio of fuel to air was critically important. If a mixture of illuminating gas and air was too rich or too lean, he found, it would not burn at all. Within the rather narrow range of combustibility a mixture too rich would ignite easily, but the shock of the explosion would be too strong for the engine; a mixture too lean would burn more gently, but ignition was difficult. Accordingly Otto's problem in the 1860's was how to get both reliable ignition and slow burning. The idea suggested by the smokestack episode was that he could have both if he controlled the intake of the gas and air in such a way that he had a mixture rich enough near the point of ignition for the ignition to occur reliably, and lean enough near the piston for the combustion to occur slowly. On this basis the exhaust gases from the last cycle, which would not burn at all, suddenly became a positive good; they might remain in a layer near the piston and help to protect it from shocks. The extra free piston shown in Otto's 1862 plan would no longer be required to absorb the shock, nor would it be needed to expel the exhaust.

To produce this special kind of charge Otto used a slide valve that was driven back and forth across the entrance to the cylinder by a half-speed control shaft [*see illustration on next page*]. The valve was a brass block with an elaborate system of passages that admitted plain air for the first half of the intake stroke and a fuel-air mixture of increasing richness for the second half of the stroke. To ignite the stratified charge at the end of the compression stroke the slide valve picked up a bit of burning gas from an external standing flame and inserted it in the combustion chamber at just the right moment. This was quite a trick, because the pressure inside was higher than it was outside. What Otto thought he had in his combustion chamber at the time of ignition was a layer of old exhaust gases nearest the piston, then a layer of air, then layers of a fuel-air mixture of increasing richness, with the richest mixture nearest the point of ignition.

Otto got into trouble with the concept of the stratified charge. In fact, he lost his patent protection in some countries when his opponents were able to persuade the courts that the charge in Otto's engine was not stratified and that an engine with a homogeneous charge would also run smoothly. Otto continued to insist that no compression engine could run smoothly unless the charge was stratified in some sense. He did not really mean, he said later, that different kinds of gas had to be in sharply defined layers. They could be in irregular swirls, like the smoke from the smokestack, but the fuel-air particles would still have to be distributed in such a way that the combustion was propagated from one fuel-air particle to another through an inert medium that did not participate in the combustion and had the function of softening the shock or slowing the burning [*see illustration on page 112*].

In those days nobody could tell exactly what was happening inside the cylinder. Even in a slow engine such as Otto's a complete four-stroke cycle took less than a second. There was plenty

of room for argument, and there *was* plenty of argument. One instrument used at that time to measure the performance of an engine and diagnose its ills was the indicator mechanism designed by James Watt for the steam engine; it recorded pressures within the cylinder as the piston went back and forth, in the form of a pressure-volume diagram.

Such an indicator diagram gives the engineer information he can use in the way the physician uses the pulse rate as evidence of what is happening inside the body. There was one important

item of information, however, that in those days could only be inferred: temperature. Here the uncertainties were so large that the most respectable experts could differ widely about what was going on in an engine. In the illustration on the preceding page are indicator diagrams Otto recorded for a Lenoir, an Otto & Langen and a Silent Otto. The oscillations in the expansion curve of the Lenoir diagram reflect the kind of shock waves Otto professed to find in competing engines without stratified charge, and the smoothly declining pressure in the expansion curves of the

1

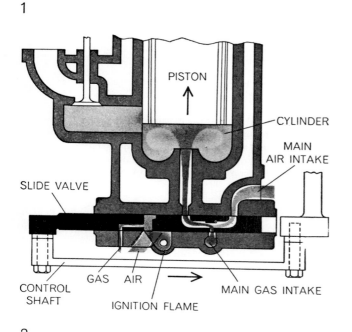

2

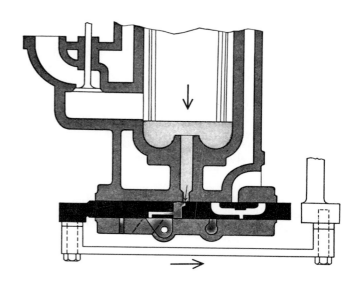

3

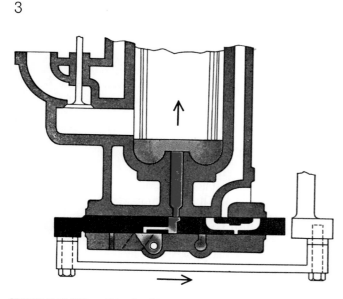

4

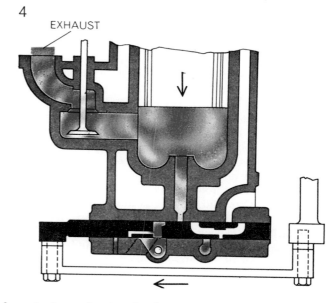

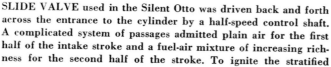

SLIDE VALVE used in the Silent Otto was driven back and forth across the entrance to the cylinder by a half-speed control shaft. A complicated system of passages admitted plain air for the first half of the intake stroke and a fuel-air mixture of increasing richness for the second half of the stroke. To ignite the stratified charge in the combustion chamber the slide valve picked up a bit of burning gas from an external flame (1) and, after equalizing the pressure in the valve and the chamber (2), inserted the flame into the chamber at the end of the compression stroke (3). A conventional poppet valve was used to release the exhaust gases (4).

diagrams of his own engines indicates the special kind of burning due to the structure of the gases that Otto claimed as the essence of his invention.

Otto never abandoned his theory of the stratified charge, but by the time of his death most experts had turned against it. It was generally agreed that the fuel-air mixture in an engine should be as homogeneous and as free of exhaust gases as possible, and that smooth running was achieved by the proper mixture and the timing of ignition. Still, not everyone was convinced, and it is curious to see how the concept of the stratified charge, or something very much like it, keeps reappearing in the history of the internal-combustion engine. Harry Ricardo, for example, reinvented the stratified charge in 1922 as a means of achieving fuel economy through improved combustion. He proposed igniting a particularly rich mixture in a small antechamber so that a hot flame would shoot out into the main combustion chamber and set fire to a mixture so lean that it ordinarily would not burn at all. (A contemporary version of the same concept appeared in a British automobile magazine as recently as last November.) Although Otto had begun by thinking of the stratified charge as a shock-absorbing device, he also saw in it a means of achieving more complete combustion. In 1877 he patented a modified shape of the combustion chamber of the Silent Otto that would concentrate the richest mixture in a narrow channel so that a hot flame would shoot out into the main charge and ignite its leanest parts.

The possibility of greater fuel economy makes such an idea interesting to automobile and oil companies today. Public awareness of air pollution has also helped to revive interest in the stratified charge as a potential way to get cleaner automobile exhaust through more complete combustion. A number of experimental programs have been launched in the past 15 years in the U.S. and abroad for the purpose of trying new shapes of combustion chambers and changes in fuel injection and ignition, all involving some kind of separation of rich and lean mixtures with the rich mixture near the point of ignition. These programs usually reflect an interest in turbulence within the cylinder that brings to mind Otto's original concept of a combustible mixture swirling about like smoke from a smokestack. One variation of the stratified charge, the Texaco Combustion Process, is said to eliminate knock, which would have been very interesting to Otto. Knock is the premature detonation of a part of the charge that is heard when low-octane gasoline is used in a high-compression engine. It was a different kind of shock that troubled Otto; he probably never tried a high enough compression to detect combustion knock as we know it. Nonetheless, he felt the same need to control combustion within his cylinder, and he tried to control it in the same way: by concentrating a rich mixture in the region of ignition.

Otto's idea of the stratified charge may not be technically respectable, but it guided him to the most important single step in the evolution of the automobile engine, and it continues to have a powerful fascination for engine designers in many countries.

MONUMENT to Otto and Langen stands in the main square of Deutz, a suburb of Cologne, near the former site of Gasmotorenfabrik Deutz. On top of the monument is a model of their atmospheric engine of 1866.

10 Rudolf Diesel and His Rational Engine

by Lynwood Bryant
August 1969

The inventor of the Diesel engine hoped it would attain the ideal of the Carnot cycle for the heat engine. He failed, but today his engine provides most of the motive power for heavy transportation

The Diesel engine has revolutionized heavy transportation in the past generation. The automobile, which is usually powered by an Otto engine rather than a Diesel, is responsible for vastly more passenger-miles of transportation than any other vehicle ever known, but the work of moving freight and passengers on a large scale on land and sea, which was accomplished in the time of our fathers by steam, is now done almost entirely by Diesel. The last stronghold of the old-fashioned steam piston engine—the railroad locomotive—has surrendered unconditionally in the U.S. and cannot hold out much longer in Europe. The revolution in heavy trucking, in earth-moving equipment and in water transportation is not so well known but is no less complete. The steam turbine competes with the Diesel in very large ships, say above 20,000 horsepower, but between that upper limit and the lower limit of the family car the Diesel engine is now supreme.

Rudolf Diesel fully expected his engine to cause such a revolution, but when he conceived the engine some 80 years ago, he had no idea the revolution would take so long. Diesel was a scientific engineer, one of a new type that began to appear in the 1880's, and he had a strong background in science and mathematics. In his view the engine he patented in 1892 was not just an improvement on existing heat engines but a machine of an entirely new kind. He called it a "rational engine," that is, an engine built on scientific, rational principles, and he fully expected it to displace the steam engine within a few years and to drive everything from battleships to sewing machines.

This new engine Diesel thought of as a Carnot engine. It was supposed to operate on the principles outlined more than half a century earlier by the French pioneer in thermodynamics Nicolas Léonard Sadi Carnot. Diesel's first calculations indicated that by following the ideal Carnot cycle his engine could convert 73 percent of the energy of coal—or almost any other fuel—into work. This was (and still is) an extremely high efficiency for a heat engine; in Diesel's day the best steam engines seldom showed a thermal efficiency higher than 7 percent. Even after making realistic allowances for the difficulties of handling the high pressures his engine required, Diesel was confident that it would need only a sixth as much fuel as a comparable steam engine would consume.

Like many another achievement in engineering, the Diesel engine we know today is quite different from the inventor's original ideal. It is the result of prolonged wrestling with practical problems, of painful and expensive incremental evolution, and not the quick breakthrough from theory to hardware that Diesel had envisioned. And it turned out to be not a Carnot engine at all. The story of the engine's development is nonetheless an unusual chapter in the history of technology because it began with an attempt to apply pure science to engineering. The inventor failed to realize his scientific ideal, but his failure was the first step in the evolution of an important new type of engine.

Rudolf Diesel was a German born in Paris who was displaced by the Franco-Prussian War and received his technical education in Germany. In 1880 he was graduated at the head of his class from the Technische Hochschule in Munich, and he entered the refrigeration-machinery business as the Paris representative of a firm founded by his teacher of thermodynamics, Carl Linde. Refrigeration was an exciting new field in those days, a field on the frontier of scientific engineering. Diesel's work gave him broad experience with heat engines and heat pumps, and a chance to work in his primary field of interest: thermodynamics. He was an extremely ambitious young man who fully intended to do for the approaching 20th century what James Watt had done for the 19th—to revolutionize energy conversion.

During the next 10 years Diesel devoted his spare time to a number of private energy-conversion projects, such as a sun-powered air engine and an ammonia bomb. As a refrigeration engineer he was familiar with the various working fluids used in refrigeration plants, such as ammonia, ether and carbon dioxide, and he made a special study of the possibility of using such working fluids in heat engines. A heat engine converts heat into work by adding heat to a working fluid—usually a gas—so that the fluid expands and exerts pressure on a piston or on turbine blades. In Diesel's time steam was the most common working fluid, but in theory any gas can serve as the medium for this kind of energy conversion. The efficiency of the process, as Carnot pointed out, does not depend on the choice of medium, but obviously some gases have more convenient properties than others. A steam power plant adds heat to the working fluid by external combustion, using a firebox to generate heat, a boiler to add heat to the working fluid and a complex system of pipes, perhaps including a condenser, to transfer the fluid between the boiler and the engine. All this equipment is in addition to the engine proper: the cylinder in which the steam does its work by expanding against a piston.

In the 1880's, when Diesel began experimenting with heat engines, hundreds of inventors were trying to find a substitute for steam that would not need such an elaborate and expensive system;

RUDOLF DIESEL (*left*) is seen with two colleagues at the time he prematurely announced successful development of his engine to the Society of German Engineers in 1897. With him are Heinrich von Buz (*center*) of the Maschinenfabrik Augsburg, one of the firms that supported Diesel, and Moritz Schröter, a professor of machine design who pronounced the engine a scientific success.

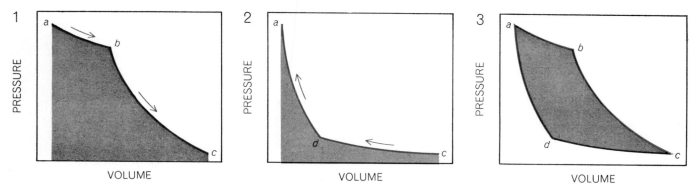

NET WORK accomplished by a heat engine is shown graphically by means of diagrams that plot pressure and volume of the "working fluid" inside the engine's cylinder as the piston first moves away from the cylinder head and then returns to its starting position. At the start (1), pressure is maximum and volume minimum (a); the curve a–b–c traces the decrease in pressure and rise in volume as the heat of the working fluid is converted into work during the fluid's expansion against the piston. The area below the curve (gray) represents "positive work." A portion of this work is need-ed to drive the piston back to its starting position (2). The increase in pressure and decrease in volume during the piston's return are traced by the curve c–d–a. The area in color below this curve repre-sents "negative work." Subtraction of the negative work from the positive work (3) produces an area indicative of the theoretical "net work" accomplished within the cylinder during a single oper-ating cycle. In a real engine this indicated area of net work will be reduced by various kinds of thermal and mechanical losses before it becomes available on an output shaft to drive an external load.

they were seeking a working fluid that would make possible more efficient or more economic power plants, particular-ly in small sizes. Air was the most popu-lar fluid tried as a substitute for steam. At this time thousands of small external-combustion engines using air instead of steam as the working fluid were in com-mon use. Another possibility that ap-pealed to Diesel was ammonia, the fluid used in Linde's refrigeration machinery, which has a number of interesting prop-erties Diesel thought he could exploit in an engine. For 10 years he pinned his hopes on a small ammonia engine that was supposed to store energy and deliver work in small quantities like a storage battery. Ammonia is so difficult to handle in an engine, however, that eventually he gave it up and turned to air.

Air has two unique advantages as a

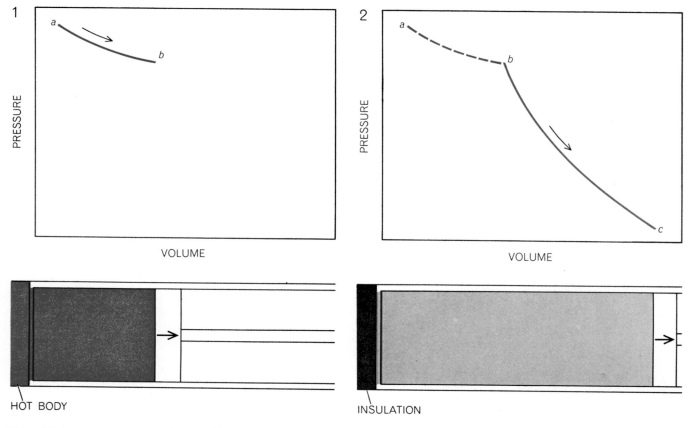

DIESEL'S DREAM, an engine that would perform according to Carnot's ideal cycle, is shown schematically; the pressure-volume diagrams at the top trace the action of working fluid and piston in an imaginary cylinder at the bottom. At the start of the cycle (1), a large heat reservoir is in contact with the cylinder head, and heat flowing from it into the working fluid causes the fluid to ex-pand (a–b) isothermally, that is, without an increase in tempera-ture. In the next step (2) the heat source is removed and the cylin-der head is insulated; the working fluid continues to expand (b–c). The expansion is adiabatic, that is, without the flow of heat to or

working fluid for an engine. One is that it is so easily available that an engine can take in a fresh supply for each cycle. Another advantage, which Diesel said occurred to him about 1890, is that it contains oxygen. This means that in an air engine combustion can take place *in-side* the working fluid, without any need for a firebox and a boiler. Air can have two functions in a heat engine: it can serve as the medium of energy conversion and at the same time it can participate in the generation of heat.

This is the essential concept of internal combustion. Diesel was by no means the first to think of it; his hero Sadi Carnot had suggested long before that air is the most promising substitute for steam because its oxygen content makes internal combustion possible, and a hundred inventors had worked on this idea before Diesel. The most successful was Nicolaus Otto, whose internal-combustion engine burning illuminating gas was well developed and widely used in the 1880's. The unique aspect of Diesel's approach was his conviction that his engine could realize the ideal Carnot cycle.

The Carnot cycle was first described in a small book titled *Reflections on the Motive Power of Heat and on Machines Fitted to Develop That Power*, published by Carnot in 1824. The book contains a number of basic insights into the nature of heat and the principles underlying the operation of heat engines. In those days the relations between heat and work were not well understood. As Count Rumford had demonstrated in his famous cannon-boring experiments, work could produce heat without limit, but a good deal of uncertainty remained about the reverse relation: the conversion of heat into work. A steam engine obviously converted heat into work, but nobody knew exactly how the work was related to the heat. In Carnot's time heat was regarded more as a kind of substance than as a form of energy. From this point of view it seemed that heat passed through a steam engine without essential change, just as water passes through a water mill. It was not generally understood that when heat is converted into work, it disappears as heat and reappears as work.

In his book Carnot showed that a heat engine can convert into work only a certain fraction of the heat supplied to it. This fraction, which is called thermal efficiency, depends on the temperature range through which the engine operates and not on the kind of working fluid used. Carnot also proved that the efficiency of a process is at a maximum when the process is *reversible*. A process, for example, that could convert heat into work and also operate in the reverse direction to convert the same amount of work into the same amount of heat would be reversible. If a water mill could extract enough energy from a certain fall of water to pump the same amount of water back up to the original level, the process would be reversible and the water mill would be perfect. Reversibility in this sense is an unattainable ideal. The concept is a way of defining the upper limit to the efficiency of energy conversion. No heat engine operating between given temperatures can possibly be more efficient than one operating on a cycle with all processes reversible.

Carnot went on to describe such an ideal cycle, with four reversible processes, and to show how it would work in an ideal engine. He imagined a cylinder containing a fixed quantity of working fluid, say air, and a piston moving back

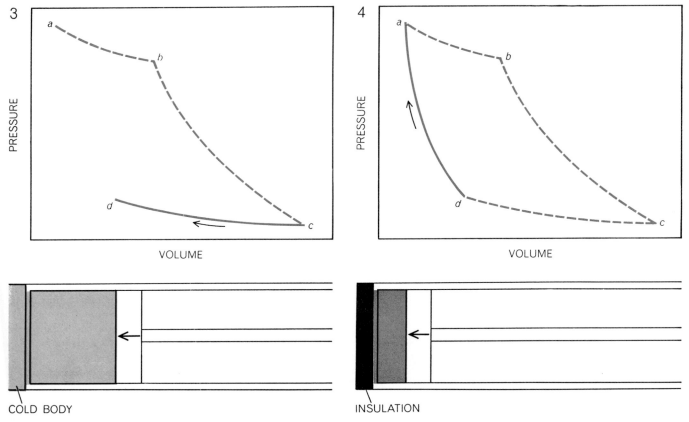

from the fluid, and the temperature of the fluid drops. Next the piston must be driven back, compressing the fluid before it. First the cylinder head is placed in contact with a cooler heat reservoir (3), so that the heat of compression flows from the fluid to the cold body; the temperature of the fluid does not rise and the compression is isothermal. Finally (4) the cold body is replaced by insulation and the piston is returned to the starting position (*d–a*). This process is adiabatic, and the heat generated by the piston's work raises the temperature of the working fluid to its original level, thereby completing the cycle. Each step of the cycle is reversible.

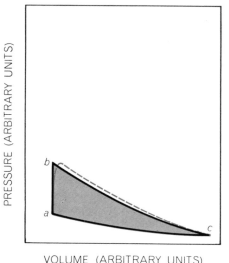
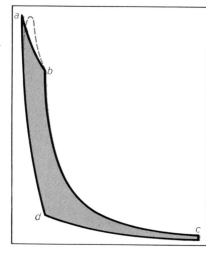

TWO PRESSURE-VOLUME DIAGRAMS, included in Diesel's first patent, compare the cycle of the Otto engine with the proposed Diesel cycle. In the Otto cycle (*left*) pressure rises sharply after ignition (*a*). Combustion is almost instantaneous, so that heat is added at roughly constant volume (*a–b*). The Diesel diagram (*right*) shows peak pressure reached solely through the compression of air (*c–d–a*). Ignition comes at *a*, and combustion continues from *a* to *b*, adding heat isothermally in accordance with the Carnot ideal. Solid lines in both diagrams show the ideal cycle; broken lines, the process expected in an actual engine.

DIESEL'S FIRST SUCCESS was this one-cylinder oil-fueled engine, which in 1897 achieved a compression of about 30 atmospheres and had a thermal efficiency of 26 percent. Diesel's compound engine, which was far less efficient to operate, was junked the same year.

and forth in the cylinder. He also imagined two very large bodies, a hot one and a cold one, that can be placed alternately in contact with the cylinder. The hot body is a reservoir of heat from which the heat to be converted into work is taken. The cold body is a reservoir into which the engine discharges the heat it cannot use.

At the beginning of the cycle the cylinder is in contact with the hot body, and the temperature of the working fluid is assumed to be very close to the temperature of the hot body. Heat flows from the hot body into the working fluid and causes it to expand and drive the piston a certain distance along the cylinder. This expansion of the fluid against the resistance of the piston is a conversion of heat into work. The transfer of heat from the hot body to the fluid is assumed to be isothermal, that is, it takes place without changing the temperature of the fluid, and the hot body is assumed to be so large that it can supply heat without any appreciable drop in temperature. Actually there has to be some difference in temperature if heat is to flow at all, but the difference can be very slight. This first process in the expansion stroke is reversible because it is isothermal. If it were a flow of heat from a hot body to a cold one, it would not be reversible, because heat will not flow from a cold body to a hot one.

Contact with the hot body is now broken to start the second process. The fluid continues to expand, but because there is no further flow of heat into it the temperature of the fluid declines until at the end of the expansion stroke the temperature of the fluid has dropped to a point only slightly higher than the temperature of the cold body. During this process some of the energy stored in the fluid is converted into the work represented by the motion of the piston. This kind of expansion, without any heat flow into or out of the working fluid, is called adiabatic. It is also a reversible process, because it can go in the reverse direction: the piston can also compress the fluid and raise its temperature.

This is the end of the expansion stroke, or power stroke. Now if the engine is to run continuously, the piston must be brought back to its original position. This return trip, the compression stroke, is going to cost something: it will take work to drive the piston back through the cylinder, compressing the working fluid ahead of it until the piston reaches its original position and the working fluid reaches its original pressure and temperature, ready to repeat the cycle. In a

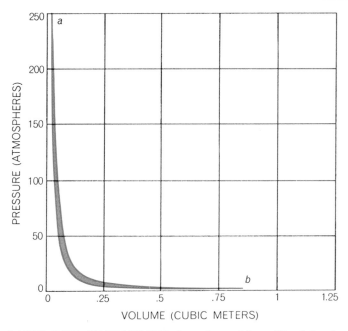

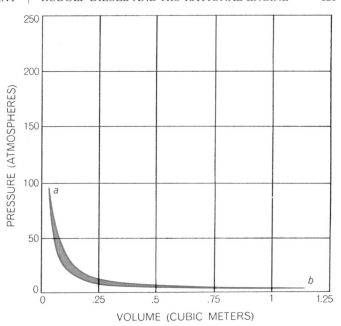

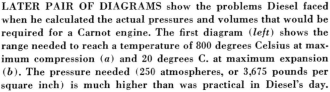

LATER PAIR OF DIAGRAMS show the problems Diesel faced when he calculated the actual pressures and volumes that would be required for a Carnot engine. The first diagram (*left*) shows the range needed to reach a temperature of 800 degrees Celsius at maximum compression (*a*) and 20 degrees C. at maximum expansion (*b*). The pressure needed (250 atmospheres, or 3,675 pounds per square inch) is much higher than was practical in Diesel's day.

Moreover, the net-work area (*color*) was dangerously slender. On recalculation Diesel found that he could cut the pressure requirement to 90 atmospheres ("*a*" *at right*) at a cost of only 5 percent in theoretical efficiency. Now he saw another problem: The long thin tail at the right in the diagram (*b*) means that complete expansion would require a very large cylinder and would make little contribution to the net work. Diesel compromised again and cut off the tail.

real engine this work will come from energy stored, perhaps in a flywheel, during the power stroke. In a Carnot engine the compression must be accomplished reversibly.

At the beginning of the compression stroke the cylinder is placed in contact with the cold body, so that during the first part of the stroke the work of compression is converted into heat, all of which flows into the cold body, so that the temperature of the working fluid does not rise. The cool reservoir is assumed to be so large that it can receive the heat without a rise in temperature. This is isothermal compression, a reversible process.

Contact with the cool body is now broken, and compression continues with no flow of heat into or out of the working fluid. In this final process of the cycle the work of compression is converted into heat, which raises the pressure and the temperature of the working fluid to its original level. This is adiabatic compression, also a reversible process.

The ideal Carnot cycle can be illustrated by pressure-volume diagrams of the kind shown in the top illustration on page 118. The vertical axis measures pressure, and the horizontal axis measures the changing volume as the piston passes back and forth in the cylinder. Engineers have used diagrams of this kind, called indicator diagrams, to study

the performance of piston engines ever since James Watt invented an indicator mechanism that would record the changing volumes and pressures inside the cylinder of a real engine.

In the ideal pressure-volume diagram of the Carnot cycle the upper line indicates the declining pressure as the piston is driven out to the right by the expanding working fluid. This is the expansion curve. The area under the curve, the product of pressure and change in volume, represents the work done by the fluid on the piston during the expansion stroke. This is *positive* work. The lower curve shows the rising pressures as the piston returns to its original position. The area under the curve represents work done by the piston on the fluid in order to bring it back to its original pressure and temperature. This is the *negative* work. In a real engine one hopes that the positive work is larger than the negative work, that is, that the expansion stroke produces at least enough power to drive the piston back. The difference between positive and negative work is the area between the two curves, which represents the *net* work. The larger this area is, the more work the engine can do.

This is the Carnot cycle, which Diesel studied in school, as all engineers still do in their thermodynamics courses, and these are the kinds of diagrams Diesel

used in his study. The Carnot cycle, however, is a theory, not an engine. The question for an idealistic, ambitious engineer such as Diesel was how the ideal could be approached in a practical engine. It would not be hard, he might have said to himself, to approximate adiabatic expansion and compression in a piston engine. The expansion and compression of air in a cylinder as a piston moves back and forth would normally be more or less adiabatic, if the piston moves fast enough, and the flow of heat through the cylinder walls could be retarded by insulation. Isothermal compression could also be achieved by known techniques. For example, water injected into a gas as it is being compressed could remove heat so that the temperature does not rise. It is not easy, however, to see how in a real engine heat can be added at the highest temperature of the cycle without raising the temperature further, as the Carnot cycle requires, particularly if the heat comes from internal combustion. Who ever heard of combustion that does not raise the temperature of the air in which it takes place?

This is precisely what Diesel thought he had discovered, a way to achieve isothermal combustion. The key idea was this: His engine would reach the highest temperature of its cycle exclusively as a result of compressing the air. Having reached that point, he would add to the

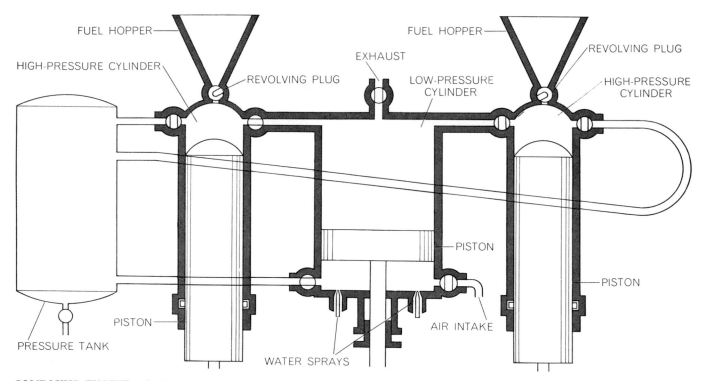

FUEL HOPPER

HIGH-PRESSURE CYLINDER — REVOLVING PLUG

EXHAUST

LOW-PRESSURE CYLINDER

FUEL HOPPER

REVOLVING PLUG

HIGH-PRESSURE CYLINDER

PISTON

PISTON

PRESSURE TANK

PISTON

WATER SPRAYS

AIR INTAKE

PISTON

COMPOUND ENGINE, splitting compression and expansion into separate low-pressure and high-pressure processes, was Diesel's first proposal for approximating the Carnot cycle. Low-pressure proc- esses occurred in the large cylinder and high-pressure processes in the smaller cylinders on each side. The operating cycle of the engine is shown in the illustration at the bottom of these two pages.

very hot air a small quantity of fuel at a controlled rate, so that the natural tendency of the air temperature to rise as the fuel burned would be exactly counterbalanced by the tendency of the temperature to fall as the air expanded with the movement of the piston. The amount of fuel would have to be very small and the amount of air very large (nearly 10 times the minimum required to support combustion), but if the idea succeeded, *all* the heat developed from the fuel would be transformed into work during the isothermal part of the expansion stroke. If this process could be achieved, an approximation of a Carnot engine was possible.

In 1892 Diesel applied for a patent on his engine, giving isothermal combustion as the essence of the invention. The patent was granted. To show the differ-

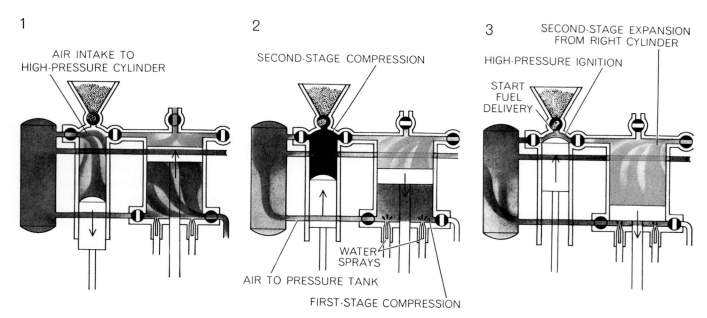

1 AIR INTAKE TO HIGH-PRESSURE CYLINDER

2 SECOND-STAGE COMPRESSION

WATER SPRAYS

AIR TO PRESSURE TANK

FIRST-STAGE COMPRESSION

3 SECOND-STAGE EXPANSION FROM RIGHT CYLINDER

HIGH-PRESSURE IGNITION

START FUEL DELIVERY

CYCLE OF COMPOUND ENGINE requires the two small high-pressure pistons (one not shown) to work 180 degrees out of phase with the double-acting, large low-pressure piston. The small pistons' cycle begins with a downstroke (1), drawing a charge of air that has already gone through first-stage compression from a pressure tank (*far left*). The following stroke (2) compresses this air further. Near the top of this stroke (3), at maximum compression, fuel begins to spill into the very hot air and ignites. Combustion continues during the first part of the following downstroke (4). This stroke effects first-stage expansion. On the next stroke (5) the expansion enters a second stage as the small piston rises and the large piston falls. The next upstroke of the large piston (6)

ence between his engine and the common Otto type, the patent has two pressure-volume diagrams. In the Otto engine combustion is practically instantaneous, so that the pressure of the working fluid rises steeply after ignition and then declines slowly as the volume of the fluid increases. In Diesel's engine, however, combustion begins at the peak pressure, and it continues, Diesel says in the patent, essentially without increase in temperature or pressure. The diagrams in the patent give no indication of scale, but Diesel's engine seems to have a much higher maximum pressure than Otto's, and the area of net work in the Diesel diagram is broad enough to suggest a good power output.

In the same year that he applied for a patent Diesel also described his ideas at greater length in a manuscript that included detailed calculations and supporting drawings. He sent copies of this study to several experts for criticism. Most of the critics were much interested in the theory, but they foresaw great practical difficulties. Diesel's original proposal was for an engine that would compress air to a pressure of 250 atmospheres and a temperature of 800 degrees Celsius. If the expansion could be carried down to a pressure of one atmosphere and a temperature of 20 degrees, as Diesel hoped, then the theoretical efficiency, on Carnot's formula, would be 73 percent. Carnot had shown that the maximum efficiency possible in a heat engine operating in a given temperature range is the difference between the highest and the lowest temperatures of the cycle divided by the highest temperature (all the temperatures being given on an absolute scale such as degrees Kelvin). In Diesel's first proposal the highest temperature would be 1,073 degrees K. (800 degrees C.) and the lowest 293 degrees K. (20 degrees C.). The difference is 780, which divided by 1,073 gives a figure for maximum possible efficiency of 72.7 percent.

The maximum pressure of 250 atmospheres that Diesel was proposing was far beyond the state of the art at the time. In a piston engine this would require a compression ratio of 60 to one. (Modern gasoline-engine compression ratios are about eight to one and modern Diesel ratios about 16 to one.) The prospect of dealing with such unprecedented pressures was naturally disturbing to technical critics and potential backers.

When Diesel plotted the pressure-volume diagram for his first proposed engine, his calculations produced an extremely narrow diagram [see illustration on page 121]. This might easily have alarmed a practical man. Not only was the pressure extreme but also the negative-work curve lay dangerously close to the positive-work curve. This meant that a large fraction of the work developed during the expansion stroke would have to be used to bring the piston back through the compression stroke.

In response to criticism along these lines, Diesel calculated the curve for an engine with more modest specifications, and he found he could reduce the peak pressure at the start of the cycle to 90 atmospheres—a reduction of almost two-thirds—at a cost of a mere 5 percent reduction in thermal efficiency. He then added a section to his manuscript describing a compromise engine with a thermal efficiency of only 68 percent, and he published the work early in 1893 under the sweeping title *Theory and Construction of a Rational Heat Engine to Take the Place of the Steam Engine and of All Presently Known Combustion Engines.* He sent advance copies to engineers and industrialists all over Europe.

From the book's detailed drawings and faultless mathematics the casual reader might easily have gained the impression that Diesel's engine already existed, but Diesel had no hardware at all, only an idea. He had to publish the book first so that he could attract the financial support he needed to develop a real engine. In order to convince manufacturers of heavy machinery that his project should be underwritten he first had to gain the endorsement of the scientific establishment. The book therefore had a strong promotional flavor, including speculation on the social and economic effects of a perfected-rational engine. The engine's efficiency, Diesel declared, would make it economic even in small sizes, which the steam engine was

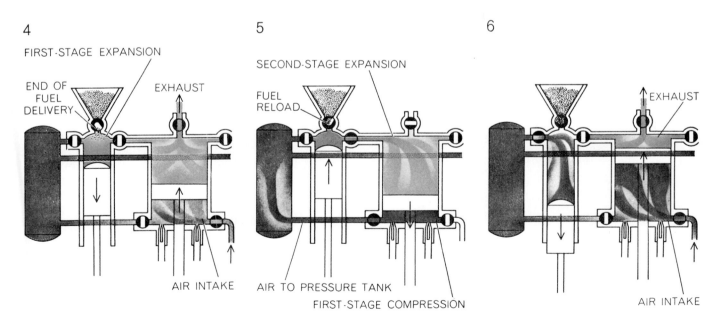

drives out the exhaust at atmospheric pressure and at the same time starts the next cycle by drawing in a new charge of air. The large central cylinder serves both high-pressure cylinders; each upstroke of its piston (1, 4 and 6) draws fresh air into the lower part of the cylinder and drives exhaust from the upper part. Each downstroke (2 and 5) performs first-stage compression in the lower part of the cylinder, forcing the compressed air through jets of water into the pressure tank, and at the same time provides room in the upper part of the cylinder for second-stage expansion of the gases from one or the other of the small cylinders. For the large cylinder this is a two-stroke cycle, for the small cylinders a four-stroke cycle and for the whole system a seven-stroke one.

not. Such an economic small power plant would enable industry to decentralize, he thought, and it would perhaps restore the small craftsman to the position of significance he had lost with the advent of steam. Diesel also outlined some of the effects the new type of power could be expected to have on railroads and ships.

Diesel's book made a good case for the Carnot engine, good enough to convince nearly all the leading European authorities on thermodynamics that his theory was sound, although there might be difficulties putting it into practice. Even the great Lord Kelvin gave Diesel's project his blessing. These endorsements, together with Diesel's eloquence, induced two major German firms—Krupp in Essen and the Maschinenfabrik Augsburg—to back the project. Both were major producers of steam engines; their support may have been a prudent hedge against the chance that Diesel's scheme would work.

The form Diesel had originally chosen for his engine was a curious three-cylinder arrangement. He realized that if he tried to accomplish all the expansion and compression that theory demanded in a single cylinder, the cylinder would have to be impossibly large. Therefore he planned a compound engine with both compression and expansion accomplished in two stages in separate cylinders. The low-pressure processes were to be accomplished in a large central cylinder and the high-pressure ones in two smaller cylinders that flanked it. Conical hoppers on top of the small cylinders would be filled with powdered coal; the fuel would be added to the air near its point of maximum compression by means of ingenious rotating plugs that would supply just enough coal at the proper rate to produce isothermal combustion.

Diesel's plan made no provision for igniting the fuel because it was assumed that almost any combustible substance would burst into flame the moment it came in contact with the very hot air. People did not understand combustion very well in those days, and Diesel found he had been much too optimistic about the possibility of using almost any fuel—solid, liquid or gaseous. He never succeeded with either coal or illuminating gas. He had a great deal of trouble with ignition in general; at one point he despaired of compression ignition altogether and began to work on various kinds of ignition devices.

Today compression ignition is considered a major defining characteristic of the Diesel engine, but Diesel always insisted that it was only an incidental feature. The essence of the invention, he held, was isothermal combustion; it made no difference how the combustion was started. One measure of his conviction is that Diesel's first plans made no provision for cooling the engine. Combustion would not raise the temperature of the working fluid, as it does in other combustion engines, and no heat would be wasted, Diesel thought, so that no cooling would be necessary. In fact, Diesel intended to insulate his cylinders to cut down heat losses.

What has been described so far is the idea for an engine and not the engine that became a reality. Diesel knew he would have to compromise with his ideal at first, and he was ready with plans for a simplified engine as soon as his backers agreed to support the development. Experimental work began in 1893 with a one-cylinder oil-burning engine that was supposed to develop 25 horsepower with a maximum compression of 70 atmospheres. To keep the size of the cylinder within reason the expansion stroke was cut short long before the pressure and temperature of the working fluid fell to the point required by theory, and isothermal compression was not attempted. Diesel was still hoping for isothermal combustion when he planned this engine, and at first he made no provision for cooling.

Early in his experimental work Diesel found that in order to get his engine to run at all he would have to use more fuel than his mathematics called for. It turns out that an internal-combustion engine cannot realize the Carnot cycle. The isothermal combustion process requires large quantities of air, much more than is necessary for combustion. The work of handling the extra air and compressing it to the high pressures required cannot be

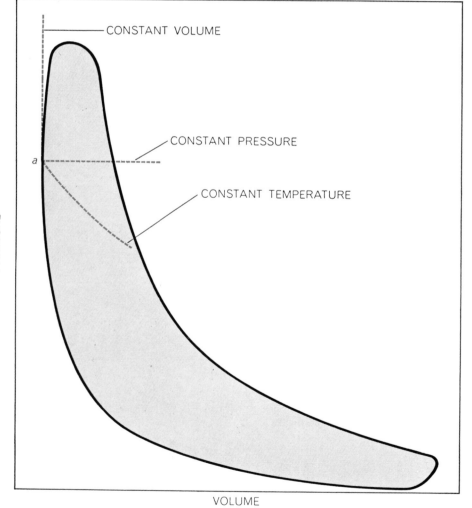

PRESSURE-VOLUME DIAGRAM of the 1897 Diesel engine (*color*) has the same shape as diagrams of modern Diesel engines. The point of ignition (*a*) indicates the start of the combustion process that adds heat to the working fluid. If Diesel's hope of adding heat isothermally could have been achieved, as it was not, the lowest of the three broken lines shown in color would have bounded the top of the pressure-volume diagram. In actuality addition of heat in a Diesel engine is a compromise between a constant-pressure process (*middle curve*) and the constant-volume process (*top curve*) characteristic of Otto engines.

done by amounts of fuel small enough to keep the combustion isothermal. The rational engine might have a very high internal efficiency, but it would not develop enough power to turn itself over. The pressure-volume diagram is simply too thin. The slightest displacement of the compression curve, such as might be caused by internal friction, would bring it right up to the expansion curve, which means that the engine could do no net work at all. The work done on the piston during the expansion stroke would scarcely be enough to bring the piston back to its starting position.

Diesel returned to his theoretical studies and wrote a long explanation of the need for making certain adjustments that would increase his engine's power. He saved as much of his theory as he could, but he acknowledged that isothermal combustion was out of the question. He saw that instead of adding heat at a constant temperature during the expansion stroke he would have to add it at approximately constant pressure.

Diesel intended to publish his explanation in a second edition of his book but there never was one. He did, however, apply for a supplementary patent that covered various devices for regulating fuel flow in his engine so that the diagram could be broadened.

The patent implies that this broadening can be achieved by an adjustment in the rate of fuel injection, but it actually requires larger amounts of fuel. This means abandoning the constant-temperature process, with isothermal combustion, and moving in the direction of a constant-pressure process, although Diesel did not say so. He clung to the ideal of the Carnot cycle as a symbol, and on ceremonial occasions he continued to refer to his engine as an approximation of the ideal Carnot engine, although he knew that the temperature of the working fluid must have risen 1,000 degrees C. after ignition. In those days uncertainties about temperatures inside engines were understandable; they could not be directly measured, and the pressure-volume diagrams used (like the ones in this article) say nothing directly about temperature. Temperatures would have to be calculated from the rough data taken from indicator diagrams, and even in a slow-moving engine the temperature would fluctuate through a range of 1,000 degrees in less than a second.

The development of the real Diesel engine is another story. It was a long and costly struggle to find ways of coping with high pressures. The problem of injecting fuel into highly compressed air

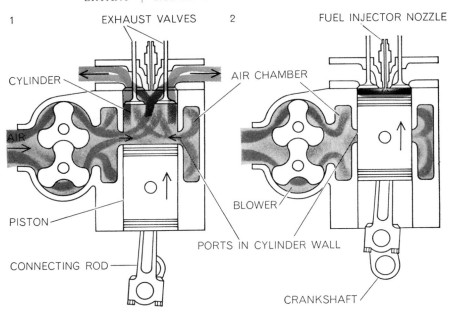

MODERN DIESEL ENGINE is seen here in a two-cycle model. A blower (*left*) is the source of high-pressure air that also helps to drive out exhaust gases after the power stroke (*1*). Near the end of the compression stroke (*2*) fuel is injected hydraulically and ignites.

proved to be extremely difficult. The first one-cylinder experimental engine never ran at all under its own power, but it gave the inventor sobering experience. It was rebuilt several times. After four years of hard work Diesel formally (and quite prematurely) announced the successful development of his engine at a meeting of German engineers in 1897. The engine he exhibited had a thermal efficiency of 26 percent with a compression of 30 atmospheres. It was a very respectable achievement, but the engine was by no means reliable or economic.

After the first year or two of practical work the Carnot cycle ceased to play any real part in the development of the Diesel engine. Diesel was nonetheless committed to it by his patent and his professional pride. As the simplified one-cylinder engine moved closer to practical reality the inventor continued to work on his scheme for the three-cylinder model. This engine, he hoped, would come much closer to the Carnot ideal because it would incorporate both the high initial pressures and the full expansion of the working fluid he had originally intended. A large compound engine was finally built, more or less according to Diesel's original plan, except that it burned oil. Its performance was far from ideal. Its fuel consumption per horsepower-hour was twice the consumption of the one-cylinder engine, and it developed only half the anticipated horsepower. Its chief problem was heat losses. The essential advantage of an internal-combustion engine is that the working fluid need not be shunted from

place to place; the entire cycle is confined to the combustion chamber. In Diesel's compound engine, however, the working fluid was first pumped from the large cylinder to the pressure tank, then on to the small cylinders and finally back to the large cylinder through pipes and valves that lost heat. In 1897, the same year Diesel announced the successful one-cylinder engine, the compound engine was junked.

The Diesel evolved slowly over the next 25 years. By 1910 a number of ships were being propelled by Diesel engines, and the submarines of World War I were mostly Diesel-powered. These first engines were heavy and spent an appreciable fraction of their output driving the air compressors then used for fuel injection. The modern Diesel, a mature and economic engine, did not appear until the 1920's, when fuel-injection devices were perfected. The Diesel of today bears little resemblance to the original rational engine. Nonetheless, three of its essential features are the same as those Diesel envisioned in 1892: it is a high-compression engine with air as the working fluid, the fuel is injected near the end of the compression stroke and the fuel is ignited by the heat of compression. The engine is economic both because the heavy oil it burns is cheap and because its thermal efficiency is high. The efficiency is up to 40 percent for the best Diesels—about the same as the figure for the most efficient large steam turbines. But the combustion is not isothermal.

T. A. EDISON.
Electric-Lamp.

No. 223,898.   **Patented Jan. 27, 1880.**

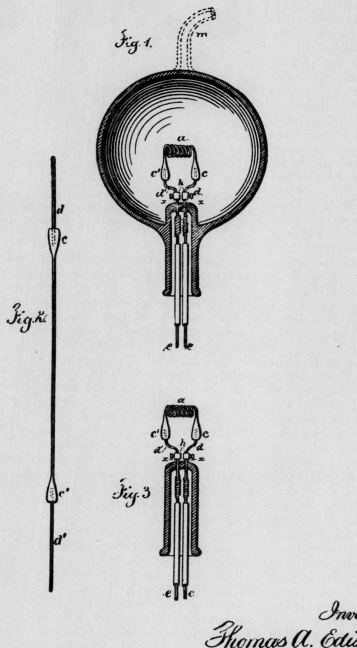

Fig. 1.

Fig. 2.

Fig. 3.

Witnesses
Chas H Smith
Geo T Pinckney

Inventor
Thomas A. Edison

for Lemuel W. Serrell
atty.

EDISON'S PATENT on the incandescent lamp was accompanied by this drawing. The labeled parts are the carbon filament (a), thickened ends of filament (c), platinum wires (d), clamps (h), leading wires (x), copper wires (e), tube to vacuum pump (m).

The Invention of
the Electric Light

by Matthew Josephson
November 1959

*It is generally assumed that Thomas Edison's
incandescent lamp was the product of inspired
tinkering. Actually it was but one element in a much
deeper invention; an entire system of electric lighting*

"I can hire mathematicians, but mathematicians can't hire me!" By such declarations in the time of his success and world-wide fame Thomas Alva Edison helped to paint his own portrait as an authentic American folk hero: the unlettered tinkerer and trial-and-error inventor who achieved his results by persistence and a native knack for things. He is said, for example, to have tried more than 1,600 kinds of material ("paper and cloth, thread, fishline, fiber, celluloid, boxwood, coconut-shells, spruce, hickory, hay, maple shavings, rosewood, punk, cork, flax, bamboo and the hair out of a red-headed Scotchman's beard") until he hit upon the loop of carbonized cotton thread that glowed in a vacuum for more than half a day on October 21, 1879. Today, in a world that relies for its artificial illumination largely on his incandescent lamp, this invention is not regarded as an especially profound contribution to technology. It rates rather as a lucky contrivance of Edison's cut-and-try methods—of a piece with his stock ticker, mimeograph machine, phonograph and alkaline storage-battery—in the esteem of a public that has come to appreciate the enormous practical significance of higher mathematics and abstruse physical theory.

If Edison's contribution to the light of the world consisted solely in the selection of a filament, this estimate of his person and achievements might be allowed to stand. But the history that is so obscured by legend tells quite another story. Edison's electric light was not merely a lamp but a system of electric lighting. His invention was an idea rather than a thing. It involved not only technology but also sociology and economics. Edison was indisputably the first to recognize that electric lighting would require that electricity be generated and distributed at high voltage in order to subdivide it among a great many high-resistance "burners," each converting current at low amperage (that is, in small volume) with great efficiency into light.

In the 15 months between the time he conceived his invention and the date on which he demonstrated it to the public, Edison and his associates designed and built a new type of electric generator, successfully adapted the then much-scorned parallel or "multiple-arc" circuit that would permit individual lights to be turned on or off separately and, last of all, fashioned a lamp to meet the specifications of his system. The laboratory notebooks of those months of frantic labor show the Wizard of Menlo Park endowed with all the prodigious capacities attributed to him by contemporary legend. They show in addition that this self-taught technologist was possessed of a profound grasp of the nature of electricity and an intuitive command of its logic and power.

It was on September 8, 1878, that Edison was inspired to devote his talents full time to the challenge of electric lighting. On that day he went to Ansonia, Conn., to visit the brass-manufacturing plant of William Wallace, co-inventor with Moses G. Farmer of the first practical electric dynamo in the U. S. Wallace showed Edison eight brilliant carbon-arc lights of 500 candlepower each, powered by a dynamo of eight horsepower. It was with such a system that Wallace and Farmer, as well as Charles Brush of Cleveland, were then beginning to introduce the electric light on a commercial scale, for street-lighting and for illuminating factories and shops. Farmer had made the first demonstration of arc-lighting in this country two years earlier, at the Centennial Exposition in Philadelphia, and John Wanamaker's store in that city was already illuminated with arc lights.

Carbon arcs are still employed in searchlights and in theater floodlights and projectors to produce light of high intensity. The current crossing a small gap between the electrodes creates an arc. Ionization and oxidation of the carbon in the heat of the arc generate a brilliant blue-white light.

In the 1870's Europe was a decade ahead of the U. S. in the technology of arc-lighting. Stores, railway stations, streets and lighthouses in Britain and France were equipped with arc lights. Shedding an almost blinding glare, they burned in open globes that emitted noxious gases, and they could be employed only high overhead on streets or in public buildings. Since they consumed large amounts of current, they had to be wired in series, that is, connected one to another in a single continuous circuit so that all had to be turned on or off together. The multiple-arc circuit, with the lights connected as in the rungs of a ladder between the main leads of the circuit, was not adapted to such systems and was considered prohibitive in cost.

Edison himself had experimented with arc lights, using carbon strips as burners. He had also investigated the

EDITOR'S NOTE

The author has based this article on material in his biography *Edison*, published by the McGraw-Hill Book Company. Copyright © 1959 by Matthew Josephson.

incandescent light, as had many inventors before him. But the slender rod or pencil of carbon or metal would always burn up, sooner rather than later, upon being heated to incandescence by the current. It would do so though substantially all of the air had been pumped out of the glass envelope in which it was contained. Edison had abandoned the effort to devote himself to a more promising invention: the phonograph.

Now at Wallace's establishment, confronted with the achievements of others in the field, he regained his earlier enthusiasm. As an eyewitness recalled, "Edison was enraptured. . . . He fairly gloated. He ran from the instruments [the dynamos] to the lights, and then again from the lights back to the electric instruments. He sprawled over a table and made all sorts of calculations. He calculated the power of the instruments and the lights, the probable loss of power in transmission, the amount of coal the instrument would use in a day, a week, a month, a year."

To William Wallace he said challengingly: "I believe I can beat you making the electric light. I do not think you are working in the right direction." They shook hands in friendly fashion, and with a diamond-pointed stylus Edison signed his name and the date on a goblet provided by his host at dinner.

From Edison's own complete and explicit notebooks and from the buoyant interviews that he gave to the press at this time we know what made him feel in such fine fettle as he left Wallace's plant. "I saw for the first time everything in practical operation," he said. "I saw the thing had not gone so far but that I had a chance. The intense light had not been subdivided so that it could be brought into private houses. In all electric lights theretofore obtained the light was very great, and the quantity [of lights] very low. I came home and made experiments two nights in succession. I discovered the necessary secret, so simple that a bootblack might understand it. . . . The subdivision of light is all right."

The Subdivision of Light

At this time there flashed into Edison's mind the image of the urban gas-lighting system, with its central gashouse and gas mains running to smaller branch pipes and leading into many dwelling places at last to gas jets that could be turned on or off at will. During the past half-century gas-lighting had reached the stature of a major industry in the U. S. It was

restricted, of course, to the cities; three fourths of the U. S. population still lived in rural areas by the dim glow of kerosene lamps or candles. Ruminating in solitude, Edison sought to give a clear statement to his objective. In his notebook, under the title "Electricity versus Gas as a General Illuminant," he wrote:

"Object: E. . . . to effect exact imitation of all done by gas, to replace lighting by gas by lighting by electricity. To improve the illumination to such an extent as to meet all requirements of natural, artificial and commercial conditions. . . . Edison's great effort—not to make a large light or a blinding light, but a small light having the mildness of gas."

To a reporter for one of the leading New York dailies who had shadowed him to Ansonia, Edison described a vision of a central station for electric lighting that he would create for all of New York City. A network of electric wire would deliver current for a myriad of small household lights, unlike the dazzling arc lights made by Farmer and Brush. In some way electric current would be metered and sold. Edison said he hoped to have his electric-light invention ready in six weeks! At Menlo Park, N.J., where his already famous workshop was located, he would wire all the residences for light and hold a "grand exhibition."

Thus from the beginning Edison riveted his attention not so much upon the search for an improved type of incandescent filament as upon the analysis of the social and economic conditions for which his invention was intended. As he turned with immense energy to expanding the facilities at Menlo Park and securing the essential financing, he continued his studies of the gas-lighting industry. In parallel he projected the economics of the electric-lighting system he envisioned.

Gas had its inconvenience and dangers. "So unpleasant . . . that in the new Madison Square theater every gas jet is ventilated by small tubes to carry away the products of combustion." But whatever is to replace gas must have "a general system of distribution—the only possible means of economical illumination." Gathering all the back files of the gas industry's journals and scores of volumes bearing on gas illumination, he studied the operations and habits of the industry, its seasonal curves and the layout of its distribution systems. In his mind he mapped out a network of electric-light lines for an entire city, making the shrewd judgment: "Poorest district for

light, best for power—thus evening up whole city." He meant that in slum districts there would be higher demand for small industrial motors. Against tables for the cost of converting coal to gas he calculated the cost of converting coal and steam into electric energy. An expert gas engineer, whose services Edison engaged at this time, observed that few men knew more about the world's gas business than did Edison.

Edison had a *homo oeconomicus* within him, a well-developed social and commercial sense, though he was careless of money and was not an accountant of the type exemplified by his contemporary John D. Rockefeller. Before the experimental work on his invention was under way, he had formed a clear notion, stated in economic terms, of what its object must be. This concept guided his search and determined the pattern of his technical decisions, so that the result would be no scientific toy but a product useful to people everywhere. By his initial calculation of the capital investment in machinery and copper for a whole system of light distribution he was led to define the kind of light he sought and the kind of generating and distributing system he needed.

Backers of the Electric Light

In the crucial matter of financing his inventive work Edison had the generous and imaginative aid of Grosvenor Lowrey, a patent and corporation lawyer well established in the financial community of Wall Street. Lowrey had fallen completely under Edison's spell and regarded him much as a collector of paintings regards a great artist whose works he believes are destined for immortality. Using his extensive connections and the favorable press-notices that he encouraged Edison to secure during late September and early October, 1878, Lowrey assembled a sponsoring syndicate of some of the most important financiers of the time. The underwriters of the Edison Electric Light Company, which was incorporated in mid-October, included William H. Vanderbilt and J. P. Morgan's partner Egisto Fabbri. This was an unprecedented development in U. S. business. Inventors had been backed in the development of inventions already achieved; Edison's financiers were backing him in research that was to lead to a hoped-for invention. In many respects the venture marks the beginning in this country of close relations between finance and technology.

"Their money," Edison said, "was in-

EDISON AND HIS PHONOGRAPH were photographed in 1878 by Mathew Brady. He had worked with electric lights but had turned to the more promising phonograph. In the year that this photograph was made, however, he resumed his work on lighting.

MENLO PARK was depicted in *Frank Leslie's Illustrated News-paper* for January 10, 1880. The barnlike "tabernacle" of Edison's laboratory is visible at the far right. In its windows passengers on the nearby railroad could see his experimental lights burning.

vested in confidence of my ability to bring it back again." The 31-year-old Edison was by now a well-known figure in Wall Street. His quadruplex telegraph system, by which four separate messages could be transmitted over a single wire, had furnished the pivotal issue in the vast economic war waged between Western Union and the rival telegraph empire of the robber baron Jay Gould. Edison's carbon microphone had transformed the telephone from an instrument of limited usefulness to an efficient system of long-range communication that was now radiating across the country. The shares of gas-lighting enterprises had tumbled on the New York and London exchanges upon Edison's announcement, in the press campaign instigated by Lowrey, that he was now about to displace gas with electricity in the lighting of homes and factories.

The alliance between Edison and his sponsors was nonetheless an uneasy one. The first rift appeared before the end of October, when the rival inventor William Sawyer and his partner Albon Man announced that they had "beaten" Edison and applied for a patent on a carbon-pencil light in a nitrogen-filled glass tube. There was a flutter of panic in the directorate of the Edison Electric Light Company. The suggestion was made that Edison should join forces with Sawyer and Man. Lowrey passed the suggestion on to S. L. Griffin, a former junior executive at Western Union whom Lowrey had hired to help Edison with his business affairs.

Griffin sent back a hasty "confidential" reply: "I spoke to Mr. Edison regarding the Sawyer-Man electric light. . . . I was astonished at the manner in which Mr. Edison received the information. He was visibly agitated and said it was the old story, that is, lack of confidence. . . . No combination, no consoli-dation for him. I do not feel at liberty to repeat all he said, but I do feel impelled to suggest respectfully that as little be said to him as possible with regard to the matter."

In view of Edison's talent for candid and salty language Griffin's reticence is understandable. After that there was no further talk of consolidation with Sawyer or any other inventor.

The Menlo Park Laboratory

In his belief that he would "get ahead of the other fellows" Edison was sustained by his unbounded confidence in his laboratory, its superior equipment and its staff. The Menlo Park laboratory was still the only full-time industrial research organization in the country, in itself perhaps Edison's most important invention. During this period the physical plant was greatly expanded; a separate office and library, a house for two 80-horsepower steam engines, and a glass blower's shed were added to the original barnlike "tabernacle." Even more important, Edison had collected a nucleus of talented engineers and skilled craftsmen, who were of inestimable help to him in working out his ideas.

The self-taught Edison thought primarily in concrete, visual terms. When he was at work on the quadruplex telegraph, he had even built a model made up of pipes and valves corresponding to the wires and relays of his system, and with running water replacing the electric current, so that he could actually see how it worked. But now he would have to depend far more on theory and mathematics.

One of the happiest effects of Grosvenor Lowrey's personal influence was the hiring of Francis R. Upton, a young electrical engineer who had worked for a year in the Berlin laboratory of the great physicist Hermann von Helmholtz. Edison jocularly nicknamed Upton "Culture," and, according to an oft-told story, put the "green" mathematician in his place with one of his scientific practical jokes. He brought out a pear-shaped glass lamp-bulb and gave it to Upton, asking him to calculate its content in cubic centimeters. Upton drew the shape of the bulb exactly on paper, and derived from this an equation for the bulb's volume. He was about to compute the answer when Edison returned and impatiently asked for the results. Upton said he would need more time. "Why," said Edison, "I would simply take that bulb, fill it with a liquid, and measure its volume directly!"

When Upton joined the staff late in October, Edison had already committed himself to the incandescent light. This, rather than the arc light, was the way to imitate the mildness of gas. But the filament glowing in a vacuum had been sought in vain by numerous inventors for half a century. In choosing the incandescent light rather than the arc light he was "putting aside the technical advance that had brought the arc light to the commercial stage." No one, including himself, had succeeded in making an incandescent lamp that would work for more than a few minutes.

Edison's first efforts in 1878 were not notably more successful. Knowing that carbon has the highest melting point of all the elements, he first tried strips of carbonized paper as "burners" and managed to keep them incandescent for "about eight minutes" before they burned up in the partial vacuum of his glass containers. Turning to the infusible metals, he tried spirals of platinum wire; they gave a brilliant light but melted in the heat. Edison accordingly devised a feedback thermostat device that switched off the current when the heat approached the melting point. The lamp now blinked instead of going out entirely. Nonetheless, with his eye on the problem of financing, Edison filed a patent application on October 5 and invited the press in for a demonstration.

As this discouraging work proceeded in the weeks that followed, Edison turned, with Upton's help, to calculating the current that would be consumed by a lighting system equipped with a certain number of such lamps. They assumed that the lights would be connected in parallel, so their imaginary householder could turn one light in the circuit on or off at will, as in a gas-lighting system. Thinking in round numbers, they assumed that these lamps, when perfected, might have a resistance of one ohm and so would consume 10 amperes of current at 10 volts. Allowing in addition for the energy losses in the distribution system, they found that it would require a fabulous amount of copper to light just a few city blocks. Such a system of low-resistance lights was clearly a commercial impossibility.

This was the gist of the objections which had greeted Edison's first announcements that he would use an incandescent bulb in a parallel circuit. Typical of the scorn heaped upon him was the opinion expressed by a committee set up by the British Parliament to investigate the crash of gas-lighting securities. With the advice of British sci-entists, the members of the committee declared that though these plans seemed "good enough for our transatlantic friends," they were "unworthy of the attention of practical or scientific men." From Ohm's law, which governs the relationship between voltage, amperage and resistance in a circuit, the report argued that if an electric light of 1,000 candlepower were divided into 10 smaller lights and connected in parallel, each of the smaller lights would radiate not one tenth but "one hundredth only of the original light." In this judgment such figures as Lord Kelvin and John Tyndall concurred. Before the Royal Institution in London the distinguished electrician Sir William Preece declared: "Subdivision of the electric light is an absolute *ignis fatuus*."

Ohm's law does indeed show that the amount of current (amperes) flowing in a circuit is equal to the electromotive force (volts) divided by the resistance (ohms) in the circuit. Edison's contemporaries reasoned that an increase in the number of lights in a circuit would increase the resistance and therefore reduce the flow of current to each. It was thought that the only way to provide these lights with sufficient current was to reduce the resistance in the distribution system. In a parallel circuit this meant increasing the thickness of the copper conductors to an impractical degree. Such were the limits on the operation of arc lights, with their low resistance and huge appetite for current. Upton's calculations showed that this conclusion also applied to Edison's first low-resistance incandescent lamps.

Edison now confounded his collaborator by proposing that he make the same sort of estimates for an entirely different kind of circuit. This time he would assume lights of very high resistance, supplied with current at high voltage and low amperage. In November and December Upton made calculations on the basis of the same number of lights, but lights with the high resistance of 100 ohms each. These lights were to operate on the low current of only one ampere. Their high resistance was to be offset, in accord with Ohm's law, by the high voltage of 100 volts in the circuit. The result was astonishing: A high-resistance system would require only one hundredth of the weight of copper conductor needed for a low-resistance system. And copper was the most costly element involved—the decisive economic factor.

The High-Resistance System

Here was the crux of Edison's insight at Ansonia. He had recognized there that the subdivision of light called for lamps of high resistance which would consume but little current; to balance the electrical equation it would be neces-

INTERIOR OF EDISON'S LABORATORY at Menlo Park was also depicted in the January 10, 1880, issue of *Frank Leslie's* *Illustrated Newspaper.* At the time of the work on the electric light the laboratory had expanded into several other buildings.

sary to supply the current at high voltage. This was the "necessary secret" that was "so simple." Today every high-school physics student learns that the power lost in transmitting electric energy varies with the square of the current. Thus a tenfold reduction in current meant a decrease of a hundredfold in the energy wasted (or a hundredfold decrease in the weight of the transmission line). It was a conception easily reached by an elementary application of Ohm's law, but it had not occurred to any of Edison's contemporaries. Even Upton did not immediately grasp the full import of Edison's idea. As he said later: "I cannot imagine why I did not see the elementary facts in 1878 and 1879 more clearly than I did. I came to Mr. Edison a trained man, with a year's experience in Helmholtz's laboratory, . . . a working knowledge of calculus and a mathematical turn of mind. Yet my eyes were blind in comparison with those of today; and . . . I want to say that I had company!"

With Upton's figures before him Edison was convinced that a new and strategic invention lay surely within his grasp. It was clear what kind of distributing system he wanted. And he knew what form of incandescent burner would serve his purpose. To offer the necessary resistance to the passage of current it must have a small cross section and so would have a small radiating surface.

By January, 1879, Edison was testing his first high-resistance lamp. It had a spiral of very fine platinum wire set in a globe that contained as high a vacuum as could be achieved with an ordinary air pump. The results were encouraging; these lamps lasted "an hour or two." He then attacked the dual problem of getting a higher vacuum and improving his incandescing element. After another trial with carbon, he returned to metals: platinum, iridium, boron, chromium, molybdenum, osmium—virtually every infusible metal. He thought of tungsten, but could not work it with existing tools. Discouraged by the problem, Edison tried nitrogen in his globe and then resumed his efforts to obtain a higher vacuum. Hearing of the new and efficient Sprengel vacuum pump, which used mercury to trap and expel air, he sent Upton to borrow one from the nearby College of New Jersey (now Princeton University). When Upton returned with the pump late that night, Edison kept him and the other men on the staff up the rest of the night trying it out.

MACHINE SHOP.

GENERATOR which Edison developed for the needs of electric lighting appears at right in this engraving from SCIENTIFIC AMERI-CAN for October 18, 1879 (at that time this magazine appeared weekly). The generator was called the "long-waisted Mary Ann."

At this stage Edison made a useful finding: "I have discovered," he noted, "that many metals which have gas within their pores have a lower melting point than when free of such gas." With the aid of the Sprengel pump he devised a method of expelling these occluded gases, by heating the element while the air was being exhausted from the bulb. The platinum wire within the bulb thereupon became extremely hard and could endure far higher temperatures. Edison later said that at this stage he "had made the first real steps toward the modern incandescent lamp."

Meanwhile the spirits of his financial sponsors had begun to droop. Their brilliant inventor, far from having achieved anything tangible, was hinting plainly that he needed more money. The first Brush arc lights were ablaze over lower Broadway, and more were being installed elsewhere with impressive effect. Edison's backers began to have serious doubts as to whether he had pursued the right course. To shore up their morale Lowrey arranged to have Edison give them a private demonstration.

In April, as one of Edison's associates recalled it, "They came to Menlo Park on a late afternoon train from New York. It was already dark when they were conducted into the machine shop where we had several platinum lamps installed in series." The "boss" showed his visitors pieces of platinum coil he was using in the lamps, pointed out the arrangement of the lights and described the type of generator he hoped to build. Then, the room having grown quite dark, he told "Honest John" Kruesi to "turn on the juice slowly."

"Today, I can still see those lamps rising to a cherry-red . . . and hear Mr. Edison saying 'A little more juice' and the lamps began to glow. 'A little more,' . . . and then one emits a light like a star, after which there is an eruption and a puff, and the machine shop is in total darkness. . . . The operation was repeated two or three times, with about the same results."

The platinum coils still consumed a lot of power for the light they gave, and they were costly and short-lived. The temporary Wallace-Farmer dynamos heated up badly, and were not powerful enough to enable Edison to connect his lamps in parallel. Edison admitted that the system was not yet "practical."

It was a gloomy gathering that broke up on that raw April evening. All of Lowrey's abounding faith would be necessary to rally the spirits and funds of Edison's despondent backers. Some

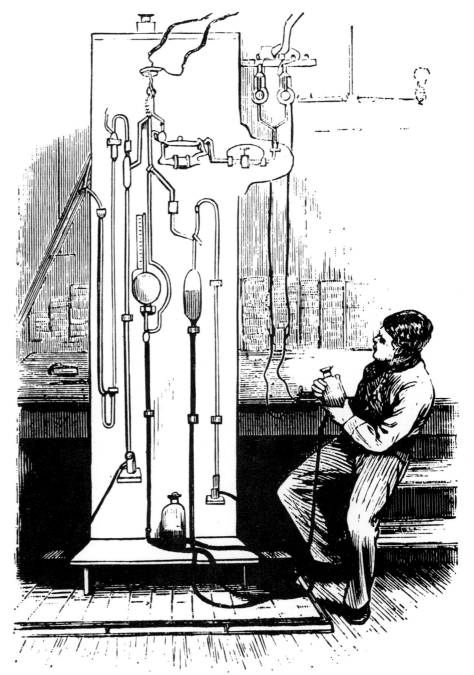

VACUUM PUMP used to remove air from lamp bulbs (*top center*) was of a new type about which Edison had read in a scientific journal. The man is holding a vessel of mercury.

rumors of the disappointing demonstration leaked out; the price of Edison stock fell sharply, while that of gas-lighting securities rose. "After that demonstration," Edison's associate relates, "we had a general house cleaning at the laboratory, and the metallic lamps were stored away."

Edison now rallied his staff to efforts on a much broader area of the front "under siege." He followed three main lines of investigation. One group he detailed to the task of developing the dynamo to supply the constant-voltage current required by his high-resistance system. He set another group to pulling down a still higher vacuum in the glass bulbs. The third team, under his watchful eye, carried out the series of experiments in which 1,600 different materials were tested for their worth as incandescent elements.

The "Long-Waisted Mary Ann"

To subdivide the electric current for numerous small lights in parallel Edison needed a dynamo which would produce

EDISON'S LIGHT.

The Great Inventor's Triumph in Electric Illumination.

A SCRAP OF PAPER.

It Makes a Light, Without Gas or Flame, Cheaper Than Oil.

TRANSFORMED IN THE FURNACE.

Complete Details of the Perfected Carbon Lamp.

FIFTEEN MONTHS OF TOIL.

Story of His Tireless Experiments with Lamps, Burners and Generators.

SUCCESS IN A COTTON THREAD.

The Wizard's Byplay, with Bodily Pain and Gold "Tailings."

HISTORY OF ELECTRIC LIGHTING.

The near approach of the first public exhibition of Edison's long looked for electric light, announced to take place on New Year's Eve at Menlo Park, on which occasion that place will be illuminated with the new light, has revived public interest in the great inventor's work, and throughout the civilized world scientists and people generally are anxiously awaiting the result. From the beginning of his experiments in electric lighting to the present time Mr. Edison has kept his laboratory guardedly closed, and no authoritative account (except that published in the HERALD some months ago relating to his first patent) of any of the important steps of his progress has been made public—a course of procedure the inventor found absolutely necessary for his own protection. The HERALD is now, however, enabled to present to its readers a full and accurate account of his work from its inception to its completion.

A LIGHTED PAPER.

Edison's electric light, incredible as it may appear, is produced from a little piece of paper—a tiny strip of paper that a breath would blow away. Through

a higher voltage than any dynamo in existence, and which would maintain that voltage constant under varying demands for current from the system. Existing dynamos were designed around the fallacious notion, held by most electrical experts, that the internal resistance of the dynamo must be equal to the external resistance of the circuit. Through study of battery circuits they had proved that a dynamo could attain a maximum efficiency of only 50 per cent. In 1877 a committee of scientists appointed by the Franklin Institute in Philadelphia had been impressed to discover that the most successful European dynamo, designed by Zénobe Théophile Gramme, converted into electricity 38 to 41 per cent of the mechanical energy supplied to it. The efficiency of the Brush dynamo was even lower: 31 per cent. These machines and their theoretically successful

contemporaries all produced current at a relatively low voltage.

Edison had concluded, however, that he must produce a dynamo of reduced internal resistance capable of generating current at a high voltage. Such a machine would not only meet the needs of his lighting system but would also convert mechanical energy to electrical energy with far greater efficiency. As his associate Francis Jehl recalled, Edison said that "he did not intend to build up a system of distribution in which the external resistance would be equal to the internal resistance. He said he was just about going to do the *opposite;* he wanted a large external resistance and a low internal resistance. He said he wanted to sell the energy outside the station and not waste it in the dynamo and the conductors, where it brought no profits." Jehl, who carried out the tests

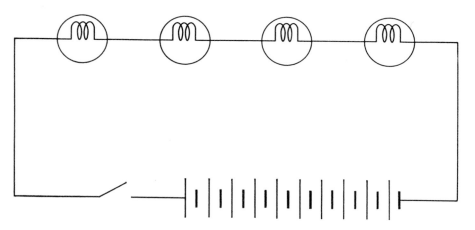

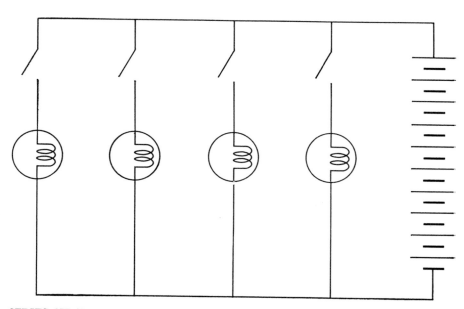

FIRST NEWSPAPER ACCOUNT of Edison's brilliant success appeared in *The New York Herald* for December 21, 1879.

SERIES CIRCUIT (*top*) requires that a number of electric lights (*circles*) be turned on or off at the same time by a single switch (*break in circuit*). Parallel circuit (*bottom*), which was adopted by Edison, makes it possible to turn lights on or off one at a time.

of resistance, also remarked that the art of constructing dynamos was then as mysterious as air navigation. All electrical testing was in the embryonic stage. "There were no instruments for measuring volts and amperes directly: it was like a carpenter without his foot rule."

Upton himself had his difficulties in this hitherto unexplored field: "I remember distinctly when Mr. Edison gave me the problem of placing a motor in circuit, in multiple arc, with a fixed resistance; and. I . . . could find no. prior solution. There was nothing I could find bearing on the [effect of the] counter-electro-motive force of the armature . . . and the resistance of the armature on the work given out by the armature. It was a wonderful experience to have problems given me by.him based on enormous experience in practical work and applying to new lines of progress."

The problem of a constant-voltage dynamo was attacked with the usual Edisonian élan. Seeking to visualize every possible structural innovation for his dynamo armature, he had his men lay out numerous wooden dummies on the floor and wind wire around them, spurring them on in their task by laying wagers as to who would finish first.

After Edison had decided upon the form of winding and type of electromagnets to be used, Upton made drawings and tables from which the real armatures were wound and attached to the commutator. Edison eventually worked out an armature made of thin sheets of iron interleaved with insulating sheets of mica; this armature developed fewer eddy currents and so produced less heat than the solid armature cores then used. When the new cores were test-run, it was Upton who made the mathematical calculations from these tests and drew up the final blueprints.

The self-effacing Upton can be given principal credit for interpreting Edison's ideas and translating them into mathematical form. A careful student of contemporary electrical knowledge, he seems to have been conversant with, and to have guided himself by, the design of a German dynamo, made by the Siemens works, that employed an auxiliary source of current to excite its field magnets.

The new Menlo Park dynamo comprised many admirable features for that period. With its great masses of iron and large, heavy wires, it stood in bold contrast to its contemporary competitors. Owing to the two upright columns of its field electromagnets, it was nicknamed "Edison's long-waisted Mary Ann."

When the dynamo was run at the correct speed, the voltage between its arma-ture brushes was approximately 110, and remained fairly constant, falling but slightly when increasing amounts of current were taken out of the machine. Edison and Upton also contrived a simple but ingenious dynamometer by which the torque of a drive belt was used to measure the work output of the steam engine that powered the dynamo. When Kruesi completed the first operating machine, Upton carefully checked the results. To his astonishment—and quite as Edison had "guessed"—the new dynamo, tested at full load, showed 90-per-cent efficiency in converting steam power into electrical energy.

Edison was as jubilant as a small boy. As was usual with him, the world was soon told all about his "Faradic machine." It was described and depicted in SCIENTIFIC AMERICAN for October 18, 1879, in an article written by Upton.

Once more there was scoffing at Edison's "absurd claims." The hectoring of Edison by some of the leading U. S. electrical experts, among them Henry Morton of the Stevens Institute of Technology, now seems traceable to their ignorance. Reading Morton's predictions of failure, Edison grimly promised that once he had it all running "sure-fire," he would erect at Menlo Park a little statue to his critic which would be eternally illuminated by an Edison lamp.

As a matter of fact, this allegedly ignorant "mechanic" was to be found reading scientific journals and institutional proceedings at all hours of the day and night. It was thus that he had learned about the Sprengel vacuum pump. This device enabled him to achieve an increasingly greater vacuum and to test a broad variety of metals, rare earths and carbon compounds under hitherto unexplored conditions.

The globe itself was also much improved, by the inventor's own design, after he had brought to Menlo Park an artistic German glass blower named Ludwig Boehm. Edison one day drew a sketch of a one-piece, all-glass globe whose joint was completely sealed, and late in April, 1879, Boehm, working skillfully with hand and mouth, fashioned it in the small glass blower's shed in back of the laboratory.

"There never has been a vacuum produced in this country that approached anywhere near the vacuum which is necessary for me," Edison wrote in his notebook. After months of effort he could say exultantly: "We succeeded in making a pump by which we obtained a vacuum of one-millionth part of an atmosphere."

In the late summer of 1879 he realized with growing excitement that a key position had been won. He had a dynamo supplying constant high voltage, and a tight glass globe containing a high vacuum. In his mind's eye he saw what might be done with an extremely fine, highly resistant incandescing substance under these conditions. His state of tension is reflected in the laboratory notebooks by such exclamations as: "S - - - ! Glass busted by Boehm!" All that remained for him was to discover a filament that would endure.

The Carbon Filament

In late August or early September—about a year after he first took up his search—he turned back to experimenting with carbon, this time for good. The rods of carbon he had tried earlier had been impossible to handle, as he now understood, because carbon in its porous state has a marked propensity for absorbing gases. But once he had a truly high vacuum and a method for expelling occluded gases he saw that he might achieve better results with carbon than with platinum.

In a shed in back of the laboratory there was a line of kerosene lamps always burning, and a laborer engaged in scraping the lampblack from the glass chimneys to make carbon cake. But lampblack carbon by itself was not durable enough to be made into fine lamp filaments. Edison and Upton had arrived at the conclusion that, given a 100-volt multiple-arc circuit, the resistance of the lamps should be raised to about 200 ohms; this meant that the filament could be no thicker than a 64th of an inch.

Through the summer months Edison and his staff worked at the tantalizing task of making fine reeds of lampblack carbon mixed with tar. His assistants kept kneading away at this putty-like substance for hours. It seemed impossible to make threads out of it; as an assistant complained one day, the stuff crumbled.

"How long did you knead it?" Edison asked.

"More than an hour."

"Well just keep on for a few hours more and it will come out all right."

Before long they were able to make filaments as thin as seven thousandths of an inch. Edison then systematically investigated the relations between the electrical resistance, shape and heat radiation of the filaments. On October 7, 1879, he entered in his notebook a report on 24 hours of work: "A spiral made of burnt lampblack was even better than the Wallace (soft carbon) mixture." This was indeed promising: the threads lasted an hour or two before they burned out. But it was not yet good enough.

As he felt himself approaching the goal Edison drove his co-workers harder than ever. They held watches over current tests around the clock, one man getting a few hours' sleep while another remained awake. One of the laboratory assistants invented what was called a "corpse-reviver," a sort of noise machine that would be set going with horrible effect to waken anyone who overslept. Upton said that Edison "could never understand the limitations of the strength of other men because his own mental and physical endurance seemed to be without limit."

The laboratory notebooks for October, 1879, show Edison's mood of anticipation pervading the whole staff. He pushed on with hundreds of trials of fine filaments, so attenuated that no one could conceive

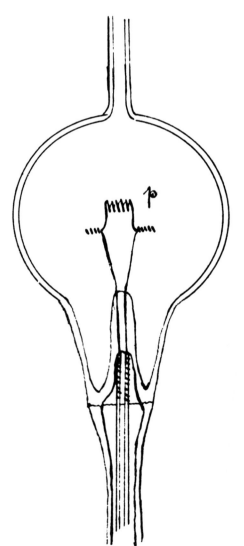

EARLY EXPERIMENTAL LAMP is depicted in one of Edison's notebooks. This lamp had a filament of platinum. It melted.

how they could stand up under heat. Finally he tried various methods of treating cotton threads, hoping that their fibrous texture might give strength to the filament even after they had been carbonized. Before heating them in the furnace he packed them with powdered carbon in an earthenware crucible sealed with fire clay. After many failures in the effort to clamp the delicate filament to platinum lead-in wires, Edison learned to mold them together with lampblack and then fuse the joint between them in the act of carbonization.

Then, as Edison later related, it was necessary to take the filament to the glass blower's shed in order to seal it within a globe: "With the utmost precaution Batchelor took up the precious carbon, and I marched after him, as if guarding a mighty treasure. To our consternation, just as we reached the glass blower's bench, the wretched carbon broke. We turned back to the main laboratory and set to work again. It was late in the afternoon before we produced another carbon, which was broken by a jeweler's screwdriver falling against it. But we turned back again and before nightfall the carbon was completed and inserted in the lamp. The bulb was exhausted of air and sealed, the current turned on, and the sight we had so long desired to see met our eyes."

"Ordinary Thread"

The entries in the laboratory notebooks, although bare and impersonal, nonetheless convey the drama and sense of triumphant resolution pervading the laboratory that night: "October 21— No. 9 ordinary thread Coats Co. cord No. 29, came up to one-half candle and was put on 18 cells battery permanently at 1:30 A.M. . . . No. 9 on from 1:30 A.M. till 3 P.M.—13½ hours and was then raised to 3 gas jets for one hour then cracked glass and busted."

As the light went out the weary men waiting there jumped from their chairs and shouted with joy. Edison, one of them recalled, remained quiet and then said: "If it can burn that number of hours I know I can make it burn a hundred." Yet all the workers at Menlo Park—Edison, Upton, Kruesi, Boehm and the rest— were completely astonished at their success. They had become accustomed to laboring without hope. "They never dreamed," as one contemporary account put it, "that their long months . . . of hard work could be ended thus abruptly, and almost by accident. The suddenness of it takes their breath away."

For once Edison tried to be discreet

and keep his momentous discoveries a secret until he could improve upon his lamp filament. At length, after experimenting with various cellulose fibers, he found that paper, in the form of tough Bristol cardboard, proved most enduring when carbonized. Edison was exultant when this filament burned for 170 hours, and swore that he would perfect his lamp so that it would withstand 400 to 1,000 hours of incandescence before any news of it was published.

On November 1, 1879, he executed a patent application for a carbon-filament lamp. Its most significant passage was the declaration: "The object of the invention is to produce electric lamps giving light by incandescence, which lamps shall have high resistance, so as to allow the practical subdivision of the electric light. . . . The invention consists in a light-giving body of carbon wire . . . to offer great resistance to the passage of the electric current, and at the same time present but a slight surface from which radiation can take place." The specifications called for a distinctive one-piece all-glass container, lead-in wires of platinum that passed through the glass base and were fused to the carbon filament, and joints that were sealed by fusing the glass.

Here were the essential features of the basic Edison carbon-filament lamp, in the form that was to be known to the world during the next half century. It was not the "first" electric light, nor even the first incandescent electric lamp. It was, however, the first practical and economical electric light for universal domestic use.

Edison had spent more than $42,000 on his experiments—far more than he had been advanced by his backers. Now he asked for more money so that he might complete a pilot light-and-power station at Menlo Park. But the directors were still uncertain about the future of the invention. Was it "only a laboratory toy," as one of them charged? Would it not need a good deal of work before it became marketable? Grosvenor Lowrey stoutly defended his protégé. He got no results until he prematurely, and over Edison's objections, made the secret of the electric lamp public.

Rumors had been spreading for several weeks. New Jersey neighbors told of brilliant lights blazing all night at Menlo Park, and railroad passengers between New York and Philadelphia also saw the bright lights with astonishment from their train windows. In Wall Street there was a flurry of speculation in Edison stock; the price rose briefly to $3,500 a share.

Then came a front-page story in *The New York Herald* on Sunday, December 21, 1879. There followed an exclusive article about the inventor's struggles for the past 14 months, told to the world, *con amore,* by Marshall Fox, who had written much of Edison before. The detailed treatment of such an adventure in applied science as a feature story was something of an innovation. Also somewhat unusual in the journalism of the time was its relative accuracy of detail, owing to help provided by Upton, who also supplied drawings for the *Herald's* Sunday supplement. The writer did his best to explain how this light was produced from a "tiny strip of paper that a breath would blow away"; why the paper filament did not burn up but became as hard as granite; and how the light-without-flame could be ignited—without a match—when an electric current passed through it, giving a "bright, beautiful light, like the mellow sunset of an Italian autumn."

In the week following Christmas hundreds of visitors made their way to the New Jersey hamlet. Edison hurried with his preparations for an announced New Year's Eve display as best he could, but was forced to use his whole staff of 60 persons to handle the crowds. He could do no more than put on an improvised exhibition, with only one dynamo and a few dozen lights.

The closing nights of the year 1879 turned into a spontaneous festival that reached its climax on New Year's Eve, when a mob of 3,000 sight-seers flooded the place. The visitors never seemed to tire of turning those lights on and off.

The inventor promised the sight-seers that this was but a token of what was in store. He was awaiting the completion

FRANCIS R. UPTON made invaluable calculations for Edison's system. An electrical engineer who had studied with Hermann von Helmholtz, he was named "Culture" by Edison.

of a new generator, he said, and intended to illuminate the surroundings of Menlo Park, for a square mile, with 800 lights. After that he would light up the darkness of the neighboring towns, and even the cities of Newark and New York.

12

The First Electron Tube

by George Shiers
March 1969

In 1904 John Ambrose Fleming wrote Guglielmo Marconi: "I have been receiving signals on an aerial with nothing but a mirror galvanometer and my device." The device was the original thermionic vacuum diode

In October, 1904, John Ambrose Fleming, a British electrical engineer, demonstrated that a high-frequency alternating current could be rectified, or converted to direct current, by what looked like a dimly glowing incandescent lamp in which a flat metal plate was supported between the legs of a carbon filament in the form of a single loop. In his master patent, filed the next month, Fleming disclosed how his primitive vacuum tube could be used as a signal detector in a wireless telegraph receiver circuit. Since his tube acted like a water valve allowing flow in one direction only, Fleming called it an "oscillation valve." In describing it to Guglielmo Marconi, in his capacity as technical adviser to Marconi's Wireless Telegraph Company, Fleming wrote: "This opens up a wide field for work.... I have not mentioned this to anyone yet as it may become very useful." Contrary to some accounts, Fleming did not casually make use of an effect first observed 24 years earlier by Thomas Alva Edison; by virtue of many careful experiments extending over two decades Fleming earned the right to be regarded as the inventor of the original electron tube.

Electric lamps with glowing filaments were developed independently by Joseph W. Swan in Britain and Edison in America at the close of the 1870's. Swan had been on the track of an incandescent-filament lamp for more than 30 years. His precocious American rival was only 32 years old when he demonstrated his version of the incandescent lamp.

Two important defects of the new electric lamp soon became apparent: the filaments broke and dark deposits formed on the inside of the bulb. Edison was particularly troubled by these defects because a reliable lamp with a reasonable lifetime was an essential element in his new electric lighting system.

He quickly assumed that the deposit resulted from molecular bombardment. Believing that electrically charged particles of carbon were being ejected from the filament, Edison decided to probe the bulb with a piece of wire to find out if it was possible to prevent "electric carrying," as he called it, and thus preserve the clarity of the glass. The deflection of a galvanometer needle when a galvanometer was placed in a circuit between the wire probe and the positive lead of the glowing filament proved the truth of his notion. This flow of current through a vacuum—the "Edison effect"—contributed nothing, however, toward helping Edison improve his lamp.

This classic experiment was performed during the infancy of the incandescent lamp. It is recorded as Experiment No. 1 and is dated February 13, 1880, in an early Edison notebook [*see illustration on page 142*]. Nothing more was done in the ensuing months, probably because the inventor's time and energy were fully devoted to the creation of his revolutionary power and lighting system.

Edison returned to the lamp problem in the summer of 1882. An entry in one of his notebooks, dated July 5, shows a wire electrode inserted through the top of the bulb. Again the experiment gave way to more urgent work. Early in 1883 Edison, pondering the strange currents within his lamps, went back to the problem and began another series of experiments. A sketch of a new kind of lamp, one with a flat plate between the legs of the filament loop, appears in a notebook entry for March 8. Another entry for March 10 illustrates a similar lamp with a list of metals to be tried as the cold electrode.

Edison found that the bulb current varied according to the temperature of the filament; it was therefore related to the voltage of the filament supply. Ignorant of the true nature of the phenomenon he had uncovered but eager to apply it, he conceived the idea of using the novel lamp as a voltage indicator. A working model was built in the summer of 1883 and was demonstrated in 1884 at an electrical exhibition in Philadelphia (the first such exhibition in America). A patent was granted to Edison for this "electrical indicator" in October of that year. Although the patent was of no commercial value, it is historic: it is the first patent in electronics.

Edison's special lamp and its peculiar behavior naturally attracted attention at the Philadelphia exhibition. The phenomenon was the subject of a note by Edwin J. Houston in the first issue of the *Transactions* of the American Institute of Electrical Engineers, organized in the same year. One may reflect that if a simple experiment had been tried by any one of several able men who attended the exhibition, a practical system of wireless telegraphy utilizing an electron tube could have been demonstrated four years before Heinrich Hertz proved the existence of electromagnetic waves.

The experiment was not done, even though the means and the experimental skills were available. Elihu Thomson had conducted experiments with electromagnetic wave phenomena as early as 1871 (experiments in which Houston played a part) and again in 1875. Edison had toyed with "etheric forces" that year, and his experiments were followed by those of Silvanus P. Thompson in 1876 and of David E. Hughes in 1879. Even then a detecting device was a major need. In most such experiments sparks occurring at some distance from a source were taken as evidence of an etheric disturbance. It remained for Hertz to design a detector consisting of a large loop

FLEMING AND HIS "OSCILLATION VALVE" were photographed in 1923, 19 years after he had discovered that his primitive electronic device could be used to rectify a high-frequency alternating electric current, thus making it suitable as a detector of wireless telegraph signals. His invention of the oscillation valve, better known today as a thermionic vacuum diode, marked the beginning of the age of electronics. At the time this photograph was made Fleming was 74 years old. He died in 1945 at the age of 96.

with a spark gap, and its sparking removed all doubt that electromagnetic radiation had traveled through space. Nonetheless, the rudimentary facts of generating, dispersing and detecting etheric radiations were known to many at the Philadelphia exhibition of 1884. They were not put to practical use until 1896, when Marconi demonstrated his wireless telegraph. The Edison effect remained a laboratory curiosity for another 20 years.

The flow of ideas in electrical science was usually from Europe to the U.S. In the case of the embryonic electron tube, however, the sequence of events was reversed. The tiny currents in the evacuated space of the carbon-filament lamp, ultimately ignored by Edison and his compatriots, became the ward of a British electrical engineer and educator.

John Ambrose Fleming was born in Lancaster in November, 1849. He began to prepare for a career in electrical engineering when he entered the University College School in London at the age of 14. He continued his studies for several years, occasionally taking time out for commercial work, and started teaching in 1871. After holding several posts as a teacher of science, Fleming entered the University of Cambridge at the age of 28, where he divided his time between his studies and laboratory work under the guidance of James Clerk Maxwell.

Both Edison and Swan formed companies in Britain to develop their electric lamps; eventually they consolidated their interests. Fleming became a consultant to the Edison Electric Light Company in 1882 and was soon involved in problems of the incandescent lamp. His first paper on the subject of "molecular radiation" was read to the Physical Society in May, 1883. He described how volatilization of the carbon filament and of copper from the lead-in clamp produced the dark deposits seen inside the bulb. He identified a narrow line free from deposit as a "shadow" of the loop, and he concluded, as Edison had, that some kind of molecular emission was going on inside the bulb.

Fleming reported similar observations in a second paper, also to the Physical Society, in June, 1885. Since ordinary lamps were used there is no mention of internal currents in these papers, nor is there any reference to the most important clue: the polarity of the leads. This omission seems surprising, but it can be partly explained by the fact that Fleming was examining used lamps and was not working with glowing lamps in a laboratory circuit.

William Henry Preece, then chief electrician of the British Post Office, visited the Philadelphia exhibition of 1884. He obtained some sample lamps from Edison and began to experiment with them soon after returning to Britain. Preece announced his results in a paper read to the Royal Society of London early in 1885. He repeated several of Edison's experiments, including one with a side extension set at right angles to the bulb, and he also tried the shielding effect of a piece of mica over the collector plate. Preece generally confirmed the results obtained by Edison, and it was he who named the basic phenomenon the Edison effect. He also remarked on a "blue glow" that appeared in the lamps at a critical temperature, which was of course directly related to voltage.

Meanwhile independent investigations were being conducted in Germany, in a continuing line of research that had

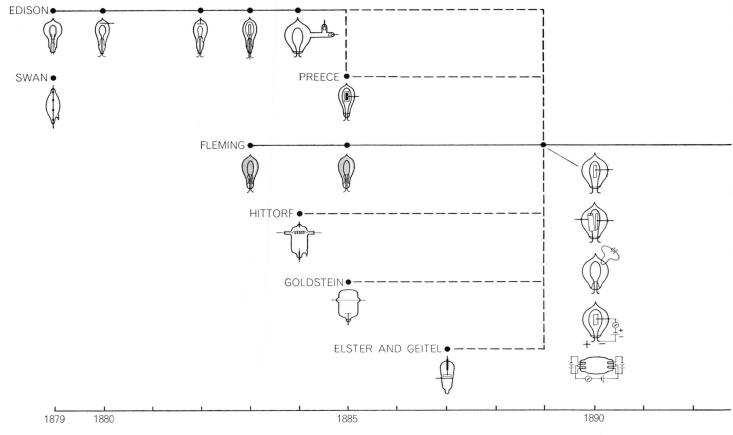

CHRONOLOGY OF INVENTION from 1879 to 1907 traces the line of work that led from the incandescent lamp of Thomas Alva Edison and Joseph W. Swan to Fleming's oscillation valve. Prior investigations into "thermionic emission" by Edison in the U.S., William Henry Preece and Fleming in Britain, and Wilhelm Hittorf, Eugen Goldstein, Julius Elster and Hans Geitel in Germany were united in Fleming's experiments of 1889. A tube made by Fleming in 1889, which was virtually the same as one devised by Edison in 1883, was

begun in the late 1860's. Wilhelm Hittorf had employed vacuum tubes with cold electrodes in his pioneer studies of gaseous conduction. In 1884 he experimented with platinum electrodes and found that the resistance of the evacuated space was reduced when the negative electrode was heated. Hittorf also observed unilateral conductivity, and he noted that the conductivity could be increased by raising the cathode temperature or by lowering the pressure inside the tube. Eugen Goldstein, another able investigator of electrical conduction in gases, carried out similar experiments the following year and confirmed Hittorf's results.

Julius Elster and Hans Geitel, later well known for their studies of photoconductivity, reported a series of experiments on electrical conduction in flames and heated gases from 1882 to 1889. One of their devices had a straight platinum filament and a flat platinum plate in a highly evacuated bulb. After trying a variety of electrode arrangements they concluded that charged particles were being dispersed uniformly from the glowing electrode; they also reported that conductivity was in one direction.

The knowledge gathered by the Germans does not appear to have been utilized by them for further research or for practical ends. Fleming, however, resumed his investigation of the Edison effect in 1889. Building on the American and German discoveries and the work of Preece, he started a line of work that eventually led him to his oscillation valve. Two different fields of endeavor came together in Fleming's hands. The Germans were scientists; their contributions were in the tradition of physical investigation. Edison was an inventor; he made his discovery through curiosity prompted by urgent technical needs. Fleming was neither an inventor nor a scientist but a practicing engineer and a teacher. His interest in the properties of a lamp with an auxiliary electrode arose through his close association with the electric lighting industry and with the training of engineers. Thus Fleming's motivation appears to have been chiefly academic; he had no practical end in view.

With some special lamps of his own design made at the Edison and Swan Lamp Works, Fleming began a series of thorough experiments to learn more

about the curious phenomenon. His results were disclosed in two papers published early in 1890, in which he gave full credit to the earlier work of Edison, Preece and the Germans. Fleming repeated some of the earlier experiments and also devised new ones in an attempt to discover the basic features of the mystifying vacuum currents.

Fleming tried a variety of electrode arrangements [see top illustration on page 144], including the use of glass and metal shielding cylinders around the filament legs. He devised lamps with extension tubes, double-ended tubes, heated carbon loops for collectors and center-tapped filaments; with them he used external circuits incorporating batteries and capacitors as well as a galvanometer. Fleming even demonstrated the Edison effect in open air by placing a metal collector plate inside an exposed carbon loop.

Through these and later experiments Fleming became familiar with the phenomenon of current flow through a vacuum. His results supported earlier findings and established unilateral conductivity as a basic property of an evacuated space containing a heated negative elec-

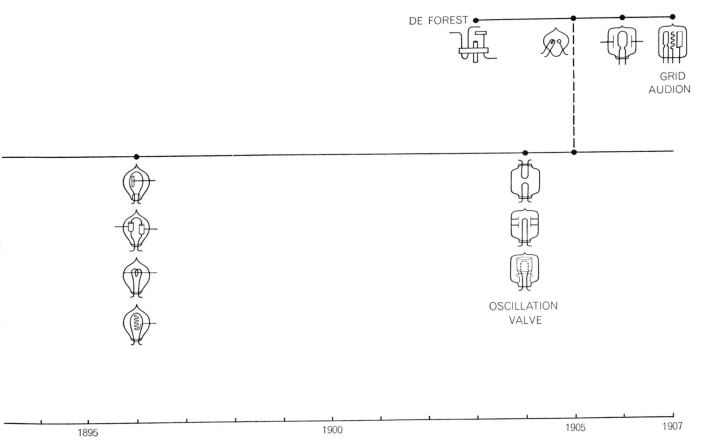

actually the first to achieve high-frequency rectification in 1904. The chart also shows the link between Fleming's 1905 paper and Lee De Forest's early work, which led to the invention of the thermionic triode. Some of De Forest's early devices incorporated open gas flames. With one exception drawings of various tube geometries were adapted from original sources. The drawing of the Goldstein tube, for which no authentic source is available, is based on contemporary tubes and represents a type likely to have been used.

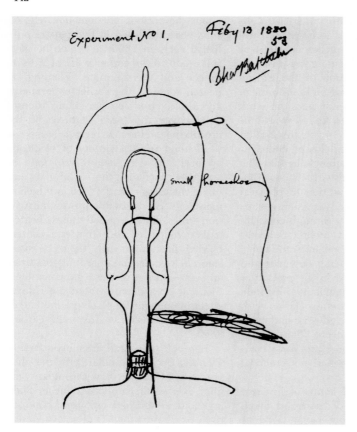

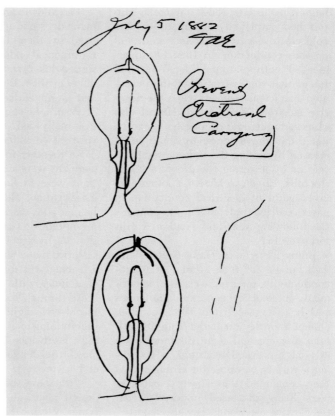

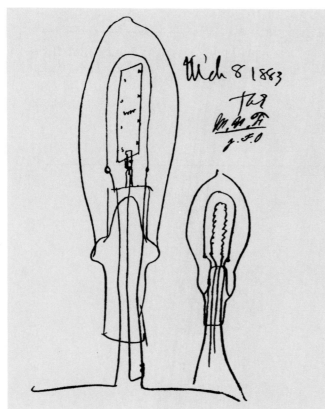

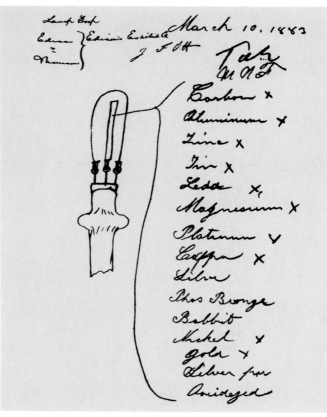

SKETCHES OF EARLY INCANDESCENT LAMPS appear in Edison's notebooks. The entry recorded as Experiment No. 1 and dated February 13, 1880 (*top left*), shows a glow lamp with a "small horseshoe" carbon filament and a wire probe designed to detect the flow of current through the evacuated space of the lamp (the Edison effect); it is probably the first illustration of the bulb that eventually became the electron tube. The sketch is signed by Charles Batchelor, one of Edison's assistants. The entry for July 5, 1882 (*top right*), shows a wire electrode inserted through the top of the bulb. The sketch dated March 8, 1883 (*bottom left*), shows a lamp with a flat plate between the legs of the filament loop. An entry made on March 10, 1883 (*bottom right*), illustrates a similar lamp with a list of metals to be tried as the cold electrode. Edison later abandoned his efforts to find the cause of the Edison effect, and the investigation was taken up by Fleming. The photographs were made at the Edison National Historic Site in West Orange, N.J.

trode and a colder positive plate. Although Fleming thus prepared himself for the invention of the oscillation valve, he did not advance any new theories or suggest any practical applications for the phenomenon or his special glow lamps. He observed in one paper that this field of research was "a region abounding in interesting facts and problems in molecular physics." Nevertheless, he chose to drop his investigations for several years and took up other matters.

Late in 1895 Fleming once more returned to his current-carrying lamps and initiated an exhaustive series of experiments in which he worked with several new models as well as earlier ones. He presented his new results ("A Further Examination of the Edison Effect in Glow Lamps") at a meeting of the Physical Society on March 27, 1896. In this lengthy paper (profusely supported with diagrams, graphs and tables) Fleming described nearly 30 experiments.

Fleming coined the term "molecular electrovection" to denote the conveyance of electric charge by moving molecules. Molecules dispersed from a glowing electrode were at that time (1896) commonly regarded as being the charge carriers in the Edison effect. It was not until a year later that J. J. Thomson demonstrated the existence of the electron. Fleming clearly described and explained unilateral conductivity, as well as the rectifier action (although he did not name it) of a lamp operated by alternating current. If the word "electron" is substituted for "negatively charged molecules," his paper is a familiar description of the elements of thermionic emission, but of course he was ahead of his time.

In the last half of the 1890's Fleming's careful experiments were scarcely noted in the press, which was full of news about the X rays that Wilhelm Konrad Roentgen had discovered in 1895 and the wireless telegraph system that Marconi had demonstrated a year later. Fleming's work seemed to be leading nowhere. With Edison's sharp intuition, or the deep perception of Thomson, or perhaps more freedom from pressing duties, Fleming could have leaped into the 20th century simply by combining the elements of his lamps. He could have suggested the possibilities of using his glow lamp as a rectifier for small currents, and since he had built lamps with two anodes, he might have demonstrated full-wave rectification as well as half-wave. Even more important, he could have placed a zigzag wire (which he had put into one of his models) between the oth-

er electrodes (just to find out what would happen, as Edison would probably have done) and thereby have anticipated Lee De Forest in inventing the three-element vacuum tube. None of these leaps, however, would have been characteristic of a man as methodical and cautious as Fleming. Although he conducted and documented a series of thorough and well-conceived experiments that apparently covered all aspects of electrical conduction in a vacuum, he did not venture beyond his own experimental limits. Once again, for the sixth time in 16 years, the curious glow lamp was put aside and forgotten.

Over the next three years wireless telegraphy was commercialized. In 1899 Fleming became technical adviser to the newly established Marconi's Wireless Telegraph Company, a post that added to his heavy obligations as professor of electrical technology at University College London. At the end of the century he was busy designing the power system for Marconi's new experimental wireless station at Poldhu in Cornwall. The detector of radio waves in those days was an erratic device known as a coherer, a metal tube filled with iron filings. This was followed by several kinds of electrolytic detector and by Marconi's magnetic detector; both were superior in some respects to the coherer, but they were not satisfactory for regular and dependable service. A new and better detecting device, or signal rectifier, was urgently needed.

Fleming, a first-rate electrical engi-

REPLICA of Edison's experimental lamp of 1883 shows the simple carbon-filament-plus-metal-plate construction of the original. Conceived of by Edison as a voltage indicator, a working model of this lamp was built in the summer of 1883 and was demonstrated in 1884 at an electrical exhibition in Philadelphia. The patent for this lamp, granted to Edison in October, 1884, was the first patent in electronics. The replica was lent by Vance Phillips.

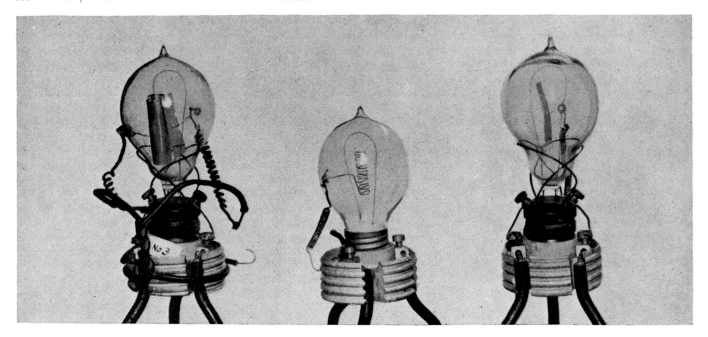

THREE EXPERIMENTAL LAMPS were made by Fleming in 1889 in the course of his investigations of the Edison effect. The assembly at left contains a metal screening cylinder around one leg of the filament as well as a metal plate inside the loop. The lamp at center has a zigzag of wire inside the loop; the one at right has a metal plate inside the loop. It was this third lamp, inserted in a wireless telegraph receiving circuit by Fleming in his crucial 1904 experiment, that was actually the prototype of his oscillation valve.

I have found a method of rectifying Electrical oscillations. That is making the flow of Electricity all in the same direction. so that I can detect them with an ordinary mirror galvanometer. I have been receiving signals on an aerial with nothing but a mirror galvanometer and my device. but at present only on a laboratory Scale This opens up a wide field for work. as I can now measure exactly the effect of the transmitter. I have not mentioned this to any one yet as it may become very useful

Yours very sincerely

J. A. Fleming.

FLEMING'S LETTER TO MARCONI, written in November, 1904, revealed Fleming's discovery of the rectifying capability of his tube. The key lines in the letter, shown here, read: "I have found a method of rectifying electrical oscillations, that is, making the flow of electricity all in the same direction, so that I can detect them with an ordinary mirror galvanometer. I have been receiving signals on an aerial with nothing but a mirror galvanometer and my device, but at present only on a laboratory scale. This opens up a wide field for work, as I can now measure exactly the effect of the transmitter. I have not mentioned this to anyone yet as it may become very useful. Yours very sincerely, J. A. Fleming." Both photographs were supplied by the Marconi Company Limited.

neer, lecturer and textbook author, rigorously applied classical methods in his work. It was therefore natural for him to ponder the general problem of making quantitative measurements in the new high-frequency apparatus. He sought ways of accurately measuring and evaluating the circuits and devices that were coming into service in wireless systems. Fleming tried the common aluminum-carbon electrolytic cell and found that it, like other detectors, was unsuitable for precision measurements with direct-current meters at radio frequencies.

In considering the problem of converting a feeble oscillatory current into a direct current that could actuate a recording instrument, Fleming later recalled that he had had "a sudden very happy thought": he perceived that his lamp was "the exact implement required to rectify high-frequency oscillations." The device he needed, neglected and all but forgotten, was sitting in one of his laboratory cupboards. Would such a lamp work at Marconi's new high frequencies? Would it be sensitive enough to operate on the extremely low voltages available in a wireless receiving circuit? Finally, could it be used to measure, as well as to detect, feeble currents oscillating at high frequencies?

These questions were soon answered. The inventive act was now nearly complete; all that remained was to unite the strange lamp with a suitable circuit and prove that it would work. Within a few hours a simple spark-coil oscillator and a resonant receiving circuit comprising one of the lamps were set up. The actual lamp that Fleming worked with was one of the 1889 models that had a flat plate supported between the legs of a single-loop carbon filament. It was virtually the same as Edison's lamp of 1883. The experiment was an immediate success; Fleming had discovered a new kind of high-frequency rectifier, one that was eminently suitable for detecting wireless signals.

This memorable event took place one afternoon in October, 1904. Fleming was considering the geometry of the collecting electrode that would be most efficient. He decided on an open metal tube (not unlike the shielding cylinders in some of his 1889 experiments) that would surround the entire filament. The following day he placed an order with the Ediswan Lamp Works for a quantity of special lamps, each with a metal cylinder around the carbon filament. These lamps, which Fleming soon called oscillation valves, were the first thermionic vacuum diodes [see bottom illustration on next page].

Early in November, Fleming wrote Marconi: "I have been receiving signals on an aerial with nothing but a mirror galvanometer and my device" [see bottom illustration on opposite page]. On November 16 he filed a provisional application for a patent on "Improvements in Instruments for Detecting and Measuring Alternating Electric Currents." This application made direct reference to the use of a galvanometer and a two-element lamp—the first vacuum tube in an electronic circuit—as a receiving instrument in wireless telegraphy. Fleming's master patent was granted on September 21, 1905 [see top illustration on next page].

Fleming lost no time in presenting his discovery to his professional colleagues. His paper "On the Conversion of Electric Oscillations into Continuous Currents by Means of a Vacuum Valve" was read at a Royal Society meeting on February 9, 1905, and was published in March. Fleming discussed the rectifying power of the valve and showed characteristic curves relating cathode-anode current and applied voltage (up to 100 volts) at three different filament voltages. He also proposed water cooling for the anode to increase the "rectifying efficiency." Fleming's hopes were only half-realized, however; although he had invented a new wireless detector, it failed him as a device for measuring high-frequency currents.

In his patent Fleming did not describe the mode of electrical conduction in his valve; he referred only to "negative electricity." Nonetheless, in his paper he employed "electron" and "free electrons" in describing how conduction took place. Fleming was the first to use the word "electron" popularly in referring to minute negative charges or particles (also called negative ions or negative corpuscles) in a paper titled "The Electronic Theory of Electricity," first published by the Royal Institution in 1902 and subsequently reprinted in The Popular Science Monthly. "Electronic" thus entered the technical literature before the advent of the first practical electron tube.

In 1906, at the peak of his career, Fleming published The Principles of Electric Wave Telegraphy, an encyclopedic treatise that became the foremost textbook for practitioners at all levels in the new communication field. In this book Fleming described his oscillation valve as one kind of "vacuum tube cymoscope," a generic name that he introduced with his invention of the cymometer, an instrument for measuring the length of electric waves. Curiously, Fleming now omitted all reference to electrons in describing the flow of current from the carbon filament to the metal cylinder in his valve; instead he spoke of a flow of "negative electricity." Nor did he refer to the theory of thermionic emission, published in 1901 by O. W. Richardson, in spite of its relevance to his own work. Fleming made various improvements in his valves early in 1905, and they went into actual service for the first time at Poldhu in the summer. The Marconi organization began manufacturing Fleming valves in 1907 and incorporated them into much of its equipment for several years thereafter.

Few inventions can have brought their inventors so much distress, disappointment and trouble as the oscillation valve gave Fleming. His patent and his interests in it immediately became the property of Marconi's Wireless Telegraph Company under the terms of his contract. Before his "glow lamp detector," as it was sometimes called, had a chance to prove its value it was supplanted by the "cat's whisker" crystal rectifier, which was first employed in 1906. (The crystal rectifier was the first solid-state device in electronic technology.) In later years, when thermionic valves were being produced by the millions annually, Fleming, perhaps understandably, showed a sour side of his nature because he had not received any financial reward for his pioneer labors. Moreover, "his valve," although it was modified and improved a thousandfold by others, was the predecessor and (in his mind) the real source of a booming industry.

The man who reaped some of the fame and fortune that Fleming considered his due was Lee De Forest, one of the most enterprising and determined radio pioneers among the Americans. In 1904, the year Fleming applied for his British patent, De Forest was 31 years old and had decided to establish a wireless system of his own free from patent domination by others, particularly Marconi. De Forest was obsessed with the idea that "a gaseous medium," maintained in a condition of "molecular activity," could serve as an "oscillation responsive device," in other words, as a detector of wireless signals.

In a series of patent applications extending from November, 1904, to the end of 1906, De Forest embodied his ideas in a variety of devices that changed from open gas flames to electrodes en-

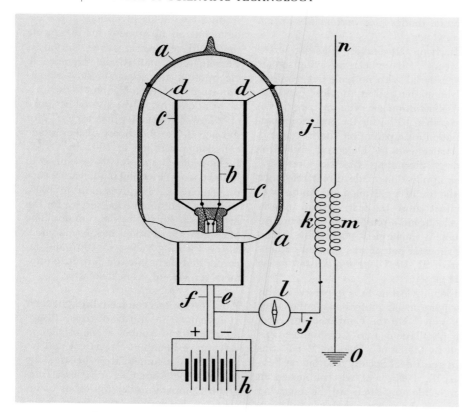

MORE EFFICIENT GEOMETRY was adopted by Fleming in the drawings of the oscillation valve that accompanied his original British patent application in 1904. The design incorporates an open metal cylinder (*c*) that surrounds the entire filament (*b*). Electrons emitted by the filament moved to the cylindrical plate when the cylinder was positively charged by a signal from the antenna (*n*). A galvanometer (*l*) was employed to indicate the arrival of a signal. The patent also included diagrams of a full-wave rectifier with two valves connected to a differential galvanometer, and three valves connected in parallel.

PRODUCTION MODEL of the Fleming oscillation valve is one of the 12 original thermionic diodes manufactured at the Ediswan Lamp Works near London in October, 1904.

closed in glass bulbs. About the middle of 1905 Fleming became aware of a De Forest device not unlike one of his own. In response Fleming called De Forest's attention to his Royal Society paper of the previous March. This move by Fleming, obviously a warning that De Forest was poaching on Fleming's preserve, was the beginning of a lifelong feud between the two men.

The situation got worse when De Forest described his experiments in a discursive and hazy paper presented to the American Institute of Electrical Engineers on October 26, 1906. This paper ("The Audion") also appeared in *Scientific American Supplement* in November and December, 1907. Unfortunately De Forest was wrong about several historical matters and slighted Fleming by making only a passing and inaccurate reference to his work. Fleming never forgave De Forest for his lack of generosity or for his "exploitation" of the basic thermionic diode. Fleming maintained to the end of his life that De Forest had merely added a piece of bent wire to "his valve."

Early in 1907 De Forest applied for an American patent on a wireless telegraph receiver that disclosed the first three-electrode tube with a control element (a zigzag piece of wire) located between the filament and the plate. The tremendous importance of this twin invention—the grid triode and its basic circuit—was not perceived by the inventor or his contemporaries for several years. Suddenly, between 1911 and 1913, a new technology emerged: triode circuits became amplifiers and oscillators, adaptable to both radio and telephone systems.

The new technology grew to maturity as radio broadcasting became a part of everyday life, followed by television, radar, the electronic digital computer and a host of other applications for the ubiquitous electron tube. In the deepest sense these technological advances were launched by Fleming's "very happy thought" and his quick verification that an old idea could serve new ends. There can be no doubt that if he had not devised the thermionic diode, someone else would have done so soon afterward. Still, Fleming was the man who took the immortal first step. In giving the world the first practical electron tube he not only justified his patient guardianship of the Edison effect but also provided an important transitional link between the age of electric power and the age of electronics.

III

THE TRIUMPH OF SCIENTIFIC TECHNOLOGY

The period since the end of the Second World War has seen an explosive growth in communication, transportation, and information. The burgeoning of electronics and the widespread use of computers, the shift in employment patterns from blue-collar to white-collar workers, and the shift from product- to service-oriented industries have contributed to a qualitatively different pattern often called a new industrial revolution. Technological innovation has become an industry in itself, and one with a vested interest in maintaining the pace of change whatever the cost; communication has become strident and universal media too easily turned to propaganda; public information swamps us, while concealed information endangers our freedom; military technology and the arms race proceed on a course we cannot seem to alter. The next eight articles describe the evolution of a technological society that has grown nearly out of control.

III   THE TECHNOLOGICAL IMPERATIVE

INTRODUCTION

After many centuries in which innovation was almost imperceptible and after a few in which all technical progress was identified with human progress, we have now reached a stage in which innovation has become compulsive—but only technological innovation. A large vested interest has been created, even apart from the military-industrial complex, embodied in the avant-garde *industries and research organizations, which believes that it must "innovate or die."*

Dennis Gabor
Innovations: Scientific, Technological, and Social

The old world came to an end in August 1914. There is a break in Western history, an enormous four-year gulf between that world and ours, that yawns so wide that our own era seems to date only from the end of the First World War, and the time before seems remote and quaint. The resentments, disillusions, frustrations, and hatreds generated in that war, the wiping out of a generation of European intellectuals and idealists and almost a whole generation of young men, left a legacy of bitterness, cynicism, and despair that forever buried the Victorian ideal of human perfectability through progress. Old beliefs were destroyed; old Empires dissolved; new ideologies enthroned; and new governments created. The momentum imparted by these forces is far from spent in our own time. The years of war, followed by recession, recovery, wild boom, depression, rearmament and war again, stretching from 1914 to 1945 have often been described as a second Thirty Years' War. The thrust of the prewar inventive age was blunted; by any of the common yardsticks of technical, social, political, or economic progress this was a period of stagnation, retrenchment, and occasional retrogression. By 1945, most of Europe and East Asia lay in ruins, and the next few years were spent in painful rebuilding of their economic-industrial base and careful testing of the new alignments of power. Since about 1950, however, the economy of the world in general, and the West in particular, has entered a period of expansion and growth unprecedented even by the most dynamic phases of the nineteenth-century industrial boom. Whether this new phase of industrial expansion and social and economic change is a continuation of the growth initiated at the beginning of the century, or whether it possesses an entirely new and qualitatively different pattern, is not yet clear.

History is possessed of an inherently inverse perspective: the closer in time an event is, the less able we are to perceive it clearly. The Edwardian period was revered as a Golden Age by the survivors who looked back across the suffering of the Great War of 1914–1918. But the war was no accident, nor was it really begun over the proper wording of a Serbian apology. The war came because the governments of Europe were ready (some even eager) for war. The frictions generated by the unleashed energies of the second phase of the Industrial Revolution had done much to create that readiness. Its products and machines had misled many of the protagonists into believing that war henceforth would be regulated by the supremacy of offensive weapons over any defense, and therefore concluded quickly. Yet, in trying to connect the present with this no longer recent past, we are hardly in a position to be more objective than our great-grandparents were. The political and social dislocations, ideologies, and cultural upheavals that were first created and then exacerbated by our own "Thirty Years' War" still remain with us, unresolved and unexpunged.

To many historians, particularly those with a Marxist orientation, the second phase of the Industrial Revolution, with its monopolies, cartels, and imperial conquests, was the climacteric of capitalism. This view is eloquently set forth by J. D. Bernal in his monumental *Science in History* (M.I.T. Press, 1971):

> Already towards the end of the [eighteen] sixties the first, simple, optimistic phase of early capitalism was beginning to draw to an end. The great depression which started in the seventies marked a transition between the era of free-trade capitalism, with Britain as the workshop of the world, and that of a new, more widely based, finance capitalism, with France, Germany, and the United States coming to the fore under the cover of protected markets. The enormous productive forces liberated by the Industrial Revolution were by then beginning to present their owners with the problem of an ever larger disposable surplus. This could not, under capitalism, be returned to the workers who made it. When invested at home it led to even greater production and to a more hectic search all over the world for markets that were soon filled. The result was colonial expansion, minor wars, and preparation for the larger wars which were to come in the next century.

Here, the Great War is seen as the beginning of the collapse of a system that had been showing preliminary patches of decay during the deflationary years before 1896 and quite clear signs during the socially and politically troubled inflationary period between 1896 and 1914. An examination of the political and social history of the Edwardian period provides much evidence to support this view. The condition of the new urban proletariat created by the pattern of nineteenth-century industrialization was very poor. The end of the century was marked by social upheaval and, at times, armed struggle between companies and unions, anarchists and police, scabs and strikers. Those who sought to preserve a supply of cheap labor came into direct conflict with those who struggled to organize the workers for higher wages and better conditions. During the thirty-year period which followed the First World War, the battle shifted from the manipulation of markets to absorb excess productive capacity to the allocation and internal distribution of these same goods and the control of their means of production. Following the Second World War, as hot peace turned to Cold War, the balance of world terror forced the relegation of military conflict to peripheral territories and proxy wars, allowing the major industrial powers to concentrate on expanding their industry and technology and to push their own growth into a completely new phase of development. The extent to which this has resolved the social problems of industrialization is less clear, even in countries where labor holds the reins of government or where socialism is the organizing philosophy. We seem no closer to the dream of a social structure in which industry is the servant of humanity rather than the forger of its chains. The chains may be more ephemeral now than they once were, but they still bind us to our machines.

Curiously enough, a parallel view of the dissolution of the old-style industrial capitalism is held by a number of influential political scientists, economists, and sociologists whose theoretical framework is decidedly non-Marxist. A collage of the views of some of the better known of these critics is provided by Victor Ferkiss in *Technological Man: The Myth and the Reality* (New York: Braziller, 1969):

> Thus Daniel Bell speaks of "post-industrial society" based on the substitution of intellectual activity for primary production as the basis of economic life, and Zbigniew Brzezinski describes essentially the same purported phenomenon as the "technetronic age." . . . But for all these prophets of the future, technology is in some fashion or other the factor that is making the radically new civilization possible and necessary and providing its organizing principle. . . . for J. K. Galbraith the "new industrial state" is dominated by the "technostructure." Psychologist Kenneth Kenniston finds the key to the alienation of many of today's brightest youth in their inability to meet the demands of the ego in a contemporary milieu that he refers to as "technological society." Marshall McLuhan sees the new civilization as based on a single aspect of technology—the new electronic media, which he holds constitute not merely new methods of communication but a totally new environment that will radically alter everything from politics to sexual behavior. Jacques Ellul concludes that technology per se will determine the future; in Ellul's opinion we are entering into a new age of "technological society," wherein technology is no longer an instrument for pre-existing human purposes but has become an end in itself, controlling both men and their society. All are agreed, however, that as a result of certain technological factors we are leaving the industrial era and bourgeois society behind and entering a radically new world.

A commonly held view of all these critics and prophets (whatever their political and philosophical orientation) is that, during the beginning of the twentieth century, the industrialization of the developed countries had proceeded far past the point of merely providing sufficient productive capacity to provide the material needs of life for their own citizens. They had developed a "surplus" that could not only turn a profit but could also have improved the material condition of every man, woman, and child on earth.

The progress of the Industrial Revolution in the nineteenth century and the early twentieth century was based largely on a certain degree of coupling between the progress of science and the advance of technology. Toward the end of the nineteenth century, however, physics, which had been the scientific breeding ground of so many of the ideas created by and for the technology of the second phase of industrialization, had solved, at least in principle, the puzzle of the perceptible, macroscopic, and basically subjective world of experience. The long history of exploring and explaining the tangible human environment, which had begun in Egypt and Mesopotamia some four thousand years before, was apparently drawing to a close. At the turn of the century, physicists were embarking on a newer and darker sea—the microscopic, statistical, and relativistic world. The technical offspring of their voyage were not to be spawned until mid-century, for the new fields of science were initially too esoteric to be exploited. The majority of the industrial developments of the 1920s and 1930s were based on the science of the previous century: synthetic dyes, the first plastics and synthetic fabrics, a metallurgy capable of designing steels for specific applications, entirely new metals, mass-produced automobiles and radios, the proliferation of broadcast stations, commercial airplanes, and the ubiquity of telephones and electric lights. These devices, which fill the technical history of the period between wars, were developments of inventions made at the turn of the century.

The emergence of the scientifically trained engineer as a numerically significant professional class was largely responsible for continuing development and innovation, which provided these devices with a history of improvement and technical refinement even during the period when demand far exceeded supply. In 1890 there were twenty-five thousand engineers in the United

States; by 1938 there were two hundred and fifty thousand. The engineer, with an education based on classical physics, chemistry, and metallurgy, had a regularized position in industry and worked toward an ideal of perfection in process or machine, toward which technique could strive quite independently of the immediate and less consistent demand of the marketplace. Much of the rapid pace of industrial development between the wars can be attributed to this inner logic for technical change. The automobile, for instance, reached what many still regard as its technical peak (within the limits of the available materials) during the 1930s, when the mass demand for these expensive and sophisticated machines was at a minimum.

The history of the technology-related industrial development of our century can, therefore, be divided into three phases. The first phase was the tail of the old Industrial Revolution, when science was still spinning off large numbers of new inventions. This was followed by a period when invention was relatively dormant and practical engineers, with their commitment to progressive evolution of existing devices, enabled industry to catch up and fully exploit those inventions. With the coming of the new industrial revolution, the engineer once again turned into applied scientist, and innovation became, not an event created by a single man, but a process carried out by thousands. In the new industrial revolution, the engineer-inventor-entrepreneur has for the most part been replaced by the engineering staff; the freelance inventor by the engineering firm. Technological change has become institutionalized, and "progress" is not only the "most important" but frequently the *only* product of many a modern industrial firm.

Walter Hossli, in his article "Steam Turbines," follows the development of one of the most important sources of industrial power through all the three phases of the century. Existing steam engines were still a factor of two or three below Carnot's limit when, in 1884, Charles Parsons adapted the water wheel, one of the earliest of power-generating technologies, to harness the power of expanding steam. Although reciprocating steam engines remained the dominant source of power for low-speed or high-specific-impulse applications such as locomotives and pile drivers, the benefits of generating continuous high-speed rotary motion by an inherently rotational method was immediately obvious. Within a few years, the turbine had become the accepted standard for ship propulsion, and the first enormous, integrated turbine-dynamo generators were being constructed to supply the flood of electrical power that Edison was so very eager to distribute. Running amazingly close to their theoretical limit of efficiency, humming smoothly and (considering their gigantic power output) quietly on their shafts, coupling rotating source to rotating load without the awkwardness of piston and crank, modern turbines are among the most esthetically satisfying of machines.

Turbine and dynamo were not yet inseparably coupled when Henry Adams, haunting the halls of the Great Exposition of 1900 for a key to the changing forces of history that were remolding the self-image of humanity during his lifetime, was drawn to the dynamo as the symbol of the new forces of industrial society. Adams contrasted the dynamo with the Virgin, whom he regarded as the symbol of the spiritual forces of the Middle Ages that had erected the cathedrals of Notre Dame and Chartres. Adams's view is best expressed in "The Dynamo and the Virgin" from *The Education of Henry Adams*:

> As he grew accustomed to the great gallery of machines, he began to feel the forty-foot dynamo as a moral force, much as the early Christians felt the Cross. The planet itself seemed less impressive, in its old-fashioned, deliberate, annual or daily revolution, than this huge wheel, revolving within an arm's-length at some vertiginous speed, and barely murmuring—scarcely humming an audible warning to stand a hair's-breadth further for respect of power—while it would not wake the baby lying close against its frame. Before the end, one began to pray to it; inherited instinct taught the natural expression of

man before silent and infinite force. Among the thousand symbols of ultimate energy, the dynamo was not so human as some, but it was the most expressive.

The comparison between dynamo and Virgin was perhaps more apt than Adams realized. The Virgin represented the force that ameliorated the harshness of medieval existence and that inspired both the building of great cathedrals and the "feminine ideal" found in chivalric romance (without which tales of knighthood would be little more than armoured Westerns). The dynamo—coupled to the steam turbine—could soften the harshness of industrialization for the factory worker. When the sources of power were limited to a single, large, stationary steam engine, huge central factories crowded with dangerous, dirty, and noisy networks of shafts, pulleys, and belts were the sole means of industrializing even those processes that had sprung from cottage industry. To keep the boilers hot, factories were run on twelve-hour shifts; many ran twenty-four hours a day, seven days a week. Electric power made possible the planning of factories where independently operable machines driven by individual electric motors could be spread out. Factories could now be made cleaner and safer and could be organized according to the flow of the product rather than the size and type of the machine. The premium on long shifts was removed. The irony is that the dynamo would also have made it possible for cottage industry itself to survive, but by the time electricity came to the cottages such industry was virtually extinct.

The new industrial revolution is nowhere so apparent as in electronics, that most characteristic of modern industries. The days when Edison hired "that engineer" (Fleming), and when Ford did not have even one engineer, have given way to the era of the modern electronics firm where, even if a product is actually manufactured on the premises, the production force is frequently much smaller than the research and development staff. In these industries, the span between creation and obsolescence has contracted from decades to years, or even months. It is not unusual for the engineering staff of an electronics firm to consider a product obsolete from the moment it reaches the production stage—that is, at that moment when research and development must halt to allow the final production jigs to be set up. The clearest example of this trend is to be found in the growth of the machine most central to the new industrial age—the electronic digital computer, whose evolution over the past two decades is described by D. L. Slotnick in "The Fastest Computer." The first practical digital computer was constructed in 1944 using electro-mechanical relays. Five years later the relays were replaced by vacuum tubes to create the first wholly electronic machine. The ILLIAC series of experimental computers described by Slotnick were begun in this first generation and have subsequently evolved at a pace not untypical even of commercial electronics in the "technetronic" period. ILLIAC I, completed in 1952, used vacuum tubes and could perform eleven thousand arithmetic operations per second; ILLIAC II, completed ten years later with the new technology of semiconductor transistors and diodes, was about fifty times faster. In 1971 the ILLIAC IV, using the most recent solid-state technology of large-scale integrated circuits, was capable of between one hundred and two hundred *million* operations per second—some ten thousand times faster than the ILLIAC I of only twenty years before and about one hundred times faster than contemporary commercial machines. The speed of the modern computer is now so great that further progress has been slowed by the definitive physical limit—the speed of light—and ILLIAC IV was designed specifically to overcome this limitation by a suitable choice of organizing principle that allows several processes to proceed simultaneously.

Progress in computer design was able to proceed so rapidly at least partially due to an interlock with progress in solid-state physics and electronics. This enabled both to advance far more rapidly than either could have advanced separately. Progress in computers called for more and more complex hardware, most of which would have been developed very slowly were it not for advances in solid-state electronics. In turn, the progress of solid-state electronics depended on computers in many ways: to design and model circuits, to augment the progress of basic research, and, in the case of large-scale integration, to provide automatic selection and testing on the production line itself. This synergistic development has become part and parcel of the rate of technical change that has so drastically altered the relationship between the new manufacturers and their markets. During the nineteenth century, it could still be argued that technological change filled clearly apparent needs. During the first half of the twentieth century, such change was governed less by the requirements of society than by its capacity to absorb new products. Innovation and change now proceed at a pace so hurried that new devices are frequently produced even before the development of their potential market (as, for instance, in the case of microwave ovens). In this respect ILLIAC IV has a history not unlike that of commercial computers. A few years ago the installation of large, sophisticated machines had proceeded so rapidly that there was a reverse "computer gap," and large blocks of very expensive computational capacity went begging due to the over-capacity of the machines for the businesses and laboratories where they had been installed. The ILLIAC should need no such justification, as it is primarily a research machine; yet, as Slotnick implicitly acknowledges, the problems that ILLIAC was uniquely capable of solving were by no means immediate and pressing. It was not the capacity of the machine that was challenged, but rather the capacity of its users to come up with problems of adequate complexity.

The idea of such challenges from the machine was apparent from the beginnings of modern computer science. Many people have been fascinated with the possibility of constructing a self-programming, self-replicating machine—an artificial "intelligence" if you will, or perhaps merely an artificial life form with some independent reasoning capacity that can be defined as "intelligence." Norbert Wiener, the father of cybernetics, was deeply involved with the implications of such a development and discussed the possible social consequences and risks by using the medieval legend of the golem as a metaphor. The golem legend has taken many forms but is best known to the modern reader as a legend about the great alchemist, Rabbi Löw of Prague. In *The Kabbalah and Its Symbolism* (Schocken, 1970), Gershom G. Scholem quotes the legend in the late Jewish form described in 1808 by Jakob Grimm:

> After saying certain prayers and observing certain fast days, the Polish Jews make the figure of a man from clay or mud, and when they pronounce the miraculous Shemhamphoras (the name of God) over him, he must come to life. He cannot speak, but he understands fairly well what is said or commanded. They call him golem and use him as a servant to do all sorts of housework. But he must never leave the house. On his forehead is written 'emeth (truth); every day he gains weight and becomes somewhat larger and stronger than all the others in the house, regardless of how little he has to begin with. For fear of him they therefore erase the first letter, so that nothing remains but *meth* (he is dead), whereupon he collapses and turns to clay again. But one man's golem once grew so tall, and he heedlessly let him keep on growing so long that he could no longer reach his forehead. In terror he ordered the servant to take off his boots, thinking that when he bent down he could reach his forehead. So it happened, and the first letter was successfully erased, but the whole heap of clay fell on the Jew and crushed him.

Rabbi Löw was reputed to have been more fortunate. Although his golem ran amok and destroyed the ghetto it was supposed to keep clean and orderly, the great Rabbi was able to tear the word from the forehead of his golem without getting caught in the collapse. Perhaps this is why the Prague form of the legend is so much more appealing to a modern, technically oriented society. It treats golem-making as an essentially reversible process.

There is no lack of examples of overly rapid technical progress. Some examples, such as the creation and testing of new drugs and industrial and agricultural chemicals, lie outside the scope of this volume; others, such as breeder reactors, will be discussed further on. One exceptionally clear cut and unambiguous example is the recent case of the supersonic transport (SST). As a purely military technic, supersonic aircraft were bound to be developed and deployed as rapidly as possible; such is the nature of the arms race. The SST, however, must be judged according to the presumably more rational standards of the nonmilitary world, where economic viability and social desirability are presumed to be nonnegligible factors. As a product *per se*, the SST is a total failure; it performs no obvious function that is either socially useful or economically profitable. Such reason as there was for its construction was primarily political, and its advocates did their utmost to exploit "gadget-worship"—the fascination of a technologically oriented society with its machines. The alienation of modern man and woman from their labor has resulted in the displacement of the sense of personal mastery over nature, of personal cleverness and ability to manipulate the inanimate world, onto the acquisition of technical goods. In mass society, participation in scientific and technical advances can be acquired by surrogate, and the desire for "participation" in the latest scientific and technological conquests is not only satisfied by the purchase of the latest technical trinkets (such as wristwatches that will let one keep an appointment to within one one-millionth of a second, traffic permitting), but also by public identification with and involvement in the success or progress of projects such as the space program. People are persuaded to accept the risks taken and the goals achieved as if they were their own. The candles we light to the dynamo are considerably more expensive than those once lit to the Virgin of Chartres.

If this attitude were confined solely to the civilian sphere, the problems would be serious enough. Its extension to the military-industrial-scientific-technological complex abets the acceptance of exorbitant fiscal costs and unnecessary new systems, which in turn threatens all efforts to stabilize and control the spread of nuclear weapons. The arms race has even had its own counterpart of the SST, the Anti-Ballistic-Missile. The battle over the ABM resembled that over the SST in many ways and was fought over many of the same issues: whether it would or could work as effectively as its proponents claimed; whether the enormous investment required was economically worthwhile; whether its acquisition constituted more of a liability than an asset; and whether it was actually being developed for technical or political reasons. The ABM case was, of course, far more serious in its implications, but the general outlines of the argument were the same. In both cases, the matter finally came down to the question of social control over science and technology, to the argument over the technological imperative: what *can* be done *must* be done; what is *possible* is *inevitable*; and the corollary, if we don't, someone else will.

The SST and the ABM aroused impassioned argument and bitter debate. But far more often the issues are less clearly delineated, the technological imperative less obviously displayed, and the claims for the new technic less obviously flawed. This is especially true in the nuclear weapons race, in which all parties involved have consistently tried to move as fast as science and technology would allow—and occasionally faster. The technological imperative has been incorporated into the arms race to such an extent that, like the purloined letter, it has become nearly invisible by virtue of being always in plain sight. The progress of this race and the attempts to halt it, the technologies and scientific advances that disturb the balance of terror and the efforts to restore it, have been regular features of *Scientific American* for many years.°

° An exemplary collection of these articles, with introductions by Herbert F. York, has recently been published under the title of *Arms Control* (W. H. Freeman and Company, 1973).

The topic selected for an example in this collection—the missile submarine—is but one of the many; yet it serves to illustrate a number of the more important general features of the new weapons systems and their development.

The missile submarine, as described in "Missile Submarines and National Security" by Herbert Scoville, Jr. is an excellent paradigm of the products of the new industrial revolution. Its components—the submarine itself, radar, the ballistic missile, inertial navigation, the nuclear warheads, even the nuclear propulsion system—are outgrowths of pre-1950 scientific and technological innovation. There have, in fact, been only two truly significant "inventions"—solid-state electronics and the maser and laser—that are attributable solely to the new industrial state; and only the former had had widespread impact. Transistors and other solid-state circuitry made possible the computers, guidance systems, electronic-warfare gear, instrumentation, and communication equipment that make the missile submarine not just a weapon, but a "weapons system." It is difficult to look back at the years when the Cold War was at its peak and criticize the decision to build the first of the Polaris submarines. The missile and technological "gaps" were then widely perceived as being real, and the need for new weapons systems was largely unquestioned. Nevertheless, when the first missile submarine was launched in 1960, with its complement of sixteen one-megaton, 1,200-nautical-mile-range missiles, prospects for arms limitation and control were diminished. The proximity of such a submarine to its target shortened the response time for a retaliatory attack and made the nuclear trigger more sensitive, while its relative undetectability raised tensions considerably. Like other military technologies, however, the missile submarine soon became an indispensable part of our "deterrent posture," and much was made of its near immunity from preemptive attack. As Scoville points out, by the end of 1966 the Polaris submarine fleet had risen to its full strength of forty-one submarines, many of which were equipped with the newer 2,500-nautical-mile-range A-3 missile with multiple warheads, which seemed more than adequate as a deterrent force. As Scoville states, "Military technology did not, however, stand still." In response to fears of possible advances in anti-submarine warfare (ASW) or the deployment of an effective ABM system, the Navy went on to develop a new generation of Poseidon missiles with multiple, independently targetable, reentry vehicle (MIRV) warheads. The projections for 1976 now show a grand total of 5,440 warheads capable of hitting 5,120 separate targets for the missile-submarine force alone—a force capable of swamping any possible ABM defense even if unsupported by other weapons systems. At 160 warheads per submarine, it would be necessary for an aggressor to destroy every one of the twenty-five to thirty submarines on station almost simultaneously in a coordinated attack, whereas it is generally accepted to be extremely difficult to track and destroy even one.

The armaments industry has long since ceased to react solely to the realities of outside threats. Advances in new weaponry are governed more by a hypothetical capability to invent new methods of destroying deterrents than by any realistically projected, actual capacity for anyone to do so. The possibilities of an effective ABM system and more advanced ASW, although they existed only as hypotheses, were enough to initiate a whole new cycle of technical advance in missile submarines. If this could have been restricted solely to research and development work, it might have been justified; but such work comes to be self-fulfilling, and Trident (as the new system came to be known) was funded as if the hypothetical threat were real. The appropriation grew from 104.8 million dollars in 1971 to nearly a billion dollars in 1972. This sum, as Scoville had predicted, constituted a nearly irrevocable commitment to full construction. One of the more persuasive arguments for increasing the Trident appropriation in 1973 was to protect the "large investment" already made in it. And so it goes. We appear once again to have fallen into "the trap of buying new military hardware just because we have made technological advances."

The MIRV-ing of the Poseidon missiles was in itself an example of this disastrous policy, and it may have done more to decrease national security than to augment it. The structure of the decision to add a MIRV system to the nuclear arsenal is the subject of "Multiple-Warhead Missiles," the latest of a number of penetrating analyses of the dynamics of the nuclear arms race by the distinguished physicist Herbert F. York. Although MIRVs always decrease the deliverable megatonnage of the missiles in which they are installed, the dispersion of the warhead actually increases the effective "kill" radius, especially against targets that have been hardened against blast. With the usual dispersion of ICBM silos, a single one-megaton warhead would have a small probability of destroying a given site, even if no missile defense is assumed. Ten warheads of fifty kilotons each would increase the probability only slightly if they were purely ballistic (MRV); if they were sophisticated MIRVs, the probability that the same single missile's load would destroy the site would be very much increased. The same holds for the soft "counter-value" targets—cities and industries—which are rarely "worth" a megaton-nage yield. Considering the size of the U.S. deterrent in 1970 (nearly 4,200 warheads), the need for a further nuclear population explosion of the retalia-tory force is not clear. Worse yet, the MIRVs satisfy the basic conditions for a preemptive attack—a "first-strike" capability—and therefore constitute a serious qualitative acceleration of the arms race and a major threat to the balance of weaponry. The deployment of MIRVs causes each side to become concerned not merely with whether they possess sufficient forces to retaliate massively in reprisal, but with whether they have a "creditable" deterrent capable of surviving a preemptive attack. One obvious consequence is the further acceleration of the design and production of additional and more sophisticated weapons; another is the more serious possibility that both sides will have an incentive to launch first in a crisis. MIRVs have also seriously delayed the progress of the arms-limitation talks, since aerial reconnaisance can only count missiles, not warheads. As the ultimate goal of disarmament, rather than just nuclear-weapons limitations, depends on restricting the num-ber of warheads, the necessity for on-site inspection (politically the most sensitive subject at the talks) has risen markedly.

The main thrust of York's article is to analyze how and why the MIRVs were designed and how the decision to deploy them was made. In York's words, the multiple course of development was more a fabric than a thread, and the severing of any single line would have done little to impede the prog-ress of the MIRV system. Although the ABM played its role by stimulating a perceived need for better penetration (and thus managed to have adverse consequences merely by becoming a subject for debate), many of the early design steps were made by technologists concerned only with solving immedi-ate problems. Much of the necessary technical development took place in programs seemingly unrelated to weapons research. As a result, MIRV capa-bility was developed almost without anyone knowing it was happening, and at the moment when actual development became feasible the technological imperative took over. As usual, any weapon that was possible became an immediate necessity; and, despite the strenuous objections of the Arms Con-trol and Disarmament Agency, the MIRVs were funded. Now they are de-ployed, and there is no way short of on-site detailed inspection to guarantee their removal. As qualitative rather than quantitative advances in nuclear weaponry, they were not subject to the restrictions of the first round of strategic-arms-limitation talks. They will probably be the major topic of the second round. As York points out, the lesson to be learned here is that many programs are too diffuse to be stopped by confronting the potential danger of a particular technology. Some decisions are woven into the very fabric of weapons research and development, and what must be stopped is not a single technology, but the arms race as a whole.

Even the most astute critics of modern technology have all too frequently overlooked the importance of such collective effects in modern society. They tie their warnings to the old individual model of the act of creation, inherited from the Judeo-Christian tradition which posits a single creator. The real golem of the modern state, however, is not any single creation—let alone the result of a single creative act—but the *process* of uncontrolled creation. When progress in scientific technology was imbued with an inner logic of its own, when *'emeth* was forged on the forehead of the machine, technological change became the golem of our time. At first the benefits outweighed the apparent costs. Technics became capable of proceeding steadily onward through political and economic crises, boom, depression, inflation, and war. But the increasing incorporation of innovation and technical intervention into the basic structure of society made continuous technological change an inexorable force —as irresistible as the golem of Rabbi Löw, which, ignoring the blows, insults, and entreaties hurled at it, continued its methodical demolition of the Prague ghetto for the sake of cleaning its streets.

The golem legend was originally a warning against the imitation of God; the golem was made of clay (*'adamah* in Hebrew), just as Adam was. Having neither mind nor soul, the golem was not only incapable of comprehending the nature of good and evil; it was inherently incapable of *being* good or evil. If it was clumsy and dangerous, if it destroyed or even killed, these were accidents of the dissonance between the logic of the golem and that of the human society in which it moved. Though molded of more abstract clay, scientific progress and technological change are also inherently neither good nor evil. They are technically rational; and rationality is a method, not a value. It is by equating the rational with the good, by failing to perceive the difference between human and technical logic, that we have allowed our golem to proceed so long unchecked. For most of this century the technological imperative has been allowed to dominate our decisions. Merely intervening at the last moment to attempt to block an undesirable technic from wide usage, or merely trying to keep track of each new technological or scientific discovery in the hope of being able to control it, has become increasingly futile as the key decisions diffuse more and more invisibly into the course of the development itself. We must alter the present system whereby technology is considered a cure for technology, and where change is taken as a remedy for change. We must stop defining progress in terms of innovation. Just as the arms race must be slowed or stopped as a whole, the unqualified, unquestioning acceptance of the benefits of ever-newer technics must cease. As we cannot long survive as human beings in a world structured to fit the logic of our machines, we must constrain our machines and make them operate according to the logic of humanity.

158

ANATOMY OF A STEAM TURBINE is exposed in the shops of Brown, Boveri & Company Limited near Zurich. When connected to an electric generator, this turboset, as it is called, is capable of producing 320 megawatts of electricity, equivalent to approximately 400,000 horsepower. It thus ranks among the larger sets built with a single shaft. Steam enters the turboset at a pressure of 2,640 pounds per square inch and a temperature of 977 degrees Fahrenheit. After passing through the high-pressure turbine, located at the extreme rear, the steam is reheated to 977 degrees and then enters a double-flow intermediate-pressure turbine, located in the housing next in line. The steam emerges from this unit divided into two streams and passes, without further reheating, into two double-flow low-pressure turbines, which are located in the large boxlike housings in the foreground. This division of flow is needed to accommodate the thousandfold expansion in volume that takes place as the steam travels through the turbine from inlet to outlet.

Steam Turbines

by Walter Hossli
April 1969

These efficient machines are the principal means of converting the heat energy released by fossil and nuclear fuels into the kinetic energy needed to drive power generators and large ships

The steam turbine ranks with the internal-combustion engine as one of the major achievements of mechanical engineering in the 19th century. Steadily increased in size, reliability and efficiency, steam turbines now account for more than 75 percent of the electric power generated in the world (most of the rest is hydroelectric) and propel most of the biggest and fastest ships. Between 15 and 20 percent of the fossil fuel consumed in the U.S. and western Europe—and essentially all the fuel now consumed in nuclear power plants—has only one purpose: to evaporate and superheat water that then passes through a steam turbine. For all their size and ubiquity steam turbines are among the least understood products of the mechanical age; they are seldom located where they can be seen by the public and the complexities of their design are familiar only to specialists. It took the recent failure of the turbines in the new Cunard liner *Queen Elizabeth 2* to remind people not only that these great machines exist but also that their design and construction is an exercise in high technology.

In principle a steam turbine is simplicity itself. It is a pinwheel driven by high-pressure steam rather than by air. It basically consists of a rotor from which project several rows of closely spaced buckets, or blades. Between each row of moving blades there is a row of fixed blades that project inward from a circumferential housing. The fixed blades are carefully shaped to direct the flow of steam against the moving blades at an angle and a velocity that will maximize the conversion of the steam's heat energy into the kinetic energy of rotary motion. Because the steam's temperature, pressure and volume change continuously as it progresses through the turbine, each row of blades has a slightly different

length, and in certain parts of the turbine the twist of the blade is usually varied along the length of the blade, from root to tip. At the inlet of the turbine the blades are stubby, with little or no twist; at the outlet the blades are much longer and the twist is pronounced.

In a typical modern power plant steam leaves the boiler, after being superheated, at a pressure of 2,500 pounds per square inch and a temperature of 1,000 degrees Fahrenheit [*see illustration on page 161*]. As it enters the high-pressure turbine (the first unit in a cascade of three or four, usually working on a single shaft) a pound of steam occupies about .3 cubic foot. A short time later, when it leaves the low-pressure turbine (the last unit in the cascade), a pound of steam occupies more than 300 cubic feet, a thousandfold expansion.

The steam enters the high-pressure turbine at a velocity of about 150 feet per second and is immediately accelerated by expansion in the first row of fixed blades. In the low-pressure turbine, steam velocity and blade velocities can exceed 1.5 times the velocity of sound under the existing steam conditions. When the steam finally leaves the low-pressure turbine, its velocity is about 600 to 1,000 feet per second. Thus the turbine engineer must design "airfoils" (more accurately blade profiles) able to operate efficiently over a range of velocities roughly equivalent to the range encountered by the designers of a supersonic airplane.

It is now 85 years since the British engineer Charles A. Parsons conceived the modern steam turbine and energetically began pushing its practical application. In an age commonly regarded as ultraconservative the steam turbine was developed and put to large-scale use with remarkable speed—a speed not

equaled later by its descendant, the aircraft gas turbine, in an age thought to be much more receptive to innovation. Parsons demonstrated his first steam turbine in 1884. Seven years later he saw the value of adding a condenser to exploit the turbine's distinctive capacity for utilizing the energy of low-pressure steam down to a near-vacuum. That same year, 1891, the first steam turbines were harnessed to generate electricity.

In 1897 Parsons installed a 2,100-horsepower turbine in the *Turbinia*, a vessel of 44 tons, and astonished the British Admiralty by driving it at 34 knots through a parade of warships in a naval review at Spithead. The fastest ship yet built, the *Turbinia* easily outraced the entire fleet, which was equipped with conventional reciprocating steam engines. By 1904 the Admiralty had installed one of Parsons' turbines in a cruiser, the *Amethyst,* and a year later decided to abandon reciprocating engines and put Parsons turbines of 23,000 horsepower in the new *Dreadnought* class of battleships and turbines of 41,000 horsepower in the 28-knot *Invincible* class of cruisers. In 1907 the Cunard line launched the *Lusitania* and the *Mauretania,* each equipped with Parsons turbines developing 70,000 horsepower. Nearly twice as powerful as the largest reciprocating engines ever employed for ship propulsion, the turbines drove the two Cunard liners at a record speed of 25 knots. Only 16 years had elapsed since Parsons had thought to add a condenser to a turbine of a few hundred horsepower.

His accomplishments made such an impact on the British that they were recounted in much detail in the famous 11th edition of the *Encyclopaedia Britannica,* published in 1910–1911. (The same edition does not mention Einstein's special theory of relativity, published in

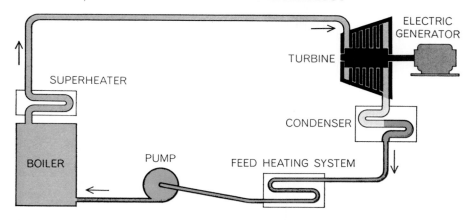

SIMPLE TURBOGENERATOR is used where power demand is light or fuel is cheap. Steam passes through a single turbine. Steam pressure is limited to about 1,500 pounds per square inch, output to 100 megawatts. Maximum thermal efficiency is around 37 percent.

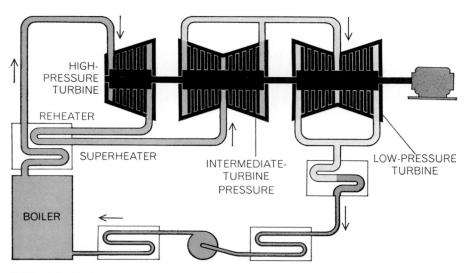

SINGLE-REHEAT TURBOSET is usually preferred when the demand for power exceeds 100 megawatts and fuel costs make a thermal efficiency above 40 percent desirable. After steam leaves the high-pressure turbine it is reheated and returned to the double-flow intermediate-pressure turbine. From there it passes to the double-flow low-pressure turbine.

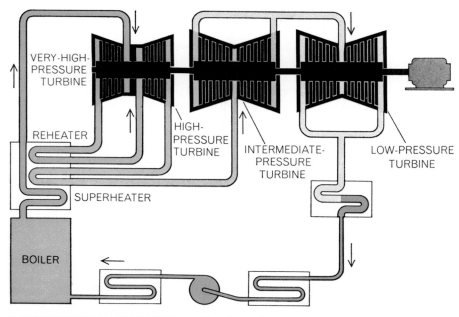

DOUBLE-REHEAT TURBOSET provides the highest power output and the highest efficiency, around 47 percent. In all three systems feedwater is heated by steam bled from the turbines. Cooling water for the condensers often comes from a nearby river, lake or ocean.

1905, or Planck's concept of the quantum of energy, published five years earlier.) The author of the entry on steam turbines has no hesitancy in describing "the invention of the steam turbine [as] the most important step in steam engineering since the time of Watt."

Steam turbines for ship propulsion have not grown very much since the days of the *Lusitania*. The turbines of the *Queen Elizabeth 2* are designed to produce 110,000 horsepower. Turbines of only 40,000 horsepower enable the largest tankers, vessels of more than 300,000 tons, to travel at 13 knots. The biggest increase in turbine size has been in the field of electric-power generation, and the end of this evolutionary trend is not in sight. In 1900 the first steam turbine used for power generation in continental Europe had a rating of 250 kilowatts; it was built by Brown, Boveri & Company Limited of Switzerland, the firm with which I am affiliated. (For purposes of comparing electric-power turbines and marine turbines, one kilowatt is about 1.3 horsepower.) Brown, Boveri is now building for the Tennessee Valley Authority the world's two largest steam turbine sets; each will have an output of 1,300,000 kilowatts, or 1,300 megawatts [*see illustrations on page 164*]. Our projections show that well before the end of the century single units of 2,000 megawatts will be required for the most efficient generation of power.

For propulsion of big ships and power generation, where large unit output and maximum efficiency are essential, the steam turbine is unchallenged. Diesel engines have a comparable efficiency but the largest units so far built are limited to about 30,000 kilowatts. Gas turbines go up to 100,000 kilowatts, but their efficiency is somewhat lower. At the other end of the size range, steam is again being considered as a power medium for automobiles; it remains to be seen, however, whether turbines or piston units will be more successful. The incentive is a reduction in air pollution. Nearly complete combustion can be achieved when fuel is used to fire a boiler instead of being burned in the cylinder of an engine.

Present-day applications of land-based steam turbines are by no means limited to electric-power production. In many industrial plants, such as refineries, sugar mills, paper mills and chemical plants, low-pressure steam is needed in large quantities for process purposes. Instead of producing this steam directly in a low-pressure boiler it is advantageous to install a high-pressure boiler at little extra cost and to drop the steam to the desired level by expanding it through a "back pressure" turbine whose outlet pressure corresponds to the pressure needed for process steam. (If steam at two pressures is desired, a second turbine can be installed ahead of the back-pressure turbine so that steam can be withdrawn between the two units as well as after the second.) When this is done, electric power can be produced as a by-product at almost no cost. In most cases, of course, an external power supply is needed to handle fluctuating electric loads. It is also common in industry to use steam turbines as a direct power source for rotating machinery, such as compressors, blowers and pumps. All told, the worldwide demand for steam turbines requires the production of units whose annual combined output exceeds 50,000 megawatts, or 65 million horsepower.

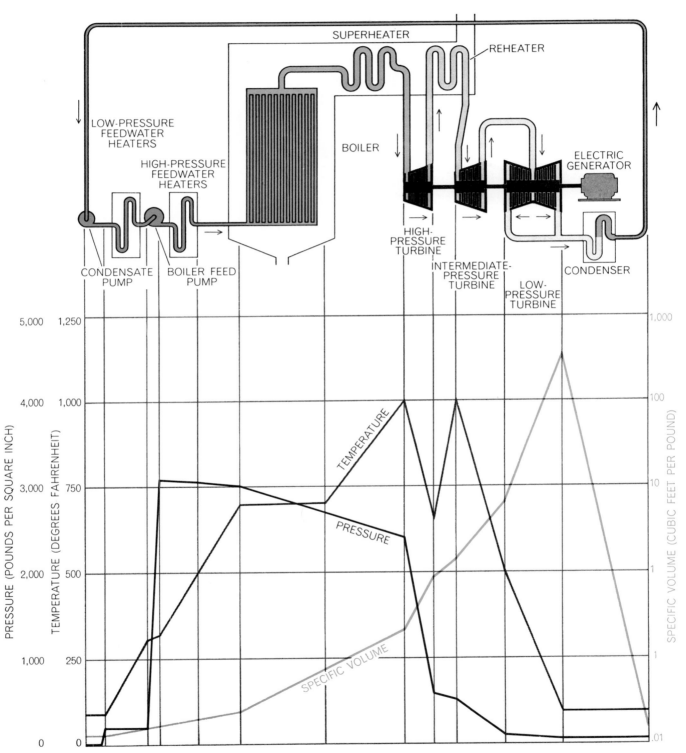

STEAM TEMPERATURE, PRESSURE AND VOLUME are the critical factors in the design of a steam power plant. The curves show how these three factors vary throughout a typical system. Pressure is highest at the exit of the feedwater pump leading to the boiler. At the entrance to the high-pressure turbine the pressure has dropped somewhat to around 2,400 pounds per square inch. Thereafter it falls rapidly as it passes through the turbine cascade. The steam temperature is raised to 1,000 degrees F. in the superheater and again in the reheater, finally plunging to about 80 degrees F. as it leaves the low-pressure turbine. The specific volume of the steam (*color*) varies over the greatest range and is therefore plotted on a logarithmic scale. At the inlet to the high-pressure turbine one pound of steam occupies about .3 cubic foot. When the steam leaves the turbine cascade, it occupies about 300 cubic feet.

The development of practical steam engines occupied many minds in the 18th and 19th centuries. The first successful steam engine had been introduced in 1698 by Thomas Savery. It was a crude affair in which steam was condensed alternately in two chambers, creating a vacuum that could be used to draw water from mines. In 1705 Thomas Newcomen built the first practical steam engine with a piston. In 1763 James Watt began making his contributions,

and in 1781 he was the first to patent methods for converting the reciprocating motion of a piston steam engine to rotary motion. With this advance the steam engine finally became a versatile prime mover. By 1804 Richard Trevithick had built the first steam-driven locomotive.

During the remainder of the 19th century piston steam engines were steadily improved. Many inventors, however, saw the advantages that would result if steam could be used directly to produce

rotary motion by means of some kind of turbine. Many devices were built in crude imitation of waterwheels. It remained for Parsons to recognize that what was needed was a device with many rows of buckets in which a small amount of the steam's kinetic energy would be extracted with high efficiency at each of many successive stages. Whereas the piston steam engine exploited only the pressure and temperature of steam as it came from a boiler,

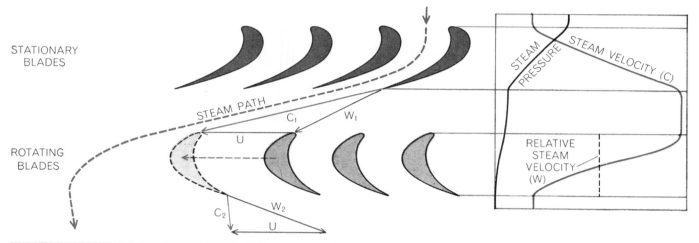

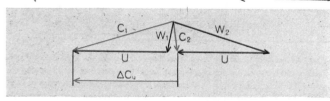

C₁, C₂   ABSOLUTE STEAM VELOCITY
W₁, W₂   RELATIVE STEAM VELOCITY
U   VELOCITY OF ROTATING BLADES

IMPULSE BLADING is one of two general methods for extracting kinetic energy from steam in a turbine. In the stationary blades steam is accelerated to a velocity (C_1) about twice that of the moving blades (U). Velocity is obtained at the expense of pressure (*curves at right*). The moving blades extract kinetic energy from the fast-moving steam, so that it leaves with essentially no tangential component of velocity (C_2). In passing through one row of fixed blades and one row of moving blades, called a stage, the amount of energy transferred to the rotor is proportional to the change in absolute steam velocity, ΔC_u (*diagram at left*).

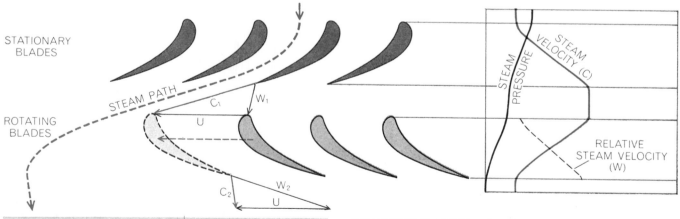

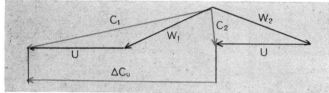

C₁, C₂   ABSOLUTE STEAM VELOCITY
W₁, W₂   RELATIVE STEAM VELOCITY
U   VELOCITY OF ROTATING BLADES

REACTION BLADING is the other concept used in steam turbine design. Here the pressure drop per stage is equally divided between fixed and moving blades (*curves at right*). In the fixed blades steam is accelerated to a velocity (C_1) only slightly greater than that of the moving blades (U). Continued expansion of the steam in the moving blades provides thrust and gives the steam a relative velocity (W_2) equal and opposite to its former absolute velocity (C_1). In reaction blading the energy, ΔC_u, transferred to the rotor in a single stage (*diagram at left*) is only about half that transferred by impulse blading. Efficiencies, however, are comparable.

the steam turbine used some of the pressure to create high-velocity jets, whose energy was then absorbed by the rotating blades.

Early in the turbine's history two concepts of blade arrangement were developed, each with its champions. Parsons favored what became known as reaction blading. Some of his competitors adopted impulse blading [see illustrations on opposite page]. In the reaction turbine the fixed blades and the moving blades that constitute one stage are practically identical in design and function; each accounts for about half of the pressure drop that is converted to kinetic energy in the entire stage. In the fixed blades the pressure is harnessed to increase the velocity of the steam so that it slightly exceeds the velocity of the moving blades in the direction of rotation. In the moving blades the pressure drop is again used to accelerate the steam but at the same time to turn it around (with respect to the blades), so that its absolute tangential velocity is almost zero as it enters the next bank of stationary blades. Thus thrust is imparted to the moving blades as the steam's absolute tangential velocity is reduced from slightly above blade speed to approximately zero. An imaginary observer moving with the steam could not tell whether he was passing through the fixed blades or the moving ones. As he approached either type of blade it would appear to be nearly motionless, but as he traveled in the channel between blades his velocity would increase steadily until he reached their trailing edges, which would then seem to be receding rapidly.

In the impulse turbine the fixed blades are quite different in shape from the moving ones because their job is to accelerate the steam until its velocity in the direction of rotation is about twice that of the moving blades. The moving blades are designed to absorb this impulse and to transfer it to the rotor in the form of kinetic energy. In this arrangement most of the pressure drop in each complete stage takes place in the fixed blades; the pressure drop through the moving blades is only sufficient to maintain the forward flow of steam. The amount of energy transferred to the rotor in each stage is proportional to the change in absolute steam velocity in the direction of rotation. This is the value labeled ΔC_u in the illustrations on the opposite page. It turns out that the value is about twice as high for impulse blading as it is for reaction blading. This means, in turn, that an impulse turbine will need fewer stages for the same pow-

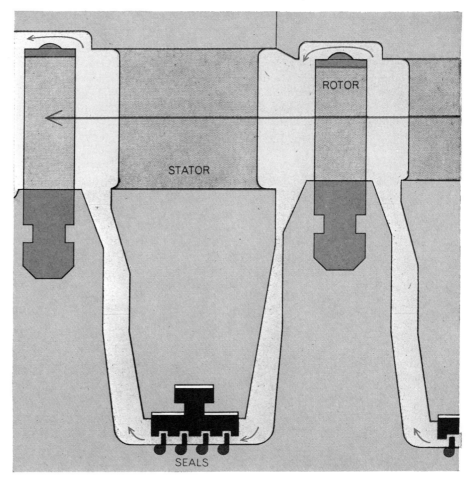

DIAPHRAGM TURBINE provides effective seals (*black*) in turbines with impulse blading, where the complete expansion of steam per stage occurs in the stator, or fixed, blades. This design allows seals to be placed on as small a diameter as possible. The moving blades are usually covered by a circumferential shroud, which may also carry seals (*not shown*).

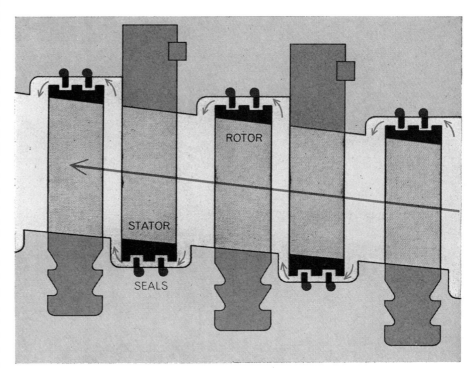

DRUM TURBINE offers the simplest arrangement of seals (*black*) when reaction blading is used. Seals on rotor and stator blades can be identical because the pressure drop across them is equal. Also the seals can be simpler than those needed for impulse blading.

er output than a reaction turbine; the efficiency, however, will be about the same for both types.

This being the case, one would expect impulse blading to have carried the day. Not so. As often happens in engineering, a design that seems clearly superior can present secondary problems of such magnitude that the choice between the alternatives becomes very nearly equal. In turbine design one of the major secondary problems is providing seals to keep the steam from leaking through the narrow spaces between the rotor and the stator. In impulse blading the complete expansion in each stage takes place in the fixed blades. It is thus desirable to place the seals on as small a diameter as possible. This has led to a turbine design known as the diaphragm type [see top illustration on preceding page]. Because the pressure differential is large the diaphragm needs considerable space in the axial direction. Therefore the width of the fixed blade must be made larger than it would otherwise have to be. A circumferential shroud is often placed around each ring of moving blades.

In reaction blading the pressure drop per stage is less than it is in impulse blading; moreover, it is divided equally between fixed and moving blades. Thus both blades can be fitted with similar seals, and the seals need not be as effective as those needed on the fixed blades in impulse blading. The result is a drum turbine [see bottom illustration on preceding page]. Another advantage of the reaction turbine is that the stationary and moving blades in each stage can have the same shape, which simplifies design and yields manufacturing economies.

For more than 50 years these two kinds of turbine, the diaphragm turbine

and the drum turbine, have been in competition without either type's demonstrating a distinctive advantage. Along the way the advocates of the two designs have moved somewhat away from pure reaction or pure impulse blading to adopt various compromise arrangements.

The efficiency that can be attained at each stage in a large modern turbine is quite remarkable: more than 90 percent. For large units with reheat systems an overall turbine efficiency of about 88 percent can be achieved. This is not the net thermal efficiency of the steam turbine as a heat engine, however. For this calculation one must introduce the limitations on the theoretical efficiency of heat engines first described in 1824 by Nicolas Léonard Sadi Carnot, who recognized that heat does work only as it passes from a higher temperature to a lower one. Specifically the efficiency is 1 minus the fraction "final temperature"/"initial temperature," both values being expressed on the Kelvin, or absolute, scale. If the working fluid is initially at 1,000 degrees F. (811 degrees K.) and is finally at 100 degrees F. (311 degrees K.), the maximum theoretical efficiency is about 60 per cent [(1 − 311/811) × 100]. Because the inherent properties of a steam cycle do not allow the heat input to take place at a constant upper temperature, the theoretical maximum efficiency is not 60 percent but closer to 53 percent. Thus the actual thermal efficiencies achieved with large reheat-steam turbines is .88 (the turbine efficiency) times 53, or somewhat better than 46 percent.

Today the chief point of competition among rival turbine manufacturers is not efficiency, since all guarantee comparable performance, but the capital cost of the unit. As units have become steadily

larger the cost per megawatt of capacity has dropped substantially because of economies inherent in size. For example, a turbine of 125 megawatts, which was considered large 15 years ago, weighed about 2.64 tons per megawatt. The 1,300-megawatt turbine Brown, Boveri is building for the TVA will weigh less than half that amount per megawatt. The cost per megawatt is roughly proportional to weight.

In the first 50 years of the steam turbine's history (roughly 1890 to 1940) turbine design was guided mainly by the intuition and ingenuity of engineers who were never far from the shop floor. Today, with the sharp increase in power density and the use of steam at higher pressures and temperatures, turbine progress depends increasingly on scientific understanding and the skillful application of new problem-solving tools such as the electronic computer. The design of a modern turbine requires the solution of difficult problems in aerodynamics, applied mathematics, metallurgy, stress, vibration and the physical behavior of steam, together with attention to many manufacturing problems (such as the production of complex blade contours at reasonable cost, the setting of permissible tolerances and size limitations imposed by transportation).

Rather than discuss such problems as they apply to the turbine as a whole I shall speak only of the design of the moving blades used in the last stage at the low-pressure end of the turbine. These are the longest blades in the turbine, hence the blades with the highest tip speed and the most acute vibrational problems. They are also the blades most subject to erosion from water droplets, which tend to appear as the steam approaches the turbine exit just before entering the condenser.

In designing such blades aerodynamic calculations reach a complexity exceeding that met in other sections of the turbine, where pressures are higher and flow patterns somewhat simpler. In passing through the low-pressure turbine the volume of steam expands about 100 times, with the result that the path of the steam has a strong radial component in addition to its forward component. The calculations must therefore deal with steam flow in three dimensions. An equilibrium condition must be maintained between the centrifugal forces that throw the steam outward and the distribution of radial pressure. In a very poor design the moving blades will tend to act as a compressor. The solution to this complicated flow problem involves

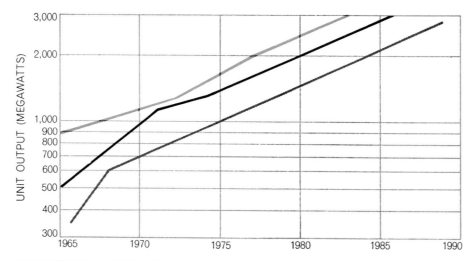

PROJECTION OF TURBINE SIZES, made by Brown, Boveri, reflects the fact that U.S. nuclear power plants (*top curve*) run larger than U.S. fossil-fuel plants (*middle curve*) and that both are larger than fossil-fuel plants contemplated in Britain (*bottom curve*).

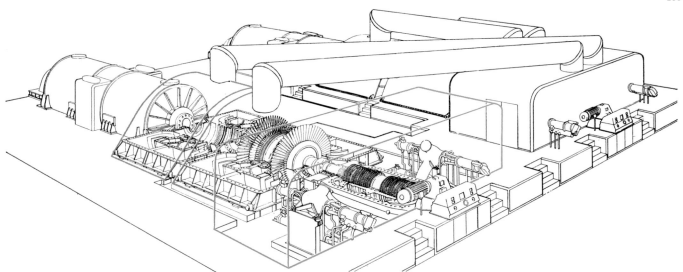

A 1,300-MEGAWATT TURBOSET, now being built by Brown, Boveri, will be the largest unit in operation when it is delivered to the TVA in 1971. It is described as a cross-compound unit, in-dicating that the steam travels sequentially through turbines of dif-ferent sizes located on two different shafts (*see flow scheme below*). The turboset will run on steam produced by burning of fossil fuel.

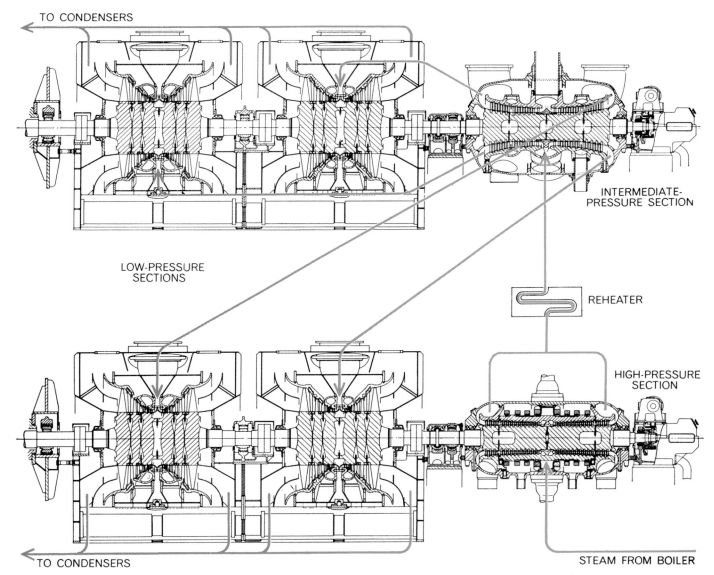

FLOW PLAN OF TVA TURBOSET shows how the steam is con-ducted through a double-flow high-pressure turbine, a double-flow intermediate-pressure turbine and finally four double-flow low-pressure turbines, all mounted on two shafts turning at 3,600 revo-lutions per minute. Steam enters the turboset at 3,500 pounds per square inch and 1,000 degrees F. Before passing to the intermediate-pressure turbines the steam is reheated to the same temperature. The flow of steam is 4,000 tons per hour, or 1.1 tons per second.

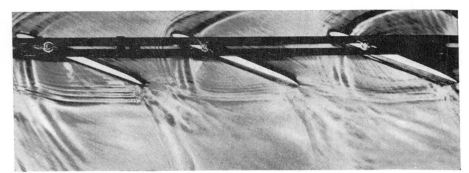

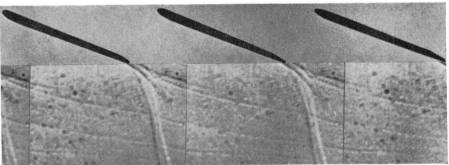

TURBINE-BLADE CHARACTERISTICS can be studied in three general ways: by water-tank analogy (*top*), by direct calculation (*middle*) and by airflow tests in a rotating cascade (*bottom*). It can be seen that the three methods give comparable results. These particular studies show patterns of transonic flow when blades are traveling at 1.4 times the speed of sound in air. In the middle picture the area tinted in color denotes supersonic flow.

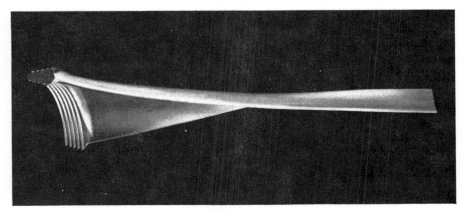

TYPICAL LAST-STAGE BLADE is 95 centimeters (37.5 inches) long, excluding the root. When installed and running, the tip of the blade will travel at 600 meters per second, or 1.6 times the speed of sound in steam under the conditions existing in the turbine's last stage.

differential equations that can be solved only with a large-capacity computer.

The results of these initial calculations provide the designer with a preliminary concept of the blade cascade, indicating the number of stages and the inlet and outlet angles of the blades in each stage. The next step is to design blades with high efficiency from root to tip. Toward this end the designer chooses profiles whose characteristics are known either from calculation or actual measurement. Even near the root, where the steam flow is subsonic, existing blade profiles may be unsatisfactory; new profiles must then be developed and evaluated. In the outer third of the blade, where the flow passes from subsonic to supersonic, things become more difficult. In calculating the line where sonic velocity is reached, mathematical equations tend to become unstable, meaning that they fail to provide a reasonable result. One must then use a combination of empirical and analytical methods. The critical passage from subsonic to supersonic velocity, known as the sonic line, must be verified by experiments in a water tank or in a rotating cascade in which the fluid is air.

In the supersonic region calculation becomes simpler again. Good methods exist for predicting the location of regions, called Mach lines, where minor disturbances will cause sudden but small changes in steam velocity and pressure. One can also compute shock lines: the lines along which high-compression waves will develop. Water-tank methods, in which water passes swiftly between model blades, provide an excellent visual representation of flow patterns and shock waves [see top illustration on this page]. Under suitable conditions flowing water behaves like a supersonic flow of air or steam. From such studies, with the help of scaling laws, one can readily determine the supersonic flow pattern as well as the sonic line. The results can be checked by passing air through a rotating cascade of blades and using optical methods to disclose the flow patterns. A critical comparison of the three methods—mathematical analysis, water-tank flow and airflow in a cascade—enables the designer to make a sound prediction of actual steam-flow conditions. In this way the aerodynamic side of the problem can be solved.

The designer is now ready to sketch a preliminary configuration for the blades in the last stage of the low-pressure turbine. Because they are highly twisted these blades are subject to unusual stresses. Under centrifugal force a last-stage blade may untwist more than seven angular degrees. To keep material stresses under control the positioning of the centers of gravity in various sections of the blade and also of the centers of inertia is very important.

Not the least of the engineer's problems is to design a root that will hold the blade in the rotor against a centrifugal force of more than 250 tons. The space available for transmitting this force to the rotor body is very limited. Tiny strain gauges and polarized-light studies of transparent plastic models provide information about stress distribution, and sample roots are stressed to destruction. Such tests can shed light on the rupture mechanism and indicate the safety factor expected when the blade is run at normal speed.

Vibration can be particularly troublesome in long blades. The failure of blades in the turbines of the *Queen Elizabeth 2* has been traced to vibrations at resonance frequencies, attributable to faulty design. In designing the last-stage blades one must calculate resonance frequencies and where the nodal points will occur. The calculation method accounts for centrifugal and torsional frequencies as well as for their coupled effect. The calculated values must then be verified by static and rotational tests; close agreement is usually obtained between calculation and experiment [*see illustration on next page*].

Even these tests are not sufficient. The distribution of steam pressure and velocity acting on the blade may vary considerably with the load, so that additional tests must be made. This is usually done in a test turbine scaled down from a complete low-pressure turbine. One can study how the last-stage blading is influenced by stages preceding the last stage under varying load conditions and also examine the back effects produced by the exit diffuser. Reliability of the last-stage blade is so important that further tests are run on turbines in commercial operation. Tiny strain gauges are attached to the blade, and their measurements are transmitted by radio.

In addition to determining resonance frequencies up to the sixth or seventh mode of vibration, it is necessary to check carefully still higher modes in a range near the exciting frequencies produced by the stationary blades. There are evidently flow disturbances, similar to wakes, behind each stationary blade that are capable of inducing resonances characteristic of a vibrating plate. Fortunately this phenomenon is nearly independent of centrifugal effects, so that stationary tests are sufficient. Laser holography has recently been used to re-

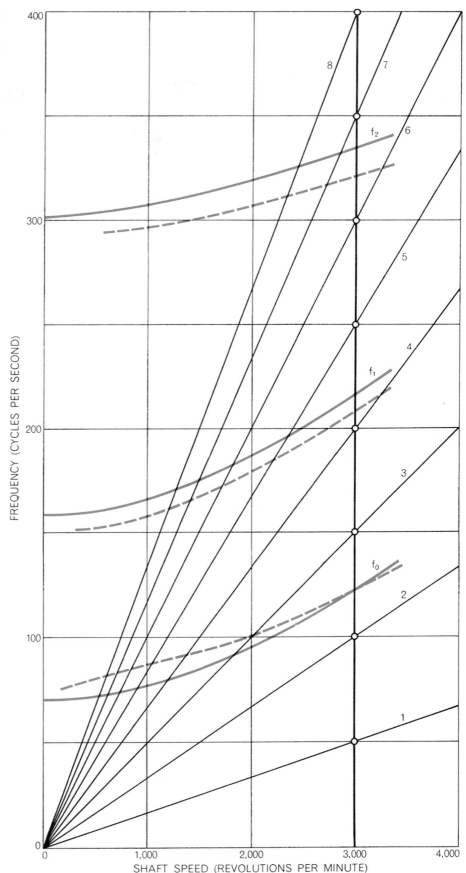

RESONANCE FREQUENCIES of last-stage blades can now be calculated with good accuracy. The slanting lines (*1 through 8*) are potential exciting frequencies. Conditions to be avoided (*open dots*) occur where these lines intersect the vertical line at the normal shaft speed, here 3,000 revolutions per minute. The solid lines in color (f_0, f_1, f_2) are resonant frequencies of the blade as obtained by calculation. The broken lines are measured values.

cord the vibration pattern in such tests.

The metallurgical requirements for a last-stage blade are, as one might expect, demanding. The steel should have a high yield point combined with good ductility in order to withstand high centrifugal stresses. It should resist fatigue as well as erosion and chemical- and stress-corrosion. (In normal operation one can expect small quantities of corrosive chemicals to be entrained occasionally in the steam.) The steel should also have high damping characteristics to minimize vibration. Finally, the steel should lend itself to machining or forging; more exotic means of shaping a blade are usually very costly.

If the steel itself does not possess sufficient resistance to erosion, the blade can be hardened or surfaced with an erosion-resistant hard alloy such as Stellite. Much study has been devoted to the mechanism by which wet steam causes erosion. Because the steam passing through the low-pressure turbine is wet, drops of water tend to collect on the trailing edge of the stationary blades. The drops are then torn away by the passing steam and hurled against the moving blades, producing tiny pits. The designer's task is to see that pitting does not impair the efficiency and reliability of last-stage blades over the 20-year minimum lifetime expected of a steam turbine. Blades in the high-pressure turbine, where the temperatures are highest, are virtually free from erosion.

I have described in some detail the development of last-stage blades not only because their design involves a representative range of engineering problems but also because they place one of the limits on the size to which individual turbine sets can grow. For maximum economy high-pressure, intermediate-pressure and low-pressure turbines should be arranged in a single line. The more steam per hour that enters the high-pressure end of the set, the larger the blades required at the low-pressure end to handle the much expanded volume of steam. "Half-speed" turbines (1,800 revolutions per minute instead of 3,600) have been introduced to provide large output in a single turbine set, but their weight and cost, megawatt for megawatt, are higher than those for full-speed turbines. To hold costs in line, therefore, it is imperative for manufacturers to learn to design full-speed turbines with larger low-pressure last stages than any built so far. These larger units will be needed to meet the growing demand for electric power at a steadily reduced cost per kilowatt-hour.

The Fastest Computer

by D. L. Slotnick
February 1971

*ILLIAC IV is made up of 64 independent processing
units that by operating simultaneously will be capable
of solving complex problems in a fraction of the time
needed by any other machine*

The computer ILLIAC IV, which is now nearing completion, is the fourth generation in a line of advanced machines that have been conceived and developed at the University of Illinois. ILLIAC I, a vacuum-tube machine completed in 1952, could perform 11,000 arithmetical operations per second. ILLIAC II, a transistor-and-diode computer completed in 1963, could perform 500,000 operations per second. ILLIAC III, which became operational in 1966, is a special-purpose computer designed for automatic scanning of large quantities of visual data. Since it processes nonarithmetical data it cannot be compared with the earlier ILLIAC's in terms of operational speed. ILLIAC IV, employing the latest semiconductor technology, is actually a battery of 64 "slave" computers, capable of executing between 100 million and 200 million instructions per second. Even that basic rate, although it is faster than that of any other computer yet built, does not express the true capacity of ILLIAC IV.

Unlike its three predecessors and all computers now on the market, which solve problems by a series of sequential steps, ILLIAC IV is designed to perform as many as 64 computations simultaneously. For such a computing structure to be utilized efficiently the problem must be amenable to parallel, rather than sequential, processing. In actuality problems of this kind constitute a considerable part of the total computational spectrum, ranging from payroll calculations to linear programming to models of the general circulation of the atmosphere for use in weather prediction. For example, a typical linear-programming problem that might occupy a large present-generation computer for six to eight hours should be solvable by ILLIAC IV in less than two minutes—a time reduction of at least 200 to one.

Subsystems for ILLIAC IV are being manufactured in a number of plants and are being shipped to the Burroughs Corporation in Paoli, Pa., for final assembly and testing. When the machine is finished a few months from now, it will be available over high-speed telephone lines to a variety of users, including the Center for Advanced Computation of the University of Illinois.

The ultimate limitation on the operating speed of a computer designed to operate sequentially [*see illustration on page 175*] is the speed with which a signal can be propagated through an electrical conductor. In practice this is somewhat less than the speed of light, which takes one nanosecond (10^{-9} second) to travel about one foot. Although integrated circuits containing transistors packed together with a density ranging from several hundred to several thousand per square inch have helped greatly to reduce the length of interconnections inside computers, designers have been increasingly aware that new kinds of logical organization are needed to penetrate the barrier set by the speed of light.

Over the past 10 years designers have introduced a number of variations on the strictly sequential mode of operation. One stratagem has been to overlap the operation of the central processing unit and the operation of input-output devices (such as magnetic-tape readers and printers). By means of a fine-grained separation of the computer's functional units a high degree of overlapping has been attained. Current efforts in "pipelining" the processing of "operands" will allow a further significant increase in speed.

Overlapping and pipelining, however, are both fundamentally limited in the advances in speed they can provide. The approach taken in ILLIAC IV surmounts fundamental limitations in ultimate computer speed by allowing—at least in principle—an unlimited number of computational events to take place simultaneously. The logical design of ILLIAC IV is patterned after that of the SOLOMON computers, prototypes of which were built by the Westinghouse Electric Corporation in the early 1960's. In this design a single master control unit sends instructions to a sizable number of independent processing elements and transmits addresses to individual memory units associated with these processing elements ("processing-element memories"). Thus, while a single sequence of instructions (the program) still does the controlling, it controls a number of processing elements that execute the same instruction simultaneously on data that can be, and usually are, different in the memory of each processing element [*see top illustration on page 176*].

Each of the 64 processing elements of ILLIAC IV is a powerful computing unit in its own right. It can perform a wide range of arithmetical operations on numbers that are 64 binary digits (bits) long, where a digit is either 0 or 1, corresponding to the two "positions" of an electronic device with two stable states. These numbers can be in any one of six possible formats; the number can be processed as a single number 64 bits long with either a fixed or a "floating" point (corresponding to the decimal point in decimal notation), or the 64 bits can be broken up into smaller numbers of equal length. Each of the memory units has a capacity of 2,048 64-bit numbers. The time required to extract a number from memory (the access time) is 188 nanoseconds, but because additional logical circuitry is needed to resolve conflicts when two or more sections of ILLIAC IV call on memory simultaneously, the minimum time between successive operations of memory is increased to 350 nanoseconds.

Each processing element has more than 100,000 distinct electronic components assembled into some 12,000 switching circuits. A processing element together with its memory unit and associated logic is called a processing unit [see illustrations on pages 172 and 173]. In a system containing more than six million components one can expect a component or a connection to fail once every few hours. For this reason much attention has been devoted to testing and diagnostic procedures. Each of the 64 processing units will be subjected regularly to an extensive library of automatic tests. If a unit should fail one of these tests, it can be quickly unplugged and replaced by a spare, with only a brief loss of operating time. When the defective unit has been taken out of service, the precise cause of the failure will be determined by a separate diagnostic computer [see top illustration on page 172]. Once the fault has been found and repaired the unit will be returned to the inventory of spares.

ILLIAC IV could not have been designed at all without much help from other computers. Two medium-sized Burroughs B 5500 computers worked almost full time for two years preparing the artwork for the system's printed circuit boards and developing diagnostic and testing programs for the system's logic and hardware. These formidable design, programming and operating efforts were under the direction of Arthur B. Carroll, who during this period was the project's deputy principal investigator.

In the course of a calculation it is frequently necessary to transfer data from one processing element to another; data paths are provided for this purpose [see bottom illustration on page 177]. In solving certain problems these data paths can be used to simulate directly the problem's geometric structure.

Although the 64 processing elements are under centralized control, only the simplest problems could be handled if the elements did not have some degree of individual control. Such control is provided by means of a "mode value," which can be set by each processing element and which depends on the different data values unique to each element. The program sets the mode value that identifies those processing elements whose state (as defined by their mode value) enables them to respond to a given instruction or sequence of instructions. The elements not in this state are turned off. As a simple example, suppose at the start of a problem all mode values are set to 1,

or "on." Now the program causes the control unit to "broadcast" to all 64 processing elements: Search your memory for X (some particular value). Each element carries out the search, and any element finding the value X sets its mode value to 0, or "off." The control unit may now issue a sequence of instructions to be performed only by those elements whose mode value is still 1, which allows them to keep operating. Similarly, the contents of two registers within a processing element can be compared, and the mode value can be set on the outcome of the comparison. Mode values are also used to determine when an iterative calculation should be terminated or when quantities have exceeded specified numerical limits. In short, mode values are the principal means for imposing a data-dependent, logical structure on a program.

In addition to the high-speed primary memories associated with each processing element, ILLIAC IV has two memories that are somewhat slower but have capacities that are considerably larger. The total capacity of the 64 primary memories is $64 \times 2,048$, or 131,072, numbers, each 64 bits in length. Thus the total high-speed storage is some 8.4 million bits. Most of the problems suitable for ILLIAC IV will require data capacities far exceeding this primary storage.

The additional data can be held either in a rotating-disk magnetic memory or in a new "archival" memory whose writing mechanism is a laser beam. The rotating-disk memory has a capacity of a billion bits, or about 120 times the capacity of the primary memory. The disk has 128 tracks, each with its own reading and recording head. The access time is determined by the time required for the disk to rotate into the position where the desired datum is under one of the fixed heads. Since the disk revolves once in 40 milliseconds, the average access time is 20 milliseconds, which is about 100,000 times slower than the access time of the primary memory. Once the disk is in position, however, data can be transferred to any of the 64 primary memories at the rate of half a billion bits per second, or roughly 100 times the rate at which data can be transmitted over a standard television channel. The archival memory, which has a capacity of a trillion bits, has a longer access time and a lower data-transfer rate [see top illustration on page 177].

These memory subsystems plus the more conventional peripheral equipment (punched cards, disk and tape units, printers, displays and so on) are under the direction of a medium-size general-purpose computer, the Burroughs B 6500

see bottom illustration on page 176]. This computer also bears the major responsibility for translating programs from the various programming languages available to the users into the detailed, hardware-determined language of the computer itself.

Let us now examine how ILLIAC IV can be used to solve a simplified problem in mathematical physics. The problem belongs to the very large class of problems whose calculation can be performed in an "all at once" manner, using either ordinary or partial differential equations. The problem we shall trace requires the solution of Laplace's partial differential equation describing the distribution of temperature on the surface of a slab. Even the reader who is unfamiliar with such equations should be able to follow this example because the method for reaching a solution relies completely on the commonsense notion that the temperature at any point on the slab tends to become the average of the

COMPUTER ILLIAC IV is nearing completion at the Great Valley Laboratories of the Burroughs Corporation in Paoli, Pa. Unlike conventional computers, which carry out logical and arithmetical operations in

surrounding values.

Laplace's equation for solving the problem is $\delta^2 U/\delta x^2 + \delta^2 U/\delta y^2 = 0$, where U corresponds to the temperature at a given position specified by the coordinates x and y on the surface of the slab. In this example we are asked to imagine that we are dealing with a rectangular slab of some material whose four edges are maintained·at different temperatures. Eventually all the points on the surface of the slab will reach a steady-state temperature distribution reflecting the way heat flows from hotter edges to the cooler ones. The temperatures at the edges of the slab, which are held constant, are called the boundary conditions. If we use an x-y coordinate system to designate the location of any point on the surface of the slab, we can say that the temperature at any point is a function of x and y. In other words, every point x,y on the slab has associated with it a temperature $U(x,y)$.

When one uses a digital computer to solve this problem, one cannot, of course, obtain the temperature at an infinite number of points. The standard procedure is to digitize the variables x and y so that the slab is covered by a mesh, each square of the mesh being h units on a side. For the sake of simplicity we shall assume that our slab is a square and that it has been digitized into 64 x,y values or mesh points [*see illustration on pages 178 and 179*].

The method of solution can now be stated very simply: The temperature at any interior mesh point is the average of the temperatures of the four closest mesh points. Thus the value of $U(x,y)$ equals the sum of four neighboring values of $U(x,y)$ divided by four. When this equation is made true for all points, there can be only one correct value for each point. This method is called "relaxation."

When the relaxation method is applied with a sequential, or conventional, computer, the usual procedure is to start at the top left of the slab and apply the basic equation at each interior point moving from left to right along each row of points and proceeding downward row by row. Since the 28 boundary points in our example are already specified, the equation would have to be applied 36 times (64 minus 28) to produce one relaxation of the relaxation method. As succeeding relaxations are performed on the set of mesh points the values of the temperatures converge to the exact solution. When values for two successive relaxations are very close to each other (within a specified error tolerance), one stops the process and says that the steady-state solution has been reached.

Let us now consider how this same problem could be solved by parallel processing on ILLIAC IV. If one stored each value of U in a separate processing element, all 36 inner values could be calculated simultaneously. A program could be written to compute new values for $U(x,y)$ not from top left to bottom right but all at once. When the first set of relaxation values for all 36 inner points has been obtained by simultaneous calculation, these values are available for the

strict sequence, ILLIAC IV will solve complex problems in an all-at-once manner by coordinating the simultaneous operation of 64 "slave" computers, or independent processing units. ILLIAC IV was conceived and developed at the University of Illinois Center for Advanced Computation.

OPEN DOORS OF ILLIAC IV reveal vertical cases holding eight of the big machine's 64 independent but centrally controlled processing units. The 12 drawers at the top of the picture hold the power-supply modules associated with the eight processing units. A group of four processing units lies behind each of the 16 bottom doors in the photograph at the left.

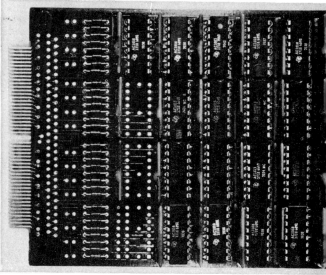

BACK-PLANE ASSEMBLY (*far left*) of one of ILLIAC IV's 64 processing elements contains up to 210 printed circuit boards arranged in six rows of 35 columns. Each circuit board (*second from left*) holds up to 20 "dual-in-line" packages (four rows by five columns) as well as some other electronic components such as resistors. Each dual-in-line package (*third from left*) contains 16 pins, which

second relaxation.

Not only are the two algorithms, or mathematical routines, different for sequential and parallel computation but also the way the temperatures converge is different [*see illustration on pages 180 and 181*]. In the sequential method the temperatures at bottom right converge faster to the exact solution than those at top left. This happens because in sweeping from top left to bottom right the last computations in each relaxation sequence contain more new data than the computations made at the start of the sequence.

When the parallel algorithm is used, the values closest to the edges converge faster than those in the center of the mesh. The reason is that the outer values are closest to the boundary values, and at each iteration they have more new data

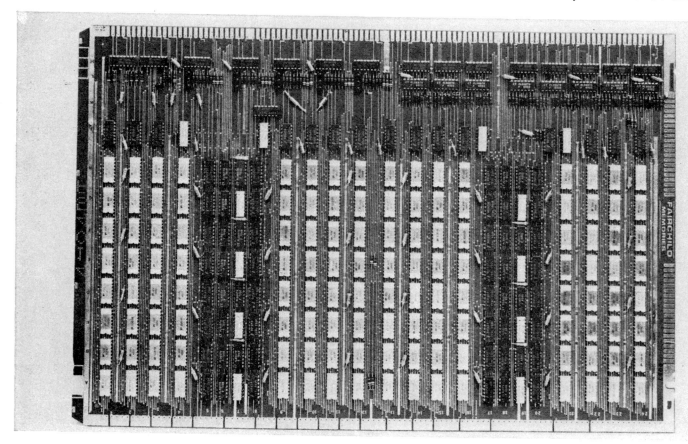

MEMORY ARRAY BOARD (*left*) is one of four that together constitute the high-speed, 131,072-bit memory associated with each of the 64 processing elements in ILLIAC IV. Each board holds up to 128 dual-in-line packages. Each package (*middle*) holds one chip and

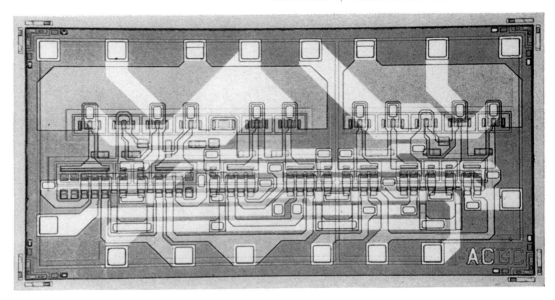

connect to an integrated circuit built up on a single chip of silicon measuring .095 by .05 inch. The integrated circuit, magnified 55 diameters at the far right, contains 34 transistors organized into seven "logic gates." The circuit chips are manufactured by Texas Instruments Incorporated. In all more than a quarter of a million chips will be used in ILLIAC IV's 64 processing elements.

available than the inner values. The convergence process can be likened to freezing. The sequential algorithm begins freezing at bottom right and proceeds to top left; the parallel algorithm begins freezing around the edges and proceeds toward the center.

The time saved by using the parallel algorithm rather than the sequential one depends on the number of iterations needed to produce convergence. If both algorithms require the same number of iterations and both compute the same number of interior values, P (or 36 in the case of the example above), the parallel process is faster by a factor of P. Since the parallel process uses less new information for each iteration, however, it will normally take more parallel iterations to produce the same degree of accuracy as a sequential calculation. Inas

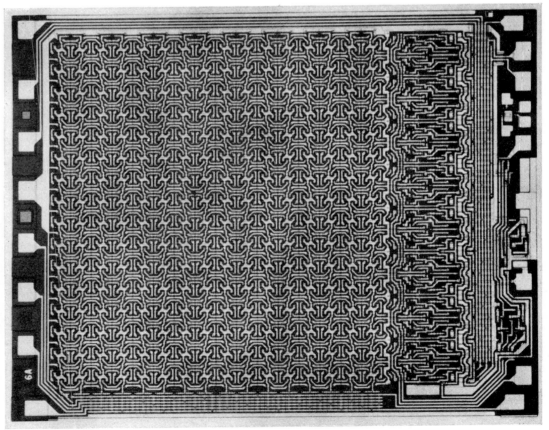

each chip contains integrated semiconductor circuits (right) with a storage capacity of 256 bits. The chips, each containing 2,485 transistors, resistors and diodes, were developed by the Semiconductor Division of Fairchild Camera and Instrument Corporation.

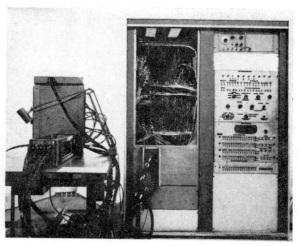

DIAGNOSTIC COMPUTERS, called exercisers, are housed in the cabinets at the right in each photograph. When one of ILLIAC IV's processors or memory units fails, it is immediately unplugged and replaced with a spare unit. The exact cause of the failure is then determined by a diagnostic computer. A defective processor has been unplugged and rolled over to the diagnostic computer in the photograph at the left; a defective memory unit is being examined by a different diagnostic computer in the photograph at the right.

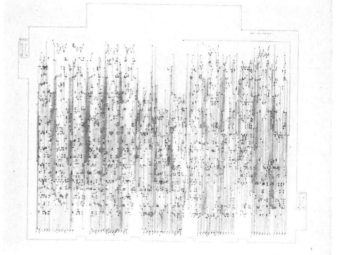

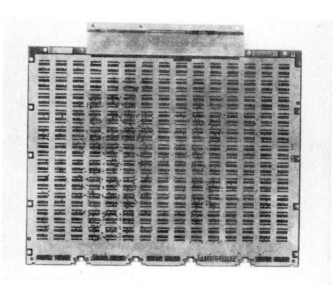

CONTROL-UNIT CARD (top left) is laminated from 12 separate layers that embody the complex wiring pattern for interconnecting several thousand electronic components. Three of the glass photographic positives of wiring patterns and etched copper wiring layers are shown in the other photographs. ILLIAC IV requires 64 control-unit cards, each of which can be removed for test or replacement.

much as ILLIAC IV has 64 "channels" available for processing iterations in parallel, it is up to 64 times faster than a sequential computer of comparable speed. This advantage in overall speed far outweighs the few extra iterations necessary to obtain a solution equal in accuracy to that produced by sequential processing.

The reader may well ask at this point: "To what purpose can a computer of this large size be applied, and, indeed, is it necessary?" Or more pointedly he may ask: "Is it worth $30 million of public funds in these days of so many identified, competing needs?" Each of us must determine the answer for himself after examining the potential value of the machine. Let us, therefore, look at some of the applications.

Among the intended tasks for ILLIAC IV is linear programming, a mathematical technique for allocating the use of limited resources to maximize or minimize a specified objective. The resource limitations (constraints) are expressed as linear inequalities in which the variables are quantities of resources. The objective is specified as a linear function of these variables. Typical linear-programming problems presented to computers involve hundreds or even thousands of variables. Examples include routing deliveries to minimize distance traveled or to maximize deliveries per trip; blending, mixing, cutting or trimming raw materials to minimize waste or to maximize output value; selecting production methods that minimize cost or maximize output; scheduling production facilities to minimize delay or to maximize throughput. ILLIAC IV will be able to solve in reasonable time much bigger problems than have been attempted in the past. This capability flows from the use of parallel computation and the high data-transfer rate of the ILLIAC IV memory disk. The linear-programming problem mentioned above that ILLIAC IV should be able to solve in less than two minutes (but which would take six to eight hours on a present-generation computer) is one that has 4,000 constraints and 10,000 variables.

A problem of this order of difficulty is now under active study at the University of Illinois under the direction of Ian W. Marceau. It involves optimizing the output of the agricultural sector of the economy, ranging in size from a large region to an entire nation. The desired objectives will reflect national policy, and they will range from production of enough food to nourish a given population to producing export crops so that a developing country can obtain

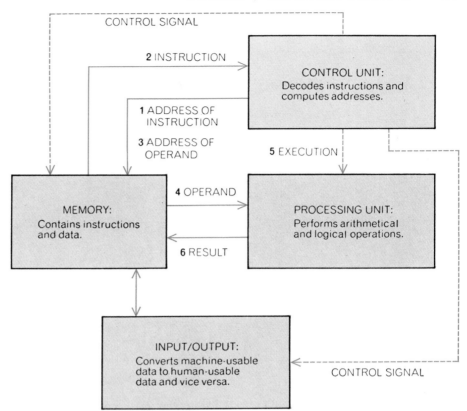

CONVENTIONAL COMPUTER is organized to carry out operations in sequence. A counter in the control unit determines the address of the next instruction in the sequence to be executed and transmits the address to the memory (1). The memory returns the instruction to the control unit (2). The instruction contains the address in the memory of the data (operand) on which an arithmetical or logical operation (also specified) is to be performed. This address is sent to the memory (3). The memory furnishes the selected operand to the processing unit (4). The control unit then transmits to the processor a sequence of electronic signals that contains the fine structure of the arithmetical or logical operation required by the program (5). The calculated result is then stored at a specified location in memory (6) for use in a subsequent operation or for conversion to printed form for the user of the machine. Advanced computers carry out this entire sequence in a few millionths of a second. Billions of repetitions may be needed to solve a complex problem.

foreign-exchange credits. The resources to be managed include land, labor, machinery, fertilizers, pesticides, herbicides, storage facilities and capital. As Marceau has demonstrated, linear-programming models of a region or nation can also recognize constraints involving social costs, for example the harm done by the intensive application of nitrogenous fertilizers, the use of certain pesticides (such as DDT) or cultivation practices with long-term deleterious effects on the productivity of the land.

It must be pointed out that in order to apply linear programming to an entire economic sector one must incur considerable expense in gathering the data to be used in the model. Here too, however, the computer can help by making experimental trials and estimating the accuracy with which various input data need to be known in order to secure answers with a given level of precision. It is also possible to simulate alternative policies on the computer and estimate their effects

on agricultural productivity. To test such policies directly "in vivo" can be very costly. There is no reason why a computer program should not be the white rat or guinea pig for a proposed cure to a social problem.

Another application contemplated for ILLIAC IV is the establishment of natural-resource inventories to be used by municipal and regional planners. A Natural Resource Information System is now being developed at the University of Illinois in cooperation with the Northeast Illinois Natural Resource Service Center. The Ford Foundation has provided funds for the initial research and development program. The system will contain a wide range of information on the natural resources of a selected area: geology, hydrology, forestry and vegetation, climate, topography, soil characteristics and current land use. Marengo Township of McHenry County in Illinois has been selected for a pilot study.

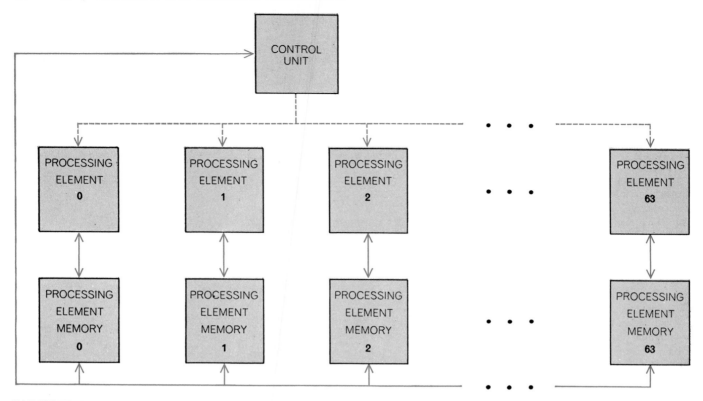

PARALLEL ORGANIZATION OF ILLIAC IV enables the control unit to orchestrate the operation of 64 processing elements, each with its own memory. There is a large class of mathematical problems that can be solved in an all-at-once manner by independent processors operating simultaneously, each about twice as fast as the single processor in an advanced sequential computer.

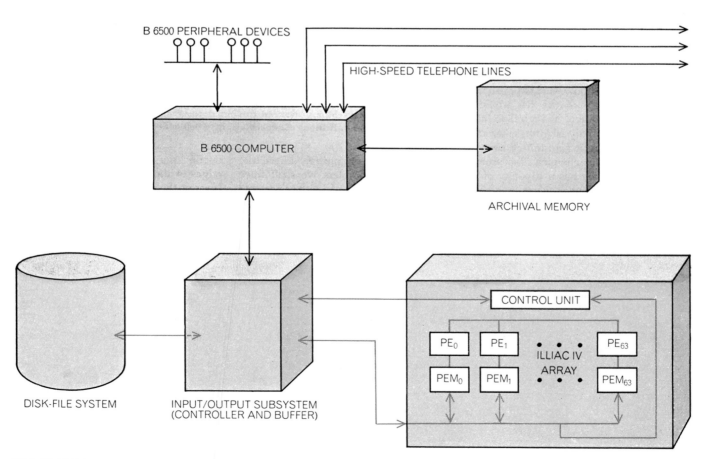

BLOCK DIAGRAM OF ILLIAC IV SYSTEM shows how the IL-LIAC's control unit, together with its 64 processors and primary memory units, will be connected to ancillary pieces of equipment. A secondary memory is provided by a disk-file system with a capacity of a billion bits (binary digits). A tertiary memory is provided by a new "archival" memory system, which uses a laser beam for reading and writing. Accessed through a medium-size Burroughs B 6500 computer, it will have storage for a trillion bits.

The system is being designed so that it can easily be used by any decision-maker (including an individual taxpayer) regardless of his technical or administrative training. For example, an individual may want to know whether or not he can have a housing subdivision (or a tennis court or a fishpond) on his land. On the other hand, county administrators may be looking for the best site for a new hospital. The search for a hospital site could be reformulated into a series of commands that could be presented to the computer. For instance, search all tracts that lie between town *A* and town *B* and that are within two miles of route *C;* the area should be no smaller than five acres and no larger than 25 acres with the following characteristics: (1) one acre of soil capable of supporting a five-story hospital, with a gradient of less than 8 percent and not subject to flooding; (2) at least four acres (for a parking lot) that can be covered with asphalt without disturbing the underground water table; (3) trees at least 20 years old. If no tracts satisfied all these requirements, one or more of the less important conditions could be relaxed until a site was located.

The output of the information system is being designed to meet three levels of need. The simplest level will consist of a concise inventory listing. The next level will be an interpretation of the computer's search in prose that should be clear to an educated layman. The third level will be a highly technical description suitable for use by a specialist, such as a geologist or an ecologist. The objective of the information system is to shorten the planning process and to improve the quality of decisions. Although the system will use existing techniques of information retrieval, ILLIAC IV, with its speed and archival memory, will be able to analyze the stored information to a far greater depth than would be possible with any earlier computer.

Our unaugmented intellectual resources have not been capable of producing satisfactory solutions to the types of large-scale planning problems just described. It is in fact evident that we are currently faced with socially debilitating aftermaths of piecemeal planning—and nonplanning—in both of these areas. A rational 20th- (or 21st-) century society will not emerge solely on the basis of universal goodwill.

A quite different assignment for ILLIAC IV is numerical weather prediction, which early computer theorists such as John von Neumann regarded as one of the important motivations for their work. Numerical techniques developed over

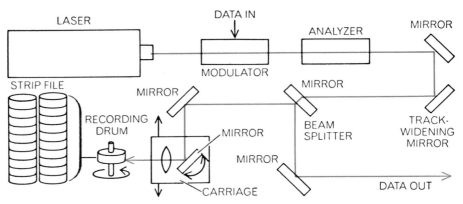

ARCHIVAL MEMORY is a new high-capacity secondary memory, developed by the Precision Instrument Company. The beam from an argon laser records binary data by burning microscopic holes in a thin film of metal coated on a strip of polyester sheet, which is carried by a rotating drum. Each data strip can store some 2.9 billion bits, the equivalent of 625 reels of standard magnetic tape in less than 1 percent of the volume. The "strip file" provides storage for 400 data strips containing more than a trillion bits. The time to locate data stored on any one of the 400 strips is about five seconds. Within the same strip data can be located in 200 milliseconds. The read-and-record rate is four million bits a second.

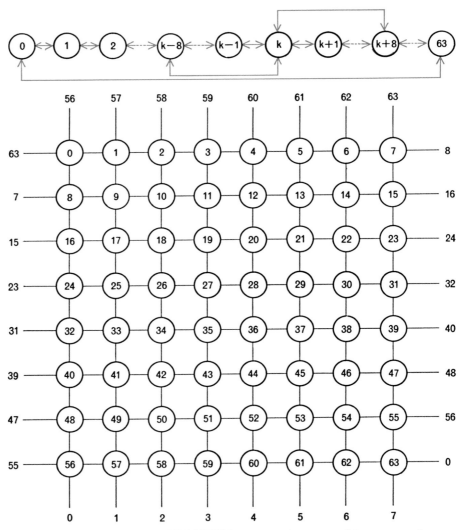

ARRAY OF 64 PROCESSING ELEMENTS in ILLIAC IV is connected in a pattern that can be regarded in either of two ways, which are topologically equivalent. The elements can be viewed as a linear string (*top*) with each processing element connected to its immediate neighbors and to neighbors spaced eight elements away. Equivalently, one can regard the processing elements as a square array (*bottom*) with each element connected to its four nearest neighbors. One can imagine the array rolled into a cylinder so that the processing elements in the top row connect directly to those in the bottom row. The last processing element in each row is connected to the first in the next row to produce a linear sequence.

the past two decades are now in daily use and yield good results for periods of from 24 to 48 hours. These techniques involve the simulation of complex atmospheric processes by a mathematical model that combines extensive knowledge of the relevant physical processes with sophisticated mathematics and advanced computer technology.

The physical basis for all numerical simulations of the atmosphere is the conservation of mass, momentum and energy. These conservation principles are embodied in sets of differential equations (Laplace's equation is an example of a differential equation describing heat distribution on a slab), which cannot be solved without a computer. The physical scales of atmospheric phenomena that are simulated on the computer range from the microphysical processes of clouds to the continental motions of frontal systems. At the upper end of the physical scale there are general-circulation models that describe the atmosphere as a heat engine driven by the sun.

The complexity of these models is illustrated by the operational model of the atmosphere used by the National Weather Service in its daily forecasts. The atmosphere over the Northern Hemisphere is represented by six horizontal slices ranging from sea level to the stratosphere. Each slice contains 3,000 points at which initial values of wind velocity, temperature and pressure are inserted. The computer then applies the appropriate equations to predict what the velocity, temperature and pressure will be in the future at 10-minute intervals. A 24-hour forecast requires about an hour of computing time on a computer that can execute 300,000 instructions per second, or more than a billion instructions in all.

If the distance between the grid points were to be halved, the number of grid points would be quadrupled and the computer time needed for a 24-hour forecast would be increased eightfold. In other words, a third of a day would be consumed merely in making a 24-hour prediction. If the model yields significantly better short-range predictions than the 3,000-point model now in use, there is a good chance that numerical forecasts can be extended to five days with an accuracy comparable to that of the 48-hour forecasts now being generated.

The actual computer techniques of weather forecasting can be advanced by testing them on ILLIAC IV. Until now investigators have been reluctant to experiment with a new predicting technique when it might involve many computer simulations, each of which could take up to 100 hours of computing time. When ILLIAC IV can reduce the running time from 100 hours to one hour, extensive experimentation will become feasible.

Mathematical models exist today for a large variety of physical systems and are in constant use as the basis for calculation aimed at prediction. Biological and biochemical systems have not been modeled with the same intensity of effort or success. There are a number of reasons for this. One can, for example,

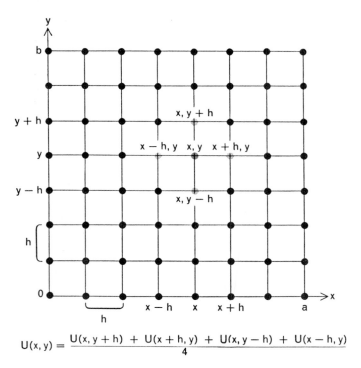

$$U(x, y) = \frac{U(x, y+h) + U(x+h, y) + U(x, y-h) + U(x-h, y)}{4}$$

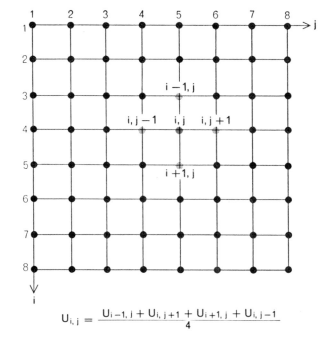

$$U_{i, j} = \frac{U_{i-1, j} + U_{i, j+1} + U_{i+1, j} + U_{i, j-1}}{4}$$

PROBLEM OF FINDING TEMPERATURES ON A SLAB must be prepared for a computer by digitizing the slab into an array of points, with some arbitrary mesh spacing (h). The purpose of this illustration and those on the next three pages is to compare how such a problem would be solved by sequential methods with a standard computer and by parallel methods using ILLIAC IV. In this problem one imagines that the edges of a slab are arbitrarily held at certain fixed temperatures. The computer is asked to calculate the temperature at a network of interior points after the slab has reached equilibrium. Two methods of identifying the points in the network are depicted above. The more familiar method at the left expresses each point in terms of x and y. The temperature U at any point x,y is the average of the temperatures at the four nearest mesh points. The equation specifies these points in terms of x and y and the mesh spacing h. In programming a computer it is more convenient to use the integers i and j as positional indicators, in which case the temperature equation is rewritten as shown at the right. The temperatures at 28 points on the perimeter of the slab are the known quantities supplied to the computer (see *illustration on opposite page*). The temperatures at the 36 interior points are the unknowns. In this example all the points on the bottom of the slab and on the right edge are held at zero degrees. The values along the top and left edge vary according to position. These boundary temperatures do not change during the calculation. The

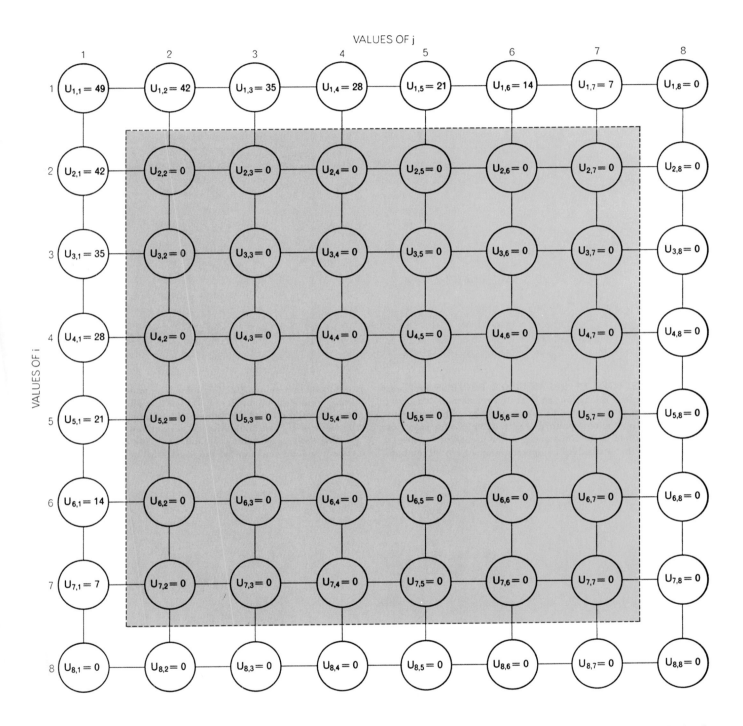

VALUES OF j

VALUES OF i

36 interior points are initially set to zero. When the problem is solved sequentially, the computer starts with the top left interior point $U_{2,2}$ and calculates its value using the numbers given above:

$$U_{2,2} = \frac{U_{1,2} + U_{2,3} + U_{3,2} + U_{2,1}}{4} = \frac{42 + 0 + 0 + 42}{4} = 21.$$

The computer then calculates the value of $U_{2,3}$ using the *new* value of $U_{2,2}$ just obtained, which is 21, instead of the initial value, 0:

$$U_{2,3} = \frac{U_{1,3} + U_{2,4} + U_{3,3} + U_{2,2}}{4} = \frac{35 + 0 + 0 + 21}{4} = 14.$$

The equation is similarly solved for the remaining 34 interior points, using at each step all the new values previously calculated. This sequence of 36 calculations is one "relaxation" of the relaxation method. If the problem were programmed for ILLIAC IV, on the other hand, each of the 36 interior points could be assigned to a separate processing element and 36 simultaneous solutions of the equation obtained. In this method the first relaxation consists of the 36 simultaneous solutions using *only* the numbers initially given. Thus the first solution of $U_{2,3}$ is $(35 + 0 + 0 + 0) \div 4 = 8.75$ rather than the value of 14 obtained in the sequential method. Succeeding simultaneous relaxations, however, can make use of values obtained previously. The way the two methods converge to yield the final answer is shown in the tables on the next page.

ONE RELAXATION

SEQUENTIAL METHOD

VALUES OF i	1	2	3	4	5	6	7	8
1	49	42	35	28	21	14	7	0
2	42	21.00	14.00	10.50	7.88	5.47	3.12	0
3	35	14.00	7.00	4.38	3.06	2.13	1.31	0
4	28	10.50	4.38	2.19	1.31	0.86	0.54	0
5	21	7.88	3.06	1.31	0.66	0.38	0.23	0
6	14	5.47	2.13	0.86	0.38	0.19	0.11	0
7	7	3.12	1.31	0.54	0.23	0.11	0.05	0
8	0	0	0	0	0	0	0	0

PARALLEL METHOD

VALUES OF i	1	2	3	4	5	6	7	8
1	49	42	35	28	21	14	7	0
2	42	21	8.75	7.00	5.25	3.50	1.75	0
3	35	8.75	0	0	0	0	0	0
4	28	7.00	0	0	0	0	0	0
5	21	5.25	0	0	0	0	0	0
6	14	3.50	0	0	0	0	0	0
7	7	1.75	0	0	0	0	0	0
8	0	0	0	0	0	0	0	0

10 RELAXATIONS

SEQUENTIAL METHOD

VALUES OF i	1	2	3	4	5	6	7	8
1	49	42	35	28	21	14	7	0
2	42	35.37	29.01	22.90	17.03	11.31	5.66	0
3	35	29.01	23.41	18.24	13.44	8.88	4.44	0
4	28	22.90	18.24	14.05	10.26	6.75	3.38	0
5	21	17.03	13.44	10.26	7.44	4.88	2.44	0
6	14	11.31	8.88	6.75	4.88	3.19	1.60	0
7	7	5.66	4.44	3.38	2.44	1.60	0.80	0
8	0	0	0	0	0	0	0	0

PARALLEL METHOD

VALUES OF i	1	2	3	4	5	6	7	8
1	49	42	35	28	21	14	7	0
2	42	34.27	27.05	20.53	14.81	9.60	4.74	0
3	35	27.05	19.87	14.08	9.48	5.90	2.83	0
4	28	20.53	14.08	9.06	5.62	3.22	1.49	0
5	21	14.81	9.48	5.62	3.09	1.61	0.69	0
6	14	9.60	5.90	3.22	1.61	0.73	0.28	0
7	7	4.74	2.83	1.49	0.69	0.28	0.10	0
8	0	0	0	0	0	0	0	0

50 RELAXATIONS

SEQUENTIAL METHOD

VALUES OF i	1	2	3	4	5	6	7	8
1	49	42	35	28	21	14	7	0
2	42	36.00	30.00	24.00	18.00	12.00	6.00	0
3	35	30.00	25.00	20.00	15.00	10.00	5.00	0
4	28	24.00	20.00	16.00	12.00	8.00	4.00	0
5	21	18.00	15.00	12.00	9.00	6.00	3.00	0
6	14	12.00	10.00	8.00	6.00	4.00	2.00	0
7	7	6.00	5.00	4.00	3.00	2.00	1.00	0
8	0	0	0	0	0	0	0	0

PARALLEL METHOD

VALUES OF i	1	2	3	4	5	6	7	8
1	49	42	35	28	21	14	7	0
2	42	35.98	29.96	23.96	17.96	11.96	5.98	0
3	35	29.96	24.94	19.92	14.92	9.94	4.96	0
4	28	23.96	19.92	15.90	11.90	7.92	3.96	0
5	21	17.96	14.92	11.90	8.90	5.92	2.96	0
6	14	11.96	9.94	7.92	5.92	3.94	1.96	0
7	7	5.98	4.96	3.96	2.96	1.96	0.98	0
8	0	0	0	0	0	0	0	0

DIFFERENT STAGES IN RELAXATION PROCESS are compared for sequential relaxations and parallel relaxations. The exact values are given in the array on the opposite page. The two methods for calculating the temperature of each of 36 interior points on a slab are described in the illustration on the preceding two pages. There one sees that a standard computer using sequential methods would obtain a value of 21 for point $U_{2,2}$ and 14 for $U_{2,3}$ in performing one relaxation. Here it is seen that after 10 relaxations by the sequential method the value of $U_{2,2}$ has climbed to 35.37 and the value of $U_{2,3}$ to 29.01. After 50 relaxations $U_{2,2}$ and $U_{2,3}$ have reached their exact values: 36 and 30. Using parallel relaxations ILLIAC IV would converge on the exact solution in a distinctly dif-

VALUES OF j

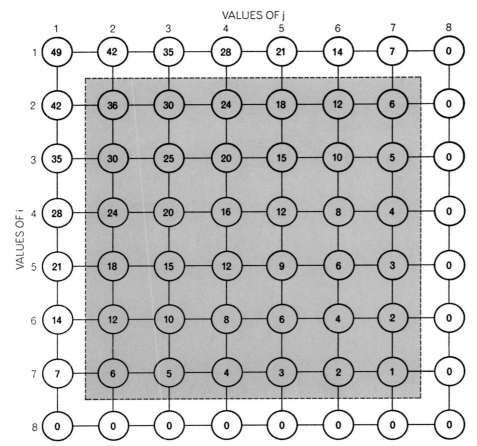

VALUES OF i

write a system of ordinary differential equations that might plausibly seem to describe the growth of a living cell. One can even measure initial concentrations with seemingly sufficient accuracy to do meaningful calculation. The number of equations in the system, however, corresponds to the number of genes in the chromosome, which is just too large a number to be handled in the cases of most interest. On the scale of real ecological systems, on the other hand, population models can be developed but measurements are extremely elusive. (How many alewives are in Lake Michigan?) Calculations would have to be performed with statistical variables to estimate a population range for each species of organism. This consumes computational capacity of a higher order of magnitude than deterministic calculation. Even the methodology of such calculation poses significant theoretical problems.

To summarize, I believe computers on the scale of ILLIAC IV will remove some of the very real barriers of capacity from certain calculations that have a direct bearing on our ability to produce a rational and enduring basis for life. Counterpoised is the computer's potential to play a significant role in the depersonalization and disordering of society. Scientists must not share the neutrality of the computer to the outcome.

ferent manner. After 10 relaxations it would obtain values of 34.27 and 27.05 for $U_{2,2}$ and $U_{2,3}$ respectively. The results after 50 relaxations, however, would be virtually the same. For this particular problem the parallel method requires a few more relaxations than the sequential method to achieve comparable results. ILLIAC IV, however, will be able to carry out 36 complete relaxations (and as many as 64 given a suitable problem) in the time that a comparably fast sequential computer would need to carry out one sequential relaxation.

RECONNAISSANCE SATELLITE

RECONNAISSANCE AND SURVEILLANCE SYSTEMS currently used by the U.S. to verify Russian compliance with the terms of the SALT I agreements are represented schematically in this illustration. U.S. photoreconnaissance satellites are typically launched (from Vandenberg Air Force Base in California) into a near-polar elliptical orbit with an orbital period of approximately 90 minutes and a perigee (lowest point) on the order of 100 miles. The latest, fourth-generation U.S. observation satellite, unofficially called Big Bird, combines the separate functions of area-surveillance photography and close-look photography and hence is required to stay aloft for a much longer period than earlier close-look satellites; orbital times to date have averaged about seven weeks. The orbit of such a satellite remains essentially fixed in space while the earth rotates, with the result that to an earth-based observer the satellite appears to move westward on each successive orbit. Hence most of the earth's surface passes under the orbital path of the satellite. U.S. early-warning satellites, in contrast, are typically launched into near-equatorial, near-synchronous "parking" orbits at altitudes of about 22,300 miles. Two such satellites, launched into identical "figure eight" orbits at the same fixed longitude over the Indian Ocean but lagging each other by 12 hours, can provide continuous infrared coverage of most of the U.S.S.R. and all of China (light colored area). The value of early-warning satellites from the point of view of arms control is that they are also capable of monitoring missile tests. The black dots indicate the locations of the major Russian missile-testing launch centers; the black line shows a typical trajectory for a Russian long-range missile test. Also shown are two types of U.S. radar used to monitor Russian missile tests: over-the-horizon transmitters (colored dots) and receivers (open colored circles), and conventional, or line-of-sight, radars (colored squares).

Missile Submarines and National Security

by Herbert Scoville, Jr.

June 1972

*Land-based missiles are giving way to submarine
missiles as a secure deterrent to a nuclear first strike.
The question now is whether or not the U.S. should
spend perhaps $40 billion on a new missile fleet*

In recent years, as the nuclear-weapons arsenals of both the U.S. and the U.S.S.R. have continued to grow, the concept of deterrence has become almost universally accepted as the key to maintaining national security and preventing the outbreak of a nuclear war. "Winning" a nuclear exchange is no longer regarded as a rational strategic objective; in such an exchange everyone, participant and nonparticipant alike, would be a loser. In keeping with the deterrence principle President Nixon affirmed in his State of the World Message of February 9 that "our forces must be maintained at a level sufficient to make it clear that even an all-out surprise

attack on the U.S. by the U.S.S.R. would not cripple our capability to retaliate." For the Russians to feel secure they must have a similar capability; only then would a stable strategic balance exist.

The primary attribute required of any deterrent force is the ability to survive a "counterforce," or preemptive, attack. Ballistic-missile submarines are almost ideally suited to satisfying this requirement. Although they are expensive compared with other strategic weapons (more than $100 million per submarine exclusive of the missiles), their mobility and invisibility make them virtually immune to destruction in a surprise attack. In contrast, land-based intercontinental

ballistic missiles (ICBM's) can readily be located with the aid of surveillance satellites, so that they must be regarded as "targetable" in the event of an enemy first strike. Attempts to "harden" such fixed missile-launchers (that is, to increase their resistance to the effects of nuclear explosions) are in the long run doomed to futility, since in the absence of qualitative arms-control agreements improvements in offensive missiles, particularly improvements in accuracy, will inevitably make fixed missile-launchers vulnerable and hence reduce confidence in their deterrence value.

The advent of multiple independently targetable reentry vehicles (MIRV's),

	SUBMARINES	MISSILES	WARHEADS		MAXIMUM TARGETS	
			BEFORE MIRV	AFTER MIRV	BEFORE MIRV	AFTER MIRV
U.S.	41	656	1,712	5,440	656	5,120
U.S.S.R.	26 (42)	416 (672)	416 (672)		416 (672)	

MISSILE-SUBMARINE FORCES of the U.S. and the U.S.S.R. are compared in this table. The submarines of both fleets are designed to carry 16 ballistic missiles each. The U.S. is currently in the process of converting 31 of its total of 41 deployed Polaris submarines to carry Poseidon missiles, each of which is capable of carrying up to 14 multiple independently targetable reentry vehicles (MIRV's); on the average each Poseidon will be able to deliver 10 nuclear warheads on 10 separate targets or at intervals on the same target. (The remaining payload can be devoted to various "penetration aids" intended to foil a potential enemy anti-ballistic-missile, or ABM, system.) Before the MIRV program began the bulk of the U.S. missile-submarine fleet was armed with Polaris A-3 missiles, which feature a multiple reentry vehicle (MRV)

capable of carrying three warheads; these warheads are not, however, widely separable and are aimable only in shotgun fashion at a single target. After MIRVing is complete the remaining 10 ships in the Polaris fleet will continue to be armed with A-3 missiles. The Russian missile-submarine fleet consists at present of 26 ships deployed and about 16 more under construction, but this program could continue. (Known future submarine and missile totals are given in parentheses.) Since its missiles carry only one warhead each, however, the number of warheads it can launch is no greater than the total number of its missiles. Moreover, in a crisis military strategists in the U.S.S.R. must add to the U.S. totals the missile-submarine forces of France and Great Britain, each of which may eventually consist of four submarines and 64 missiles.

which are currently being deployed on a large scale by the U.S., creates a situation in which the "exchange ratio" strongly favors the attacker. Thus a single missile with, say, six warheads can potentially destroy six enemy ICBM's if they are caught in their silos. Moreover, strategic bombers are extremely vulnerable while they are on the ground and would therefore be very susceptible to annihilation in a surprise missile attack. Attempts to avoid this weakness by maintaining aircraft on continuous airborne alert have proved to be expensive and potentially dangerous. Even the current 15-minute ground alert is not completely satisfactory, since adequate warning would be more difficult to obtain if fractional-orbital-bombardment systems (FOBS) or depressed-trajectory missiles launched from submarines were used to attack the bombers.

Hence given the present state of military technology and reasonable anticipated advances, the primary element in the strategic-deterrent forces of both the U.S. and the U.S.S.R. will continue to be the ballistic-missile submarine. All other strategic systems will remain secondary. Moreover, it seems likely that any agreement that may emerge in the near future from the strategic-arms-limitation talks (SALT) will further enhance the relative importance of the missile-submarine forces. Since the chances that

MIRV's will be limited by a SALT agreement are extremely low, ICBM's will become increasingly vulnerable. The more likely limitations on anti-ballistic-missile (ABM) systems, on the other hand, would guarantee the retaliatory capability of even a comparatively small number of submarine-launched ballistic missiles (SLBM's). The expected failure to limit antiaircraft defenses and to restrain qualitative improvements in offensive-missile systems would further decrease the value of strategic bombers. Although there will probably not soon be restrictions on antisubmarine-warfare (ASW) measures, the technology in this area is so far behind that it could not possibly threaten the submarine deterrent, if it can threaten it at all, until far in the future. In sum, the Navy will increasingly be the principal military guardian of our national security.

What characteristics must an SLBM force have in order to fulfill its function as a deterrent against the initiation of nuclear warfare by the U.S.S.R.? (Since China is so far behind both the U.S. and the U.S.S.R. in this respect, the same forces would be more than adequate to deter China as well.) First of all, the submarines should be designed to operate in, and fire their missiles from, large enough ocean areas in a variety of directions around the U.S.S.R. so as to

decrease their vulnerability to ASW detection and tracking and to facilitate the penetration of any ABM system. The closer these areas are to ports in the U.S., the less will be the time lost in moving to and from operational stations and the less will be the need for overseas bases. Higher submarine speeds will also reduce this travel time and increase the ability to break contact with a trailing ASW submarine or surface vessel. The gains here may be marginal, particularly since tracking vessels will probably be faster than any missile submarine. The faster a submarine moves through the water, however, the more noise it will produce, and in countering ASW measures quietness may be much more important than speed. The reduction of submarine noise is the most critical element in preventing detection and continuous covert tracking, both of which must rely on passive acoustic sensors.

If an ABM defense is a realistic possibility, then the submarine missiles must have enough payload capacity to allow the use of multiple warheads and other penetration aids. The entire submarine force should be large enough so that the destruction of a few submarines by a concerted enemy attack, by slow attrition or perhaps by a series of accidents does not seriously degrade its overall capability. If continuous tracking by antisubmarine submarines or other ASW

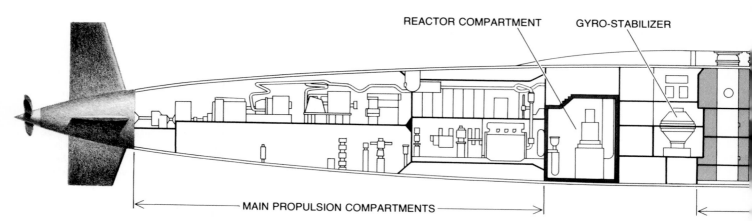

TYPICAL POLARIS SUBMARINE, in this case an advanced model belonging to the Lafayette, or 616, class is shown schematically in the cutaway drawings on these two pages. The nuclear-powered submarine has a length of 425 feet, a beam of 33 feet and a submerged displacement of 8,250 tons. In addition to its 16 ballistic-missile silos it is equipped with four bow torpedo tubes. Each of its

REACTOR COMPARTMENT   GYRO-STABILIZER

MAIN PROPULSION COMPARTMENTS

vessels ever becomes a realistic operation on a large scale, then the more vessels there are in the missile-submarine
fleet, the harder it will be for this tactic
to be successful in destroying the entire
force. Ballistic-missile submarines cannot be used to attack other submarines
and are no threat to the SLBM deterrent
of the other side.

In addition to an adequate number of
submarines, missiles and warheads, it is
essential to have secure and reliable
communications between these vessels
and their command authorities. It is not
enough to send the submarines to sea
with sealed orders. Controls to prevent
inadvertent or unauthorized firing are
an absolute necessity, and reliable methods for ordering retaliation in the event
of a surprise attack are required. These
communications must be jam-proof; the
potential attacker cannot be allowed
to hope that a communications failure
might prevent a retaliatory strike.

The submarines must also be able to
navigate accurately, so that after they
have moved through the oceans into
their operational areas they will always
be in a position to fire their missiles at
predetermined targets. High navigational accuracy is not as great a requirement
for a retaliatory strike against cities as it
would be if the submarines were to be
used in a counterforce role for destroying
such "hard" targets as enemy missile

sites. In fact, if one side wishes to use
missile submarines only as deterrent
weapons, then it is important that the
accuracy-yield combination of the system not be so great as to give the other
side concern that the submarines have a
first-strike capability against land-based
ICBM's; otherwise the position of mu-
tual stable deterrence will be eroded.

With these general principles in mind,
let us examine how the U.S. missile-submarine forces have developed
over the years. The U.S. launched the
first nuclear-powered submarine, the
Nautilus, in 1955, but it was not until
the late 1950's that development of long-
range missiles had proceeded to the
point that these could be installed in
such submarines. The first ballistic-missile submarine, the George Washington,
became operational in November, 1960.
It was armed with 16 solid-fuel Polaris
A-1 missiles, which could be fired at a
rate of about one per minute. The range
of this missile was about 1,200 nautical
miles and the warhead yield about one
megaton. The submarines were designed
to fire their missiles while submerged,
using compressed gases to expel the missile; the rocket engine is then ignited
after the missile has cleared the surface.
By 1963, 12 more Polaris submarines
were operational.

Meanwhile the development of more

advanced missiles continued. The next-
generation missile, the A-2, had a range
of about 1,500 nautical miles. The first
test of the A-3 missile, with a range of
2,500 nautical miles and a "triplet" reentry vehicle, was conducted from a submerged submarine in the fall of 1963.
The triplet reentry vehicles, which could
carry three individual nuclear warheads
each, did not have independent guidance; the three warheads were intended
to reenter the atmosphere in a shotgun
pattern with the target at the center.
Since such warheads cannot be aimed at
separate targets, they do not alter the
exchange ratio and do not provide any
first-strike advantage. Their advent was
therefore not in itself regarded as de-
stabilizing.

In the early 1960's there was considerable debate over the appropriate size
of the Polaris fleet. The Navy originally
sought 48 ships, and the final decision
was to build 41. One factor limiting the
number of submarines is the problem of
manpower recruitment. Nuclear-submarine duty, which involves 60-day underwater cruises, calls for a certain type of
person who is not easy to find and who
must be highly trained. Normally each
vessel has two crews of about 140 men
who go on alternate patrols.

By the end of 1966 all 41 Polaris submarines were operational; eight carried
A-2 missiles and 33 were eventually

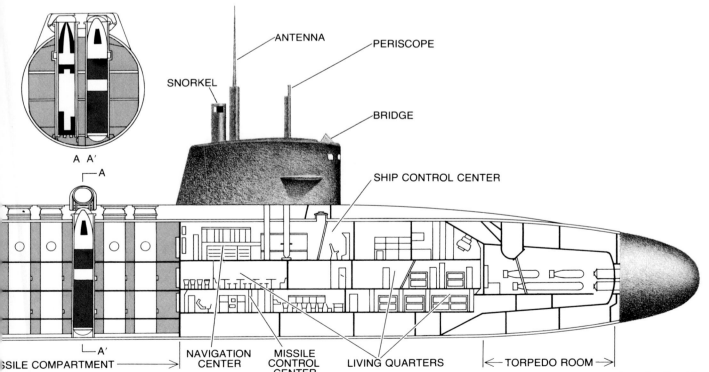

alternate crews consists of 14 officers and 126 enlisted men. The
transverse section (AA') shows the relative sizes of the Polaris A-3
missile (left) and the Poseidon C-3 missile (right). All the 31 missile submarines in this class will be converted to the Poseidon
MIRV system by 1976. The Russian Y-class submarines, although
not as advanced, are believed to be roughly similar in design.

fitted out with *A-3*'s. Thus the force carried a total of 1,712 warheads, but since the triplet warheads cannot be aimed separately, 656 was the maximum number of separate targets that could be hit. Of course, not all of these submarines can be kept at operational stations at all times. In general a submarine spends 60 days at sea and 30 days in port for maintenance. In addition the submarine might take five or more days to move from the U.S. to its launch point and the same period to return. If a submarine wished to avoid detection by moving

quietly and therefore slowly, the travel time would be even greater. Thus the number of submarines at launch stations at any one time could be reduced to some 20 to 25 ships. The situation is improved by using forward bases (on Guam, at Holy Loch in Scotland and at Rota in Spain), which reduces the time needed to reach launch stations from five or six days to one or two days.

With a range of 2,500 nautical miles, a submarine-launched missile can hit Moscow from most of the North Atlantic (inside an arc extending from the tip of

Greenland to North Africa), from the Mediterranean and even from some parts of the Indian Ocean, a total sea area of about six million square miles [*see illustration below*]. The sea area from which a submarine-launched missile could hit important targets other than Moscow, say targets only 200 miles inside the U.S.S.R., is even larger [*see illustration on opposite page*]. One high-ranking Navy officer reported in 1964 that a Polaris submarine equipped with the *A-3* missile could operate in 15 million square miles of ocean area while

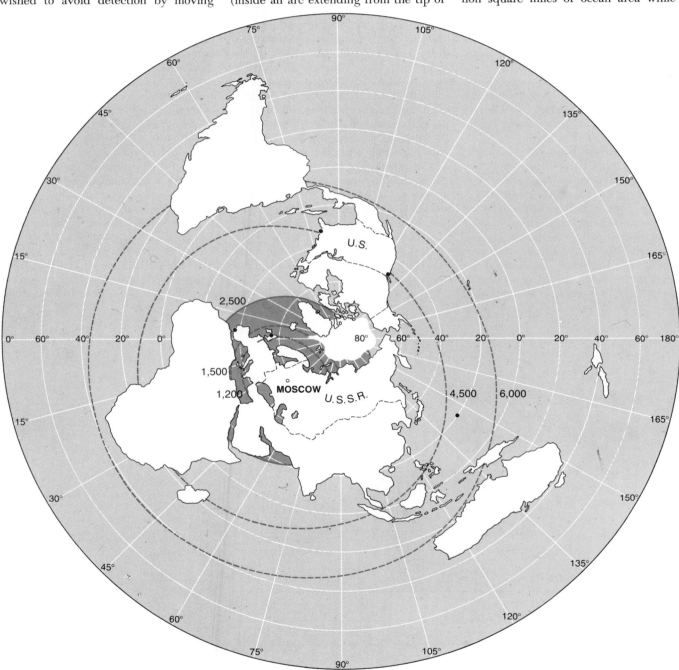

RANGE OF U.S. SUBMARINE MISSILES defines the operational sea areas from which submarine-launched ballistic missiles (SLBM's) could hit strategically important targets in the U.S.S.R. In the map at left the target is assumed to be Moscow; in the map at right the targets are assumed to include population centers and industrial complexes within 200 miles of the border of the U.S.S.R. (*gray areas*). The contours in the map at left are actually circles; their somewhat misshapen appearance is caused by the polar projection used in plotting the map. The contours in the map at right are concentric with the border of the U.S.S.R. The solid colored

covering its targets in the U.S.S.R. No land target anywhere in the world is inaccessible from attack by the A-3 missile.

Although mobility provides a submarine with the tremendous advantage of improving its survivability, it creates a new problem: the determination of its location at the moment when the missile is to be launched toward its target several thousand miles away. Unless the missile is provided with some means of determining its position during flight or with a terminal-homing capability, the accuracy at the impact point can never be better than the uncertainty in the launch point. To determine the launch position calls for accurate submarine navigation, which is made more difficult by the requirement that in order to avoid disclosing its presence the submarine should not surface to determine its location. The attitude of the ship with respect to the vertical and true north at the time of the launch is also needed. When the missile force is being used for deterrent purposes, an accuracy greater than a few thousand feet is not needed; it is only necessary to be able to hit a large urban complex. Today this order of accuracy in locating the position and attitude of a submarine can be readily achieved. The U.S. has made tremendous advances in the development of inertial-navigation systems in recent years, and reasonably accurate position fixes can be obtained even after the submarine has been submerged at sea for many days.

The inertial-navigation system in a Polaris submarine is a complex system of gyroscopes, accelerometers and com-

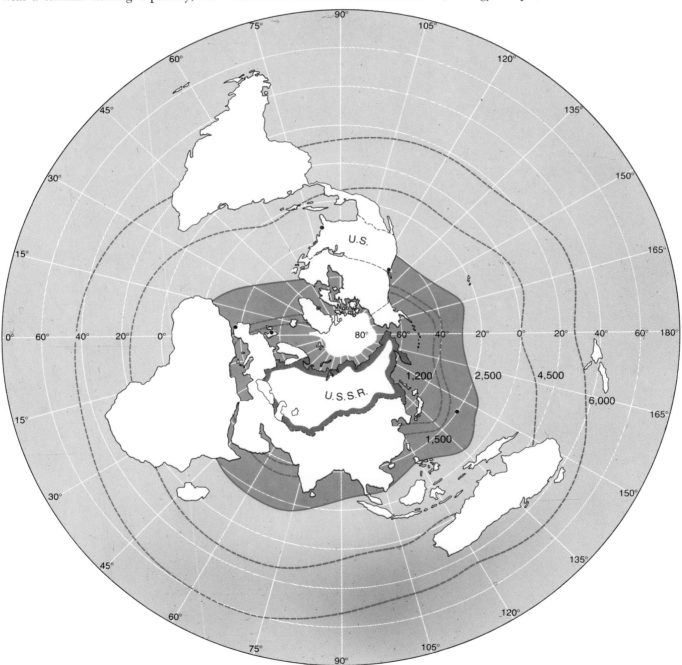

contour in each map shows the range of the Polaris A-3 and the nominal range of the Poseidon C-3 missiles (2,500 nautical miles). The two inner broken colored contours in each case show the ranges of the older Polaris A-1 and A-2 missiles (1,200 and 1,500 nautical miles respectively). The two outer broken colored contours show the estimated ranges of the Navy's proposed undersea long-range missile system, which envisions two new generations of missiles, the ULMS 1, with a range of 4,500 nautical miles, and the ULMS 2, with a range of 6,000 nautical miles. Black dots denote home ports and forward bases of the U.S. missile-submarine fleet.

puters that relate the movement and the speed of the ship in all directions with respect to true north. If an initial position is known, then this system will provide continuous data on the ship's position. For an absolutely stationary submarine, or one whose motion can be corrected for, inertial sensors can determine without external data the vertical, the true north, the latitude and all velocity components by inertially sensing the earth's gravitational and rotational vectors, but there is no way of determining the longitude by inertial means. Subma-

rines that have been voyaging at sea for protracted periods and whose inertial-navigation errors may have become unacceptably large can, by trailing an antenna while they are still submerged, get a radio "position fix" from navigation satellites or land-based transmitters. It may also be possible to locate a submarine by reference to accurately known geographical landmarks on the ocean bottom such as seamounts. In sum, the present technology has advanced to the point where the location and attitude of the submarine could in principle no long-

er be the critical factor in obtaining missile accuracies down to less than an eighth of a mile.

A deterrent force must also be able to receive communications from national command centers. Direct command and control originating with the President and with many verification checks is vital to prevent unauthorized launching; it is also essential that command authorities be able to communicate in times of crisis with the submarine captains without fear of interference by the other side. Otherwise communication would

RANGE OF RUSSIAN SUBMARINE MISSILES is estimated in these two maps in order to show the operational sea areas from which their SLBM's could hit targets anywhere in the U.S. (*left*) and targets within 200 miles of the U.S. border (*right*). Alaska and Hawaii are excluded from consideration. Again the solid colored contour in each map shows the present range of the Russian Y-class missiles (1,300 nautical miles). The inner broken colored contour in each case shows the range of the previous generation of Russian SLBM's (700 nautical miles). The outer broken colored contour in each case shows the range of a new Russian missile that

be the Achilles' heel of the submarine deterrent. There are a number of means of communicating with a submarine, at least one way from shore to ship, that do not require the submarine to surface. Very-low-frequency (VLF) radio waves can penetrate a short distance into the water, so that a receiving antenna does not need to be exposed at the surface. Moreover, the submarine can operate at a considerable depth, since it can trail an antenna as much as several hundred feet above its deck. The U.S. has a number of land-based VLF transmitters at

various locations around the world for communication with Polaris submarines, and more recently an airborne VLF system has been devised in order to eliminate the possibility that the fixed land-based stations would be destroyed in a surprise attack.

The use of satellites for relaying messages to submarines provides an alternative means of communication. Recently much research has been devoted to extremely-low-frequency (ELF) waves, which can penetrate even deeper into the water. The Navy project named

Sanguine proposed to set up a vast antenna for this purpose in Wisconsin. The data rate of such a system would be quite low, but it would be adequate for command-communication purposes. The project has run into difficulties with local residents because of the large antenna currents and the potential hazard to living things. For communication from a submarine to the command center the problem is more difficult; it calls at the very least for trailing an antenna close to the surface and must in any case be avoided in order to prevent disclosure of

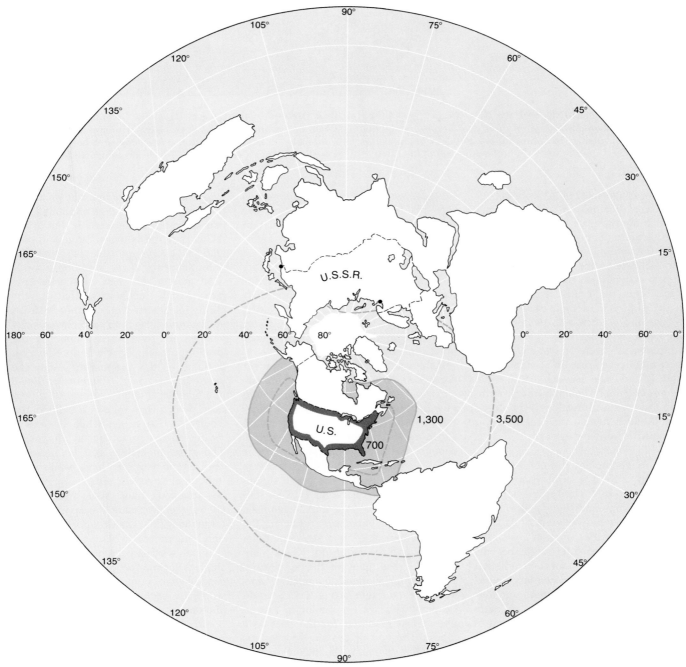

has recently been tested; this missile, however, is believed by U.S. strategic planners to be too long to fit readily in the present *Y*-class missile submarines. The black dots indicate the shipyards of the Russian missile-submarine fleet; the U.S.S.R. has no forward bases for its missile submarines. In general the greater distance from the

Russian missile-submarine fleet's home ports to their operational launching areas, combined with their lack of forward bases, means that their submarines waste more time in getting "on station." Hence the U.S.S.R. requires more missile submarines than the U.S. to maintain the same deterrent force operational at any one time.

U.S. MISSILE SUBMARINE, the U.S.S. *Lafayette*, namesake of the advanced 616 class of Polaris vessels, was photographed while she cruised in the Atlantic Ocean during her builder's sea trials in April, 1963. The *Lafayette* was built in the shipyards of the Electric Boat Division of the General Dynamics Corporation in Groton, Conn. As originally deployed, this submarine was armed with 16 solid-fueled Polaris *A-2* missiles; it is scheduled to be converted to Poseidon *C-3* missiles with MIRV warheads sometime before 1976.

the submarine's location to listening enemy radios. Fortunately such communication is not essential to the viability of the submarine deterrent force.

By the end of 1966 the U.S. submarine-missile force together with its support systems was by itself more than adequate to deter any nuclear attack on the U.S. It had more than enough missiles and warheads to devastate the U.S.S.R. even when only a fraction of the submarines were on station. It could operate in ocean areas on all sides of the U.S.S.R., and the Russian ASW capability was quite rudimentary, with virtually no ability to "draw down" the size of the U.S. fleet. At that time the Russians had no ABM system deployed.

Military technology did not, however, stand still. The need to operate in restricted sea areas close to the northern coast of Europe and the Mediterranean in order to reach Moscow and other interior Russian cities created fears that someday ASW measures might become a threat. More important, concern that the U.S.S.R. might deploy a large ABM system capable of coping with our missile-submarine force was becoming more acute every day. The Russians were in the process of deploying an ABM defense around Moscow, using a large interceptor missile estimated to have a single-warhead yield large enough to destroy all three warheads of the Polaris *A-3*. In addition they were deploying radars and defensive missiles in the "Tallinn system" widely throughout the

U.S.S.R. Some "worst case" analyses of U.S. planners, particularly during the early phases when factual information was limited, postulated that these facilities were for ABM defense. Later, as more data became available, the predictions were scaled down to the effect that the Tallinn system was an antiaircraft defense that could perhaps be upgraded to provide an ABM capability. (Now even this upgrading is not considered practicable by most experts.)

As a result research and development proceeded on a next-generation missile for the Polaris submarine that would give increased future assurance of penetrating any ABM defense and would at the same time give the submarines enough flexibility to operate at greater distances from the U.S.S.R. To increase either the payload or the range significantly called for a larger missile, and the resulting Poseidon missile required enlarging the launching tubes in the Polaris submarines, a costly and time-consuming task requiring 13 or more months. Since many of the submarines were due for overhaul in any case, however, the two shipyard activities could be combined with a minimum loss in operational readiness for the fleet as a whole. The cost of converting a Polaris submarine to carry the advanced Poseidon missiles is on the average $29 million, with another $38 million for normal submarine overhaul and replacement of the nuclear fuel.

The new Poseidon missile is about twice as heavy as the Polaris *A-3* and has

a payload about four times as great. Although its nominal range of about 2,500 nautical miles is the same as that of the *A-3*, a trade-off between range and payload is always possible, so that the potential range of the Poseidon is somewhat greater. The new missile incorporates MIRV technology, that is, the ability to disperse many warheads aimed at separate targets. The technique developed for this purpose employs the "bus" approach, in which shortly after burnout of the propulsion stages the missile's final stage (the bus) is aimed at a first target point and releases a warhead, which then follows a ballistic trajectory to that target while the bus is redirected toward a second aim point. The same procedure can be repeated until all the warheads have been sent to individual targets. If a single target is to be attacked, then MIRV technology allows the warheads to approach the target at widely spaced intervals and on different trajectories so that no more than one warhead can be destroyed by a single ABM interceptor. The Poseidon is reported to be capable of carrying 14 warheads, each with a yield of about 50 kilotons, several times the yield of the bomb that destroyed Hiroshima. Warheads can be traded off for either ABM penetration aids or increased range. The nominal complement is usually taken as being 10 MIRV warheads. Department of Defense officials have repeatedly stated that these warheads do not have the accuracy-yield combination to provide a first-strike capability against hardened silos (a "circular-error probability" of about an eighth of a mile or better would be needed with that yield), but the Russians might still be concerned on this score.

As with the Polaris A-3, a Poseidon missile with a range of 2,500 nautical miles can launch warheads at Moscow not only from large areas of the North Atlantic and the Mediterranean but also from some parts of the Indian Ocean, a total area of about six million square miles. Targets within 200 miles of the border of the U.S.S.R. can be reached from some 15 million square miles. These large ocean areas present great problems for any possible future ASW system. To deploy detection and tracking systems throughout these waters is a prodigious, if not impossible, task. Furthermore, on short notice these areas might be somewhat enlarged if it ever became critical by reducing the Poseidon payload, either by eliminating penetration aids or by cutting down the number of warheads in each missile.

The first Poseidon missile was tested in August, 1968, and the development of the entire Poseidon system was completed two years later. The first Polaris submarine went to sea with Poseidon missiles in March of last year. At present about 10 submarines have been converted. The program calls for modifying 31 submarines to carry these new missiles, leaving 10 to be equipped with the older A-3's. When the program is com-

pleted in 1976, the U.S. submarine force will be able to launch 5,440 warheads at 5,120 separate targets. It should be possible to keep considerably more than half of these submarines on station at all times, and in times of crisis the operational readiness can be stepped up if necessary. This is an awesome force, capable of overwhelming even a massive ABM defense system. There is, of course, no evidence that the Russians have any intention of building a large ABM system with nationwide coverage, and it is highly likely that such a system will be precluded in the first stage of a strategic-arms-limitation treaty.

Even one missile submarine can launch 160 warheads at separate industrial centers in the U.S.S.R., an attack that the Russians could not afford even if the U.S. had been annihilated. This means that any ASW system would have to be able to eliminate almost instantaneously every single submarine, a herculean task. Today it is difficult, if not impossible, to destroy even a single submarine that follows skilled evasion tactics. Yet if ABM defenses were forbidden by treaty, an ASW system that has still to be devised would be the only threat to the submarine deterrent. That is one reason why an ABM agreement at SALT would by itself be such an important gain to the security of both the U.S. and the U.S.S.R.

The U.S.S.R. has always lagged considerably behind the U.S. in the devel-

opment of nuclear submarines and SLBM's. The first Russian nuclear submarine was built about four years after the *Nautilus*, and Admiral Hyman G. Rickover, the director of the U.S. nuclear-submarine program from the beginning, has made it clear he believes the Russian submarines are technically inferior to the U.S. ships. The first Russian ballistic-missile submarines were diesel-powered and therefore had limited endurance and cruising range. Their first nuclear-powered missile submarine carried only three missiles with a 300-nautical-mile range, which in later models was extended to about 700 nautical miles.

By the late 1960's it must have been obvious to military planners in the U.S.S.R. that their land-based ICBM's would become increasingly vulnerable to the U.S. MIRV's, which were then under development and which had been publicly justified as providing an improved counterforce capability. The Russian deterrent needed shoring up with a more effective SLBM force, whose value had been demonstrated by the U.S. In 1966 the U.S.S.R. launched its first Y-class submarine, which carries 16 missiles with a reported range of 1,300 miles. This class of vessels was similar to the Polaris submarines, which the U.S. had put into operation seven years earlier. All the Russian SLBM's deployed so far have had storable liquid fuels,

RUSSIAN MISSILE SUBMARINE, a representative Y-class vessel, is remarkably similar in appearance and general dimensions to the earlier U.S. Polaris designs. The ship carries 16 liquid-fuel missiles with a reported range of about 1,300 nautical miles. Each missile is armed with a single nuclear warhead with a yield of about one megaton. This photograph was released by the U.S. Navy in 1970.

whereas all the U.S. missiles have had solid fuels. The Russians apparently decided to continue with the Y-class design and began building submarines at a rapid pace, initially at the rate of six to eight per year and currently at nine per year. Two shipyards are engaged in this work, one at Severodvinsk on the Arctic Sea and one in Siberia on the Pacific. At present the Russians have about 26 Y-class submarines operational and another 16 under construction.

Although the Russians have tested a new missile with a range of about 3,500 miles, John Stuart Foster, Jr., chief research scientist for the Department of Defense, recently reported that this missile was so long that he did not believe

the Y-class submarine could be modified to launch it. The Russians have never even tested multiple warheads on their submarine missiles, let alone MIRV's. Their missiles are each armed with a single warhead with a yield of about one megaton. These missiles have no capability for attacking our ICBM silos, but it has been postulated that they might be employed to attack our bombers on the ground and our command and control centers, using a depressed trajectory to achieve the necessary surprise. The missiles have not, however, been tested in this mode, and this approach would, in any case, entail a reduction in their already limited range.

The U.S.S.R. will have a slightly

larger ballistic-missile submarine fleet than the U.S. when it completes those vessels now under construction (42 to 41). Even then, however, the capabilities of the Russian fleet will be far inferior to those of the U.S. Polaris fleet. President Nixon in his 1972 State of the World Message said that "our missiles have longer range and are being equipped with multiple independently targetable warheads. Moreover, our new submarines are now superior in quality." The shorter range of the Russian missiles requires that their submarines operate fairly close to the U.S. coast in order to be able to strike inland U.S. targets; this makes the Russian submarines potentially more vulnerable to a U.S. ASW system [see illustration on page 188]. On the other hand, the population centers and industrial complexes on the east and west coasts of the U.S. can be reached from much larger ocean areas, and these targets would be quite satisfactory if the SLBM's were to be used for deterrent purposes [see illustration on page 189]. The restricted range would be a serious factor only if the SLBM's were to be used against our bomber bases or missiles in the interior of the U.S.

There are other reasons why the parity between the U.S. and the U.S.S.R. in operational submarines cannot be evaluated on numerical grounds alone. Since bases in the U.S.S.R. are farther from the operational launching areas and since the Russians have no locations available for forward bases, more time is wasted getting submarines on station and it takes more submarines to maintain the same deterrent force operational at any one time. It would take Russian submarines a minimum of six days in the Atlantic and eight days in the Pacific to reach the nearest launch stations, so that the transit time to and from home ports, in many cases a quarter to a third of the duration of the patrol, seriously degrades the operational readiness of the Russian fleet. This disadvantage can be only partly alleviated by using submarine tenders for maintenance and crew exchange at sea. Moreover, in any East-West comparison the small British and French missile-submarine fleets, each of which may eventually consist of four submarines and 64 missiles, must be added to the U.S. total. Thus, whereas the Russians now have an adequate missile-submarine deterrent, their fleet is markedly inferior to that of the U.S. and its allies and provides no threat to the U.S. deterrent.

As the first phase of SALT is drawing to an end, then, it is becoming universally recognized that ballistic-missile

ARRANGEMENT OF MISSILE SILOS in a Lafayette-class Polaris submarine is revealed by the open silo doors in this view. The submarine, the U.S.S. *Sam Rayburn*, was photographed at Newport News, Va., at about the time of her commissioning in December, 1964.

CONVERSION of the Polaris submarine U.S.S. *James Madison* to accommodate the new, larger Poseidon missile was undertaken in February, 1969, and was completed in June, 1970. This photograph shows the *Madison*, the first of 31 such missile submarines to be converted, being test-fitted for the new Poseidon missile system at the Groton shipyards of the General Dynamics Electric Boat Division.

submarines are the essential foundation of a secure and stable strategic balance. Under these circumstances it is only natural to investigate ways to still further improve submarine missile systems. The U.S. has had a research and development program in this area for several years, so that the developments at the frontiers of technology could be incorporated in the successor to the Polaris-Poseidon system. The particular system proposed by the Navy for this role has been called ULMS, for undersea long-range missile system.

One obvious way to improve the present submarine missile would be to extend its range, making possible the launching of missiles from larger ocean areas in all directions around the U.S.S.R. Increasing the missile's payload would allow the incorporation of more warheads per missile or additional ABM-penetration aids. Payload can, of course, always be traded

off for range. A longer-range missile would reduce the time required to move from U.S. ports to launching areas and thereby reduce the need for overseas basing in order to maintain the submarines on station for a larger fraction of their cruising time. With a range of 4,500 nautical miles a missile could reach Moscow shortly after the submarine leaves the U.S., whereas with a range of 2,500 nautical miles at least three days' travel would be required. Thus for a 60-day cruise lengthening the missile range by 2,000 miles could increase the period of operational effectiveness by about 10 percent if forward bases were abandoned.

More advanced guidance systems employing terminal control (the ability to change the path of the warhead during reentry) are being developed to avoid interception by ABM systems or to improve accuracy. Higher accuracy ob-

tained by this means or others is not required if missiles are to be used as deterrent weapons. Indeed, it might be construed by the other side as an attempt to attain a counterforce capability for a first strike. Although it is more difficult to acquire such a capability with a submarine missile system than with an ICBM because of the inherent limitations on the payload and the nuclear explosive yield and because of navigational complications, there are no scientific barriers to its achievement. The greatest technical restriction on the use of SLBM's for a counterforce first strike may lie in command and control. It may be feasible to preprogram and command the initial launchings, but there will inevitably be failures. The difficulties of directing subsequent firings to destroy the silos missed the first time appear to be virtually insurmountable.

The submarine itself can be improved

by making it quieter as it moves through the water, thereby rendering detection and tracking by passive acoustic techniques more difficult. Although increasing the speed of the submarine will make it somewhat harder for enemy ASW ships to follow it, the higher speed will also raise the level of noise produced by the submarine. In any case it is probably a losing proposition for a missile submarine to try to outrun an ASW vessel, which can always be designed to move faster. High speed will enable the submarine to reach its launch area more rapidly and thus reduce the time it spends in a nonoperational condition, but again the potential gains are not large, and they may be outweighed by the disadvantages. If all other factors are equal, it will require a bigger power plant and larger submarine, both of which will increase cost and detectability. Increasing the depth at which a submarine can operate is not particularly significant, at least for the depths that are likely to be achieved in the next generation of submarines; submarines can be detected acoustically and destroyed by nuclear depth charges or homing torpedoes at any reasonable depth.

If space for more and larger missiles is needed to increase the destructive capacity and the ABM-penetrability of any single submarine, then larger submarines with bigger power plants will be required. The larger the submarines, however, the fewer the ships that will be available for the same investment. Therefore if funds are limited (and they always are), this means a smaller fleet, which is more vulnerable to being wiped out in a simultaneous surprise attack. Thus there are many trade-offs in system design, and the final decision on a successor to the Polaris-Poseidon system should be based only on the nature of the specific threat.

In 1971 $104.8 million was appropriated for the advanced development of ULMS. Although this expenditure still left options open, it was a major step on the path toward procuring a specific new submarine system. This year the Department of Defense is seeking $977 million for ULMS. If that amount is authorized, the U.S. will be irrevocably committed to a large and very expensive new shipbuilding program. Unclassified details of the proposed ULMS program are still scarce, but it appears that the submarine in the system would be quite large (more than twice the size of the Polaris ships) and that it would be capable of launching 20 to 24 large missiles equipped with MIRV's. It is proposed to have a higher maximum speed and to incorporate the latest available silencing techniques, although these two objectives are competitive.

The ULMS program has been divided into two parts. The first stage (ULMS 1) would involve a new missile with a range of about 4,500 nautical miles capable of being deployed in the present Polaris submarines as well as in any new vessel. The second stage (ULMS 2) would include the development of the new submarine and a still more advanced missile with a range of about 6,000 miles that would be too big to be

FIRST SUCCESSFUL TEST LAUNCH of a Poseidon missile from a submerged submarine was carried out in October, 1970. The missile was launched from the *James Madison*. The Poseidon missile, like the earlier Polaris models, is expelled from its silo by compressed gas; its rocket engine is then ignited after the missile has cleared the surface of the water.

substituted for the Poseidon in the existing Polaris ships. A maneuvering reentry vehicle (MARV) is also being developed for the new ULMS missiles. According to one estimate, the total cost of a program for 30 such ULMS vessels would be $39.6 billion [*see illustration at right*].

So far no convincing case has been made for the need to proceed with a replacement for the Polaris-Poseidon system and for making a commitment to a new large, high-speed submarine. Russian construction of SLBM's is no justification for ULMS; the Russian missile submarines do not in any way threaten the Polaris deterrent. Numerical superiority in launchers is meaningless; all authorities agree that the U.S. is far ahead qualitatively and can deliver from submarines about 5,000 warheads to fewer than 700 for the U.S.S.R. Even if we foolishly choose to race the Russians in the number of SLBM's, ULMS is certainly not the way to do it; each ULMS system will probably cost five or more times per missile launched than the Russian Y-class system.

The Poseidon with 10 or more MIRV's on each missile has a far greater capability than is needed to overwhelm any Russian ABM system that can be foreseen at present. Admiral Thomas H. Moorer, chairman of the Joint Chiefs of Staff, testified in February that "the Moscow ABM system even with improved radars and more and better interceptors could still be saturated by a very small part of our total missile force. In any event, the programmed Minuteman III and Poseidon forces, with their large number of reentry vehicles, provide a hedge against a future large-scale Soviet ABM deployment." Since such a large-scale ABM deployment will almost certainly be precluded by a first-stage SALT agreement, there is nothing in the ABM area that would require replacement of the Poseidon; in fact, even the Poseidon MIRV's will not be needed if a SALT ABM treaty is realized.

Therefore it is necessary to examine antisubmarine warfare to determine if there is anything that would currently justify the major ULMS step. Without going into a detailed analysis of possible ASW measures and countermeasures, suffice it to say that no evidence has yet been presented that the Russian ASW program could present a threat to the Polaris deterrent in the next decade. [See "Antisubmarine Warfare and National Security," by Richard L. Garwin, SCIENTIFIC AMERICAN Offprint 345]. Admiral Levering Smith, director of the

RESEARCH, DEVELOPMENT, TEST AND EVALUATION	MILLIONS OF DOLLARS
SUBMARINE	$800
MANEUVERING REENTRY VEHICLE (MARV)	$600
ULMS 1 MISSILE	$2,100
ULMS 2 MISSILE	$1,500
PROCUREMENT	
SUBMARINES (30 AT AVERAGE COST OF $400 MILLION EACH)	$12,000
ULMS 1 MISSILES (ENOUGH FOR 500 LAUNCHERS AT AVERAGE COST OF $8 MILLION EACH)	$4,000
ULMS 2 MISSILES (ENOUGH FOR 600 LAUNCHERS AT AVERAGE COST OF $11 MILLION EACH)	$6,600
SUPPORT	
ULMS REFIT FACILITIES (ONE PER 15 SHIPS AT COST OF $1,800 MILLION FOR FIRST, $1,200 MILLION FOR SECOND)	$3,000
OPERATION AND MAINTENANCE (FOR 30 SHIPS AT AVERAGE COST OF $30 MILLION PER YEAR FOR 10 YEARS)	$9,000
TOTAL	$39,600

ESTIMATE OF TOTAL COST of the Navy's proposed ULMS program was prepared by Members of Congress for Peace through Law. The first stage of the ULMS program (ULMS 1) would involve a new missile with a range of about 4,500 nautical miles capable of being deployed in the present Polaris submarines as well as in any new vessels. The second stage (ULMS 2) would include the development of a new, larger submarine and a still more advanced missile with a range of about 6,000 miles. A special maneuvering reentry vehicle (MARV) is being developed for these new, larger missiles. The entry under research, development, test and evaluation for the submarine includes the cost of all nonmissile subsystems and integration work. The procurement estimates assume that the new program would eventually involve 30 new submarines. The support estimates assume a need for two special refit facilities for the ULMS ships and include crew-related costs for 10 years. The cost of the ULMS 1 missile program by itself is estimated to come to about $6.7 billion.

Navy's Strategic Systems Project, testified in 1969 that even the new generation of Russian ASW submarines will not be able to follow our Polaris submarines, and that the U.S.S.R. has no specific new ASW methods that would make the Polaris fleet vulnerable to attack. That is still true today. The U.S. has spent tens of billions of dollars on ASW efforts over the past 20 years and still does not have any system that could even begin to approach the kind of capability that would be needed to eliminate 20 to 30 missile submarines almost simultaneously. The Russians are far behind the U.S. in this area, and they have the serious geographical disadvantages of remoteness from and unavailability of land areas contiguous to the oceans in which their ASW systems would have to operate. Since the nature of the potential ASW threat to the Polaris-Poseidon system cannot even be foreseen at this point, ULMS, if built now, may be designed to cope with the wrong threat. The most obvious improvement to Polaris-Poseidon would be to increase the range of the missile in order to enlarge the ocean

areas from which missiles could be launched. The deployment of such a new long-range missile might cost nearly seven billion dollars, however, and in any case, as the maps on pages 6 and 7 show, the Poseidon system already has a tremendous operational flexibility and is not threatened in its present launch areas.

Thus there are strong arguments for keeping both the ULMS missile and the ULMS submarine options in the research and early development stage. This would allow the exploration of all approaches, including smaller, slower but quieter submarines, and would avoid the making of a premature commitment to a large, expensive submarine and missile program. We must not fall into the trap of buying new military hardware just because we have made technological advances; there is no quicker way to price ourselves out of the security market. The submarine missile force is the backbone of our deterrent; its present strength and invulnerability obviate the need for its replacement for at least a decade.

16

Multiple-Warhead Missiles

by Herbert F. York
November 1973

MIRV's increase the number of strategic nuclear weapons and now threaten the stability of the nuclear balance of power. Their history shows why they present special problems of arms control

From 1945 to 1970 the number of nuclear warheads in the U.S. strategic arsenal went from zero to about 4,000. From 1970 to mid-1975 the number will increase to almost 10,000. This increase, about half of which has already been achieved, is made possible by a single development in military technology: the multiple independently targeted reentry vehicle (MIRV).

The MIRV system enables a single rocket to launch several warheads, which can be aimed at separate targets or made to approach the same target on different trajectories. Two of its characteristics may have important consequences for military strategy. First, MIRV's are better able to penetrate an anti-ballistic-missile (ABM) system than missiles with a single warhead. Second, because MIRV warheads are numerous and can be guided with great accuracy, MIRV's could lead to an effective "counterforce" weapon, one capable of destroying a very large part of an adversary's retaliatory forces if used in a surprise attack. Counterforce weapons are usually considered the most disruptive to the strategic balance, because they threaten the deterrent on which national security is presumed to depend. A measure of how threatening MIRV's can seem was provided in August, when the U.S. Department of Defense announced that the U.S.S.R. had made its first successful test of a MIRV missile. The Secretary of Defense, James R. Schlesinger, cited the flight test as evidence that the Rus-

sians "are seeking a strategic advantage."

In actuality it is highly unlikely that MIRV's could affect the ultimate outcome of a nuclear war between the U.S. and the U.S.S.R. It seems clear that each has ample power to destroy the society of the other, with or without multiple-warhead missiles and whether or not a surprise attack destroys the opposing force. By generating fears of the other side's intentions, however, MIRV's could affect the likelihood that a nuclear war will occur. Thus in the peculiar logic of the nuclear arms race the possession of a potent weapon may diminish national security rather than enhance it. In this context I shall discuss what the MIRV system is and how it works, what its antecedents were in the U.S. arms program and why the decision to build it was made. Finally I shall examine why efforts to stop its development and deployment failed.

Few details of the operation of multiple-warhead missiles have been made public. The dimensions and mass of the individual warheads, for example, have not been revealed, nor has the maximum separation of targets been stated. A general, nonnumerical explanation of the MIRV system can nevertheless be given. It is based on statements made by officials of the Department of Defense, on testimony given before Congressional committees and on data published by nongovernment organizations such as the Stockholm International Peace Re-

search Institute (SIPRI). In addition certain characteristics of the MIRV missile can be inferred from the operation of related devices known to have preceded MIRV's in the weapons and space programs. With this information it is possible to describe a hypothetical flight of a MIRV missile (*see illustration on pages 198 and 199*).

For this purpose let the missile be a land-based intercontinental ballistic missile (ICBM) carrying three independently aimed warheads. (Minuteman III is such a missile.) In the initial stages of flight the main rocket motors put the missile on a trajectory calculated to terminate near the first of the three selected targets. When the fuel of the last of the main rocket stages is exhausted, the final stage containing the warheads and their associated apparatus separates from the rocket and coasts on toward apogee, the point of greatest altitude. The apogee for ICBM's is typically about 800 miles.

Once the stage containing the warheads has separated from the larger stages, propulsion and guidance are provided by the post-boost control system (PBCS), the most important component of the MIRV system [*see illustration on next page*]. The final stage, consisting of the warheads and the PBCS, is often called the "bus"; it carries the warheads as passengers to be discharged at intervals. The bus has an inertial guidance system and small rockets that can modify its velocity and attitude.

After a brief period of coasting the bus refines its trajectory until it is aimed as precisely as possible at the first target. It then gently ejects one of the warheads. In satellite programs that were precursors of MIRV this ejection was accomplished with small springs in compression; the MIRV system could have a similar mechanism.

When the first warhead has been released, the bus alters its course in preparation for releasing the second. It can do this in several ways. First, the bus can increase its speed in the direction of its original orbit, causing the second warhead to impact "downrange," or farther away than the first. A decrease in speed, brought about by firing the rocket in a direction opposite to the direction of motion, will cause the warhead to land "uprange," or closer to the launching site. Another mode of deflection is perpendicular to the plane of the original trajectory. Impulses in this direction will aim the warhead at targets in an arc on each side of the initial target. Finally, by giving the bus an impulse in the plane of the original trajectory but roughly perpendicular to its direction, the second warhead can be aimed at the same target as the first. The second will approach from a higher or lower angle, however, and its arrival will be delayed by as much as several minutes. Ordinarily there would be movement on all three axes.

When the second weapon has been released, the bus is reoriented once again, this time on a new trajectory terminating at the third target. If more than three warheads are incorporated in the bus, the process is repeated until all have been launched at their targets. Several kinds of decoys and other "penetration aids" might also be released with some or all of the warheads.

Numerous engineering compromises are necessary in the design of the bus. The maximum separation between the targets, for example, depends primarily on the total impulse available from the bus propulsion system and therefore on the payload weight allotted to it. MIRV's now in use by the Navy and the Air Force can evidently reach targets separated by distances on the order of a few hundred miles, or a few percent of the total range of the missiles.

A more obvious compromise is that between the number of warheads and their aggregate explosive power, or "yield." Some yield is always lost in the transition from single to multiple warheads. Even in multiple-warhead missiles that do not provide separate guid-

ance for each reentry vehicle much weight, and therefore yield, must be sacrificed to extra heat-shielding for reentry into the atmosphere and to the diseconomies of smaller scale.

For true MIRV's the ratio is even less favorable, because the weight of the post-boost control system as well as that of the shielding must be subtracted from the payload available for the weapons themselves. The engineering is also more complicated, because the missile-maker has more options: he can choose to emphasize wider separation of targets, higher multiplicity of warheads, greater overall range, more penetration aids or higher yield. Presumably, optimum compromises have been found for all these choices; they depend on the nature of the target being attacked and on the defenses to be overcome.

The Department of Defense has not disclosed the yield of the warheads carried by the U.S. missiles now equipped with MIRV's. Estimates have been made, however, that for the purposes of this discussion are sufficiently reliable. (The figures given here are those published by SIPRI.)

Even if the precise yield of a warhead were known, it would be difficult to calculate the amount sacrificed in converting to multiple warheads, since in each case MIRV's were introduced in new missiles that had never carried single warheads. The Navy's Poseidon, for example, is a larger missile than the Po-

laris it is replacing, and it has twice the payload. Nevertheless, it is estimated to have a smaller yield. The last Polaris to carry a single warhead is reported to have had a yield of about one megaton. The Poseidon is usually said to carry 10 warheads of 50 kilotons each, for a total yield of about half a megaton.

A better indication of the compromises required may be given by the Minuteman III, an Air Force ICBM. The single-warhead versions of Minuteman had yields estimated at from one megaton to two megatons; Minuteman III is said to have three independently aimed warheads of 200 kilotons each, suggesting again that the aggregate yield is one-half or less.

It is important to note that a reduction in total yield does not necessarily imply a reduction in destructive effect. Megatonnage and "throw weight" are not reliable indicators of the destructive capability of nuclear weapons. A better measure is the circular area within which a given warhead will cause some specified degree of damage; this in turn is a function of the "blast overpressure." For overpressures high enough to be effective against military targets the radius of destruction increases roughly as the cube root of the explosive yield; the area of destruction, therefore, increases as the two-thirds power of the yield. For example, if a one-kiloton device could destroy a particular target within an area of one square mile, then to produce

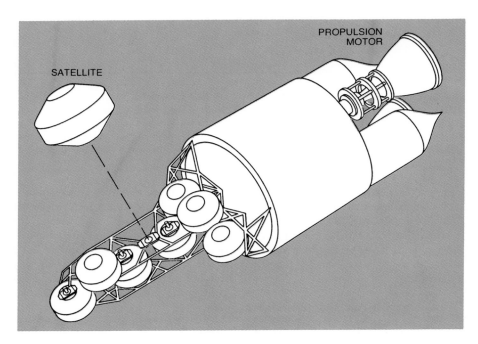

TRANSTAGE, a post-boost control system used with the Titan III booster rocket, is shown carrying defense communication satellites, which it was used to launch beginning in 1966. The satellites are mounted in a tubular frame and are ejected by small springs. The Transtage was the immediate predecessor of the MIRV system used in the Minuteman III ICBM and has been said to incorporate all the essential technology of a MIRV warhead.

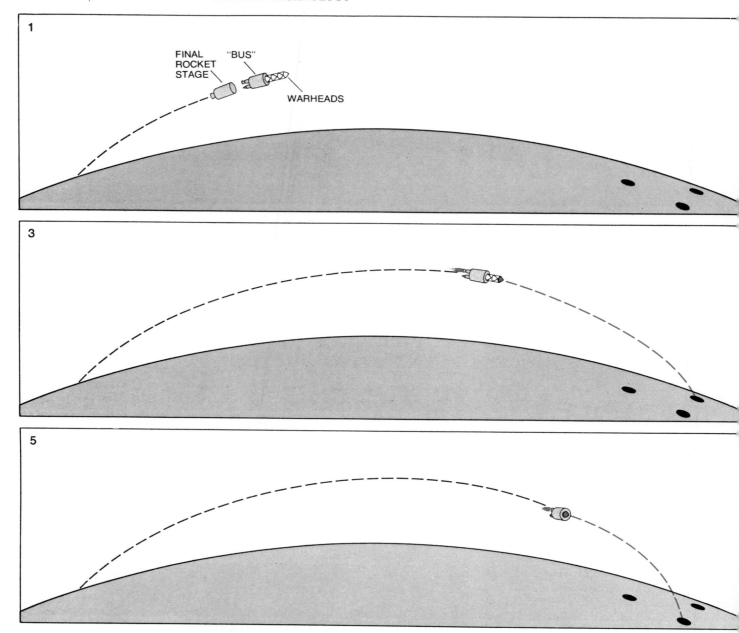

POST-BOOST CONTROL SYSTEM, or "bus," is at the heart of MIRV; its operation is illustrated schematically in these drawings, which show a hypothetical flight of a bus carrying four warheads. When the last of the main rocket stages is exhausted (1), the bus, with the warheads attached, separates from it. After coasting brief- ly the bus adjusts its trajectory until it is aimed as accurately as possible at the first of the selected targets. The first of the warheads is then gently ejected (2) and will continue to follow the ballistic trajectory to the target. After firing its rocket motor to add an increment of speed the bus ejects another warhead (3), which will

an area of destruction of two square miles would require a warhead of about 2.8 kilotons. Because of this exponential relation the potential for destruction is greater with many small weapons than it is with a single large one of the same total yield.

The history of how multiple-warhead missiles came to be developed in the 1960's and how they came to be deployed in the early 1970's provides an interesting lesson in the structure and operation of the military and military research organizations. The technology necessary for MIRV's evolved from re-

search directed toward several independent and quite disparate goals. Ideas and personnel were exchanged among the various programs, so that the course of development became not a thread but a fabric. It could have been cut in any number of places without seriously impeding the progress of the MIRV system.

Once the technology was developed MIRV assumed a momentum of its own; the chances of halting it were by then slim. In addition a number of quite different arguments were presented in favor of deployment, and apparently any one of several might have been sufficient

to gain the necessary Congressional and Department of Defense approval. Many of the development decisions could have been countermanded on several occasions and the result would have been about the same: MIRV's on U.S. missiles at the beginning of the 1970's.

A convenient moment at which to begin an examination of the history of MIRV's is the launching of the first Russian artificial satellite in October, 1957. About a month after the satellite went into orbit, and at least partly in response to the launch, William M. Holaday, director of guided missiles in the Office of the Secretary of Defense, established

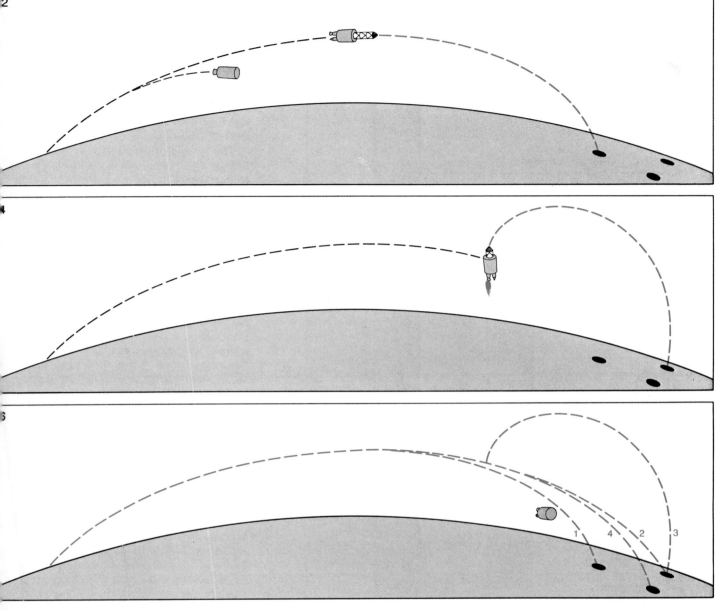

strike a target farther downrange. The bus next increases its velocity in a direction roughly perpendicular to the direction of the trajectory (4), placing the third warhead on a path that will lead it to the same target as the second but will delay its arrival by as much as several minutes. The last warhead is released after the bus has executed a final maneuver, adding a velocity increment in the plane that in this illustration is perpendicular to the page (5). Many other combinations of these movements are possible. Finally, the bus disintegrates on reentering the atmosphere and the warheads reach their assigned targets (6), although not simultaneously.

the Reentry Body Identification Group, with representatives from several agencies of the Department of Defense, from industry, from academic research departments and from such consulting firms as the Rand Corporation. The committee was formed to determine whether or not the designers of offensive ballistic missiles should consider seriously the possibility that defensive missiles might be built by the opposition. In early 1958 the committee reported that missile defense should be given consideration; it also described, however, a number of countermeasures available to the offense.

All but one of the countermeasures proposed by the Reentry Body Identification Group were intended to confuse the radars of the defenders. They included decoys, objects that to radar would resemble a warhead; booster fragments, pieces of the rocket-motor fuel tanks used as decoys but available at no weight penalty; chaff, small lengths of wire dispersed in space to act as a radar reflector; a reduced radar cross section for the warhead, and radar blackouts produced by exploding thermonuclear devices in the upper atmosphere. The remaining proposal, and the most important for this discussion, was the use of multiple warheads. Rather than confuse the defense,

multiple warheads simply exhaust it. These countermeasures and others are what are collectively called penetration aids.

Decoys, even cheap and light ones, can approximate the characteristics of a warhead as long as both the decoys and the real weapons are in space, where all objects, regardless of mass and shape, follow ballistic trajectories. Once the decoys enter the atmosphere, however, they are slowed by air resistance to a greater extent than the heavier warheads, and the difference in velocity becomes progressively greater as the reentry bodies reach denser regions of

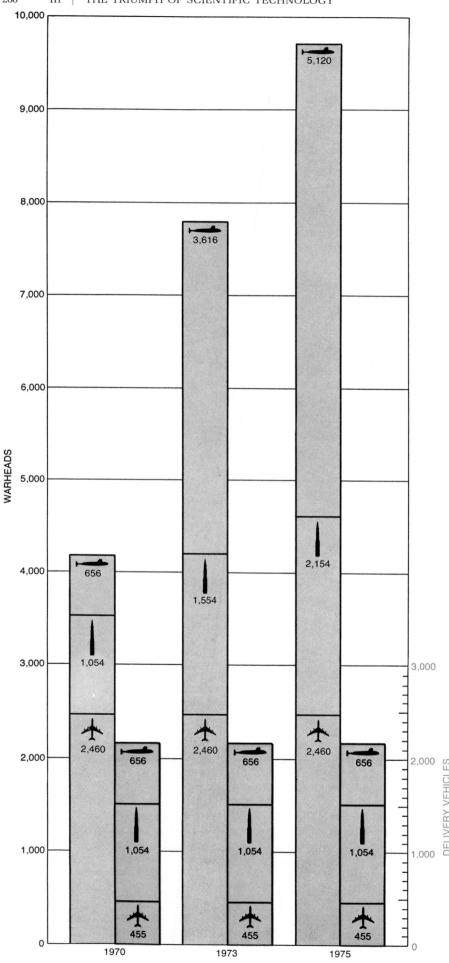

the atmosphere. Booster fragments, chaff and such light decoys as balloons covered with metal foil will disintegrate and burn up in the upper atmosphere. Thus the differential effects of the atmosphere on the reentry bodies enable the defense to discriminate between decoys and weapons (if it is willing to wait until well after they have entered the atmosphere) and to allocate defensive missiles only to the real weapons.

The solution for the offense is to use heavier decoys that will mimic the flight characteristics of real warheads to lower altitudes. As the weight of the decoys approaches that of the warhead itself, however, it becomes more efficient simply to use several warheads.

This was in fact the strategy adopted for the Polaris A-3, the last of the submarine-launched ballistic missiles (SLBM's) in the Polaris series. The A-3 is a multiple-reentry-vehicle (MRV) system with three warheads. The cluster is launched as a unit, on a trajectory chosen to guide it as accurately as possible to the single target. When the rocket has burned out, the three weapons are separated and given small additional impulses. They continue on separate but close trajectories and impact in a triangular pattern, presumably centered on the target. The missile might be compared in principle to a shotgun.

The dimensions of the triangle formed by the impact points of the warheads have never been made public, but the military situation as it was perceived in the late 1950's set obvious limits. The separation had to be more than a few tenths of a mile or all three warheads could have been destroyed by the explosion of a single antimissile missile. On the other hand, they could be no farther

NUMBER OF WARHEADS in the U.S. strategic force will increase almost two and a half times in the first half of the 1970's, even though the number of delivery vehicles (bombers, ICBM's and SLBM's, indicated in the graphs by symbols) will remain constant. The statistics are those published by the Stockholm International Peace Research Institute. The graph marked 1970 represents the force level before any MIRV's were installed; 1973 bars are for early in that year; figures for 1975 show the number of warheads when all Poseidon and Minuteman III MIRV's now planned are deployed. In these charts it is assumed that the Minuteman III carries three warheads and the Poseidon 10 warheads. MRV's are considered single warheads since each MRV assembly could be directed to only one target. The figures given for weapons carried by bombers are approximate and can change quickly.

apart than a few miles or the dimensions of the pattern would have exceeded the size of most cities in the U.S.S.R. The yield of each warhead is estimated to have been about 200 kilotons.

The first Polaris A-3's were deployed in 1964. Even before they joined the strategic-weapons force it was recognized that the MRV warhead would not be able to cope with improved ABM systems. By 1962 or 1963 progress in U.S. defensive missiles and the knowledge that Russian defenses were also being improved made it clear that the separation of warheads in the A-3 MRV was much too small to thwart any but a primitive ABM. Moreover, it was seen that the solution could not be found by increasing the horizontal spread of the impact points; to do so would make the pattern larger than most targets. Another solution was soon found: the MIRV.

The launching of the first Russian satellite and the launching of the first Russian ICBM (which had preceded the satellite by two months) stimulated in the U.S. an outburst of ideas about how to make and use satellites and missiles. The use of penetration aids, as we have seen, was one of the results of this process. Another was the concept of launching more than one satellite with a single rocket.

The earliest proposal of multiple satellite launchings of which I am aware was directed to defense against missile attack. This was the ballistic antimissile boost interceptor (BAMBI). The objective was to intercept an enemy's missiles during the first few minutes of flight, while the booster motors were still operating. Missiles were thought to be particularly vulnerable during this period because simply puncturing their propellant tanks could make them fall thousands of miles short of their target.

Two versions of BAMBI were proposed; both would have placed large numbers of satellites in orbits from which missiles could be detected and destroyed in the early stages of flight. One version, the random-barrage system, would have used many thousands of small satellites, launched by, say, hundreds of boosters. The other, called space patrol air defense (SPAD), would have deployed the small "killer" satellites in a "mother ship" equipped with central guidance and detection devices. On command the mother ship would have oriented itself and determined when, at what rate and in what direction to launch its subsatellites.

Neither of these systems was built,

but they were studied on paper by the Advanced Research Projects Agency (ARPA) of the Department of Defense with help from the Rand Corporation. During the early years of ARPA (it was founded in 1958) there was an active interchange of technical personnel between the agency and industry and between the various companies most heavily involved in missile and space technology. As a result many of those who helped to fashion these early proposals were later members of the organizations that designed real multiple-warhead devices.

A quite unrelated development whose basic technology was later adapted to MIRV's was the Able-Star, a second-stage vehicle designed to be used with the Thor booster. It was the first spacecraft where the main propulsion rocket could be shut off and later restarted. The Able-Star used hypergolic propellants (substances that ignite on contact) and incorporated restart, guidance and control devices, a programmer and an accelerometer—all necessary to the operation of MIRV's.

The Able-Star was first tested in space in April of 1960. Two months later it was used in the first multiple satellite launch, in which a Transit II-A satellite and a Naval Research Laboratory solar radiation satellite were placed in near-circular orbits 500 miles above the earth. Once the Able-Star achieved the proper orbit the satellites were detached and separated by a compressed spring, giving the smaller satellite an additional velocity of 1.5 feet per second.

In a subsequent launch the Able-Star was used to place three satellites in similar orbits, although the procedure was only partly successful. In 1963 the Atlas-Agena rocket was used in a more difficult maneuver: placing a pair of satellites in very different orbits. Later versions of the Agena second stage, like the Able-Star, could be stopped and restarted during flight. The satellites, called Vela, were used to monitor compliance with the Limited Test-Ban Treaty of 1963. They were placed 180 degrees apart in orbits from 62,000 to 72,000 miles high.

The immediate technological ancestor of the Air Force version of MIRV was Transtage, a highly flexible post-boost control system. It was crucial in the development of the components and techniques used in MIRV's, yet it was devised for reasons unrelated to the effort to improve missiles and missile warheads.

Transtage was used with Titan III, which in the early 1960's was the largest of the U.S. booster rockets. Transtage

had a propulsion system capable of coasting and restarting, like the Able-Star and the Agena, but it carried a larger payload and was capable of more complex and·more extensive maneuvers. It was conceived without a specific mission in mind, and it was first used to launch a series of defense communication satellites called IDCSP (for initial defense communication satellite program).

The special requirements of defense communication demanded that the satellites be many and that their orbits be quite high. On June 16, 1966, a Titan III-C and Transtage placed eight 100-pound satellites in eight different equatorial orbits, all at an altitude of about 21,000 miles.

The operation of Transtage was comparable in almost all respects to that of the MIRV bus. Using its ability to coast and restart, it first achieved a near-circular orbit at the proper altitude with a period of 1,334.2 minutes. It gently nudged off one of the subsatellites with compressed springs. Then, with four vernier motors of 50 pounds' thrust (whose main purpose was controlling pitch and yaw), it added a small increment of velocity and ejected a second satellite.

This one would orbit at essentially the same altitude, but with a period of 1,334.7 minutes. The maneuver was repeated for each satellite, until the last was dropped off three minutes after the first in an orbit with a period of 1,347.6 minutes.

Three more groups of IDCSP satellites were launched, using the same technique, at intervals of about six months. The importance of the program in the development of MIRV's was indicated in 1968 by John S. Foster, Jr., the director of Defense Research and Engineering. Asked during his testimony before a Senate subcommittee why he was so confident the Minuteman III and Poseidon MIRV's would work, he cited the successful operation of Transtage as proof that all the essential engineering problems had been solved.

Another way to pursue the relation between these projects and the later MIRV missiles is through the contractors who produced them. For example, Agena was designed by the Lockheed Missiles and Space Company (a subsidiary of the Lockheed Aircraft Corporation), which later designed the Poseidon missile. Similarly, systems engineering for the Titan III and Transtage was done by the Aerospace Corporation, which later did the concept engineering for the

Minuteman III MIRV. Another company involved in the MIRV program for Minuteman was the Space Technology Laboratory of the Thompson Ramo Wooldridge Corporation, which had earlier participated in the Able-Star and Vela satellite programs. In the Air Force itself the Space and Missile Systems Office, which supervised the Transtage program, was soon to begin development of the Minuteman MIRV.

In addition to all these programs, bits and pieces of MIRV technology were invented or reinvented independently in the course of unrelated endeavors. One of them was a study by the Rand Corporation of what advantage the U.S.S.R. might gain from its relatively large missiles. Among other conclusions of the report was the possibility that such missiles might be used to deliver multiple warheads; a MIRV-like device was hypothesized as the means for doing so. Another was a study of orbital offensive weapons, conducted for the Air Force by seven competing firms. In an orbital bombardment system nuclear weapons would be placed in permanent orbit and brought down on their targets from space on receipt of coded instructions.

Some of the proposals involved what was essentially a one-passenger bus. The system was never developed, but if it had been, the technology of MIRV's might well have been derived from it.

Another program that could have been used as a point of departure for MIRV was the sequential-payload-delivery system (SPD), used to deliver unarmed warheads from California to Kwajelein Atoll in the Pacific, where they served as test targets in the ABM program. The justification for this system was economic: it is cheaper to attack several targets with a single rocket than to use a separate rocket for each target. The design and construction of the sequential-payload-delivery system was supervised by the Aerospace Corporation, which had been responsible for Transtage.

Almost all the mechanisms and techniques used in MIRV's could also have been derived from the civilian National Aeronautics and Space Administration manned space program, particularly the systems used in lunar exploration. I have described here, however, only those programs that were addressed to some military purpose.

By the mid-1960's it was clear to military planners that a MIRV missile could be built. Even before then, in 1962 and 1963, two independent arguments had been put forward in support of deployment.

One was embodied in the "counterforce" speech given by Secretary of Defense Robert S. McNamara in Ann Arbor, Mich., in 1962. The notion of counterforce did not begin with McNamara; it was a part of military strategy before nuclear weapons were invented. It holds that one should plan to attack an adversary's weaponry, the "counterforce targets," rather than his cities and industries, the "countervalue targets." It is the strategy usually adopted by those who favor expanded deployment of weapons, and it is necessarily the strategy of those who contemplate making a preemptive surprise attack.

McNamara subsequently modified his position on counterforce, but the speech nevertheless stimulated the proposal of MIRV's as a means of increasing the number of targets that could be attacked. Indeed, there can be no doubt that in a counterforce strategy MIRV is a powerful weapon. In a preemptive at-

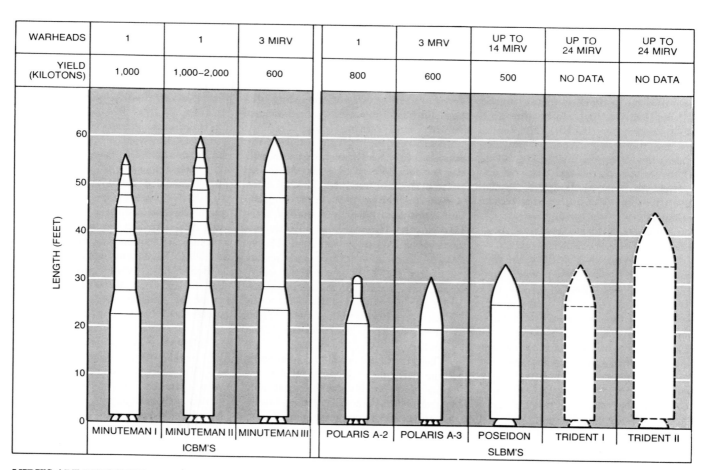

WARHEADS	1	1	3 MIRV	1	3 MRV	UP TO 14 MIRV	UP TO 24 MIRV	UP TO 24 MIRV
YIELD (KILOTONS)	1,000	1,000–2,000	600	800	600	500	NO DATA	NO DATA

MINUTEMAN I	MINUTEMAN II	MINUTEMAN III	POLARIS A-2	POLARIS A-3	POSEIDON	TRIDENT I	TRIDENT II
ICBM'S			SLBM'S				

MIRV'S ARE DEPLOYED on two of the six U.S. missiles that are now operational, the Air Force Minuteman III and the Navy Poseidon. Although both have larger payloads than the missiles that preceded them, the total yield of their warheads is less. The Trident I, which could carry as many as 24 warheads, is expected to be introduced by 1978; the Trident II is planned for the mid-1980's.

tack almost perfect system reliability is essential, since the failure to destroy even only a few of the enemy's missiles would result in great harm to the aggressor. Because the reliability of individual missiles is not likely to even approach 100 percent, the first-strike force must be substantially larger than the force being attacked. Even if the adversary offers no defense, several warheads must be allotted to each of his missile silos. If missiles with single warheads are used, the military and economic advantage is clearly with the defender.

MRV's offer no advantage over single warheads; they merely scatter weapons over the target area and give no assurance of hitting a small, specific site. MIRV's, however, substantially improve the chances of the offense. With three or more warheads per missile, designed so that each can be assigned a separate target, a nation could attack an opponent's fixed-base forces without first enlarging its own arsenal. By aiming many weapons at each silo it could greatly increase the probability that all the missiles would be destroyed.

MIRV's also satisfy another major requirement of a first-strike force: high accuracy. If a missile is to be destroyed inside a hardened silo (a buried tube armored with steel and concrete), a warhead must be exploded at quite close range, perhaps even within the "cratering radius" of the weapon. Because the radius of destruction against a hardened target increases very slowly with increasing yield it is generally considered more profitable to improve accuracy than to increase explosive power.

The customary measure of accuracy is the "circular error probable," a circle centered on the target with a radius such that half of the warheads aimed at the target will fall outside the perimeter. To be useful as a counterforce weapon a warhead should have a radius of destruction against its particular target somewhat larger than the circular error probable. For example, if a given warhead has a circular error probable of one mile and a radius of destruction against a given silo of half a mile, far fewer than half of the warheads could be expected to destroy the missiles they are aimed at. To achieve 50 percent reliability the yield could be increased eightfold; the same result could be obtained by a twofold improvement in accuracy.

The second argument favoring the deployment of MIRV emphasized its ability to penetrate missile defenses. Russian nuclear tests at high altitude in 1961 and 1962 and Premier Khrushchev's famous boast that "You can say our rocket hits a fly in outer space" led many U.S. planners to ascribe to the Russian ABM performance at least as good as that which could in theory be achieved by the U.S. system. In particular it was concluded that a single large antimissile missile could simultaneously destroy all three warheads of the Polaris A-3.

For the Navy there were at least two additional motives for building MIRV's. First, the Polaris A-3 development program was nearing the end, and it was almost automatically assumed that there would be a new generation of sea-launched missiles. Second, improvements in submarine navigation for the first time made high accuracy possible in SLBM's. (In order to launch a missile with a specified accuracy the submarine must know its own position and orientation to at least that degree of accuracy.)

The new missile was at first designated the B-3, but the name was later changed to Poseidon. From the beginning it was evident that it would be a more powerful rocket than the A-3. The increase in power was obtained largely by making the rocket bigger.

At the outset it was not at all clear that MIRV's would be used in the Poseidon. The desire to compete with the Air Force in the counterforce role led some on the Navy staff to favor a large, accurate, single warhead, whereas others, including Harold Brown, who was then the director of Defense Research and Engineering, sought continued emphasis on MRV's as a means of ABM penetration.

The Navy Special Projects Office soon proposed a compromise, a device that was never built but that served as a link between MRV and MIRV. In this concept a bus would have delivered several warheads to a single target but on trajectories that had different apogees. Each warhead would be aimed at the target as accurately as guidance technology would allow; they would arrive at different times, however, and would be spaced as much as 100 miles apart. They would have the same capacity for exhausting a newer, more sophisticated ABM as the Polaris MRV had against a crude ABM. Such a delivery system requires a bus capable of making modest changes in its velocity in the plane of the trajectory and roughly perpendicular to its direction. To transform this proposed bus into one suitable for MIRV's it is necessary only to provide a means of adding small increments of velocity in the other two directions.

Continuing progress in missile guidance and anticipation of even more improvement soon led the Special Projects Office to propose the high-multiplicity MIRV deployed today. The Poseidon is almost twice as heavy as the Polaris A-3 and carries about twice the payload. Its nominal range is about the same, 2,500 miles, but this could be increased by sacrificing payload. It is reported to be capable of carrying as many as 14 warheads, but in most analyses it is usually considered to carry 10. The warheads are estimated to have a yield of 50 kilotons each.

The first boatload of Poseidon missiles was deployed at sea in April, 1971, aboard the missile submarine *James Madison*. Converting from Polaris to Poseidon requires the installation of new launching tubes to accommodate the greater length and girth of the new missile. At least 17 boats have been converted so far, and eventually 31 of the nation's 41 missile submarines will carry Poseidons. Thus if each missile does indeed carry 10 warheads, the number of independently targeted reentry vehicles in the nation's sea-launched fleet will increase from 656 to 5,120.

Although the Poseidon will not be fully deployed for another two years, its successor is already being planned. It is the Trident, a still larger missile that in its final version will have a range of 6,000 miles. Each missile will be armed with as many as 24 independently aimed warheads. The Navy intends initially to build 10 Trident submarines, carrying 24 missiles each.

In the Air Force the argument over single v. multiple warheads persisted into the mid-1960's. Concerned, like the Navy, over progress in missile defenses and over the Russian high-altitude test series, the Air Force began to consider MIRV's as a means of improving penetration. At the same time a steady increase in the perceived number of military targets in the U.S.S.R. stimulated interest in the use of multiple warheads to improve force effectiveness.

On the other hand, the Air Force had for some time been committed to large, single warheads protected against interception by decoys, chaff and other penetration aids. At one point the factional dispute reached the missile-industry press when a contract for the Minuteman III warhead was delayed following the intercession of the director of Defense Research and Engineering.

Early in the debate two versions of MIRV were considered. One employed a bus virtually identical with the Transtage already under development. In the alternative system a rocket booster would have launched a cluster of small, single-stage missiles, each incorporating a propulsion and guidance system. After the cluster as a whole was placed on a trajectory leading to the area of the targets the individual missiles would adjust their velocities separately to impact on their particular targets. The extra weight and extra cost of the individual guidance and control systems made this proposal less attractive than the bus, and its development was never authorized.

The bus type of post-boost control system was selected for Minuteman III in 1964; the first flight of 10 missiles was turned over to the Strategic Air Command in June of 1970 at Minot Air Force Base in North Dakota. Minuteman III carries three warheads, usually said to have a yield of about 200 kilotons each.

The decision to deploy MIRV's was made all but inevitable by the decision to develop them. Weapons systems, once proved feasible, assume a momentum of their own, under the rationale of "If it can be done, it must be done, because if we don't, they will."

Even so, the deployment of MIRV's

was debated after the development decisions were made, and the matter was not settled until the missiles were in place in 1970 and 1971. As we have seen, the arguments usually cited in favor of MIRV deployment are improved penetration of missile defenses and an increase in the number of targets that could be attacked. The penetration of an anti-ballistic-missile system was stressed by those who perceived a tradition of defensive measures in Russian history and who therefore believed that the U.S.S.R. would construct such a system. The proliferation of potential targets was emphasized by those who advocated a counterforce mission for U.S. nuclear weapons.

Secretary McNamara had other, primarily political motives for the deployment of MIRV's. Multiple warheads, he contended, offered a less costly way than the addition of more missiles to expand the strategic force and maintain at least some counterforce capability against growing Russian forces. Thus the potential powers of MIRV's were invoked in the arguments of McNamara and his

staff against strategic-force expansion.

McNamara also mentioned MIRV's in arguing against deployment of missile defenses. He doubted that the proposed anti-ballistic-missile network would work and believed it might bring on a new cycle in the arms race. He opposed its deployment in the U.S. and tried to persuade Premier Kosygin (at the conference in Glassboro, N.J.) that the U.S.S.R. also should forgo antimissile systems. A U.S. commitment to deploy MIRV's was among his arguments, since MIRV's represent a relatively inexpensive means of overcoming any conceivable antimissile system. Thus, from the point of view of McNamara and some of his immediate associates in the Office of the Secretary of Defense, the deployment of MIRV's could benefit the cause of arms control.

Indeed, the limitation of ABM systems achieved as part of the agreements made in the strategic-arms-limitation talks (SALT I) of 1972 might in part be attributed to the existence of MIRV's. It could be argued that the U.S. and the U.S.S.R. were willing to renounce extensive antimissile systems because MIRV's

STIMULI TO DEVELOPMENT

FIRST RUSSIAN SATELLITE AND ICBM

LIMITATIONS OF DECOYS

RUSSIAN HIGH-ALTITUDE NUCLEAR TESTS

COUNTERFORCE CAPABILITY

RUSSIAN ABM AND MRV VULNERABILITY

PRECURSOR DEVICES AND PROGRAMS

ABLE-STAR

AGENA

TRANSTAGE

SEQUENTIAL PAYLOAD DELIVERY

MRV

NASA

PARTICIPANTS IN DEVELOPMENT

REENTRY BODY IDENTIFICATION GROUP

ADVANCED RESEARCH PROJECTS AGENCY

AIR FORCE SPACE AND MISSILE SYSTEMS OFFICE

NAVY SPECIAL PROJECTS OFFICE

DIRECTOR OF DEFENSE RESEARCH AND ENGINEERING

AEROSPACE CORPORATION

SPACE TECHNOLOGY LABORATORY

INDUSTRIAL AND CONSULTANT COMPANIES

ADVISORY COMMITTEES TO ALL OF THE ABOVE

JUSTIFICATIONS FOR DEPLOYMENT

ABM PENETRATION

PROLIFERATION OF TARGETS

COUNTERFORCE STRATEGY

LIMITATIONS OF FORCE EXPANSION

DISCOURAGEMENT OF ABM

MULTIFARIOUS PATHS OF DEVELOPMENT, culminating in MIRV warheads, are suggested by a chart showing some of the motives for development, the projects and devices that preceded the weapons, the organizations that participated and the justifications proposed for deployment. The elimination of any one, or even several, of these programs would not have halted MIRV, since develop-

ensured the futility of building them. MIRV's are themselves excluded from the regulation of the SALT I agreements. They are considered a "qualitative" refinement in weaponry and therefore are not subject to restriction. They are expected to be a major topic in the second round of the talks (SALT II), scheduled to conclude by the end of 1974.

If McNamara believed that the deployment of MIRV's could slow the arms race, officials of the U.S. Arms Control and Disarmament Agency saw it differently. As early as 1964 Herbert Scoville, Jr., and George W. Rathjens noted ways in which the deployment of MIRV's could upset the "balance of terror." In particular, they predicted that an adversary could construe the deployment of MIRV's as possible preparations for making a preemptive strike. If a MIRV force is capable of destroying an adversary's ICBM's, that is, if it is an effective and reliable counterforce weapon, it could be considered a threat to the nation's deterrent and to that extent would be provocative. Of course, even a perfect

MIRV force could not make a first strike a rational policy, since sea-launched missiles and perhaps bombers would still be able to retaliate, but MIRV's do contribute some increment of instability. For example, in a crisis where nuclear war seemed imminent MIRV's might be perceived as giving an advantage to the aggressor and therefore could encourage a first strike.

Whether or not MIRV's are now sufficiently accurate and reliable to serve as counterforce weapons, they are considered menacing by those responsible for military planning. The MIRV missiles tested in August by the U.S.S.R., for example, were described by Secretary of Defense Schlesinger as leading the U.S.S.R. to "a clear advantage in counterforce capability." Under the SALT I agreement the U.S.S.R. is allowed about 25 percent more offensive missiles than the U.S.; with the deployment of MIRV's, Schlesinger suggested, this superiority in launchers and throw weight could be "married" to equality in technological sophistication.

During most of the period in which MIRV's were under development the program was kept secret and the controversies it entrained were unknown to the public and to many members of Congress. It did not become a political issue until late in 1967, when public debate over deployment began. In the 1968 Presidential campaign Senator Eugene McCarthy echoed the view of the Arms Control and Disarmament Agency, noting that "the introduction of sophisticated anti-ballistic-missile systems and new missiles equipped with multiple warheads threatens to make the situation unstable. With the deployment of such weapons systems, each side will become concerned as to whether in the event of a preemptive attack it will be able to inflict sufficient damage in retaliation—if not, its deterrent will not be credible. The arms race will thus be impelled to a new intensity. In crises, there could be an incentive to launch a first strike."

Later Senator Edward W. Brooke introduced a resolution calling for the suspension of testing of multiple-warhead missiles. Some other senators and a number of other people supported the resolution, but the effort was in vain. The first deployment of MIRV's took place while the issue was being debated.

The MIRV program had many roots and branches. Decisions were made by many people, some only loosely connected with one another, over the course of more than a decade. Of all the stimuli that led to the development of MIRV the

most important was the perceived need to penetrate ABM systems whose theoretical capabilities were slowly but steadily improving. Even without the stimulus of ABM, however, MIRV's would probably have been devised, and probably at about the same time. All their essential technology was developed in unrelated programs and for unrelated reasons: Able-Star, Agena, Transtage, sequential payload delivery. During the early phases of development the progress of MIRV was determined largely by the decisions of technologists who were attempting to solve problems presented by nature or responding to their perception of the technological challenges of the Russian missile and space programs.

Once proved feasible, MIRV's were also proved necessary, at least to the satisfaction of those who made the decision to deploy them. That decision was compelled by three factors: (1) the participants in and sponsors of the development program urged application of the new weapon; (2) MIRV's promised to thwart any ABM system the U.S.S.R. might construct and at the same time served as an argument against the deployment of missile defenses by either nation; (3) MIRV offered a relatively inexpensive way to increase the number of Russian targets that could be attacked, and thus ended debate over strategic-force expansion.

Opposed to these arguments were the predictions of the Arms Control and Disarmament Agency that deployment of MIRV's would be perceived as seriously disturbing the strategic balance and in crises would make the most dangerous policy the most profitable one. The first of these predictions has already turned out to be right. The number of people holding such views was negligible, however, and their voices were not powerful. Plans to deploy proceeded while the arguments went on.

In the development of multiple warheads there is a lesson for those who would reduce the world's armaments. Some programs, such as the B-70 bomber and ABM, are expensive, are addressed to a clearly evident and single purpose and depend on what might be called a unitary decision-making process. In principle they can be stopped by direct confrontation. Other programs, however, and MIRV is an example, evolve from many independent and seemingly unrelated goals and decisions. They are too diffuse, too protean, too difficult to define and delimit to be stopped by confrontation. They can be slowed or stopped only by slowing or stopping the arms race as a whole.

MIRV

ment could have taken alternative pathways. Similarly, had some arguments for deployment been refuted, others would have served.

III THE SOFT REVOLUTION

INTRODUCTION

The third revolution began gathering momentum after World War II, with the massive entrance into college of ex-GIs; with the concomitant explosion in higher learning and the growing acceptance of the social primacy of education; with the union of national power and modern science crowned by the harnessing of nuclear energy and the federal government emerging as a major sponsor of scientific investigation; with the sudden birth of rapid continental communication, ranging from the world's most modern and developed highway system, through rapid air passenger transport, to a uniquely effective instant transcontinental telephone system, and finally to a nation-wide television intimacy; with the transformation in managerial techniques wrought by the appearance of computers and other electronic devices that conquer complexity, distance, and even the diffusion of authority; and with the fading of industry as the most important source of employment for most Americans.

Zbigniew Brzezinski,
Between Two Ages

Technical and scientific advances are often directly or indirectly responsible for extensive social changes. Indeed, the significance of a new discovery is more often judged by a social than a scientific standard. The printing press, antisepsis, synthetic dyes, and the longbow, for example, are all judged important for their profound social effects. If the sole effect of the post-war scientific-technological-industrial complex had been merely to create such hardware, then its impact would have been significant, but hardly revolutionary. It is when a new technic arises primarily as a consequence of extensive social change or synergistically interacts with it that we come to speak of a "revolution" in society. Such revolutions are infrequent, but happen with enough regularity to be catalogued. In a few instances, the social and psychological effects of a new technic have been so overwhelming as to completely remold the self-image of the human race and to totally alter its relation to its environment.

The first of these revolutions, predating all possibility of accurate records, occurred some hundreds of thousands of years ago when an ancestral species, possibly not even of genus *Homo*, learned to use language and to form social structures. Although the direct connection between these two developments is purely hypothetical, they are frequently lumped together as the cultural revolution. Language and culture are technics only in the most general sense, but mechanical technics have also had their effects. The agricultural revolution, which involved "hard" technics to some extent, was the next great event. The existence of a guaranteed food supply was in large part responsible for the creation of formal social organization and the apportionment of labor. It shifted *Homo sapiens* from the common role of environmental adaptor to the extraordinary one of environmental manipulator. Lewis Mumford has emphasized the role of a balanced and secured food supply and relative freedom from famine on the development of culture in *Technics and Human Development* (Harcourt Brace Jovanovich, 1966):

> With this security, it was possible to look ahead and plan ahead with confidence. Except in the tropical areas, where soil regeneration was not mastered, groups could now remain rooted in one spot, surrounded by fields under permanent cultivation, slowly making improvements in the landscape, digging ditches and irrigation canals, making terraces, planting trees which later generations would be grateful for. Capital accumulation begins at this point: the end of hand-to-mouth living. With the domestication of grains, the future became predictable as never before; and the cultivator not merely sought to retain the ancestral past, but to expand all his present possibilities: once the daily bread was assured, those wider migrations and transplantations of men, which made the country, town and the city possible, speedily followed.

Thus agricultural technics—and by this is meant not only the introduction of a specific implement or technique, but also the conceptualization of controlled farming and husbandry—were responsible for the second "software" revolution, a reorganization of the basis of human society and the pattern of human interaction around the ownership and control of land.

This primal "soft" revolution continued to exert its influence over thousands of years. As villages grew, so did the specialization of labor, and with it a rough barter economy. The securing of food and shelter increased the probability of survival, which in turn encouraged population growth; and as the spreading villages became closer together a regular intervillage trade developed. Eventually the exchange between such villages began to harden into more formal ties, and some villages grew into trade centers, cities, and civilizations. The verbal contract, more binding to an isolated village culture than any subsequent body of written law, is less useful in a more diversified urban one. Actual barter would come to be replaced by symbolic barter, by money and some form of record. It is not at all surprising that the oldest written records in existence are inventories and account books. With the advent of writing, culture acquired a degree of permanence, and now sprang up the famed cities of antiquity. Babylon, Tyre, Athens, and Rome were all born of the lowly cereal grain, beside which the legendary acorn was an underachiever. For over ten thousand years, agriculture continued to be the primary activity of the human race. Even in our own century, wars were usually begun in the fall, after the harvest, when the food supply was secured and the men could be spared to fight.

Society remained by and large agricultural and rural for most of human history; not until the mid-nineteenth century did this pattern shift appreciably. Kingsley Davis has shown that until 1850 no single society in recorded history could be correctly described as being predominantly urbanized (see "The Urbanization of the Human Population," *Scientific American*, September 1965, Offprint 659). The urbanization of the industrial countries with the attendant creation of an urban proletariat was the sociological "software" accompanying the Industrial Revolution. The bulk of the productive labor had rested for some thousands of years on the backs of the slave, the peasant, and the serf; the great cities of history, from Babylon to Venice, were but the cultural tip of the agricultural iceberg. In Mumford's terminology, the first "megamachine" was built of humans and fueled by rice and corn and beans. Not until the Industrial Revolution was well underway did the economic base begin to shift from farm to factory, from country to city; only when machines began to run on coal and oil was there enough power available to rebuild the megamachine from nonorganic components.

This is not to say that society no longer has roots in agriculture. The newer technical base was at first superimposed on top of the agricultural base, not alongside of it. Among the first of its products were new machines to accelerate the shift of the population. Agricultural workers left the fields to man the factories to build the machines that enabled still fewer farmers to cultivate still larger fields. The continuous improvements in agricultural technology thus received an enormous boost just at the beginning of the Industrial Revolution. This boost further accelerated the shift of the labor force from the farm to the city. The rate of urbanization of the population of the industrial countries, which had been increasing steadily from 1850 to 1950, rose sharply between 1950 and 1960, when the growth of the new technological society began to shift the pattern of employment again—this time from the factory to the office. The agricultural worker had dominated the labor force for nearly ten thousand years, but the numerical superiority of the blue collar worker will certainly not extend more than two centuries from beginning to end.

The most important software effect of the new industrial state and of the emergence of the scientific and technological elites (who, as Daniel Bell suggests, will be its new managers, its technocratic bourgeoisie) is the transfer of employment and economic growth from production and distribution of goods to the creation, promotion, and distribution of *services*. The scientific and technological elites are most geared to the management of software, such as communication, information, transportation, and control. The hardware they produce is oriented primarily toward supplying these services. The cultural revolution provided us with language, the agricultural revolution with food, and the industrial revolution with material goods. The "technetronic" revolution now seems to be supplying us with "life style," an ephemeral product whose necessity was less than apparent to the generations preceding ours.

The decade during which the major part of this shift occurred is discussed in the article by Anne P. Carter, "The Economics of Technological Change." The input-output method of economic analysis developed by Wassily Leontief was necessary for dissecting the increasingly complex interlock of industries and technologies that prefigured the shift from product-oriented to service-oriented economy and the growth of the polymorphous-diverse conglomerate. In retrospect, this method appears to have been more revealing of the shift in the U.S. economic base in the early 1950s than even Carter realized. Having just analyzed the years 1947–58, Carter was still too close to perceive how clearly her results pointed to the developments of the next decade. Carter's table, covering the key transitional years, shows a large relative increase of nonmaterial inputs that is balanced by a decrease in the material inputs from which tangible goods are produced. The relatively small size of the computer and electronics industries in 1958 masks the equally important shift from tangible to intangible outputs, which was at that time just beginning. The trend to ephemeral goods has not been altered by the recent revival of handcraft industry, for what is marketed by craftsmen these days is more the handcrafting itself than the object crafted. Similarly, what the computer industry actually markets is the capacity to compute, not the machine itself. A scan of Carter's table reveals some quite clear trends toward a far more rapid increase in services than in goods. Less obvious, due to the complexities of the interlocks of the table, are the factors that measure the shift in employment patterns. Electronics, for instance, the largest percentage gainer in the input list, is under metalworking; but electronics is a "soft" industry with a very high ratio of white-collar to blue-collar workers. The growth of such technically based soft industries and the increasing interdependence of industries (which, as Carter points out, is a principal consequence of technological and scientific change over the past few decades) have reduced the market for "labor" and correspondingly increased the demand for coordinators and integrators—that is, industrial bureaucrats who administer not only to production and distribution but to advertising, interlocking, counting, accounting, management, and the management of management.

The evolution of the human economy may be compared with that of an organism. From an amorphous primeval soup, society evolved into a multicellular structure with functional, structural differentiation. The Industrial Revolution marked a shift toward centralized control and highly specialized organ functions. The post-war scientific and technological explosion, with its emphasis on transportation, communication, and information, may be said to correspond to the development of a "central nervous system" and "brain." As with an organism, the separate parts of the economy became increasingly interdependent as they grew more complex. They could now perform more efficiently by eliminating duplication of function, but the whole became more vulnerable as each sector came to be less capable of taking over for another in case of failure. Beyond a certain level of complexity, increasing vulnerability begins to outweigh increased efficiency unless the individual and collective functions of the separate parts can be coordinated and regulated. This requires an effective net of internal and external communication and a comprehensive information storage system with rapid access and retrieval. Without the explosive growth of communication and information in the past few years, the new industrial state could not possibly have developed the highly technologized, tightly interwoven social complex required to sustain it.

John McCarthy explains the importance of the electronic computer in this transformation in the opening sentences of his essay, "Information": "The computer gives signs of becoming the contemporary counterpart of the steam engine that brought on the Industrial Revolution. The computer is an information machine." The steam engine processed energy, converting it from a raw state as the locked-in chemical energy of fuel to a finished form as usable mechanical work. The computer processes raw information, accumulating and storing inputs and data and shaping them into usable forms—tables, charts, predictions, even decisions. The information used by Carter, for instance, to set up her input-output table has always been more or less obtainable. The processing of such information by the method of Leontief, however, would have had only a limited value without the availability of digital computers to perform the inversion of the enormous matrices required. Input-output economics itself is a piece of software derived from the computer, and mathematical economists are part of the new technological labor force. There is another reason why McCarthy's analogy is well-chosen. The steam engine was the keystone of the Industrial Revolution, not for what it was, but for what it made possible. The computer industry occupies a similarly narrow economic niche in itself; but the primary role of the computer lies in its use, not its production. Those factors that qualitatively differentiate the new industrial state from the pre-war period are directly attributable to the enormous increase in the ability to handle, store, and process information, which the computer has made possible. Automation, for example, depends on the ability of a machine to store and retrieve large quantities of information concerning the position of the work and the tool, an evaluation of the progress of the work at each stage, and a system for handling various common malfunctions in an orderly manner. Many industries now use processes so complex and rapid they would not have existed without computer control. The information is delivered to the controller too quickly and in quantities too great for a human to assimilate, process, and react to the data in time. Communication systems, from the post office to the telephone, depend on simple computers for rapid sorting and routing and optimal use of channels. Even such mundane institutions of modern society as credit-card services would be nearly paralyzed without the ability to store and retrieve information quickly and efficiently.

Another new software technique made possible by the computer is systems analysis, a method of modeling social and other problems not inherently suited to mathematical treatment in a form that makes them amenable to investigation and optimization by computer techniques. The general approach is first to create a model that can be programmed and then to vary the parameters corresponding to different parts of the system in order to test its response to various possible actions that might be taken. For any reasonably complex social system, the model itself is so enormously complicated that experiments can only be performed on a high-speed digital computer. Systems analysis has often been called a solution that distorts its problem, a method that causes only those questions to be asked that have readily quantifable answers. The basic difficulty is that not all components of a human society are reducible to mathematical statement, and the tendency all too often is simply to disregard all values not amenable to cost-benefit analysis.

Both the benefits and the limitations of systems analysis are quite apparent in the article by William F. Hamilton II and Dana K. Nance, "Systems Analysis of Urban Transportation." The work of Hamilton and Nance is of particular interest because of its attack on a transportation problem of considerable social importance. Such luxuries as super-airliners, SSTs, and high-speed intercity ground transportation are technologically fascinating; but the major transportation problem in the United States is intracity transport. Despite the increasing urbanization of the population, the amount and quality of transport in most cities has suffered a marked decline over the past twenty years. The private automobile has not only caused the cities to sprawl out and the population to disperse, but has actively undercut public transportation by totally dominating all thinking about transportation problems. With the growing acceptance of the reality of the strangulation of the city by its automobiles and the impending shortages of fuel, those agencies that had calmly let public transportation systems wither or die have come to recognize that the lack of urban transport is a tragedy not only for the poor but for the entire population. The authors set out to analyze both old and new transportation technologies in an attempt to estimate whether allocated transportation funds would be better spent on revamping older systems or designing new ones. The analysis indicated that new technologies—personal-transit and personal-capsule systems—would be the most efficient technical solution.

It is characteristic of systems analysis that the technically optimal solution almost always seems to lie with innovation. This is not surprising, since a new system can usually be designed to optimize the analysis, whereas the old systems just grow. The experience of skyrocketing costs and many years of delay on the Bay Area Rapid Transit system in the San Francisco area should serve as a warning to those who imagine that the costs of a new technology can be accurately predicted, as should the regular experience of military procurements, which frequently have large cost overruns. Hamilton and Nance, however, took their cost-benefit analysis one step further to determine whether the social benefits of the new modes of transport would outweigh the social liabilities. With the inclusion of this very important factor, their analysis indicated that cities with existing rapid-transit might spend their money better by upgrading and modernizing the old system. Their procedure is most interesting precisely because of their careful screening of the meaning of their results; they attempt to include the subtle social issues and refuse to bow to the technological imperative. They have not fallen into the trap of leaving the larger part of the decision-making process to the computer. The analysis they present may be deficient in some ways, but it does not have that central flaw. Not all systems analysis has been done so cautiously, and not all systems analysts are so firmly in control of their results. The computer is an intimidating machine, and its behavior with sophisticated programming often causes the metaphor of "electronic brain" to be taken far too seriously.

As with the hardware technologies, the dangers of the computer have been considerably—and mistakenly—anthropomorphized. The golem legend, transmitted via Frankenstein's monster, robots, and HAL the anthropomorphic computer in *2001, A Space Odyssey*, continues to predominate in human fears. But only a handful of the tens of thousands of computers in the world are even remotely capable of approaching a behavior that could, even in the most limited sense, be considered "intelligent." Again it is the process, not the product, that poses the greatest potential danger. The vast amount of information that can be stored and made readily available is all too easily subject to misuse by whoever has access to it. The right to privacy is not so easily defined or defended in a society where occasional bits and pieces of information—from credit purchases to educational records to personnel data— can be easily collected, collated, indexed, and retrieved; and an extensive file can be kept on virtually anyone who is active in society. Access to what should properly be privileged or confidential information has been too easily obtained by private credit firms and public security agencies alike. Such files, which were once restricted in their scope by the amount of laborious hand work necessary to maintain and search them, can now be manufactured *en masse*, either automatically or on demand.

The information explosion is, of course, not solely attributable to the computer. Through paperback books, magazines, newspapers, radio, and, above all, television, we have access to more information than any single mind can assimilate. Through microfilm, microfiche, dry copying, and instant duplicating, even rare, handwritten, or roughly sketched work can be copied, circulated, stored, indexed, and cross-referenced. Journals have given birth to abstracting journals, which themselves are now being abstracted in second generation journals. At times, we seem to be well on our way to drowning in a sea of information, being buried alive in a mountain of paper. Yet, it is not merely the quantity or quality of the information available that has so altered our lives. Without the accompanying scientific-technological advances in communication, there would be little point in developing such sophisticated information systems. Information, to be useful, must be transmitted; and communication is, by definition, the transmission of information.

The analytic theory of communication, as discussed by John R. Pierce in his article "Communication," was developed in close conjunction with the theory of computers. A brain, be it animal or mechanical, must have a communication system as well as a data store in order to function. The work of Claude E. Shannon on mathematical communication theory was to the computer what the Carnot cycle was to the steam engine. Comparable to the second law of thermodynamics, the quantification of an upper limit to the information-transmitting capacity of a communication channel established a ground rule for eliminating the impossible and also served as a guideline for further development by showing that most channels were working far below theoretical maximums. Thus *communication* has become somewhat more than a general descriptive term; it is also the subject of a field of analytic study. Some of the offshoots of this quantification, such as cybernetics or the investigation of syntactic structure, have themselves blossomed into the most intellectually challenging of modern research fields. These fields in turn depend on the new technologies that spawned them to analyze their own disciplines.

Communication is an extremely difficult phenomenon to describe in a quantitative fashion, because, as Pierce points out, the transmission of data already possessed by the receiver is not a communication in the analytic sense, since no new information is received. The message is then redundant, as its content is not information but duplication. This may be illustrated by considering a well-known object—Adams's beloved Chartres, for instance. It is unlikely that one who has not seen the cathedral of Chartres could be given the "information" that describes it over the telephone or even by television; it would be faster to fly to France and see it. The greater the bandwidth of the channel, the higher the rate of transmission of information. But suppose you have seen the cathedral. A telegram saying only "Chartres" would suffice to trigger a recall of the stored information and give you access to it. Therefore, the most efficient transmission of a message depends not only on its inherent complexity, but also on the extent to which its content is anticipated. In Shannon's terms, a message conveys some entropy, some piece of unexpected information. But *unexpected* refers not to the sender, but to the receiver, and a message to a friend can be much briefer than one to a stranger. Any theory concerning communication must take this into account. Furthermore, communication modifies both sender and receiver by bringing them into a greater degree of correspondence.

The sheer complexity of the ethnic, cultural, and geographical diversity of our world, however, has rendered many messages unintelligible. To return to the previous example, a global announcement concerning Chartres will make sense only to those who contain an index "Chartres" by which some stored information can be called up for examination. Pierce calls this a "community of interest." A new idea or discovery in physics, neurophysiology, economic theory, or literary criticism is similarly a "communication" only to the community of interest possessing the correct library of information for understanding it. Such communities may now be global, cutting across all boundaries of race, nation, or geography, but they have generally been narrowly defined and limited in scope. At a time when facile and inexpensive communication has made possible the creation of a global society that could transcend the artificial borders and boundaries that have so long divided the race and set one person against another, the store of information necessary for making the messages intelligible has increased more than enough to compensate.

The first industrial revolution supplied both power and a promise. The power of the machine could be harnessed to do the drudging and dangerous labor of humanity; the promise was a life free of hardship, poverty, hunger, and deprivation. The abundance of industrial capacity could alleviate the condition of the ill-housed, the ill-clothed, and the ill-fed without strain. Badly used, the power has been corrupted, and its promise rarely kept. The soft revolution, with its power to promote free exchange of information and make explicit the common condition and destiny of the human race, has exposed some new communities of interest to which we all properly belong. In matters of nuclear war, radioactive poisoning of the atmosphere, or destruction of the protective ozone blanket of the stratosphere, there are no disinterested observers, no outside referees. The misuse of our hardware has given us a capacity for total destruction and global poisoning that must be watched over forever.

The maldistribution of goods has done much to create a political tension that makes the exercise of that destructive capability all too probable. The use or misuse of the new power of communication and information can either ease those tensions or exacerbate them. Such power can be used to eliminate ignorance, xenophobia, narrow nationalism, and fear. It can also be used for the manipulation of public opinion, propaganda, vilification, deceit, and disguise. It is a power coming far closer than that of the Industrial Revolution to being the power that corrupts absolutely. Technology, whether software or hardware, is and always has been a double-edged sword; the choice as to whether it is wielded or where it strikes must remain under human control. We can afford neither to abdicate the power of decision, leaving it to the machines, nor to pretend to an age of innocence, as if the apple of nuclear destruction had never been tasted.

The Economics of Technological Change

by Anne P. Carter
April 1966

The effects of such change are brought out by a comparison of input-output tables listing the transactions among all sectors of industry in the U.S. for the years 1947 and 1958

Technological change in the U.S. economy currently evokes expressions of both satisfaction and anxiety. According to the National Commission on Technology, Automation, and Economic Progress, reporting to the President and Congress earlier this year, the "vast majority" of people recognize that technological change "has led to better working conditions by eliminating many, perhaps most, dirty, menial and servile jobs; that it has made possible the shortening of working hours and the increase in leisure; that it has provided a growing abundance of goods and a continuous flow of improved and new products." At the same time people are assailed by fears and concerns: "Perhaps the [concern] most responsible for the establishment of the Commission has arisen from the belief that technological change is a major source of unemployment..., that eventually it would eliminate all but a few jobs, with the major portion of what we now call work being performed automatically by machine."

The members of the commission, for their part, concluded "that technology eliminates jobs, not work" and attributed current unemployment to more or less "normal" cyclical processes. This distinguished group of industrialists, labor leaders and economists nonetheless concurred in the recommendation ("directed to making it possible, or easier, for people to adjust to a fast-changing technological and economic world with-

out major breaks in the continuity of employment") that Congress "examine wholly new approaches to the problem of income maintenance [and] give serious study to a minimum income allowance or a negative income tax program."

Clearly there is need for an objective and consistent way to identify and measure the economic consequences of technological change. There is concrete evidence of change all around us. Developments such as the replacement of the steam locomotive by the diesel engine, the transformation of electronics by solid-state physics and the substitu-

tion of aluminum for steel or polyethylene for aluminum are explicit and visible enough. Their economic impact, however, is not easy to gauge. The citing of a single example does not exhaust the possible applications of a new technique, and a spectacular example may give an exaggerated idea of its importance. Moreover, for all the evidence of change, the internal-combustion engine under the hood of the 1966 automobile bears considerable resemblance to the engine of 20 or 30 years ago, 50-year-old blast furnaces are still reducing iron, and machinists and secre-

	AGRICULTURE	INDUSTRY	FINAL DEMAND	TOTAL OUTPUT
AGRICULTURE	25 / .25 .... 1.46	20 / .40 .... .66	55	100
INDUSTRY	14 / .14 .... .23	6 / .12 .... 1.24	30	50
VALUE ADDED	61 / .61	24 / .48	85	

INPUT-OUTPUT TABLE depicts a hypothetical economy broken down into two sectors: "Agriculture" and "Industry." Reading across the row for one of these sectors, the large figures in each cell show the distribution of its output of intermediate products to itself and to the other sector inside the "interindustry" matrix (*two columns at left and two top rows*), and its delivery of finished products to final demand. Reading down a column, the figures show the input of intermediate products required by the sector plus its "value added": its inputs of labor, depreciation and profit. The final demand and value added for the system as a whole sum to the same gross national product (85 arbitrary units, for example billions of dollars). The total output for each sector redundantly adds its interindustry deliveries to its contribution to the gross national product. The "input-output coefficients" in small figures at bottom left of each cell express the ratio of the input shown to the total ouput of the sector in whose column it appears. Figures at bottom right are "inverse coefficients," showing the direct and indirect requirement for the input per dollar of delivery to final demand.

taries have not all been driven from the labor force by "automation." It is, in fact, difficult to relate the effects of specific innovations to changes in the national indexes of output and productivity; individual changes may reinforce or offset one another. What is wanted to make sense out of the fragmentary and conflicting evidence about change and lack of change in the economy is a technique that will allow us to organize this piecemeal information in the context of the structure of the system as a whole.

One useful approach to the structure of an economic system is provided by "input-output" or "interindustry" analysis. This technique takes account of the fact that the division of labor in a modern industrial economy is embodied in a diversity of highly differentiated and mutually dependent technologies. The production and delivery of the output of any one industry to its final market requires inputs of raw and semifinished materials, components and services from other industries. For any

		FOOD AND TOBACCO		TEXTILES AND APPAREL		DRUGS, CLEANING PREPARATIONS, COSMETICS, PAPER		FURNITURE		CONSUMERS' APPLIANCES (INCLUDING AUTOMOBILES)		CONSTRUCTION		PRODUCERS' DURABLE GOODS (NON-ELECTRICAL)		ELECTRICAL EQU[…]	
		DOLLARS	PERCENT	DOLLARS	PERCENT	DOLLARS	PERCENT	DOLLARS	PERCENT	DOLLARS	PERCENT	DOLLARS	PERCENT	DOLLARS	PERCE[…]
GENERAL INPUTS	BUSINESS SERVICES	1427	72	111	22	84	9	26	23	841	198	1983	97	347	172
	COMMUNICATIONS (INCLUDING RADIO AND TELEVISION)	474	158	81	75	84	87	27	128	179	158	462	131	135	207
	ELECTRICITY, GAS AND WATER	590	87	169	84	86	95	49	117	218	100	775	127	181	112
	FINANCE, INSURANCE, REAL ESTATE AND RENTALS	703	17	281	38	292	114	82	65	556	123	900	37	466	185
	PAPER PRODUCTS AND CONTAINERS	362	16	155	27	118	6	125	155	241	52	549	43	50	16
	PETROLEUM REFINING	336	24	1	0	50	37	10	19	9	3	419	26	63	42
	WHOLESALE AND RETAIL TRADE	520	12	47	4	186	67	191	114	1045	152	2187	40	647	154
	PRINTING AND PUBLISHING	370	43	77	40	72	2	17	42	248	125	554	72	115	111
	OTHER SERVICES	29	2	76	34	40	44	28	75	152	96	218	16	117	131
	METAL CONTAINERS	238	19	9	25	31	47	1	14	22	43	34	28	5	19
	TRANSPORTATION AND WAREHOUSING	508	11	15	2	40	9	15	6	149	13	301	7	167	31
	MAINTENANCE CONSTRUCTION	20	2	85	33	2	3	3	8	22	9	23	0	13	10
	COAL MINING	211	52	60	50	34	47	12	36	76	30	172	32	36	24
	WOODEN CONTAINERS	229	53	18	59	12	74	5	55	31	47	27	34	19	73
CHEMICALS	CHEMICALS AND SELECTED CHEMICAL PRODUCTS	434	27	373	61	251	36	42	55	127	29	781	91	77	35
	DRUGS, CLEANING AND TOILET ITEMS	143	75	10	6	23	155	3	52	27	71	80	122	20	113
	PAINTS AND ALLIED PRODUCTS	3	2	26	45	12	34	2	3	59	29	9	2	3	7
MATERIALS	PLASTICS AND SYNTHETIC MATERIALS	89	33	462	61	34	76	41	96	11	3	241	103	34	42
	STONE AND CLAY PRODUCTS	28	16	17	48	9	29	4	15	0	0	1496	40	89	64
	RUBBER AND PLASTICS PRODUCTS	91	20	199	94	66	149	103	211	348	28	335	72	34	13
	LIVESTOCK AND LIVESTOCK PRODUCTS	2750	16	382	53	104	52	0	0	12	17	16	7	27	75
	GLASS AND GLASS PRODUCTS	193	23	8	24	111	50	46	102	98	48	105	29	7	14
	AGRICULTURAL PRODUCTS (OTHER THAN FOOD)	2375	12	373	26	122	61	9	13	16	14	295	72	22	44
	PRIMARY NONFERROUS METAL MANUFACTURING	175	30	44	28	50	38	2	2	502	26	825	22	143	16
	LUMBER AND WOOD PRODUCTS, EXCEPT CONTAINERS	163	28	35	21	31	24	91	13	125	38	1401	21	46	30
	PRIMARY IRON AND STEEL MANUFACTURING	444	25	80	34	52	25	72	18	1004	26	1167	16	511	17
METALWORKING	ELECTRONIC COMPONENTS AND ACCESSORIES	25	372	11	640	6	513	5	777	591	173	83	423	157	941
	SCIENTIFIC AND CONTROLLING INSTRUMENTS	15	66	16	210	1	5	38	75	102	60	193	120	33	2
	HARDWARE, PLATING AND VALVES AND WIRE PRODUCTS	54	18	31	39	12	12	57	31	255	29	73	4	7	2
	HEATING, PLUMBING AND STRUCTURAL METAL PRODUCTS	49	31	13	41	5	29	9	38	39	18	790	16	33	17
	ELECTRIC INDUSTRIAL EQUIPMENT	28	52	0	1	7	70	5	70	86	13	317	64	97	23
	BATTERIES, X-RAY AND ENGINE ELECTRIC EQUIPMENT	20	24	2	22	0	7	0	3	48	16	22	17	13	19
	ELECTRIC-LIGHTING AND WIRING EQUIPMENT	10	14	2	19	2	32	5	131	55	31	27	3	13	26
	STAMPINGS AND SCREW MACHINE PRODUCTS	93	24	6	15	1	2	14	25	368	26	140	21	117	23

TECHNOLOGICAL CHANGE IN THE U.S. has led to increases or decreases in the inputs required of selected producing sectors listed vertically at left to satisfy 1958 final demand for the categories of end products specified at head of each column. Figures in cells show the difference, in millions of dollars (*cell at left*) and in percentages (*cell at right*), between inputs required to satisfy the same 1958 final demand computed with the coefficients of a 1947 input-output table for the U.S. economy and those provided

EQUIPMENT	SERVICES (EXCLUDING UTILITIES)	TOTAL OUTPUT REQUIREMENTS TO MEET 1958 FINAL DEMANDS ($)		DIFFERENCE IN REQUIREMENTS (1958 TECHNOLOGY-1947 TECHNOLOGY)			
				1958 TECHNOLOGY	1947 TECHNOLOGY		
	643	189	1654	16	24366	17120	7246	42
	177	156	1126	16	11278	8499	2779	33
	311	105	1418	54	20204	15498	4706	30
	561	125	4294	6	88088	79241	8847	11
	147	36	684	16	13032	12377	655	5
	18	5	239	9	17310	16641	669	4
	1101	187	1972	3	94584	90663	3921	4
	235	169	1082	19	12582	12102	479	4
	134	85	425	1	49320	48324	996	2
	16	36	74	33	2090	2057	33	2
	196	18	1001	5	32608	34523	1916	6
	24	9	4108	34	16855	21408	4553	21
	69	26	800	71	2742	4589	1847	40
	23	69	43	28	446	879	433	49
	148	39	225	16	11790	9025	2766	31
	39	134	422	69	6609	5853	754	13
	42	22	177	24	1868	2212	344	16
	70	28	148	39	4222	2993	1230	41
	58	23	11	2	7484	5854	1629	28
	246	23	108	10	6820	6146	674	11
	9	13	218	12	22621	20539	2082	10
	89	51	85	20	22444	25703	3259	13
	16	14	670	24	2137	2479	342	14
	230	13	626	39	9150	11881	2730	23
	7	2	716	47	7890	10606	2716	26
	1050	21	1963	57	19080	26081	7002	27
	171	486	284	276	2646	1458	1189	82
	40	15	60	14	3511	3020	492	16
	320	39	243	27	6420	6129	291	5
	42	21	588	51	8012	7991	21	0
	127	56	42	13	5068	5440	372	7
	72	23	209	44	1529	1845	317	17
	90	113	289	61	2284	2868	584	20
	143	12	250	42	3683	4812	1129	23
CENT	DOLLARS	PERCENT	DOLLARS	PERCENT	1958 TECHNOLOGY	1947 TECHNOLOGY	DOLLARS	PERCENT

by the 1958 table. Producing sectors are grouped as "General inputs," "Chemicals" (excluding plastics), "Materials" and "Metalworking." Negative numbers are in color.

given industry a detailed accounting of its purchases from other industries tells exactly what inputs it used to make its product and thus describes its production process. Correspondingly the record of interindustry transactions for the entire economy, displayed in the square matrix of an input-output table [see illustration on page 213], describes the structure of the underlying technological order.

Input-output tables have now been prepared for the economies of more than 50 nations by the statistical agencies of their governments. With the advent of computers that can manipulate the information quickly and easily these tables are finding a wide range of uses. They have been applied in setting the priorities of capital-investment programs of developing nations, in ordering trade relations among the Common Market countries and in planning the market-development campaigns of large industrial corporations. Comparison of input-output tables for two different countries highlights differences in their industrial systems.

For planners and forecasters an input-output table provides what is in effect a working model of the technological apparatus that helps to make and test economic projections. Investigators in this field, beginning with Wassily W. Leontief of Harvard University, the originator of the technique, have also been intrigued by the insights that input-output analysis might yield into the economics of technological change. Comparison of two input-output tables compiled for the same economy at different times should reveal the structural changes corresponding to the changes in technology that occurred in the interim.

During the past year the author and her colleagues at the Harvard Economic Research Project have been engaged in just such a comparative study of two input-output tables for the U.S. economy. The tables are based on detailed accounting of the interindustry transactions for the years 1947 and 1958, conducted respectively by the Bureau of Labor Statistics of the Department of Labor and the Office of Business Economics of the Department of Commerce. Ideally one could wish for a terminal year other than 1958, because the business recession of that year may have distorted some of the interindustry product flows; furthermore, the passage of eight years would seem to make the comparison descriptive of "history" rather than of current developments. The work has nonetheless prepared the ground for studies of more up-to-date tables that are to come. The present study has also brought out patterns of technological change that persist in the system today.

Comparison of the technologies of 1947 and 1958, as portrayed by the two tables, shows a relative increase in what can be described as "nonmaterial" or "general" inputs. This increase has largely been balanced by decreases in the input of the materials and semi-finished goods out of which the system makes its tangible products. The category of general inputs embraces those used in all, or almost all, sectors of the economy: energy, communications, trade, packaging, maintenance construction, real estate, finance, insurance and other business services, printing and publishing, and business machines and their related information technologies. Increased energy consumption results, of course, from the increasing mechanization of productive processes. It also reflects mechanization of office functions and heavy expenditures for air conditioning. Increases in other general inputs are explained by the growth of the relative importance of the coordinating functions required by the logistics of a larger and more complex industrial system. As demand for general inputs rose from 1947 to 1958, diverse sectors of the system developed increasingly similar demands for these inputs.

In the pattern of relatively declining materials inputs there appears a second broad trend of equal significance. The classical dominance of single kinds of material—metals, stone, clay and glass, wood, natural fibers, rubber, leather, plastics and so on—in each kind of production has given way by 1958 to increasing diversification of the bill of materials consumed by each industry. This development comes from interplay between keenly competitive refinement in the qualities of materials and design backward from end-use specifications.

All these changes in the relative importance of sectors imply change in the kinds and numbers of jobs available. Translation of the input figures into man-years shows, in fact, that the sectors producing the general inputs had come to employ more than half of the labor force by 1958.

In order to compare the tables for 1947 and 1958 it was necessary first to make them reasonably compatible. The 1947 table presents a full accounting of interindustry transactions with the U.S. economy broken down into 450 sectors. For intelligibility on the printed page this table was consolidated to 192 sectors and, for publication in the pages of this magazine, to 50 sectors [see "Input-Output Economics," by Wassily W. Leontief; SCIENTIFIC AMERICAN Offprint 610]. The 1958 table is based on analysis of primary data at a comparable level of detail but presents the information conservatively aggregated to 81 sectors [see "The Structure of the U.S. Economy," by Wassily W. Leontief; SCIENTIFIC AMERICAN Offprint 624]. Alignment of the two tables was further complicated by the very process of technological change that was the subject of our study. The waning of older industries and the burgeoning of

new ones had necessitated a complete revision of the Standard Industrial Classification by the Office of Statistical Standards in 1957. As a result 45 of the 450-order sectors of the 1947 table had to be split, each into two or three parts, in order to create a set of 1947 81-order industries that would conform to the new "S.I.C." code employed in the 1958 table. To improve the comparability of

sectors between the two years the 81-order classification was finally aggregated to a 73-order system. Needless to say, reconciliation of the two tables also required price adjustments (inflation of the 1947 transactions to 1958 prices) and many other accounting adjustments, particularly those occasioned by change in the conventions for defining and measuring outputs in certain sectors and, inevitably, differences in the treatment of taxes.

With the interindustry transactions for 1947 and 1958 arrayed in two roughly compatible 73-order tables, the next step was to derive the input-output coefficient matrix for each of the two years. An input-output, or "direct," coefficient expresses the ratio of a given input to the total output of the industry receiving the input. The full column of direct coefficients for an industry shows just how much that industry must draw from each of the other industries in order to produce a unit of its product. Change in a coefficient, from one input-output table to the next, shows the effect of changes in an industry's technology on its requirements for a particular input. Often changes in input coefficients will reflect qualitative as well as quantitative changes in input requirements.

When it comes to comparing changes in the system as a whole, however, direct comparison of two tables, coefficient by coefficient, would be a cumbersome task. Moreover, some coefficient changes are of much greater interest than others, and some may reflect not real changes but minor differences in the techniques of constructing the tables of transactions. In the present study we treated the two complete matrices of coefficients as working models of the system, one for 1947 and one for 1958. We asked each "technology" in turn what outputs of raw materials, intermediate products and services it would have to produce in order to yield the same gross national product.

The total of such outputs, the "gross domestic output," exceeds the gross national product, which is the value of goods and services consumed in their final markets, because each industry delivers more or less of its total output in the form of primary and intermediate inputs to other industries. Gross national product may also be reckoned as the total of the inputs of the prime factors of production, or the "value added," of all the industries [see illustration on page 213]. To the gross national product, reckoned either way, the gross

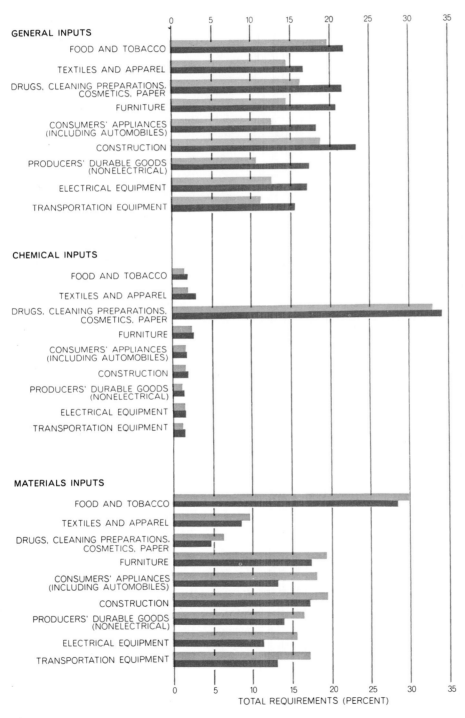

DIFFERENT REQUIREMENTS are made on producing sectors because of changing interindustry relationships. Graph at top shows that "General inputs" contribute proportionately more to output of end-product groups in the 1958 economy (*dark bars*) than they did in the 1947 economy (*light bars*). Chemical industries also play a larger role in the 1958 economy. Bottom graph shows that industries producing materials, however, play a smaller role.

domestic output adds the redundant values of the interindustry transactions. Since all of the product of the system is ultimately delivered to final demand, it is the gross national product that is taken as the conventional index of total economic activity. For the purposes of our study we employed the gross national product for 1958, specified by a bill of final demand showing preliminary estimates of actual deliveries to the ultimate markets represented by households, industries (on their inventory and capital accounts), government agencies and so on. (The official estimates for the 1958 final demand, which were published after we had done our work, would change some of the numerical results but not the general conclusions of our study.)

To facilitate our comparison of the 1947 and 1958 technologies, we had the Harvard Computation Center "invert" the two 73-order coefficient matrices and furnish us with matrices of what are called Leontief inverse coefficients. These coefficients express the total value of each industry's output required directly or indirectly to satisfy an additional dollar's worth of final demand for the product of a given industry. The inverse coefficient for the output of coal occasioned by the final demand for shoes, for example, tells the total increase in the consumption of coal by the shoe industry, the leather industry, the chemicals industry—by all industries in the economy—that would be required to increase final consumption of shoes by one dollar. Given the inverse coefficient matrix, one can derive the total outputs required to produce any stipulated bill of final demand by simple multiplication and summation.

With the matrix for the 1947 economy, therefore, we derived the outputs of all the goods and services that would have been required to satisfy the 1958 final demand on the basis of 1947 technology. These hypothetical outputs could then be compared with the actual 1958 outputs that satisfied the 1958 demand on the basis of 1958 technology. We found that the 1958 gross national product of $444 billion required a gross domestic output of $786 billion by 1947 technology and $800 billion by 1958 technology. It would appear that technological change (or progress!) had actually added about $14 billion to the task of satisfying the same final demand. Because the difference is accounted for entirely by increase in the interindustry transactions one can say that the pro-

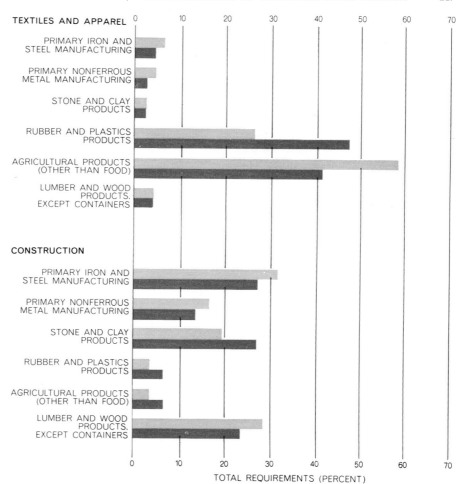

COMPETITION OF MATERIALS is evident in graphs comparing the percentage of materials required by two end-product groups according to 1947 technology (*light*) and 1958 technology (*dark*). The inputs tend to level out, the large 1947 inputs becoming smaller and the small ones becoming larger. In textiles and apparel, natural fibers ("Agricultural products") yield traditional predominance to synthetics ("Plastics, rubber"). In construction, iron and steel ("Ferrous metals") yield to cement and concrete ("Stone and clay products").

ductive and distributive process has become somewhat more "roundabout."

Our figures also showed that the productivity of labor had increased markedly in practically all industries between 1947 and 1958. Roughly 33 percent more labor would have been required to produce the 1958 bill of final demand with 1947 technology.

The differences between the technologies for the two years become more substantial when the gross changes in output are added up, ignoring the plus and minus signs. The gross change comes to $87 billion, or roughly 10 percent of the gross domestic output computed for either system. About half of the total swing is accounted for by the preponderantly positive changes in the general inputs. The principal offset to these increases is provided by the decline in the materials inputs, leaving the small net positive change of about $16 billion. Expressed as percentages of the gross domestic output none of

these movements, up or down, appears very large. Clearly the system has considerable inertia; technological change must be regarded not as a revolutionary process but as an evolutionary one.

When the changes are expressed as percentages of the major classes of inputs, however, they loom larger. The gross change of $39 billion in general inputs represents 11 percent of the total inputs from these sectors computed for 1947 technology. The net of the positive and negative changes is an increase of $22 billion, or 6 percent. For materials the gross changes add up to $21 billion, or 19 percent of the total 1947 inputs from these sectors, and the changes net out in a decline of $10 billion, or minus 9 percent.

To calculate the corresponding changes in employment we multiplied the total output for each industry by a "labor coefficient," that is, the man-years required per unit of output in each of the two years. The total swings in the job market estimated in this fashion reflect increases in productivity as well as changes in the requirements for the several classes of inputs considered in our study. Thus the smaller inputs of materials required by 1958 technology generated only about half as many jobs as the hypothetical materials requirements of the 1947 matrix [see illustration below]. The larger 1958 total of general inputs required 83.6 percent of the jobs computed to meet the 1947 demand for these inputs. The movement of jobs into the general input sectors indicated by these findings is fully substantiated by comparison of the 1950 and 1960 census figures, which show that the "providers of services" increased from 47 percent of the labor force to 54 percent.

Consideration of the swings in the outputs of a representative sample of individual producing sectors, as shown in detail in the table on pages 214 and 215, brings one still closer to the ferment of technological change and to the level at which change is experienced by the people and the institutions involved. The biggest dollar-and-cents changes occur in sectors that are large to begin with. On a percentage basis one finds large changes affecting relatively small sectors and smaller percentage changes in large sectors. Communications (including radio and television broadcasting), plastics and electronic components show increases ranging from 33 to 82 percent; these fast-growing sectors are still comparatively small, and the absolute increases in the requirement for their products and services range around $1 billion. On the other hand, some large sectors show substantial declines: coal mining by 40 percent, iron and steel by 27 percent and stampings, screw-machine products and bolts by 23 percent. Further disaggregation of the sectors would reveal other significant changes. The decline of nearly $3 billion, or 23 percent, in nonferrous metals represents the net of a substantial increase in the use of aluminum and of declines in other nonferrous metals. In the case of iron and steel the decline reflects not only substitution by plastics and aluminum but also improvements in the performance of various steels and design changes that take advantage of these improvements to reduce the total amount of material used.

Such change in the relative importance of industries brings changes in the distribution of employment. Comparatively rapid growth in the demand for chemicals, plastics and drugs tends to offset reductions in their man-year requirements per unit of output and to moderate the decline in their total labor requirements. In industries such as coal mining, steel and wooden containers "technological" unemployment stems as much from relative decrease in demand for their output originating elsewhere in the system as from decline in their labor coefficients.

Even the largest swings in output and employment disclosed by our study assume a gradual slope when averaged over the 11 years from 1947 to 1958. The replacement of a metal bearing by a nylon one may look like an abrupt change, but it takes time for such substitutions to percolate through the system. Changes in input-output coefficients therefore occur gradually. As it becomes possible to plot them from one input-output table of the U.S. economy to the next it will become increasingly feasible to project the diffusion of new technologies.

Demand for an industry's product may increase in some markets and decline in others. The net change in the output of a sector may therefore understate the gross change in the re-

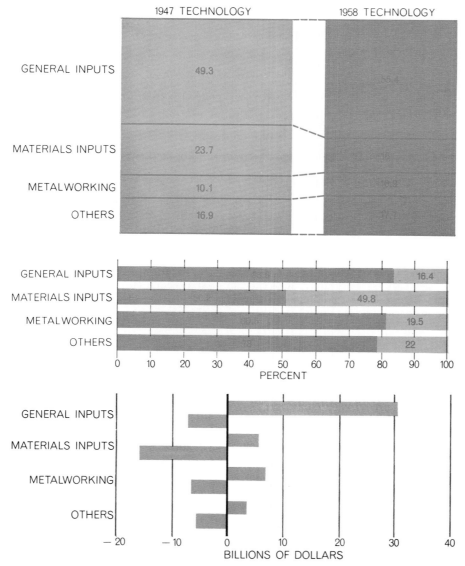

EFFECT OF CHANGE ON LABOR FORCE is presented in three graphs. Graph at top shows manpower required to satisfy 1958 final demand according to 1947 technology (*bar at left*) and 1958 technology (*right*). Bars show the percentage of workers employed in each industry group. Width of each bar is proportional to the number of workers called for by the technology of each year. Bars of middle graph show the percentage of workers needed in 1958 compared with 1947 to produce the 1958 bill of goods in each major inputs category. The graph at bottom expresses the changed requirements for major groups of inputs. Extension of bars to the right of zero represents a positive change, extension to the left represents a decline. The positive overall change in general inputs (production of which employs more than half of the 1958 labor force) is accompanied by an overall decline in materials inputs.

quirements of the sectors that consume its product. To show how technological change has affected detailed input requirements for different types of final products, we subdivided the 73-order "vector," or bill of final demand, into 10 "subvectors of final demand," or end-product groups. We then computed changes in direct plus indirect requirements for the various inputs generated by final demand for "Food and tobacco," for "Construction," for "Transportation equipment" and so on when 1958 technology is substituted for that of 1947. The results of this detailed analysis are shown in the table on pages 214 and 215. Reading down the columns, one sees the changes in the input structures for each of the 10 end-product groups. The requirements for the production of durable goods, particularly those involving electrical and electronic technology, change most. The relatively old and mature food and textile groups show correspondingly less change. Reading across the rows in the table, one sees the changes in requirements for a given input. As the figures for the system as a whole suggest, the requirement for general inputs, particularly for services, increases in almost all end uses.

In the energy rows coal requirements decline and the demand for petroleum, natural gas and electricity increases in keeping with the common elements of fuel technology in all sectors. Growth in the packaging sectors reflects the fact that the U.S. economy now packages almost everything except bulk raw materials. The use of paper and paper containers in packaging apparently increases at the expense of wood; metal containers similarly displace glass in the delivery of food, beverages, drugs and cleaning preparations. The general rise in the use of plastic film in packaging is unfortunately obscured in the aggregation of plastics products with rubber.

Although the entries in the materials rows reflect the gross trend downward for these inputs, there are significant departures from the general pattern. The increase in demand for synthetics reflects the parallel diffusion of several technologically distinguishable developments, particularly the new dominance of synthetic fibers in textiles and the increased use of plastics in durable goods. One is impressed to find, however, that the decline in steel requirements shows up in every subvector of final demand and adds up to a cumulative decline six times as great as the rise in plastics and synthetics. The growth of aluminum offsets the decline

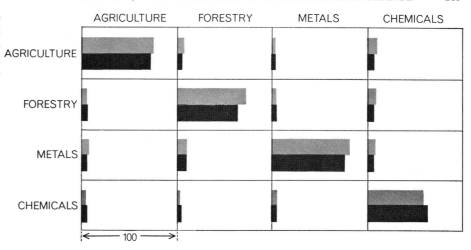

GROWING INTERDEPENDENCE of its parts has characterized the U.S. economy in recent years. In this matrix for four materials-oriented industrial complexes each cell shows the percent of total direct and indirect requirements on the complex listed at left by the complex listed at top. Gray bar gives figure for 1947 technology; black bar, for 1958 technology. All but the chemicals bloc need a rising percentage of inputs from other sectors.

of other nonferrous metals only in the production of nonelectrical producers' durables.

In the producing sectors grouped under metalworking two major trends can be discerned. The more traditional sectors, such as stamping and screw-machine products, lose markets as fabrication technologies become more specialized and fewer parts are assembled to make a given product. Losses for this group are offset by increases in demand for electrical apparatus and motors, electronic components and instruments in the production of most kinds of consumers' and producers' durable goods.

As between the 1947 and 1958 technologies, the requirement for general inputs in all 10 end-product groups increases from an average of 15 percent to an average of 18 percent of total inputs. Concurrently the requirements for these inputs tend to become still more uniform among the different end-product groups [see illustration on page 216]. The inputs of materials show a nearly offsetting decline in all end-product groups. In effect the rapid development of materials technology has brought the replacement of old inputs with new ones of lower value. The changing input pattern shows, however, that these developments have had the more important effect of making materials increasingly interchangeable. In all 10 end-product groups the larger materials inputs tend to become smaller and the smaller to become larger; in the textile industry synthetics displace natural fibers from first place and in the construction industry steel yields to concrete [see illustration on page 217]. As

a result the familiar materials-oriented industrial complexes—agriculture, forest products, metals and chemicals—show less self-sufficiency and increasing interdependence. The chemicals bloc is an exception to this generalization; it is the least self-sufficient in the 1947 technology and develops increasing reliance on its own, rather than agricultural, raw materials [see illustration above].

By and large technological change from 1947 to 1958 tended to reduce the differences in input structure distinguishing the major groups of industries. This may seem an improbable consequence of the increasing specialization and complexity of technology. The fact remains that the proliferation of new materials and new methods tends to increase the variety of inputs to each sector; with greater variety the input columns show more elements in common. The diversification of materials breaks down the primary identity of major industrial blocs. The increase in general inputs that render the same services and deliver such indistinguishable products as kilowatt-hours to all customers makes input structures more alike. Thus as a principal consequence of technological change the diverse major industries in the U.S. economy tend to become interlocked in increasing interdependence. In the job market there is declining demand for people in the "productive" functions, as traditionally defined, and increasing demand for people who can contribute to the coordinative and integrative functions required by the larger and more complex system.

Information

by John McCarthy
September 1966

*The moral of this article: computers, far from robbing
man of his individuality, enable technology to adapt
to human diversity*

The computer gives signs of becoming the contemporary counterpart of the steam engine that brought on the industrial revolution. The computer is an information machine. Information is a commodity no less intangible than energy; if anything, it is more pervasive in human affairs. The command of information made possible by the computer should also make it possible to reverse the trends toward mass-produced uniformity started by the industrial revolution. Taking advantage of this opportunity may present the most urgent engineering, social and political questions of the next generation.

A computer, as hardware, consists of input and output devices, arithmetic and control circuits and a memory. Equally essential to the complete portrait is the program of instructions—the "software"—that puts the system to work. The computer accepts information from its environment through its input devices; it combines this information, according to the rules of the program stored in its memory, with information that is also stored in its memory, and it sends information back to its environment through its output devices.

The human brain also accepts inputs of information, combines it with information stored somehow within and returns outputs of information to its environment. Social institutions—such as the legislature, the law, science, education, business organizations and the communication system—receive, process and put out information in much the same way. Accordingly, in common with the computer, the human brain and social institutions may be regarded as information-processing systems, at least with respect to some crucial functions. The study of these entities as such has led to new understanding of their structures.

The installation of computers in certain organizations has already greatly increased the efficiency of some of the organizations. In the 15 or 20 years that computers have been in use, however, it has become clear that they do not merely bring an increase in efficiency. They induce basic transformation of the institutions and enterprises in which they are installed.

In the first place, computers are a million to a billion times faster than humans in performing computing operations. This follows from the fact that their working parts now change state in a few millionths or billionths of a second. Why should this quantitative change in speed produce a qualitative change in human activities that are facilitated by a computer? It might seem that there is no way to use such speeds outside of the missile business and other exotic undertakings. The answer is that the increase in speed has meant the building of computers with the capacity to handle information on a correspondingly larger scale. The interaction of high-speed, high-capacity computers with their environment is often continuous, with many input and output devices operating simultaneously with the ongoing internal computation.

The computer is, furthermore, a universal information-processing machine. Any calculation that can be done by any machine can be done by a computer, provided that the computer has a program describing the calculation. This was proved as a general proposition by the British mathematician A. M. Turing as early as 1936. It applies to the most rudimentary theoretical system as well as to the big general-purpose machines of today that make it possible, in practice, to write new programs instead of having to build new machines.

MICROELECTRONIC CIRCUITS of the kind shown on the opposite page can be regarded as the nerve tissue of the next generation of computers. The circuits, which are enlarged about 200 diameters, are part of a "complex bipolar array chip" made by Fairchild Semiconductor. Each of the eight complete circuits shown (*dark gray*) is a functional unit consisting of 18 transistors and 18 resistors. These units are connected by a larger microelectronic network (*white*); there are 28 units in the entire chip. Some recent computers incorporate microelectronic circuits, but the circuits are not connected microelectronically. Possibly microelectronic circuits will be used not only as logic elements but also as memory elements.

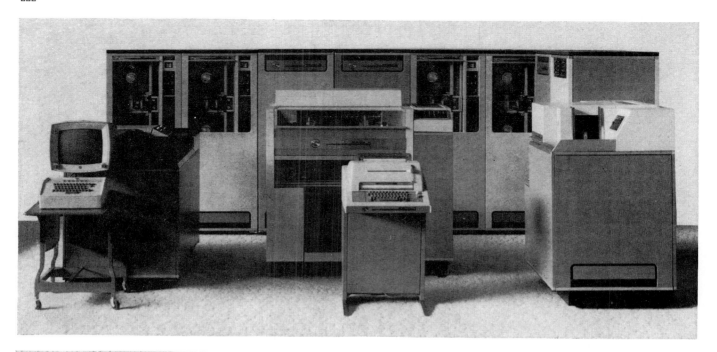

TYPICAL COMPUTER INSTALLATION includes components of the kind shown here in front and top views; the components are identified in the diagram at right. The heart of the system, which is a computer in the Spectra 70 series of the Radio Corporation of America, is the central processor and memory unit; the other units serve for input, output and storage of data. The input devices are

The speed, capacity and universality of computers make them machines that can be used to foster diversity and individuality in our industrial civilization, as opposed to the uniformity and conformity that have hitherto been the order of the day. Decisions that now have to be made in the mass can in the future be made separately, case by case. To take a practical example, it can be decided whether or not it is safe for an automobile to go through an intersection each time the matter comes up, instead of subjecting the flow of automobiles to regulation by traffic lights. A piece of furniture, a household appliance or an automobile can be designed to the specifications of each user. The decision whether to go on to the next topic or review the last one can be made in accordance with the interests of the child rather than for the class as a whole. In other words, computers can make it possible for people to be treated as individuals in many situations where they are now lumped in the aggregate.

The quality of such individual response and attention is another matter. It will depend on the quality of the programs. The special attention of a stupid program may not be worth much. But then the individual can write his own program.

The future that is contemplated here has come into view quite abruptly during the past few years. According to a report published by the American Federation of Information Processing Societies (AFIPS), there were only 10 or 15 computers at work in the U.S. in 1950. Today there are 35,200, and by 1975 there will be 85,000. Investment in computers will rise from $8 billion to more than $30 billion by 1975. Present installations include 2,100 large systems costing about $1 million each; in 1975 there will be 4,000 of these. Even the medium and small systems that are in use today have a capacity equal to or exceeding that of the 1950 generation.

A scientific problem that took an hour on a big 1950 machine at 1,000 operations per second can be run on the fastest contemporary computers in less than half a second. Allowing another 3.5 seconds to transfer the yield to an external storage memory for later printing, it can be said that program running time has been reduced from an hour to three or four seconds. This reflects the impressive recent progress in the design and manufacture of computer hardware.

Big computers are currently equipped with internal memories—the memory actively engaged in the computation under way—that usually contain 10 or 12 million minute ring-shaped ferrite "cores" in three-dimensional crystalline arrays. Each core is capable of storing one "bit," or unit, of information. Along with the replacement of the vacuum tube by the transistor and now the replacement of the transistor by the microelectronic circuit [see illustration on page 220], there has come a steep increase in the speed of arithmetic and control circuits over the past 10 years. The miniaturization of these circuits (from hundreds of circuits per cubic foot with vacuum-tube technology to hundreds of thousands and prospectively millions of circuits per cubic foot with solid-state technology) has speeded up operations by reducing the distance an impulse has to travel from point to point inside the computer.

As increases in speed and capacity have realized the inherent universality of the computer, expenditures for programming have been absorbing an increasing percentage of total installation costs. The U.S. Government, with a dozen or so big systems serving its military and space establishments, is spending more than half of its 1966 outlay of $844 million on software. Without doubt the professions in this field—those of system analyst and programmer—are the fastest-growing occupations in the U.S. labor force. From about 200,000 in 1966 it is estimated that their numbers will increase to 500,000 or 750,000 by 1970. Courses in programming are now offered in many universities and even in some high schools. In a liberal education an exposure to programming is held to be as bracing as an elementary course in mathematics or logic.

Calculating devices have a history that goes back to the ancient Greeks. The first mechanical digital calculators were made by Blaise Pascal in the 17th century. In the mid-19th century Charles Babbage proposed and partially constructed an automatic machine that would carry out long sequences of calculations without human intervention. Babbage did not succeed in making his machine actually work—although he might have, had he used binary instead of decimal notation and enjoyed better financial and technical support.

In the late 1930's Howard H. Aiken of Harvard University and George R. Stibitz of the Bell Telephone Laboratories developed automatic calculators using relays; during World War II, J. Presper Eckert and John W. Mauchly of the University of Pennsylvania developed ENIAC, an electronic calculator. As early as 1943 a British group had an

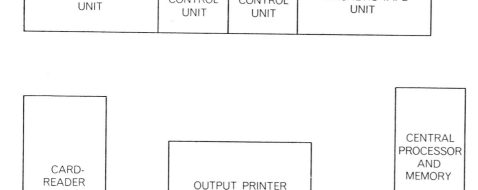

the typewriter, with which the operator communicates with the machine, and the card-reader, for which the card-punch is an adjunct. The main output devices are the printer and the video data terminal, which employs a cathode ray tube. The tape units store data.

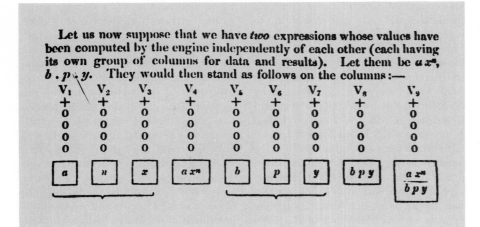

EARLY PROGRAM was written for Charles Babbage's "analytical engine" by Lady Lovelace, who was the daughter of Lord Byron. She wrote the program, which was for computing the number series known as Bernoulli numbers, to show what the engine could do. The program, of which this is a fragment, was published in 1840 in *Taylor's Scientific Memoirs*.

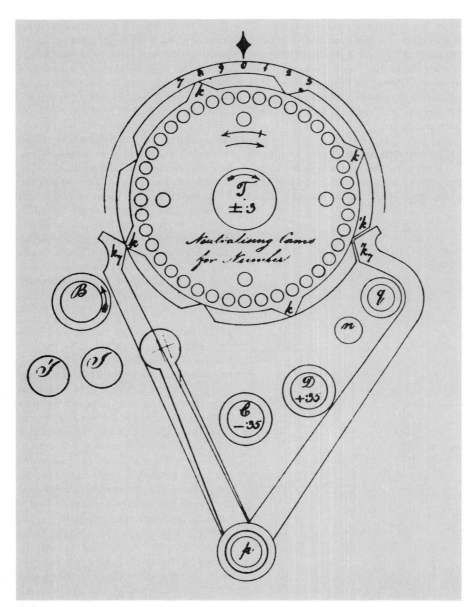

PART OF ANALYTICAL ENGINE was drawn by Babbage. This drawing, one of many that he made for the engine, bears the inscription "Neutralising Cams for Number." Financial and technological difficulties prevented Babbage from completing the machine.

electronic computer working on a wartime assignment.

Strictly speaking, however, the term "computer" now designates a universal machine capable of carrying out any arbitrary calculation, as propounded by Turing in 1936. The possibility of such a machine was apparently guessed by Babbage; his collaborator Lady Lovelace, daughter of the poet Lord Byron, may have been the first to propose a changeable program to be stored in the machine. Curiously, it does not seem that the work of either Turing or Babbage played any direct role in the labors of the men who made the computer a reality. The first practical proposal for universal computers that stored their programs in their memory came from Eckert and Mauchly during the war. Their proposal was developed by John von Neumann and his collaborators in a series of influential reports in 1945 and 1946. The first working stored-program computers were demonstrated in 1949 almost simultaneously in several laboratories in Britain and the U.S. The first commercial computer was the Eckert-Mauchly UNIVAC, put on the market in 1950.

Since that time progress in the electronic technology of computer circuits, the art of programming and programming languages and the development of computer operating systems has been rapid. No small part in this development has been played by the U.S. Government, which is always in the market for the latest and biggest systems available. It is not too much to say that the systems designed for the industry's biggest customer have been the prototypes for each major advance in computer hardware. The creation of the high-speed computer has been as central to the contemporary revolution in the technology of war as the intercontinental missile and the thermonuclear warhead.

The basic unit of information with which these machines work is the bit. Any device that can be in either of two states, such as a ferrite core or a transistor, can store a single bit. Two such devices can store two bits, three can store three bits, and so on. Consider a five-bit register made of five one-bit devices. Since each device has two states, represented, say, by 0 and 1, the five together have 2^5, or 32, states. The combinations from 00000 to 11111 can be taken to represent the binary numbers from 0 to 31. They can also be used to encode the 26 letters of the alphabet, with six combinations left over to represent word spaces and punctuation. This would permit representation

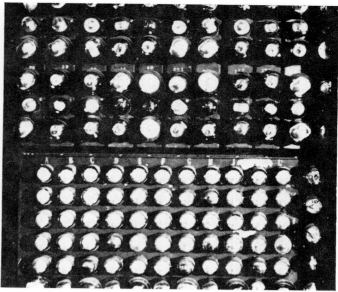
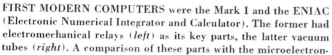

FIRST MODERN COMPUTERS were the Mark I and the ENIAC (Electronic Numerical Integrator and Calculator). The former had electromechanical relays (*left*) as its key parts, the latter vacuum tubes (*right*). A comparison of these parts with the microelectron-ic circuits illustrated on page 64 will indicate how far miniaturization has progressed. Both the Mark I and the ENIAC have been disassembled. The portion of the Mark I shown here is now at Harvard University; panel from ENIAC is at U.S. Military Academy.

of words and sentences by strings of five-bit groups. (Actually, to accommodate uppercase and lowercase letters, the full assortment of punctuation marks, the decimal digits and so on it is now customary to use seven bits.)

For many purposes, however, it is better not to be specific about how the information is coded into bits. More important is the task of describing the kinds of information to be dealt with, the basic operations to be carried out on them and the basic tests to be performed on the information in order to decide what to do next. For bits the basic operations are the logical operations $\cdot$, $+$ and $-$ (the last usually placed above a symbol, as in $\bar{A}$), which are read "and," "or" and "not" respectively. Operations are defined by giving all the cases. Thus $\cdot$ is defined by the equations $0 \cdot 0 = 0$, $1 \cdot 0 = 0$, $0 \cdot 1 = 0$, and $1 \cdot 1 = 1$. The basic decision concerning a bit is whether it is 0 or 1. Designers of computers do much of their work at the level of bits [see illustration on next page]. They have systematic procedures, as the following article by David C. Evans shows, for translating logical equations into transistor circuits that carry out the functions of these equations.

At the next level above bits come numbers. On numbers the basic operations are addition, subtraction, multiplication and division [see illustration on page 227]. The basic tests are whether two numbers are equal and whether a number is greater than zero. Programmers are generally able to work with

numbers because computer designers build the basic operations on numbers out of the logical operations on bits in the design of the circuits in the machine.

Another kind of information is a string of characters, such as A or ABA or ONION. It is well to include also the null string with no characters. A basic operation on characters may be taken to be concatenation, denoted by the symbol *. Thus ABC*ACA = ABCACA. The other basic operations are "first" and "rest." Thus first(ABC) = A and rest(ABC) = BC. The basic tests on strings are whether the string is null and whether two individual characters are equal.

Out of one kind of information, then, more elaborate kinds of information can be built; numbers and characters are built out of bits, and strings are built out of characters. Similarly, the operations and tests for the higher forms of information are built up out of the operations and tests for the lower forms. One can represent a chessboard, for example, as a table of numbers giving for each square the kind of piece, if any, that occupies it. For chess positions a basic operation gives the list of legal moves from that position. A picture may be similarly represented by an array of numbers expressing the gray-scale value of each point in the picture. The *Mariner IV* pictures of Mars were so represented during transmission to the earth, and this representation was used in the memory of the computer by the program that removed noise and enhanced contrast. Christopher Strachey shows in

this issue how programmers put together the basic operations and tests for a given class of information in designing a program to treat such information ["System Analysis and Programming," by Strachey, *Scientific American*, Sept. 1966].

What computers can do depends on the state of the art and science of the programming as well as on speed and memory capacity. At present it is straightforward to keep track of the seats available on each plane of an airline, to compute the trajectory of a space vehicle under the gravitational attraction of the sun and planets or to generate a circuit diagram from the specifications of circuit elements. It is difficult to predict the weather or to play a fair game of chess. It is currently not clear how to make a computer play an expert game of chess or discover significant mathematical theorems, although investigators have ideas about how these things might be done [see "Artificial Intelligence," by Marvin L. Minsky, SCIENTIFIC AMERICAN Offprint 313].

Input and output devices also play a significant part in making the capacity of a computer effective. For the engineering computations and the bookkeeping tasks first assigned to computers it seemed sufficient to provide them with punched-card-readers for input and line-printers for output, together with magnetic tapes for storing large quantities of data. To fly an airplane or a missile or to control a steel mill or a chemical plant, however, a computer must receive inputs from such sensory organs

LOGIC

YES	NO
1	0

NUMERATION

0	0	0	0
1	1	1	1
2	10	2	10
4	100	3	11
8	1000	4	100
16	10000	5	101
32	100000	6	110
64	1000000	7	111
128	10000000	8	1000
256	100000000	9	1001

SEVEN-BIT CODE

COLUMNS

b_4	b_2	b_3	b_1	Row	$b_7 b_6 b_5$ = 000 (0)	001 (1)	010 (2)	011 (3)	100 (4)	101 (5)	110 (6)	111 (7)
0	0	0	0	0	NUL	DLE	SP	0	`	P	@	p
0	0	0	1	1	SOH	DC1	!	1	A	Q	a	q
0	0	1	0	2	STX	DC2	"	2	B	R	b	r
0	0	1	1	3	ETX	DC3	#	3	C	S	c	s
0	1	0	0	4	EOT	DC4	$	4	D	T	d	t
0	1	0	1	5	ENQ	NAK	%	5	E	U	e	u
0	1	1	0	6	ACK	SYN	&	6	F	V	f	v
0	1	1	1	7	BEL	ETB	'	7	G	W	g	w
1	0	0	0	8	BS	CAN	(	8	H	X	h	x
1	0	0	1	9	HT	EM	)	9	I	Y	i	y
1	0	1	0	10	LF	SS	*	:	J	Z	j	z
1	0	1	1	11	VT	ESC	+	;	K	[	k	{
1	1	0	0	12	FF	FS	,	<	L	~	l	¬
1	1	0	1	13	CR	GS	–	=	M	]	m	}
1	1	1	0	14	SO	RS	.	>	N	∧	n		
1	1	1	1	15	SI	US	/	?	O	—	o	DEL

(ROWS)

BINARY NUMBERS serve computers in logic, arithmetic and coding functions. The array of binary numbers at left under "Numeration" shows that the system, which is based on 2, represents each new power of 2 by adding a 0. The same arrangement reappears at right in the binary version of the numbers 1 through 9; it shows, for example, that 111, representing 7, can be read from the left as "one 4, one 2 and one 1." The seven-bit code (*bottom*) is widely used to accomplish the printing done by computers in issuing results and communicating with operators. On receiving pulses representing 1011001, for example, the computer would print *Y*. Columns 0 and 1 contain control characters; *BS*, for example, means "back space."

as radars, flowmeters and thermometers and must deliver its outputs directly to such effector organs as motors and radio transmitters. Still other input and output devices are demanded by the increasing speed and capacity of the computers themselves. To keep them fully employed they must be allowed to interact simultaneously with large numbers of people, most of them necessarily at remote stations. This requires telephone lines, teletypewriters and cathode-ray-tube devices. For many purposes a picture on the cathode ray tube is more useful than the half-ton of print-out paper that would deliver the underlying numerical information. Simultaneous access to the computer for many users also calls for new sophistication in programming to establish the time-sharing arrangements described in the article by R. M. Fano and F. J. Corbató [see "Time-sharing on Computers," by R. M. Fano and F. J. Carbató, *Scientific American*, Sept. 1966].

It is possible to describe at greater length the perfection and promise of the new technology of information. This discussion must go on to certain pressing questions. To put the questions negatively: Will the computer condemn us to live in an increasingly depersonalized and bureaucratized society? Will the crucial decisions of life turn on a hole punched in Column 17 of a card? Will "automation" put most of us out of work?

Experience with the computerized systems most people have so far encountered in governmental, business and educational institutions has not tended to dispel the anxiety that underlies such questions. One can ascribe the bureaucratic ways of these systems to their computers or to the greed, stupidity and other vices of the people who run them. I would argue three more direct causes: one economic, one technical and one cultural.

In the first place, computers are expensive. When a computer is first installed in an organization, the impulse of the authorities is to use the new machine to cut corners, to do the old job in the old way but more cheaply, to achieve internal economies even at the expense of external relations with citizens, customers and students. Secondly, the external memories that store the data for most large organizations are inherently inflexible. Between runs through a magnetic-tape file, for example, there is no possibility of access to the account that generates today's complaint. Finally, most practitioners in the expanding software professions were beginners; it was all they could do to get the systems going at all.

In my opinion the opportunity to cure these faults is improving steadily. Computers are cheaper, and competition between systems should soon compel more attention to the customers. (The effect is not yet noticeable at my bank.) Secondly, high-speed memory devices such as magnetic-disk files, now used as internal memories, are taking up service in external data-storage. They make access to any record possible at any time. Finally, although there are a lot of young fogies who know how things are done now and expect to see them done that way until they retire in 1996, programmers are acquiring greater confidence and virtuosity.

All of this should encourage the development of systems that serve the customer better without offending either his intelligence or his convenience. In particular, organizations such as schools should not have to ask people questions the answers to which are already on file.

The computer will not make its revolutionary impact, however, by doing the old bookkeeping tasks more efficiently. It is finding its way into new applications that will increase human freedom of action. No stretching of the demonstrated technology is required to envision computer consoles installed in every home and connected to public-utility computers through the telephone system. The console might consist of a typewriter keyboard and a television screen that can display text and pictures. Each subscriber will have his private file space in the computer that he can consult and alter at any time. Given the availability of such equipment, it is impossible to recite more than a small fraction of the uses to which enterprising consumers will put it. I undertake here only to sample the range of possibilities.

Everyone will have better access to the Library of Congress than the librarian himself now has. Any page will be immediately accessible, although Ben-Ami Lipetz holds that this may come later rather than sooner [see "Information Storage and Retrieval," *Scientific American*, Sept. 1966]. Because payment will depend on usage, all levels and kinds of taste can be provided for.

The system will serve as each person's external memory, with his messages in and out kept nicely filed and reminders displayed at designated times.

Full reports on current events, whether baseball scores, the smog index in Los Angeles or the minutes of the 178th

ADDITION

```
 1 11
 111          7
  11        + 3
1010         10
```

```
 11   1
 110101       53
  11001     + 25
1001110       78
 64  8 4 2
```

SUBTRACTION

```
 11
1101         13
 111        − 7
 110          6
```

```
 1 1
110101        53
 11001      − 25
 11100        28
 16 8 4
```

MULTIPLICATION

```
1001          9
 101        × 5
1001
1001
101101       45
 32  8 4  1
```

DIVISION

```
11000 ÷ 110      24 ÷ 6

           100
      110) 11000         4
```

BINARY ARITHMETIC involves only the manipulation of 0 and 1 and hence is the basis of the extremely rapid calculating done by computers. The superscript colored numerals represent carries; subscript colored numerals show how binary numbers are read decimally.

meeting of the Korean Truce Commission, will be available for the asking.

Income tax returns will be automatically prepared on the basis of continuous, cumulative annual records of income, deductions, contributions and expenses.

With the requisite sensors and effectors installed in the household the public-utility information system will shut the windows when it rains.

The reader can write his own list of assignments. He can do so with the assurance that various entrepreneurs will try to think up new services and will advertise them. In this connection the Antitrust Division of the Department of Justice should see to it that companies set up to operate the computers are kept separate from companies that provide programs. Competition among the programmers will intensify and diversify demand on the public-utility systems. Anyone who has a new program he thinks he can sell should be free to put it in any computer in which he is willing to rent file space and to sell its services to anyone who wants to use it.

As for the conformities currently imposed by mass production, consider how the computer might facilitate the purchase of some piece of household equipment. In the first place, the computer could be asked to search the catalogues and list the alternatives available, together with appraisals from such

institutions as Consumers Union. If the consumer knows how to use an automatic design system such as that described by Steven Anson Coons [see "The Uses of Computers in Technology," *Scientific American*, Sept. 1966], he might design the desired equipment himself. The system will deliver not only drawings but also the findings of a simulation study that will show how well the equipment works. The consumer could also consult a designer, who will be able to render his service through the computer at less cost, together with firm estimates from prospective suppliers. With more or less elaboration, the procedures sketched here could do the paper work for the building of an entire house.

Apart from the physical construction of the public-utility information system, the full realization of these possibilities will require new advances in programming. No application illustrates the virtues and limitations of present-day programming so well as do efforts to use computers to aid teaching in elementary and secondary schools. In principle, one computer can give simultaneous individual attention to hundreds of students, each at his own console, each at a different place in the course or each concentrating on a different topic. The treatment of the student can be quite individual because the computer can remember the student's performance in every preceding session of instruction. The pace and the range of study can be entirely determined by the student's progress.

The teaching programs that have been written so far, however, put the student in a passive role. They are extremely pedantic. They have no understanding of the student's state of mind; they decide what to do next only in accordance with rather stereotyped sets of rules. As Patrick Suppes concludes, these programs do not compare too unfavorably with the performance of a teacher who has a large class [see "The Uses of Computers in Education," *Scientific American*, Sept. 1966]. Particularly where practice and repetition are the dominant ways of learning, the computer may even prove superior. The present programs fail in subjects that ought to cultivate the student's capacity for generating new ideas.

For the future it would be well, perhaps, to think of computers as study aids rather than teachers. The aim of the program should be to place the system under the control of the learner. He should be able to select from a list of topics the one he wants to work on; he should decide whether he prefers to read an exposition or to try to solve a problem. Best of all, he should be able to use the computer as a tool for testing his own ideas.

Reflection on the power of computer systems inevitably excites fear for the safety and integrity of the individual. In many minds the computer is the ultimate threat. It makes possible, for instance, a single national information file containing all tax, legal, security, credit, educational, medical and employment information about each and every citizen. Certainly such a file would be the source of great abuses. The files that exist today are abused. Security files, for example, have provided material for politically motivated persecutions. Credit files, to which access is wide open in the business community, have been used for purposes irrelevant to credit decisions. Accordingly it can be expected that more centralized files will facilitate even greater abuses.

On the other hand, citizens could seize the creation of centralized files as the occasion to cure existing abuses and to establish for each individual certain rights with respect to these files. Such a "bill of rights" might specify the following:

No organization, governmental or private, is allowed to maintain files that cover large numbers of people outside of the general system.

The rules governing access to the files are definite and well publicized, and the programs that enforce these rules are open to any interested party, including, for example, the American Civil Liberties Union.

An individual has the right to read his own file, to challenge certain kinds of entries in his file and to impose certain restrictions on access to his file.

Every time someone consults an individual's file this event is recorded, together with the authorization for the access.

If an organization or an individual obtains access to certain information in a file by deceit, this is a crime and a civil wrong. The injured individual may sue for invasion of privacy and be awarded damages.

At present an organization that claims to be considering extending credit to a person can learn a lot about his financial condition. In the new system no such information will be available without authorization from the person concerned. The normal form of authorization will allow no more than a yes-or-no answer to the question of whether he meets a particular definite credit criterion—whether he meets credit condition C1, for example, and can be expected to manage the installment purchase of a television set.

To establish such rights people must revise their ideas about the source and nature of their freedom. Most individual rights now recognized are based on the claim that the individual always had them; the safeguards of the law are said to be designed to prevent their infringement. Technology is advancing too fast, however, to allow such benevolent frauds to work in the future. The right to keep people from keeping files on us must first be invented, then legislated and actively enforced.

It may be supposed that, as happened with television and then color television, the enthusiasts and the well-to-do will be the first to install computer consoles in their homes. Eventually, however, everyone will consider them to be essential household equipment. People will soon become discontented with the "canned" programs available; they will want to write their own. The ability to write a computer program will become as widespread as the ability to drive a car.

Not knowing how to program will be like living in a house full of servants and not speaking their language. Each of the canned programs will be separately useful. It will be up to the individual, however, to coordinate them for his own fullest benefit. People will find, in fact, that console control of a process leads directly to the writing of one's own programs.

At first the computer says in effect: I can do the following things for you, which do you want? You reply. Then it says: In order to do this I need the following information. You respond and the dialogue continues. After you get used to using a particular facility, the computer's questions become annoying. You know in advance what they will be and you want to give the answers without waiting for the questions. Next you want to be able to give the entire sequence of actions a name and bring forth the sequence by typing only the name. As you become bolder you will want to make a later action conditional on the results of earlier actions and to provide for the repetition of actions until a criterion is reached. You are then already programming in full generality, albeit awkwardly.

As a skill, computer programming is probably more difficult than driving a car but probably less difficult than flying an airplane. It is more difficult than arithmetic but less difficult than writing good English. It does not require long study. Many people can write simple programs after an hour or two of instruction. Some success ordinarily comes quickly, and this reward reinforces further effort. Programming is far easier to learn than a foreign language or algebra.

Success in writing a program to do a particular task depends more on understanding the task and less on mastery of programming technique. To program the trajectory of a rocket, for example, requires a few weeks' study of programming and a few years' study of physics.

Writing a program to carry out some activity requires that an individual make explicit what he wants. The public-utility computer will do exactly what it is told to do within limitations imposed to protect other people's interests. A person who has experienced the unexpected and sometimes unpleasant consequences of the faithful execution of his wishes is usually ready to reexamine his preferences and premises. Fortunately programs can be readily changed. As people acquire greater control over their environment by explicit programming they will discover greater self-understanding and self-reliance. Some people will enjoy this experience more than others.

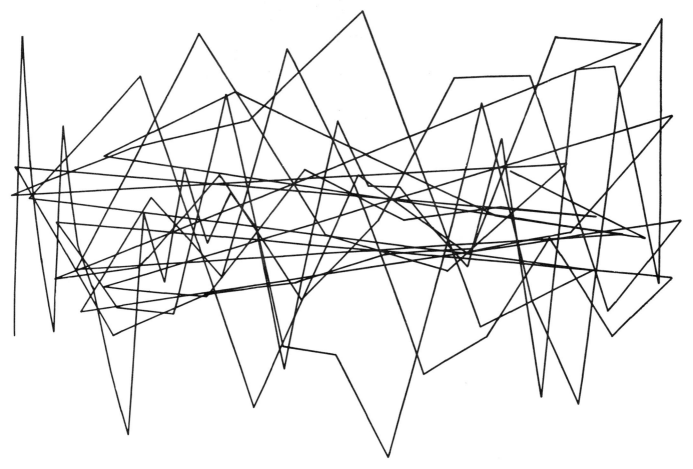

COMPUTER-GENERATED ART includes two works devised by A. Michael Noll of Bell Telephone Laboratories. At top is "Gaussian-Quadratic." The end points of each line have a Gaussian random distribution vertically; the horizontal positions increase quadratically. The pattern begins at left and is "reflected" back from right. At bottom is a portion of "Ninety Parallel Sinusoids with Linearly Increasing Period." The top line was mathematically specified as such a curve; the computer then repeated the line 90 times.

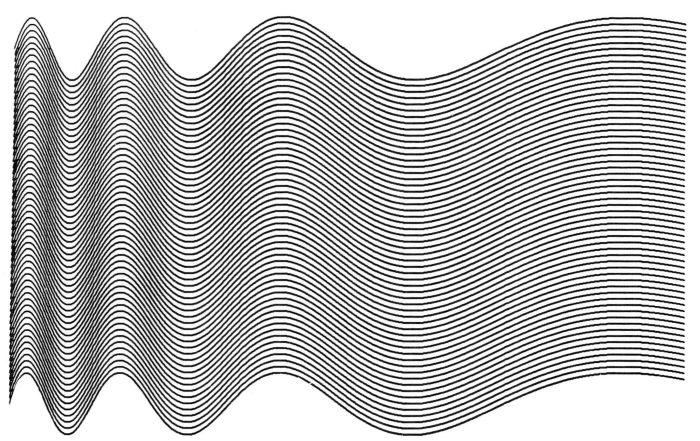

Systems Analysis of Urban Transportation

19

by William F. Hamilton II and Dana K. Nance
July 1969

*Computer models of cities suggest that in certain
circumstances installing novel "personal transit"
systems may already be more economic than building
conventional systems such as subways*

There is a growing recognition that many of the ills of U.S. cities stem from the problem of transportation within the metropolis. Although the automobile has endowed the American people with unprecedented mobility, the long-range trend toward movement by private automobile rather than by public transit has created a new complex of difficulties for urban living. The price being paid for the privacy and convenience provided by the automobile is enormous. It includes the engulfing of the city by vehicles and expressways, congestion, a high rate of accidents and air pollution. The automobile has brought another consequence that tends to be overlooked but is no less serious: by fostering "urban sprawl" it has in effect isolated much of the population. In the widely dispersed metropolis, much of which is not served by public transit, those who cannot afford a car or who cannot drive are denied the mobility needed for full access to the city's opportunities for employment and its cultural and social amenities.

These "transportation poor" constitute a far larger proportion of the population than is generally realized. Half of all the U.S. families with incomes of less than $4,000, half of all Negro households and half of all households headed by persons over 65 own no automobile. Even in families that do own one it is often unavailable to the wife and children because the wage earner must drive it to work. The young, the old, the physically handicapped—all those who for one reason or another cannot drive must be counted among the transportation poor in the increasingly automobile-oriented city. The generalization concerning the mobility made possible by the private automobile must be qualified by the observation that 100 million Americans, half of the total population, do not have a driver's license, and the proportion of nondrivers in the big cities is higher than in the country at large.

The gravity of the urban transportation problem prompted Congress three years ago to direct the Department of Housing and Urban Development (HUD) to look into the entire problem. HUD awarded 17 study contracts to a wide variety of groups: transportation experts, university laboratories, research institutes and industrial research organizations. Our group, the General Research Corporation of Santa Barbara, Calif., which is experienced in the discipline known as systems analysis, was assigned to apply such analysis to the transportation problem, considering the entire complex of transportation facilities for a city as an integrated system. In analytic method our study resembled a number of earlier ones devoted to this subject. It is nonetheless unique in that it weighed not only existing systems of transportation but also future systems. Furthermore, it carried cost-benefit accounting to a new breadth and depth of coverage.

We set out to build a mathematical model of urban transportation and to test with the aid of a large computer the effectiveness and the costs of various possible networks. Systems analysis is a general approach that consists in examining a complex system by exploring the interactions of its many parts. One "wiggles" each part in order to see what will happen to the whole when all the parts are taken into account. When the system does not exist and would be too expensive or too risky to build for testing by direct experimentation, the analyst tries to construct a model representing it and does the experiments on the model. Most often the model turns out to be a set of equations that can be solved together. For a system of any complexity the model usually is so complicated that the experiments can only be performed with a high-speed computer.

Our goal was to model all the significant modes, actual and potential, of transporting people in an urban area. We were not trying to design a particular optimal system; rather, we undertook to examine various combinations of the possible modes to see how the system as a whole would work.

To make our model as realistic as possible it was plainly desirable to use data from actual cities rather than from a hypothetical average city. We therefore decided on a case-study approach, selecting four representative cities as models. On the basis of an elaborate factor

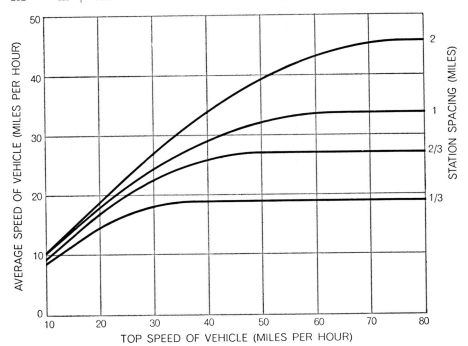

AVERAGE SPEED OF SUBWAY IS LIMITED by the spacing of stations and the accelera-tion that passengers can tolerate. It is assumed that the maximum tolerable acceleration is three miles per hour per second and that stops are 20 seconds long. Thus regardless of what the top speed of the train is, it can only average (if stations are a mile apart) 33 miles per hour. Improved equipment cannot overcome this limitation of conventional transit.

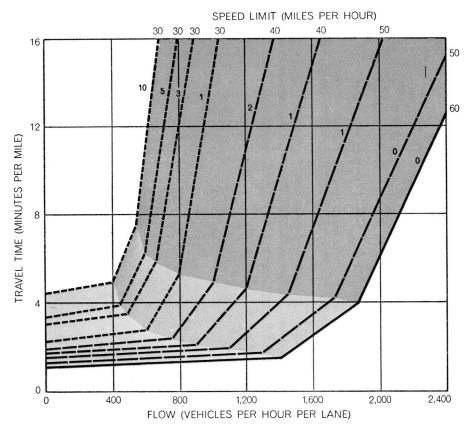

STREET CAPACITIES were represented in mathematical models of a city's transportation by the equations of these lines. At low traffic flow (*light color*) the speed and number of signal-marked intersections per mile (*numbers within grid*) are governing factors. The region above (*medium color*) is mainly governed by car density. Where flow exceeds street capacity (*dark color*) the slope was calculated from queuing theory. Data for particular streets are from city maps. The number of signal-marked intersections is an approximation.

analysis of census data we chose Boston as a typical example of a large city that was strongly oriented to public transit, Houston as a large city oriented to the private automobile and New Haven and Tucson as corresponding representatives of smaller cities (between 200,000 and 400,000 in population). These four cities offered the valuable advantage that de-tailed studies of their traffic flows had recently been made in each of them, so that they provided data not only for building our model but also for validat-ing the results of experiments with the model.

The formulation of the model for each city started off with a description of present transportation facilities and con-sidered the travel needs of its people both now and in the future. We de-scribed for the computer the streets, freeways, bus service and rail service (if any). For evaluation of the present system and of possible future improve-ments the model had to take into account a great deal of demographic and techni-cal detail: the population density and the average family income in each area of the city, the location of residential and business areas, the traffic flows over the transportation routes at typical peak and off-peak hours, how the speed of flow over each route would be affected by the number of vehicles using it, the amount of air pollution that would be generated by each type of vehicle and a great many other factors that must en-ter into the measurement of the costs and benefits of a transportation system.

Starting with computer runs that eval-uated how well the existing system per-formed, we went on to model pro-gressively more advanced systems and compare their performance. All together we tested some 200 models, each loaded with a tremendous amount of detail and each taking about an hour for the run-through in our computer. The project occupied a large team of specialists: en-gineers, city planners, mathematicians, sociologists, economists and computer programmers. A measure of the amount of work entailed is the fact that our final report, written by 17 authors, ran to 500,000 words—and we tried hard to be brief!

As our study proceeded, the results of the experiments showed that the possible strategies for the improvement of urban transportation fell into two sharply different categories from the standpoint of effectiveness. One of these was an approach we called "gradualism." It consisted in building improvements into the existing methods of transporta-

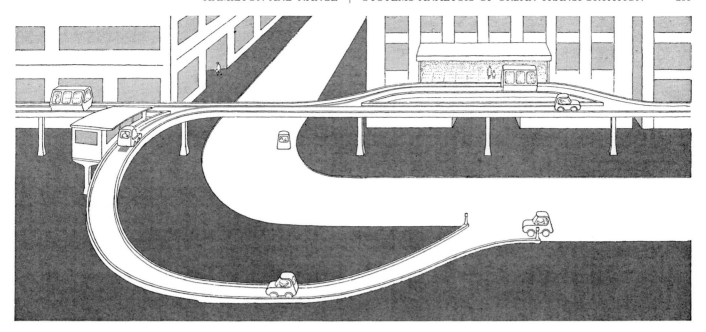

HYPOTHETICAL PERSONAL-TRANSIT SYSTEM would combine the speed and privacy of the automobile with the advantages of rail transit. A passenger entering the automated guideway network at a station would be carried by a small vehicle nonstop to his destination at speeds of up to 60 miles per hour. Specially equipped automobiles could enter the guideway by ramp, affording the driver swift, safe and effortless transport. Such dual use would make it feasible to extend the system to urban fringe areas.

tion. These, for example, included modernizing and extending old subway lines and building new ones, redesigning buses to make them quieter and easier to enter and speeding up their movement, equipping automobiles with devices to minimize air pollution, and so forth. The other approach, which we labeled "new technology," consisted in a jump to entirely new modes of transport, involving the creation of new kinds of vehicles and interconnections. Our tests of models indicated, as we shall show, that the gradualistic approach could not meet the future transportation needs of the cities, whereas innovations already in sight promise to do so.

Let us briefly examine some of the most promising of these new concepts. Engineers have described a system called "personal transit" that will operate like a railroad but will transport individual passengers or small groups nonstop to stations of their own selection. Its cars will be small, electrically propelled vehicles, with a capacity of two to four passengers, running on an automated network of tracks called "guideways." All stations will be on sidetracks shunted off the through line [see illustration above]. A passenger will enter a waiting car at a station, punch his destination on a keyboard and then be carried to the designated station with no further action on his part—no transfers, no station stops, no waiting, no driving. It appears that such a system could take the passenger from starting point to destination at an average speed of 60 miles per hour, as against the present average speed of 20 miles per hour counting station stops in U.S. subways.

The guideways could be designed to carry private automobiles as well as the public-transit cars, so that a driver coming into the city could mount the guideway at a station and ride swiftly to a downtown destination. Transport of the automobile by the guideway could be arranged either by providing flatbed vehicles that carried ordinary automobiles "piggyback" or by building into automobiles special equipment that enabled them to be conveyed by the guideway itself. The dual-mode use of guideways—by automobiles as well as by passengers in the small public vehicles—could make it financially feasible to extend the guideway system to outlying districts of a metropolitan area.

In some of our models of transportation systems incorporating new technology we also postulated entirely new automobiles designed from scratch for maximum safety and minimum air pollution. Such steam-engine automobiles are a feasible alternative to vehicles that could be combined with a personal-transit system. In contrast to gradualistic improvements, such as the padded dashboard or the smog-control device added to an internal-combustion engine, all-new automobiles would dramatically reduce accident casualties and fatalities and essentially eliminate air pollution. On the other hand, these cars would not help to defray the cost of personal-transit facilities nor would they automate any part of the burdensome task of driving.

For the suburbs, to transport people between their homes and local guideway stations or ordinary railroad stations, a promising possibility is a system known as "Dial-A-Bus." It would employ small buses (for eight to 20 passengers) and provide door-to-door service at a cost substantially less than that for a taxi. A commuter preparing to go into town would simply dial the bus service and be picked up at his front door in a few minutes to be taken to the nearest rapid-transit station. As calls for the bus service came in, a computer would continuously optimize the routes of the buses in transit for speedy responses to the developing demand. The computer technology to make such a system work is already developed, and the system could be tried out on a large scale immediately in connection with present suburban railroads. The Dial-A-Bus system would be most effective, however, in conjunction with a guideway network for personal transit.

For short-distance travel in the dense central areas of cities something is needed that would be faster than buses and cheaper than taxis. One classic proposal is the moving sidewalk. Unless someone can think of a better way of getting on and off than any yet proposed, however, the moving-sidewalk idea would work only for those who are content to travel at about two miles per hour or for people with a certain amount of athletic agility. A small-scale version of the personal-transit guideway looks like a more practical solution to the problem. The tracks for this system would stand above street level, to avoid interference with other traffic. The passenger would enter a personal "capsule" (which might hold one or two people) at a siding, dial his destination and travel to it at a speed of about 15 miles per hour. Such a system could be very compact and quiet.

Engineers generally agree that these innovations, specifically the personal-transit and personal-capsule systems, are already within the realm of feasibility. There are problems of safety and reliability to be solved, and decisions have to be made as to the best methods for propulsion, suspension and control. There is little doubt, however, that a system based on the innovations here described could be operating within a few years.

The big question is not whether such a system *could* be built but whether it *should*. The new system would take several years to develop, and there can be no guaranty that it would live up to its promise when it was completed. Meanwhile cities are hard-pressed for immediate relief from their transportation crisis. Would it not be wiser to adopt the gradualistic approach, to invest in improved buses, in better scheduling and perhaps in rapid-transit networks, than to invest millions of dollars in an untried system that in any case could not bring any help to our cities until years hence? This was the major question our computer tests of the various alternatives sought to answer. Our systems analysis attempted to compare the alternatives as fairly as possible in terms of the measurable costs and benefits—social as well as financial.

The heart of our model was a network representing a city's transportation. Network-flow analysis is an outgrowth of the mathematical theory of graphs. In the abstract, the question it deals with is this: Given a set of "nodes" (points) connected by a set of "arcs" (lines), with a specified cost associated with each arc, how can each shipment from node to node be routed at minimum cost, taking into account all other shipments? In our network each node represented a district, or "zone," in the city under study (for precision the node was defined as the center of population in the district) and each arc represented the capacity of the collection of streets that carried traffic from one node to the next. Besides the city streets we added separate arcs to represent expressways, rail lines, bus routes and walking and waiting for a conveyance, all of which had to be taken into account in order to calculate the minimum cost of travel from one node to another. Our basic measure of "cost" was the time required to traverse an arc, which depends not only on the length of the arc but also on how many other users are on the arc at the same time. We assumed, as could reasonably be done, that people usually take the fastest route (not necessarily the shortest in distance) between points.

A city's transportation system involves thousands of places to go, dozens of ways to get there and thousands of possible choices by an individual. As powerful as a large computer is, it can handle only so many calculations an hour. For our program the computer was limited to dealing with a maximum of 200 zones, 1,500 nodes and 5,000 arcs. Hence we had to divide each of our model cities into no more than 200 zones. We varied the zones in size from just a few blocks in the dense central city to substantially larger sections in areas away from the center. The criterion for zone size was that travel within a zone be negligible compared with travel among zones. We also had to make certain other simplifications.

The most crucial simplification had to do with the expected behavior of individuals in choosing their routes and means of travel. For a precise prediction of the traffic flows from zone to zone in the network we would have needed answers to a number of specific questions. Would a given resident going downtown take the bus, drive his car or have his wife drive him to the subway? How far would a $5,000-a-year male worker living in Zone 27 in Boston walk on an average winter day to save a 25-cent fare? How heavy would the traffic have to get before a person contemplating a nonessential trip decided not to go at all? If we had had detailed information such as this, we could have computed who went how by routing each person in the way that cost him the least in time, money and trouble—or, as economists say, "minimized the disutility to him."

Lacking sufficiently detailed data on such questions, we developed a general basis for predicting behavior that turned out to be reasonably reliable. First, we applied a simple formula, which had

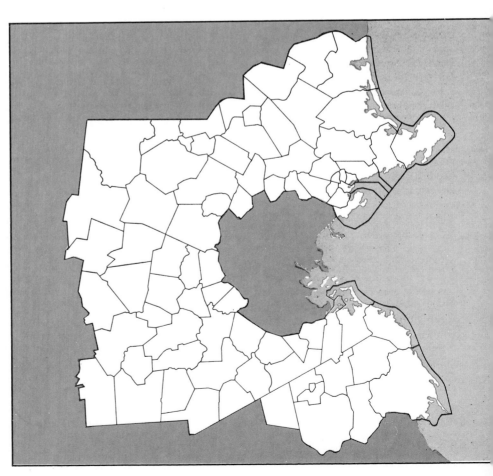

TRAFFIC-ZONE BREAKDOWN of the Boston area formed the basis of a model (*see illustration on page 236*). **Boston represents a typical example of a large city strongly oriented to**

been developed by the Traffic Research Corporation and had been found valid in traffic studies in several cities, to determine what proportion of the people in any given home zone would choose public over private transportation. (The formula computes this "modal split" on the basis of the average family income in the home zone, the relative amounts of time needed to reach a target zone by the two transportation methods and the relative "nuisance time" spent in walking and waiting.) Second, we assumed that within either of the two modes, public or private, each traveler will simply choose the route that minimizes his total travel time.

After thus working out a program for computing the expected zone-to-zone traffic flows in a city network under given conditions, we fed our data for each city into the computer to calculate the flow in the network with given demand for travel. The procedure was "iterative," employing a series of trials to arrive at the final allocation of flows. The program first calculated what the travel time for each arc would be if there were no congestion. Then it considered the

destination zones one at a time and calculated the quickest route to each destination from all the other zones. Next it introduced, for each route, the complicating factor of the relative numbers of travelers who would use the public mode and the private mode respectively. When this had been done for all the arcs, the program went back to the beginning and recomputed the travel times on the basis of the traffic flows indicated by the foregoing trials. It took about five such iterations to produce a stable picture of traffic flow that did not change in further trials.

For a quantitative assessment of what benefits could be brought about by improvements of the system, we modeled not only the existing modes of transportation but also various possible future systems with entirely different flow characteristics. The program included a number of subroutines that measured the costs and benefits of each system, in social terms as well as in terms of travel speed and money. Among the factors we introduced into the calculation were air pollution, the intrusion of automobiles into the city, the accessibility of key areas and the mobility of ghetto resi-

dents. Thus the transportation system judged to be "best" was not necessarily the one that was simply the cheapest or the fastest.

Obviously no model or program is worth much if it overlooks crucial factors or if its key assumptions are wrong. How much confidence could we place in the general model we finally developed? Fortunately it passed every validation test we could apply.

In the first place, as the work proceeded we took a skeptical view of the model's basic assumptions, trying out different assumptions to see how they would affect the results and encouraging each expert to criticize the others' work. We had some lively conferences and threw away a lot of computer printouts before we settled on a model we felt we could trust.

As it happened, the representation of traffic flow that we developed on the basis of our experience in studying quite different systems turned out to be very similar to flow models that had been devised by transportation engineers for use in traffic planning. Since we had had no prior knowledge of the traffic engineers' ideas, the fact that we had arrived

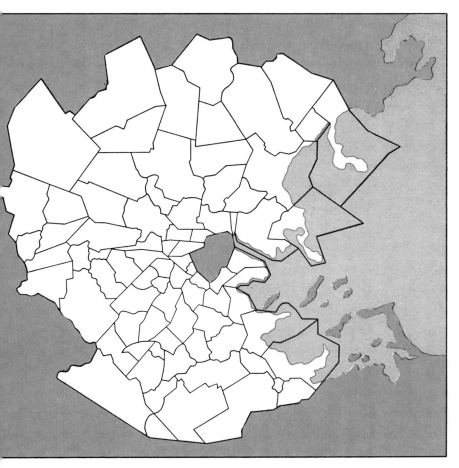

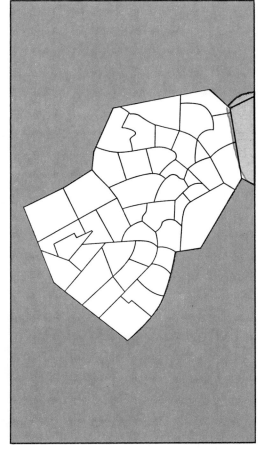

public transit. The size of a zone varies with population density and relative contribution to total traffic. In the dense core of the city (*right*) a zone often comprises only a few blocks. The total number of zones (200) represent an area of 2,300 square miles.

at much the same method of predicting traffic flow gave us considerable confidence that we were on the right track. Furthermore, we found that our network-flow program reproduced a faithful picture of the known flow in specific situations. As we have mentioned, each of the four cities we modeled had recently undergone a detailed traffic survey. These studies had recorded the average speed of traffic movement on the major streets, the numbers of people using public-transit facilities, the times for various trips in the city and so forth. To test the prediction ability of our network-flow program, we fed into the program the characteristics of the city's population and transportation network as of the time of the survey and let the program route the flow according to its own rules. In each case the results in the computer print-out corresponded so closely to the actual flow pattern as the direct, on-the-spot survey had described

it that we were satisfied our model could do a realistic job of representing a city's traffic flows.

Further validation of the general usefulness of our model emerged when we came to testing the alternative approaches for dealing with the urban transportation problem. In all four of our model cities the results of the analysis pointed to the same major conclusion: the best hope of meeting the cities' future needs lies in developing new transportation systems rather than in merely improving or adding to present systems.

A summary of our tests of various systems in Boston will serve to illustrate our findings. The story begins with the situation in 1963 [*see illustration on page 238*]. In that year the average door-to-door speed of public-transit travelers in the peak hours was nine miles per hour; the average automobile traveler's speed in the city was 16.4 miles per

hour; 32 percent of the people used public transit at peak hours, and the downtown streets were heavily burdened with automobile traffic. We next projected what the situation would be by 1975 if there were no change in the transportation facilities and traffic reached the level predicted by the Boston Regional Planning Project. Our calculations showed that public-transit travel would slow to 7.8 miles per hour and automobile travel to 15.7 miles per hour; the use of public transit would fall to 23 percent, and the intrusive concentration of automobile traffic downtown would rise by more than 15 percent. (One disastrous day in 1963 automobile traffic in downtown Boston reached a level of congestion that stopped all movement for several hours.)

We then proceeded to consider the effects of improvements in the transportation network. Addition of the costly freeways and extensions of rapid transit

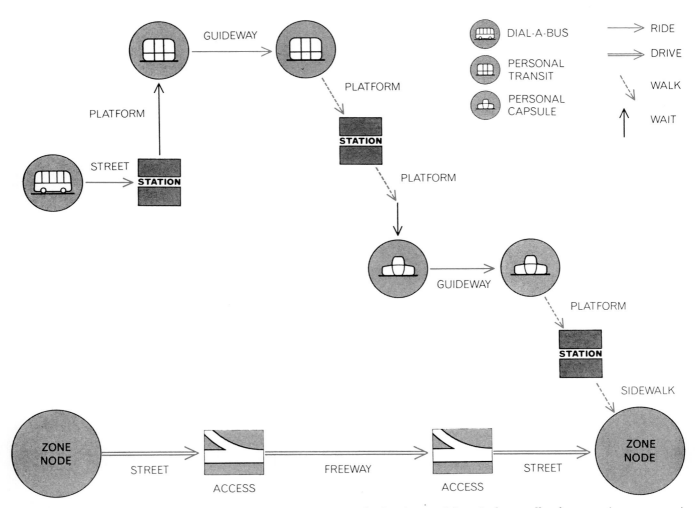

NETWORK simulates a city's transportation in terms of nodes (points) and arcs (lines). A zone node represents the center of population of a traffic zone and the point at which any trip begins or ends. Other nodes represent transfer points. Here two trips are represented in diagrammatic form, both beginning at the zone node at left and ending at the zone node at right. One trip utilizes transport by "Dial-A-Bus" (a hypothetical door-to-door system where the bus is routed by telephone calls of prospective passengers), personal transit and personal capsule, a version of the personal-transit guideway that could serve a central urban area, traveling at a speed of about 15 miles per hour. The second trip is by automobile. The parking and walking time at the end of this trip are not indicated. The relative lengths of the lines are not significant. Boston's transportation was modeled in terms of network flow.

that metropolitan Boston planned to build by 1975, it turned out, would bring about some improvement in speed over 1963 (to 10 miles per hour for public transit and 20.7 miles per hour for private automobiles) and somewhat reduce the crush of automobile traffic in the downtown streets. Replacement of buses by personal capsules for short-distance travel downtown would produce modest additional improvements, at a small net reduction in transit cost. In order to see what effects might result if public transit were considerably speeded up by improvements in the conventional system, we fed into the program an arbitrary assumption of a 50 percent rise in speed (disregarding the cost). On this assumption (which represents the maximum speedup that is likely to be attained on the basis of any current proposal) we found that automobile travel also would speed up substantially, because more people would be drawn to public trans-

portation and congestion on the freeways and in the streets would be relieved. The percentage increase in the use of public transit was only moderate, however, which suggests that an investment in speeding up conventional public facilities will not pay for itself unless it can be done very cheaply.

When we came to testing systems incorporating a network of personal transit by means of guideways, we saw really striking improvements in service. Speeds took a jump, particularly in the public mode, and more riders were attracted to public transportation. Had our calculations taken into account the comfort and privacy that personal transit offers in relation to conventional transit, the fraction of public-mode travelers would doubtless have been considerably higher. Furthermore, the introduction of a guideway network reduced the intrusion of vehicles and congestion in the downtown streets to less than half the 1963

level. Installation of a 400-mile network for personal transit in the Boston area would speed up travel to an average of 24.6 miles per hour in public facilities and 25.7 miles per hour in private automobiles, and 38 percent of the city's travelers (in 1975) would use public transit. If the network were extended to 600 miles and provided for the transport of automobiles as well as transit cars on the guideways, the average speeds of travel and the use of public transit would increase still further.

More important than these gains is the great improvement a personal-transit system would provide in mobility for the transportation poor or disadvantaged populations in the city. The 400-mile network we postulated for 1975 would make some 204,000 jobs in outlying areas of the metropolis accessible within half an hour's travel to people living in the city center; at present these

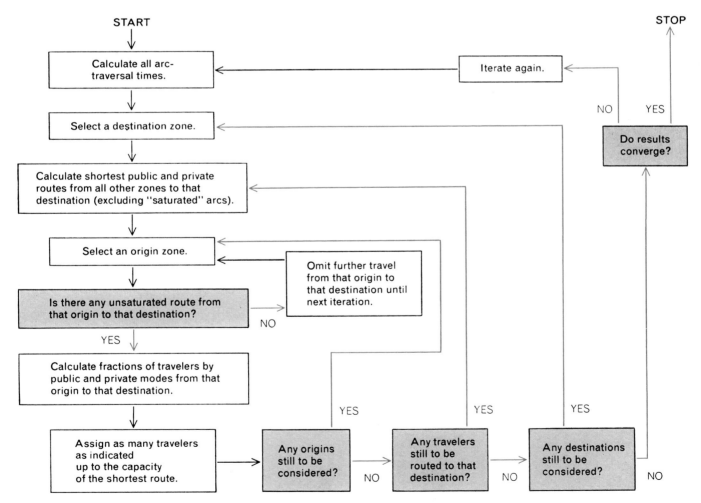

CALCULATION OF NETWORK PERFORMANCE utilized a computer program that employs a series of trials, or iterations, to compute the flow in the network. The program first calculates what the travel time for each arc would be if there were no traffic congestion. After the quickest route to each zone from all the others is calculated, the program introduces, for each route, the complicating factor of "modal split," namely the proportion of people traveling by public mode. If these numbers cannot be handled within the capacities of the shortest routes, the program goes back and computes the next-shortest routes. After the first iteration the program computes travel times as they are influenced by the flow assigned on earlier iterations. The exclusion of "saturated" arcs is an artifice to keep all the flow from following a few routes on early iterations. It speeds the convergence of the iteration process.

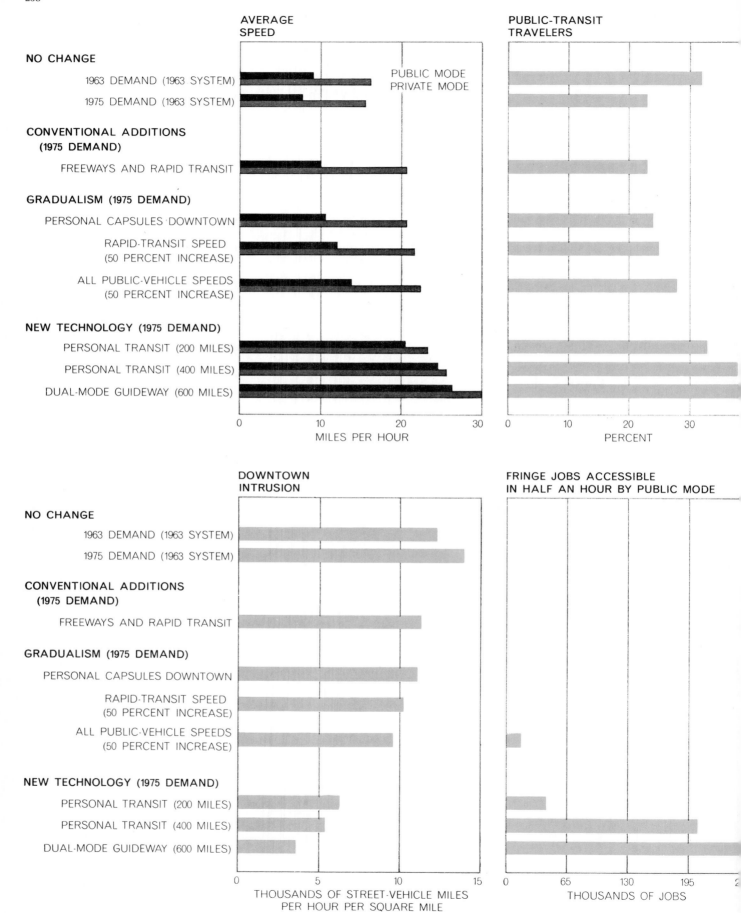

AVERAGE
SPEED

PUBLIC-TRANSIT
TRAVELERS

NO CHANGE

1963 DEMAND (1963 SYSTEM)

PUBLIC MODE
PRIVATE MODE

1975 DEMAND (1963 SYSTEM)

**CONVENTIONAL ADDITIONS
(1975 DEMAND)**

FREEWAYS AND RAPID TRANSIT

GRADUALISM (1975 DEMAND)

PERSONAL CAPSULES DOWNTOWN

RAPID-TRANSIT SPEED
(50 PERCENT INCREASE)

ALL PUBLIC-VEHICLE SPEEDS
(50 PERCENT INCREASE)

NEW TECHNOLOGY (1975 DEMAND)

PERSONAL TRANSIT (200 MILES)

PERSONAL TRANSIT (400 MILES)

DUAL-MODE GUIDEWAY (600 MILES)

0   10   20   30
MILES PER HOUR

0   10   20   30
PERCENT

DOWNTOWN
INTRUSION

FRINGE JOBS ACCESSIBLE
IN HALF AN HOUR BY PUBLIC MODE

NO CHANGE

1963 DEMAND (1963 SYSTEM)

1975 DEMAND (1963 SYSTEM)

**CONVENTIONAL ADDITIONS
(1975 DEMAND)**

FREEWAYS AND RAPID TRANSIT

GRADUALISM (1975 DEMAND)

PERSONAL CAPSULES DOWNTOWN

RAPID-TRANSIT SPEED
(50 PERCENT INCREASE)

ALL PUBLIC-VEHICLE SPEEDS
(50 PERCENT INCREASE)

NEW TECHNOLOGY (1975 DEMAND)

PERSONAL TRANSIT (200 MILES)

PERSONAL TRANSIT (400 MILES)

DUAL-MODE GUIDEWAY (600 MILES)

0   5   10   15
THOUSANDS OF STREET-VEHICLE MILES
PER HOUR PER SQUARE MILE

0   65   130   195   2
THOUSANDS OF JOBS

PERFORMANCE MEASURES from the authors' cost-benefit summary show how different transportation systems behaved in the case of the Boston model. All figures except public-mode cost refer to travel at a peak hour. In the full cost-benefit summary there

PUBLIC-MODE
COST

2   4   6   8
CENTS PER PASSENGER MILE

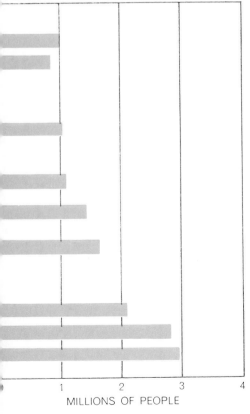

DOWNTOWN ACCESS
IN HALF AN HOUR BY PUBLIC MODE

1   2   3   4
MILLIONS OF PEOPLE

were 229 performance measures. The term
"downtown intrusion" refers to automobiles.

job areas are beyond that range of accessibility. The system would also make the downtown area, the airport, universities and hospitals quickly accessible even in peak hours to millions of people in the suburbs. Our full "cost-benefit" survey of the system indicated other benefits such as reductions of traffic accidents and air pollution.

In general, the results of our analysis made clear that, even with the most optimistic view of what might be achieved through improvement of the existing methods of transportation, such improvement could not satisfy the real needs of our cities in terms of service. Automobiles, even if totally redesigned for safety and smog-free steam propulsion, have the irremediable drawbacks that they must be driven by the user and are unavailable to a substantial percentage of the population. Buses and trains, however fast, comfortable and well scheduled, are unavoidably limited in average speed by the necessity of making frequent stops along the line to let riders on or off. All in all, our study suggested strongly that the course of gradualism is not enough: at best it is merely an expensive palliative for the transportation ailments of the cities.

On the other hand, our tests of the new-technology approach, particularly the personal-transit type of system, showed that it could provide really dramatic improvements in service. The personal-transit system would offer city dwellers a degree of convenience that is not now available even to those who drive their own cars. The city and its suburbs could be linked together in a way that would bring new freedoms and amenities to urban living—for the ghetto dweller now trapped in the city's deteriorating core as well as for the automobile-enslaved suburban housewife.

One must take account of the probability that drastic alteration of a city's transportation will bring about changes in the structural pattern of the city itself. We tested certain structural variations, such as concentration of the city population in a few dense nuclei, and found that the personal-transit system still offered striking advantages.

How would a personal-transit system compare with improvement of the existing system in the matter of financial cost? In Boston the cost of building and operating a personal-transit system would be somewhat more expensive per passenger-mile than a conventional rapid transit even if the city built an entirely new subway system from scratch. Remember, however, that we are talking about a personal-transit network of 400

miles, whereas Boston's rail system with its planned extensions by 1975 will consist of only 62 route-miles. Nevertheless, Boston and other cities that already have rather extensive rail rapid-transit systems may well think twice before scrapping the existing system to replace it with personal transit, even though personal transit offers benefits that rail rapid transit cannot approach.

For most of our large cities, now lacking rapid transit, personal transit looks like a much better bet than a subway. In automobile-oriented Houston, for example, personal transit in our calculations came out far cheaper than rail rapid transit, as well as far more effective. In such a city personal transit is clearly a best buy.

For smaller cities such as Tucson or New Haven personal transit looks less attractive. Because of their limited extent and lower density of population (and consequently smaller use of the system) the cost of personal transit per passenger-mile would be about three times the cost for a large city. It appears that a personal-transit system (as well as rail rapid transit, for that matter) would be too costly for cities with a population of less than half a million. Such cities, however, do not have to contend with the congestion that is overwhelming large cities.

To sum up, the installation of a personal-transit system, perhaps serviced by Dial-A-Bus feeders (the performance of such vehicles is not yet predictable), and designed from the start for eventual expansion to dual-mode service, seems well worth considering for the immediate future of many U.S. cities. We estimate that a personal-transit system could be developed and tested on a fairly large scale within five years at a cost of about $100 million. Compared with the cost of any sizable subway this development cost is insignificant. (A rail rapid-transit system recently proposed for Los Angeles, and rejected by the voters, would have cost about $2.5 billion.)

On the basis of the reports HUD has received from the groups it commissioned to study the urban transportation problem the department has submitted a number of recommendations to Congress, giving prominence to the proposed systems for personal transit, dual-mode transit and Dial-A-Bus. If the funds for development of these systems were made available immediately, the systems could be ready for installation in cities five years hence. Our study has convinced us that no time should be lost in proceeding with these developments.

SCIENTIFIC AMERICAN

September 1972   Volume 227   Number 3

Communication

It is not only the essence of being human but also a chief
property of life. Technological advances in communication
scope society and make its members more interdependent

by John R. Pierce

Communication

by John R. Pierce
September 1972

It is not only the essence of being human but also a vital property of life. Technological advances in communication shape society and make its members more interdependent

Our existence depends on communication in more ways than one can easily enumerate. Without our initial backlog of genetic messages we would not be who we are, and without our internal communication system we could not continue to live and function as we do. Our internal communications are handled by a network with nervous and metabolic subsystems [see "Cellular Communication," by Gunther S. Stent; SCIENTIFIC AMERICAN, September, 1972]. Nerve impulses are like telephone calls that are switched to a particular recipient and heard nowhere else. Hormones are like messages addressed to individuals or groups but sent out broadcast; only those concerned need respond. Furthermore, certain cells in the body behave with seeming autonomy, seeking out invaders and destroying them. Indeed, the falure of these cells to destroy cancer cells has been attributed to a blockage of the mode of communication by which foreign or abnormal cells are recognized.

When we think of communication, however, we usually think of external communication, of those processes by which we communicate with others. Without external communication we might live, but we would be ignorant, lonely individuals. We would have neither the inspiration of accumulated skill and knowledge nor the support of a society. That society, which communication makes possible, supplies us with necessities we would otherwise have to obtain for ourselves. Moreover, communication with others conveys rewards far beyond the basic necessities of life.

Animals live without knowing how they live, and they communicate without knowing how they communicate. By and large, so do we. Unlike animals, however, we speculate about how we live and how we communicate. Our better brains and our unique means of communication—language—make such speculation possible. Occasionally we learn something incontrovertible, and such knowledge can be very powerful in our lives.

Our clearest and most fruitful knowledge is not knowledge of ourselves or of how we communicate. Rather it is knowledge of various physical and chemical processes. That knowledge has enabled us to make powerful tools that have changed our lives greatly. We understand these tools far better than we understand ourselves.

In part that is because the tools are simpler than we are. A television set, a computer or even a national communication network such as the telephone system is simpler than a human being. We also understand our tools better because those tools have been built according to our understanding. It is easier to understand a computer than it is to understand art or the weather because the man who built the computer wanted to build something that would work. In order to build something that would work he had to build something he could understand.

No one man understands all the devices in a big airplane, a tall building or a telephone system, but some men have at least a working knowledge of each. That is not true of man and his means of communication. We have learned some remarkable things, but chiefly we live and communicate not through our understanding of these processes but in spite of our ignorance concerning them.

In our puzzlement about man and his communication it is natural to turn for enlightenment to the sure knowledge and deep insight that we have concern-

ONE KIND OF COMMUNICATION is print. This page, as it originally appeared, was reproduced from the "form" shown on the opposite page. The columns of type (and the caption) are set by Linotype. The headline ("Communication") is a copper photoengraving, and the subheading is type set by hand. These items and the incidental type at the top of the page are locked into position by a rectangular set of "bearers," which are beyond the margins of the picture. The complete form is used to make a nylon replica that is mounted on one cylinder of a high-speed rotary press and does the actual printing. Running at full speed, the press delivers 20,000 copies of SCIENTIFIC AMERICAN per hour. Photograph was made at The Lakeside Press, R. R. Donnelley & Sons Company in Chicago.

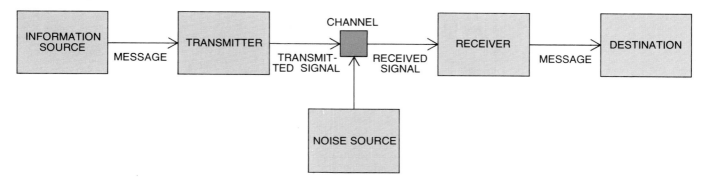

UNIVERSAL COMMUNICATION SYSTEM symbolically represented in this block diagram was originally proposed by Claude E. Shannon in his influential 1948 paper "The Mathematical Theory of Communication." The usefulness of Shannon's theory to the communication engineer is based in the broadest sense on the fact that it reduces any communication system, however complex, to a few essential elements. Modern communication theory is derived in large part from the ideas outlined by Shannon in his paper.

ing the nature, behavior and limitations of some of man's machines. Frequently it seems that what we have learned about machines should apply in some way to ourselves and our activities. We feel on occasion that the scales have fallen from our eyes and that the means for understanding are in our hands. At last we shall know.

In 1948 two publications created an intellectual stir about communication that has not yet subsided. These works were a paper titled "The Mathematical Theory of Communication," which Claude E. Shannon published in the July and October issues of the *Bell System Technical Journal,* and Norbert Wiener's book *Cybernetics: Control and Communication in the Animal and the Machine.* For the purposes of this discussion I should like to add to these two works a third: *Syntactic Structures,* published by Noam Chomsky in 1957.

Shannon's communication theory (also known as information theory) and Wiener's cybernetics created a broad and lasting interest and enthusiasm. In 1949 Shannon's paper was reprinted in book form together with exposition and comment by Warren Weaver. Weaver's contribution was based in part on an article by him in the July 1949 issue of *Scientific American.* These books alerted the scientific world. Conferences were held here and abroad. Scientists, shallow and profound, tried to extend the ideas of Shannon and Wiener to all aspects of life and art. I remember Shannon's telling me how he sat in a conference watching a man who spoke with his eyes turned toward the heavens, as if he were receiving divine revelation. Shannon was disappointed that he could not make any sense out of what the man was saying. Cybernetics and communication theory have been with us for nearly a quarter of a century. What have they

taught us? I looked through the books to find out.

Shannon's "The Mathematical Theory of Communication" summarizes and clarifies the task of the communication engineer in a remarkable way. The system Shannon studied is a truly universal communication system. As shown in the first figure of his paper [*see illustration above*], it consists of an information source, a transmitter, a communication channel, a noise source, a receiver and a message destination.

In such a communication system the information source and the message destination are usually human beings. As an example, a human being may type a message consisting of the letters and spaces on the keyboard of a teletypewriter. The teletypewriter serves as a transmitter that encodes each character as a sequence of electrical pulses, which may be "on" or "off," "current" or "no current." These electrical pulses are transmitted by a pair of wires to another teletypewriter that acts as a receiver and prints out the letters and spaces, which are in turn read by the human being serving as the message destination.

As the pulses travel from the teletypewriter that serves as the transmitter to the teletypewriter that serves as the receiver, an intermittent connection or an extraneous current may alter or mutilate them so that the receiver teletypewriter prints out some wrong characters. Shannon sums up such alterations or mutilations of the signal by including a noise source in his diagram of the communication system. Ordinarily the signal is an electrical signal, and the noise is an unpredictable interfering signal that is added to the desired signal.

The noise can cause the received message to differ from the message the source wished to transmit. The greater the ratio of the power of the signal to the power of the noise is, that is, the

greater the signal-to-noise ratio, the less effect the noise will have on the operation of the communication system.

This example of a communication system based on the use of teletypewriters is a very simple one. In contrast let us consider a voice communication system based on the automatic speaking machines known as vocoders. The transmitter part of such a system analyzes a speech signal and obtains data to control a speaking machine at the remote receiver. The control signals so obtained are the signals that are sent from the transmitter to the receiver. These control signals operate the distant speaking machine. Excessive noise in the transmission path may cause errors of articulation in the speech the message destination hears.

In this example we come on another concept introduced by Shannon. The vocoder and its control signals must be complex enough to produce speech that meets a fidelity criterion set by the message destination. If we fail to meet the fidelity criterion, the purposes of the communication system will not be fulfilled. If we produce better speech than is necessary, the system will be inefficient.

Thus Shannon's diagram of a communication system can apply to a system that is very complex indeed. Moreover, Shannon's theory can tell the communication engineer something meaningful and useful about any communication system. It gives him a measure of the commodity he is trying to transmit. That measure, a unit of uncertainty or choice, is called a bit. A bit (a contraction of "binary digit") is the uncertainty between "yes" or "no" or "heads" or "tails" when both are equally likely; it is the choice we exercise in choosing "left" or "right" most unpredictably (that is, with equal probability for each). The symbols 0 and 1 (that is, the digits of a binary number) can specify yes or no, heads or tails, left or right.

Shannon showed how in principle to measure the information rate of a message source—a speaker, a person writing, the output of a television camera—in bits per message or bits per second. This quantity is called the entropy or entropy rate of the source.

Shannon also showed how to measure the capacity of a communication channel. Again the measure is bits per second. The entropy rates of different message sources vary widely. Studies conducted by Shannon show that the entropy of English text is probably about one bit per character. The entropy rate of speech is not known. Experiments with vocoders indicate that it is less than (and probably much less than) 1,000 bits per second; in contrast we routinely use a channel (the telephone) capable of transmitting 60,000 bits per second of high-quality speech. A channel capable of transmitting some 60 million bits per second is generally used in transmitting television pictures, but the entropy rate of the picture source must surely be far less than that.

One thing is clear: It is necessarily more costly to transmit a television picture than to transmit speech, and more costly to transmit speech than to transmit text. Transmitting a detailed picture, whatever its worth, is bound to be more than 1,000 times more costly than efficiently transmitting 1,000 words of text.

Shannon contributed a great deal in establishing the bit as a measure of the complexity of message sources and the capabilities of channels. In addition he proved an extraordinary theorem. He showed that if the entropy rate of a message source does not exceed the capacity of a communication channel, then in principle messages from the source can be transmitted over the channel with a vanishingly small probability of error. This can be achieved only by storing an exceedingly long stretch of message before transmitting it [see "Communication Channels," by Henri Busignies; SCIENTIFIC AMERICAN, September, 1972].

Shannon's communication theory has had two chief values. One is rather like the value of the law of conservation of energy. Shannon has made it possible for communication engineers to distinguish between what is possible and what is not possible. Communication theory has disposed of unworkable inventions that are akin to perpetual motion machines. It has directed the attention of engineers to real and soluble problems. It has given them a quantitative mea-

sure of the effectiveness of their systems. Shannon's work has also inspired the invention of many error-correcting codes, by means of which one can attain error-free transmission over noisy communication channels.

Information theory and coding theory are widely pursued areas of research. It is true that many real and significant problems seem too difficult to solve, and that many of the problems that are solved seem made up because they can be solved rather than because they are very significant. Nonetheless, communication theory as Shannon conceived it has enduring importance.

Shannon is an engaging and lucid

writer. In his exposition of what he had done he used amusing as well as enlightening examples. One example was a message source in which the probability of the next letter depended on the preceding letters. Such a source produced "words" such as *grocid, pondenome, demonstures* and *deamy.* Shannon's choice of technical terms, on the other hand, was evocative as well as apt; it included *information, entropy, redundancy* and *equivocation.* Surely a theory that dealt with such wide-ranging concepts must have wide-ranging application.

So thought Weaver, as he expressed himself in "Recent Contributions to the Mathematical Theory of Communica-

CHANNEL		CHANNEL BANDWIDTH (HERTZ)	CHANNEL CAPACITY (BITS PER SECOND)
TELEPHONE WIRE (SPEECH)		3,000	60,000
AM RADIO		10,000	80,000
FM RADIO		200,000	250,000
HIGH-FIDELITY PHONOGRAPH OR TAPE		15,000	250,000
COMMERCIAL TELEVISION		6 MILLION	90 MILLION
MICROWAVE RELAY SYSTEM (1,200 TELEPHONE CHANNELS)		20 MILLION	72 MILLION
L-5 COAXIAL-CABLE SYSTEM (10,800 TELEPHONE CHANNELS)		57 MILLION	648 MILLION
PROPOSED MILLIMETER-WAVEGUIDE SYSTEM (250,000 TELEPHONE CHANNELS)		70 BILLION	15 BILLION
HYPOTHETICAL LASER SYSTEM		10 TRILLION	100 BILLION

CAPACITY OF VARIOUS COMMUNICATION CHANNELS can be measured according to Shannon's theory in bits per second, as indicated in the column at the extreme right in this chart. The numbers given for each entry are more or less rough estimates for a particular system (either currently in operation, proposed or merely envisioned); in many cases it is difficult to ascribe even an approximate value for capacity in bits per second to an analogue channel because of the variability of parameters such as signal-to-noise ratio.

tion," his part of the 1949 book. Weaver plausibly divided the problem of communication into three levels: (A) the technical problem: How accurately can the symbols of communication be transmitted? (B) the semantic problem: How precisely do the transmitted symbols convey the desired meaning? (C) the effectiveness problem: How effectively does the received meaning affect conduct in the desired way?

Weaver recognized that Shannon's work pertained to level A, but he argued persuasively that it overlapped and had significance for levels B and C as well. Weaver was intrigued by the idea of using the powerful body of probability theory involving Markoff processes in connection with both languages and semantic studies. Moreover, in pointing out the pertinence of information theory to cryptography, he argued that information theory "contributes to the problem of translation from one language to another."

This brings us to the fascinating story of machine translation. Since most of the details of the story are irrelevant to communication, I shall simply sketch the broad outlines. In brief, when simple machine "dictionary look-up" produced ambiguous, unreadable texts, the proponents of computer translation turned to grammar for a cure. They found that existing grammars were unhelpful. In fact, they found that no attainable grammar cured the persistent ills of machine translation: inaccuracy, ambiguity and unreadability. As Victor H. Yngve, once an associate of Chomsky's, wrote in 1964: "Work in mechanical translation has come up against a semantic barrier. ... We have come face to face with the realization that we will only have adequate mechanical translation when the machine can 'understand' what it is translating and this will be a very difficult task indeed.... 'Understand' is just what I mean.... Some of us are pressing forward undaunted."

Both the Shannon-Weaver book and the search for a syntactical cure for the ills of machine translation led to a reexamination of syntax. In *Syntactic Structures* Chomsky sought an avenue toward a grammar of a language. Letting L stand for a language, he wrote: "The grammar of L will thus be a device that generates all of the grammatical sequences of L and none of the ungrammatical ones. One way to test the adequacy of a grammar proposed for L is to determine whether or not the sentences that it generates are actually grammatical, i.e., acceptable to a native speaker."

Chomsky rejected the Markoff process, which Weaver had mentioned, as being inappropriate and inadequate. Part of his argument was statistical. Although a statistical process can generate sentences, the likelihood that a sentence will occur has nothing to do with whether or not it is grammatical. Indeed, in 1939 Ernest Vincent Wright published a novel, *Gadsby*, in which the letter *e* simply does not occur. Although the text violates the statistics of English, it violates neither grammar nor sense [*see illustration on page 8*].

Another part of Chomsky's argument was that neither the Markoff process nor any existing grammar gave simple, sensible explanations for many obvious features of English, including the passive voice. To this end Chomsky proposed a new grammatical concept: the transformational grammar.

Chomsky's work was important in two ways. It led a number of linguists, Yngve among them, to try to produce transformational grammars of English. This work produced grammars that became larger and larger and less and less intuitive; moreover, none of the grammars produced could generate all grammatical sequences and no ungrammatical sequences.

Another outcome of Chomsky's work was to inspire certain psychologists,

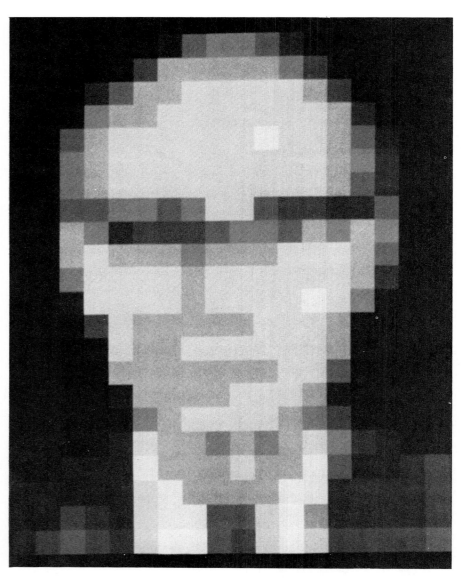

CUBIST PORTRAIT of John R. Pierce, the author of this article, was made by Leon D. Harmon of the Bell Telephone Laboratories as part of a computer experiment designed to determine the least amount of visual information a picture can contain and still be recognizable. The picture is divided into about 200 squares, with each square rendered evenly in one of 16 intensities of gray. For best results the portrait should be viewed from a distance of 15 feet or more; jiggling the picture, squinting or removing one's eyeglasses also helps. Such studies of the information content of a visual image may be useful in devising future systems for storing, transmitting and displaying pictures with the aid of computers.

notably George A. Miller, then at Harvard, to do experimental work on human response to language. At first this line of inquiry seemed very fruitful. For one thing, response to the active voice appeared to be quicker than response to the passive voice, a finding consistent with the view that the passive voice involves an additional transformation. As such work has proceeded it has become complicated and unclear in implication, except for one thing: the importance of meaning in communication [see "Verbal Communication," by Roman Jakobson; SCIENTIFIC AMERICAN, September, 1972].

Remember that Chomsky proposed as a criterion for a sentence's being grammatical that it be "acceptable to a native speaker." When in fact naïve native speakers are asked whether or not a sentence is grammatical, they frequently reject it as being ungrammatical unless it makes sense to them. Conversely, ingenious people will spontaneously and gleefully invent tortured circumstances in which a seemingly meaningless or ungrammatical sentence will actually be acceptable to a native speaker. We are inclined to reject "I is a boy," but what is our reaction to be if we are visiting a school in which all the children are assigned pronouns rather than being called by their names?

Indeed, whether or not a sentence is grammatical or meaningful seems to depend ultimately on the ingenuity of the person who puzzles over it. Even then we are left with Weaver's exception: "Similarly, anyone would agree that the probability is low for such a sequence of words as 'Constantinople fishing nasty pink.' Incidentally, it is low, but not zero; for it is perfectly possible to think of a passage in which one sentence closes with 'Constantinople fishing,' and the next begins with 'Nasty pink.' And we might observe in passing that the unlikely four-word sequence under discussion has occurred in a single good English sentence, namely the one above."

Weaver therefore stresses meaning and effectiveness in communication. We find this search for meaning and effect outside the laboratory both in the myths and magic of antiquity and in our own daily experience. We find meaning in text or speech even when the text is corrupt or the speech scarcely audible. William James had observed this truth in 1899: "When we listen to a person speaking or read a page of print, much of what we think we see or hear is supplied from our memory. We overlook misprints, imagining the right letters, though we see the wrong ones; and how

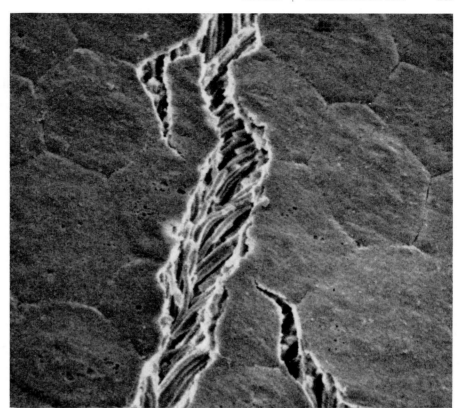

VISUAL-INFORMATION RECEPTORS are represented in this scanning electron micrograph by the outer rod segments seen through a tear in the membrane at the back surface of a rabbit's retina. The micrograph was made by Thomas F. Budinger and Thea Scott of the Lawrence Berkeley Laboratory. The magnification is approximately 4,000 diameters.

AUDITORY-INFORMATION RECEPTORS are shown enlarged some 6,000 diameters in this scanning electron micrograph of the ciliary tufts of hair cells in the sacculus, or inner ear, of a frog. The micrograph was made by James Pawley of the University of California at Berkeley working with a sample prepared by Edwin R. Lewis in his laboratory at Berkeley.

little we actually hear, when we listen to speech, we realize when we go to a foreign theatre; for there what troubles us is not so much that we cannot understand what the actors say as that we cannot hear their words. The fact is that we hear quite as little under similar conditions at home, only our mind, being fuller of English verbal associations, supplies the requisite material for comprehension upon a much slighter auditory hint."

This passage from James gives us a strong clue to understanding where the meaning of language really resides. It is useless to seek the meaning or utility of a book between its covers, or the intent or effect of a speech in the sounds that are uttered. A textbook on physics or mathematics exists in the context of the physics and mathematics of its time. It is not a complete exposition of the physics or mathematics it is intended to teach. Such a text is addressed to a person who already knows a language, something about the world and something about physics and mathematics. It is intended to enable him to learn, understand and use knowledge, skills and insights that the author himself has acquired. Taken out of its environment of language, such a textbook would be completely unintelligible. Taken out of its environment of physics or mathematics, such a textbook might be interpreted symbolically, construed as magic or dismissed as boring nonsense.

Our everyday communication has meaning in a much more restricted context. An order on a battlefield communicates only to a soldier who knows the situation, and the situation cannot be re-created by an examination of the words uttered. Phrases passed back and forth in a discussion may earn John Smith a raise or cost him his job, but the phrases have meaning only with respect to what those who are discussing Smith already know about him. Communication can take place only between people with a common aim, a common problem, a common curiosity, a common interest; in other words, something in common that is meaningful and important or fascinating to both. The process of communication is not one of imparting entire areas of knowledge or of drastically changing views. That is the process of education or training, which makes use of communication but goes far beyond communication as we commonly construe it. Communication in everyday use is a process of adjusting understandings and attitudes, of making them congruent or of ascertaining how and where they agree or disagree. A common language is of extreme advantage in our efforts to communicate, but it is not as important as a common interest and some degree of common understanding.

Yet if we had everything, or almost everything, in common, no communication, or very little communication, would suffice. That is the case with certain animals, such as bees [see "Animal Communication," by Edward O. Wilson; SCIENTIFIC AMERICAN, September, 1972]. The need for communication arises because something unguessable must be imparted concerning our understanding or actions. A little must be added to what we already know or as a basis for modifying what we would otherwise do. It is this element of the unguessable that Shannon measures as entropy. And it is the unguessable, the surprising, that is an essential part of communication, as opposed to the mere repetition of gestures, incantations or prayers.

Meaning can exist only through what we have in common in our lives, minds and language. It is concerning what we share that communication takes place. We cannot be certain that reassurance will be offered, and we are upset if it is not. We are excited by a new discovery or by a tidbit of gossip about someone we love or hate. We spring to a task we understand and feel impelled to undertake. And, as I have observed, we con-

---

XXIX

GADSBY WAS WALKING back from a visit down in Branton Hills' manufacturing district on a Saturday night. A busy day's traffic had had its noisy run; and with not many folks in sight, His Honor got along without having to stop to grasp a hand, or talk; for a Mayor out of City Hall is a shining mark for any politician. And so, coming to Broadway, a booming bass drum and sounds of singing, told of a small Salvation Army unit carrying on amidst Broadway's night shopping crowds. Gadsby, walking toward that group, saw a young girl, back towards him, just finishing a long, soulful oration, saying:—

". . .and I can say this to you, for I know what I am talking about; for I was brought up *in a pool of liquor!!"*

As that army group was starting to march on, with this girl turning towards Gadsby, His Honor had to gasp, astonishingly:—

"Why! Mary Antor!!"

"Oh! If it isn't Mayor Gadsby! I don't run across you much, now-a-days. How is Lady Gadsby holding up during this awful war?"

[201]

IMPROBABLE TEXT, represented by this page copied from the 1939 novel *Gadsby*, by Ernest Vincent Wright, violates the statistics of English, although it violates neither grammar nor sense. Wright's entire novel of more than 50,000 words was written without a single word containing the letter *e*. This extraordinary counterexample serves to symbolize the difficulties associated with a purely statistical approach to the analysis of language. Such problems came into prominence in the course of early attempts at computer translation, leading certain linguistic theorists, such as Noam Chomsky of the Massachusetts Institute of Technology, to undertake a reexamination of syntax in linguistic communication.

tinually search for meaning, whether or not meaning is there.

I believe it is a mixture of surprise and the search for meaning in a familiar context—the English language—that lends charm to Shannon's statistical words *grocid, pondenome, deamy*. They are so like English that we wonder what they mean, or perhaps we feel that we sense some elusive meaning or connotation. I think that it is our straining toward some such extraordinary context or meaning that gives charm to computer-produced "poems" such as those of Marie Borroff:

The river
Winks
And I am ravished.

Dangerously, intensely, the music
Sins and brightens
And I am woven.

These poems are not so much out of the world as enticingly on the fringes of it.

We are challenged by what seems almost within our comprehension, but we reject or are bored with whatever completely eludes our grasp. Using computer-generated patterns, Bela Julesz of the Bell Telephone Laboratories has shown that a given degree of order or redundancy (in the sense of information theory) can be detectable or undetectable to the eye, trained or untrained. Here is one point of departure between information theory and problems of human communication.

The information theorist asks: What is a general measure of order, or rather of disorder, in a message source, and how can I take advantage of the order that is there in transmitting messages from that source efficiently? To do this, however, one must adapt the communication equipment to the order that lies in the source. Our eyes, ears and brain, adapted as they may have been through evolution, are with us throughout our lives. We can make use of order only if we can perceive it. Some order we can learn to perceive; some must escape our senses, to be detected only through statistical analysis.

Many artists and musicians who have been inspired by information theory have sought to produce works with an optimum combination of order and randomness. Such a criterion antedates information theory. Beethoven is said to have declared that in music everything must be at once surprising and expected. That is appealing. If too little is surprising, we are bored; if too little is expect-

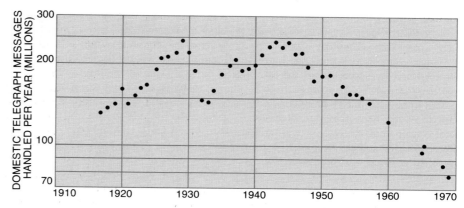

RISE AND FALL of a form of communication, the telegraph, is traced in this graph of the total number of domestic telegraph messages handled in the U.S. from 1917 to 1970.

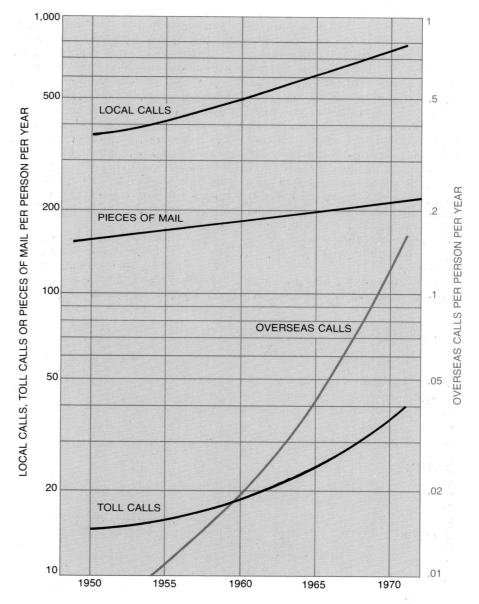

ACCELERATING USE of two other forms of communication, the postal service and the telephone, in recent years has contributed to the decline of telegraphy. The rate of increase for pieces of mail per person is currently about 3.5 percent per year, for local telephone calls per person about 5 percent per year, for toll telephone calls per person about 10 percent per year and for overseas telephone calls per person about 25 percent per year.

ed, we are lost. Communication is possible only through a degree of novelty in a context that is familiar.

Music communicates at once by reassurance and by surprise, but we can be neither reassured nor surprised by what passes unnoticed. Sir Donald Tovey, the British composer and writer, characterized certain contrapuntal devices as "for the eye only" and so implied that they must go unheard and unappreciated by the listener.

Order undetected is order in vain. One contemporary Canadian composer, Gerald Strang, has said that much random music and much totally organized music sound alike: uniformly gray. This would indeed be the consequence of an order undetectable by the ear and brain.

As to what is perceivable order and what is not, there will be long arguments. Some will say that the occurrence of the diatonic musical scale in unrelated cultures argues that it is physically or psychologically suited to man, whereas the only basis for 12-tone music is the fact that the piano has 12 black and white keys per octave. Others will argue that what we expect and what surprises us in older music is entirely cultural; that with sufficient training and effort on the part of the listener today's gray music will come to shimmer with color and startle with surprise, intellectual if not acoustical. It is plausible that what has moved many men may move others; it is less plausible that men will necessarily come to be moved by what has moved no one.

So far in this article I have associated communication entirely with man and communication among men. Another kind of communication is growing up around us: communication among machines. Here we are back to the process of devising, understanding and improving tools. We do not care about the working conditions or interests of computers. In the end we do not care how our credit-card transactions are processed as long as the process is accurate, cheap and prompt. What we do care about is providing a simple, easy-to-use, compact, reliable terminal by means of which a man can send messages to a machine and receive messages from it.

It seems ridiculous to speak a credit-card number into a telephone and have the computer operator key the number into the machine and relay the machine's yes or no answer by voice. It *is* ridiculous, and people are working on machines to read credit cards and communicate with a distant computer directly.

All of us look toward communication with distant computers from the home or the office, but as yet there are no adequate terminals. Teletypewriters are large, clanking and costly. Cathode ray terminals are large, limited and costly. The problem is that we have no cheap and adequate means for display. Keyboards are compact, reliable and easy to use. I for one would rather command a computer through a keyboard than talk to it, even if that science-fiction dream were possible.

The widespread use of computers is held up by lack of cheap, compact, reliable terminals, and by other things as well. The field of data communication and interacting computer systems is powerful and fascinating. It calls into play not only communication theory but also the construction of artificial languages, complete with grammars and the devising of computer programs of great complexity. Such communication is a growing part of our world, but it is remote from the average user of communication, who sees and cares to see only the terminals by means of which he uses such computer systems. As the computer and terminal arts advance, we shall have all kinds of new services in our homes, whether for consulting advertisements, purchasing, making reservations, learning algebra or learning Japanese. All of this and more will come, but when it will come is hard to judge [see "Communication Terminals," by Ernest R. Kretzmer; SCIENTIFIC AMERICAN, September, 1972].

Whatever one may say about communication among machines, human communication takes place within a community of interest made up of human beings. Within that community people think and act, and they communicate in moving others toward or away from what they regard as appropriate thoughts and actions. In terms of information theory only certain messages among all possible messages occur in communication within a community of interest, because only certain messages are appropriate, make sense and will be understood.

In a simple society, an isolated band or a village with a subsistence economy, a community of limited size and complexity provides a community of interest common to all. Speech supplies ade-

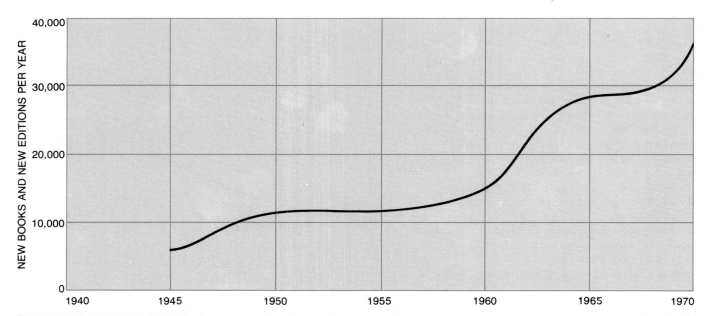

IRREGULAR UPWARD TREND characterizes this plot of the number of new books and new editions of old books issued in the U.S. each year from 1945 to 1970. An initial rise after the end of World War II was followed by a plateau (perhaps associated with a period of postwar practicality and conformism), which in turn gave way to a wavering but very steep rise beginning in about 1960.

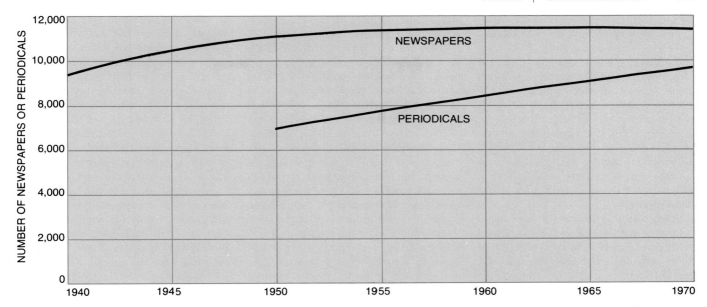

INTERESTING CONTRAST is evident in this graph, which shows that the number of periodicals published in the U.S. over the period 1950–1970 has increased steadily, whereas the number of newspapers published over the same period has become fairly stable.

quate communication. The family or the clan may be a subsidiary community of interest. Other places, other people and other times must somehow be taken into account in myth and custom. Nonetheless, to a degree all of life (that is, all that is noted) is accessible and intelligible to all members of the community.

Our world has far outgrown the reach of the voice or the comprehension of any one mind. It is divided into countless communities of interest. Increasingly these communities are intellectual rather than geographical. A physicist or a mathematician or a biologist may have more in common with a colleague in a foreign land than he does with his next-door neighbor.

Our multiple communities of interest and the institutions that serve them overlap. A mathematician may be interested in the stock market and in riding unicycles. An engineer may be interested in spectator sports, skiing and music. A science-fiction writer may be interested in local politics. That each man is a member of many communities of interest makes them no less real.

Clearly such communities of interest could not exist without adequate communication. I shall not argue that advances in communication and travel have brought these communities of interest into being, or that conversely wide-ranging interests have stimulated the technologies of travel and communication. What is clear is that communication plays a complex role in a complex way of life. In so serving us modes of communication interact with the people

they serve in many ways [see "Communication and Social Environment," by George Gerbner; SCIENTIFIC AMERICAN, September, 1972].

Some forms of communication serve many communities of interest, ranging from the individual and the family to large business and cultural organizations. The postal service, the telegraph and the telephone are among these forms. The telegraph provides an interesting example of the rise and fall of a form of communication. When one plots the domestic telegraph messages handled from 1917 to 1968, one finds a faltering peak from 1930 to 1935 [see top illustration on page 247]. Anyone who has sent or received telegrams can read a variety of economic and social observations into these data. On the one hand one sees competition from the telephone, through direct distance dialing and decreasing long-distance rates. On the other hand one sees that a dispersal of population into suburbia and rising labor costs have combined to make the physical delivery of telegrams uneconomic. Those with long memories may recall the knock of the telegraph boy at the door. They may even recall office switches that would summon telegraph boys to take messages. Such services have vanished, and Western Union is seeking a new mode of life.

Mails and telephony, in contrast, are still on the increase, with the telephone winning out [see bottom illustration on page 247]. There are more local calls per person than pieces of mail per person, and the rate of increase is faster. Furthermore, toll calls increase twice as fast as local calls and overseas calls increase

more than twice as fast as toll calls. In such statistics we see a society in which human relations and business transactions both have a shorter time scale. Life has become less leisurely, less planned, more immediate. The increases in toll calls and overseas calls also tell us that communities of interest are becoming more far-flung.

These increases in part reflect advances in technology that have made such calls cheaper and better. Quality is important to the use and success of communications. Overseas calls increased rapidly after 1954, when the first undersea cable made overseas calls more intelligible and useful than they had been in the days of shortwave radio. Cost is equally important. Surely a reduction in the cost of toll calls, and particularly of calls during certain hours, has stimulated such calling.

Although a decrease in cost can stimulate the use of communications, too high a cost can result in no use whatever. Invitations are no longer carried by footmen. And, technological marvel though it is, the Bell System's Picturephone service seems unlikely to sweep the country. The cost is greater than people care to pay, particularly for long-distance calls. So it seems to be with many ambitiously proposed communication systems. The cost is more than the traffic will bear. In time videotelephone and home communication centers will be available to all, but only when radical advances in technology bring the cost down. That time will not be tomorrow.

Telegrams, the postal service and the telephone serve almost all communities of interest. Some modes of communication serve large communities of interest only. Books are among these. Nothing is more diverse than books. With rare exceptions (the Bible is among them) each book reaches a small fraction of the total population. Yet a successful book may reach a large fraction of those deeply interested in a particular area or subject, be it yoga, tennis, the stock market or general relativity. In the last instance a successful book can blanket the real community of interest with a few thousand copies.

A plot of the number of new books and new editions of old books issued each year from 1945 to 1970 reveals that there was an initial rise after the war and then a plateau. We can associate the plateau, if we wish, with postwar practicality and conformism [see illustration on page 248]. About 1960 a wavering but very steep rise set in. Does this rise reflect an increasing diversity and a growing number of communities of interest? Or does it reflect a reduction of cost and an increasing number of outlets during the paperback revolution? I am sure that it reflects both, and that both are a part of the interaction of communication and society.

A book can serve a rather narrow community of interest, for example the worldwide community of specialists in some field of science. In general periodical publications must have a larger readership in order to survive. Nonetheless, the number of periodicals published over the period 1950–1970 has increased steadily [see illustration on preceding page]. This increase reflects a growth in communities of interest of substantial

size, including the community of interest that is served by this magazine.

I believe that such statistics actually underestimate the true number of periodical publications, and that they omit many that serve very real communities of interest. Particularly in technology and business, there are a large number of highly specialized periodical publications: monthly, quarterly or yearly. They are directed to subjects as special as digests of commercially available computer hardware and analyses of the financial condition and prospects of certain companies. The subscription price for some is extremely high; this factor seems to show that they serve particular communities of interest better than more general technical or financial journals. Other highly specialized and probably uncounted periodicals include the mimeographed science-fiction fan magazines called "fanzines" and endless club and society bulletins and the like. All serve real communities of interest.

It is interesting to note that whereas the number of periodicals is rising, the number of newspapers is static. I think that here too one can read meaning into statistics. As we progress from the mails and the telephone through books, through periodicals and on to newspapers we are considering successively media that necessarily serve larger and larger communities of interest. With the newspaper we have arrived at what survives of the geographical and the national communities of interest. Newspapers must appeal to such interests as we all have in common because we live in a particular community, state or nation.

What is true of newspapers is also true

of radio and television [see illustrations below]. Here we see that whereas virtually every home has a radio and a television set, the number of stations is still rising somewhat. With the advent of the communication satellite, television has become a truly international medium. President Nixon's trips to China and the U.S.S.R. and the astronauts' trip to the moon have been hit shows. With the growth in diversity, however, the geographical, the national and indeed the international communities of interest have come to occupy an equivocal place in our lives. Mass communications have come to have an uncertain role in the functioning of society. It can be argued that mass communications are not effective eyes and ears on the world, and that they do not ensure popular participation in government.

Here I come back to Wiener's *Cybernetics,* to which I have scarcely referred since I mentioned it at the beginning of this article. Unlike Shannon's "The Mathematical Theory of Communication," *Cybernetics* does not expound a novel, well-defined and well-implemented body of theory. Rather it is a discussion, partly mathematical and partly popular, of the relation between certain concepts and man and his world.

One prime concept expounded by Wiener is homeostasis: the functioning of an organism or a system so as to correct for adverse disturbances. Homeostasis involves two processes: the detection of deviation from the desired state, and negative feedback, by which the discrepancy between the desired state and the present observed state is cor-

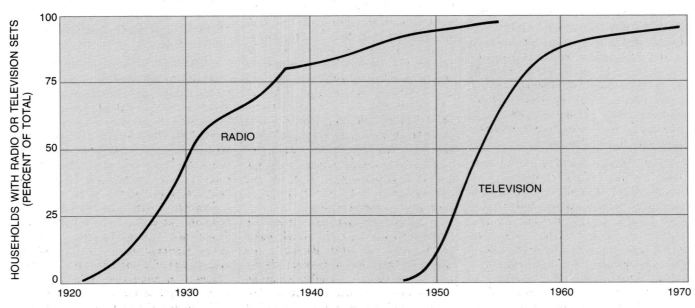

ASCENDANCY OF MASS COMMUNICATIONS is reflected in the graphs on these two pages, which show that whereas virtually every home in the U.S. now has both a radio and a television set (*left*), the number of radio and television stations is still rising

rected. The process of homeostasis may fail in many ways.

For the overall process to function, we must satisfactorily sense that some correction needs to be made. Beyond that, however, negative feedback may itself result in instability. This happens when there is too much negative feedback, or when it acts too quickly in a system in which some responses are necessarily slow. The normal human being can observe the fallibilities of feedback by trying to carry a shallow pan full of water without spilling it. He had best keep his eyes off the pan as he walks along, however. Efforts guided by the eye ordinarily heighten rather than reduce the sloshing of the water.

Wiener observes that "small, closely knit communities have a very considerable measure of homeostasis; and this, whether they are highly literate communities in a civilized country, or villages of primitive savages. Strange and even repugnant as the customs of many barbarians may seem to us, they generally have a very definite homeostatic value, which it is part of the function of anthropologists to interpret. It is only in the large community, where the Lords of Things as They Are protect themselves from hunger by wealth, from public opinion by privacy and anonymity, from private criticism by the laws of libel and the possession of the means of communication, that ruthlessness can reach its most sublime levels. Of all of these antihomeostatic factors in society, the control of the means of communication is the most effective and most important."

No doubt greed, self-interest and cal-

lousness do account for failures of communication in our overall community of interest. We may also observe, however, that the world in which we live and which affects our daily lives has become overwhelmingly complex. We cannot possibly understand it in the sense that we might understand a small, closely knit community. If we do not disregard the larger world, we must perceive it as it is represented by the media of mass communication. Is this, however, a real world, or is it a myth that stands in the place of something we cannot possibly understand?

It seems common that those with special knowledge, be it of science or publishing, distrust mass-media presentations of their own fields while accepting the pictures the media present of other fields. Even in the large a civil war in Nigeria attracted notice and aroused passion, whereas one in the Sudan went unnoticed and left us cold. Can the mass media present a useful picture of an overall community of interest? To what degree does such a community of interest exist? And whose interests are taken into account?

Whatever may be the reality of the overall community of interest people observe, they do react to it strongly and sometimes violently. Even if this reaction were a reaction to the true state of affairs, it would not necessarily result in a return to the desired state, whatever that might be. I think all societies perceive this difficulty of perception and reaction. In totalitarian societies great efforts are made to narrow and control the channels of communication. In open societies the channels of communication

are open, but constraints such as due process and checks and balances are put on the speed and magnitude of reaction. Yet, as society becomes ever larger and more complex, the problem of communication and control in the overall community of interest becomes progressively more difficult [see "Communication and Freedom of Expression," by Thomas I. Emerson; SCIENTIFIC AMERICAN, September, 1972].

At times governments show a public awareness of this. One response is to try to make society simpler and more uniform through centralization and drastic restrictive legislation concerning whatever is deemed to be antisocial or unfair or simply unpopular. Another response is to try to make communication and control operate meaningfully and effectively by pushing power toward communities of interest, whether these are intellectual, technological or geographical in the sense that regions, states and cities are.

Long before we took deliberate steps in this latter direction, men's communities of interest had somehow multiplied, differentiated themselves and had in many cases broken their geographical bonds. The amount of communication we use, the distances over which we communicate and the increasing diversity of books and periodicals show this clearly. Still, geography governs some aspects of our lives. However universal our understanding of science or our enjoyment of music may be, we prefer stoves in the Arctic and air conditioners in the Tropics. The world that communication and transportation have built is exceedingly complex and very difficult to govern wisely.

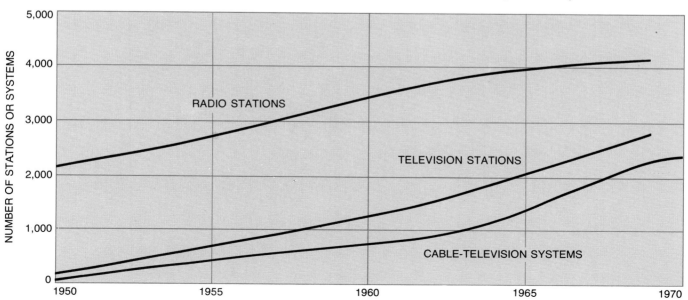

somewhat (right). The graphs are derived mainly from two sources: *The American Almanac*, which is published by Grosset & Dunlap   in 1972, and *Historical Statistics of the United States: Colonial Times to 1957*, published by the U.S. Department of Commerce.

IV

ENERGY: THE ULTIMATE RESOURCE

The first law of thermodynamics states that energy is conserved. The energy we use must come from somewhere, and it must also be disposed of somewhere. The energy economy of an advanced society is a giant machine for transforming high-grade energy into waste, while obtaining some useful work and heat in the process. The next six articles describe where the energy is found, how it is produced, the processes of conversion, and some of the problems of disposing of both the degraded energy and the noxious by-products of that conversion.

IV

ENERGY: THE ULTIMATE RESOURCE

INTRODUCTION

The lightning is his slave; heaven's utmost deep
Gives up her stars, and like a flock of sheep
They pass before his eye, are numbered and roll on!
The tempest is his steed, he strides the air;
And the abyss shouts from her depth laid bare
Heaven, hast thou secrets? Man unveils me; I have none.

Percy Bysshe Shelley
Prometheus Unbound

From prehistoric time, the goal of technology has been to control both matter and energy, whereas the goal of science has been to comprehend them. The control of matter by the use of tools and the harnessing of energy by the use of fire antedate the origin of our species. Matter and energy have been our constant servants, and in order to maintain the current state of human society, copious amounts of both must continue to be diverted to human uses. The world once appeared to have an abundance of resources; if shortages occurred, they were generally shortages of material, not power. The Romans denuded most of central and southern Italy of firewood by the time Rome fell; the British came near to deforesting their island to maintain the fleet upon which their power rested. Some materials have always been scarce. Gold and silver, for instance, would have been a marvellous choice for household utensils had these metals been abundant; their value as metals has been obscured by the high price set by their scarcity. Energy, on the other hand, has come to be considered a finite resource only in recent years, when the creation of a highly energy-intensive economy through scientific technology began to rapidly deplete existing fossil-fuel resources. Not so many years ago, the U.S. government was stockpiling comparatively rare vital materials such as tungsten and manganese, for fear that shortages might develop and cripple industry. At the same time, the energy companies assured us that their new explorations guaranteed an ever rising reserve of fuel, and we were encouraged to use it freely. But energy is the more fundamental resource. Problems concerning materials are frequently dealt with by clever technological-scientific improvisation, the so-called "technological fix," which ameliorates the immediate crisis without solving the underlying substantive problem. Such fixes, however, usually depend on increased technical sophistication, and increased sophistication generally implies an increased consumption of energy. Thus steel takes more energy to make than bronze, and aluminum requires far more than steel; synthetic fabrics consume copious supplies of energy when compared to wool and cotton; urban living uses far more energy per capita than rural. Intervention via the technological fix resembles nothing so much as throwing a ball in the air and then maneuvering endlessly to keep it aloft with one trick after another; what goes up can be kept up, but it costs. In order to evaluate what it costs, we now turn to the questions of what energy is, and how and in what forms it is used.

Matter and energy together make up the universe, and to our limited perception they appear quite distinct. Albert Einstein's celebrated discovery of the equivalence of matter and energy affects this division very little, however. Only in the realm of elementary particle physics, where particles and energies appear and disappear like wraiths until the boundary between matter and energy fades into argument, is the equivalence clearly evidenced. Ordinary, tangible matter is, in general, conserved. The exceptions to the rule of conservation of matter are of great practical importance; but despite the enormous amount of energy released by nuclear fission and nuclear fusion, only an imperceptible fraction of the mass of the earth has been annihilated in such processes. The sun itself consumes only about one ten-billionth of its mass each year to fuel the furnace of the solar system.

Conventionally defined in mechanical terms, energy is the capacity to do work. Although an attempt to follow this definition through to first principles leads to difficulties, it is nevertheless comfortable in that it connects energy with effects that are familiar and tangible. The potential energy locked in a mountain snowbank is most easily visualized in terms of the power of an avalanche; the chemical energy locked in a shovel of coal can be readily perceived in terms of horses pulling a load.

Energy, like matter, is also, in general, conserved. The exceptional case of the nuclear reactions is, however, relatively far more important in dealing with energy than in dealing with matter. This is true largely because life as we know it exists in a zone where the mass-energy of the universe is almost totally condensed, and very little of it remains free as radiant energy. Energy, the concomitant of all motion and the source of all life, comes and goes, ebbs and flows for the most part imperceptibly, for the greatest part of the energy in our island universe is gravitational, serving as cement to hold the parts together (see "Energy in the Universe," by Freeman J. Dyson, *Scientific American*, September 1971, Offprint 662). Next in the hierarchy is the energy of motion, by which the parts avoid collapsing on to one another. Far below in the hierarchy are the thermonuclear reactions that radiate the energy that supports life. The earth intercepts but a little of this radiant energy, and life itself uses even less.

How small a fraction of the energy flow of our planet is directed to the immediate support of life is discussed by M. King Hubbert in his essay, "The Energy Resources of the Earth," the keynote of this section. Of the total incident flux of solar energy (which contributes about 99.98 percent of the total energy input to the earth's surface), only one three-thousandth is used for the direct support of living things by photosynthesis. In the cyclic energy economy of the plant world, the chlorophyll factories process this solar energy and store it as chemical energy in organic molecular bonds. As the plants grow, die, decay, and break down, the energy from the sun is assimilated, stored, and ultimately released again as heat. Animal life, parasitic on the plant world, obtains its energy by robbing the chemical storehouses, but it undergoes the same cycle of decay and energy release. The fraction of the energy retained in organic residue is valuable, especially for building soil, but relatively small. Life processes on earth, then, operate by interposing themselves into the solar energy cycle and diverting a small portion of the flow to their own uses. The arrival of early technical man on this pleasant prelapsarian scene did little to alter the flow of energy at first. The interventions became more pronounced with increasing control of nature; but such processes, while of the utmost import for human life, were inconsequential perturbations of the energy flow. The processes were speeded up a little here, slowed down somewhat there; some of the energy was locked away in bricks or steel by technology. But taken as a whole, the energy cycle of civilization remained on a hand-to-mouth basis until the nineteenth century.

Not all the solar energy has been returned as heat. About six hundred million years ago, plants of the early Cambrian began to lay down tiny amounts of organic material under conditions that prevented their complete breakdown. These pockets supply the fossil fuels—coal, oil, and natural gas—to power machines and industry and heat homes. It was perhaps inevitable that the first power technicians should turn to this buried treasure to feed the ever-hungry machines. With the opening of the ancient seams of coal and pockets of oil, the chains of the short-term energy cycle were broken, and the generation of power took a giant leap. Hubbert's data on U.S. energy production show a precise correspondence with our earlier discussion of the phases of industrialization. From 1850 to 1910 energy production rose at a rate of almost 7 percent annually. (By the "rule of 70" for compound interest—doubling time equals 70 divided by the interest rate in percent—this means a doubling every ten years.) During the troubled decades that followed, the rate fell to less than 2 percent. When the new industrial revolution began, expansion was somewhat less energy-intensive, but the rate of energy production increased rapidly to about 4.5 percent, and the rate is still rising. Electricity, which is a better indicator for the new scientific-technological society, is rising at better than 7 percent.

The result of such exponentially rising consumption of a finite resource is usually not well understood. It is the paradox of Achilles and the Tortoise in reverse. In one famous example, if a lake is being covered by weeds whose area doubles every day, it may take a month for the lake to be half covered, but the weeds will finish the job and cover the other half the next day. In practice, diminishing supplies will reduce consumption, so that a graph of the consumption of a finite resource as a function of time will be a bell-shaped curve rising to a relatively narrow peak followed by a rapid fall-off as the supplies are exhausted. We are already near the peak of supply for natural gas and oil, and 80 percent of the six-hundred-million-year-old inheritance will be burned during the peak fifty years of consumption. Since total energy demand will continue to rise throughout this period, it is clear that oil and natural gas are, as a percentage of our total energy resources, more abundant now than they ever will be again. The coal situation is less immediately serious, but fundamentally the same. Acceleration of coal production to fill the hole will only step up the process. The fossil fuels are rapidly vanishing, and, even from such a cosmically limited viewpoint as the recorded history of the human race, the reign of the fossil fuels will be a brief one. We have placed ourselves and our culture in a position where an energy supply much greater than can be obtained from the natural cycles of plant fuels, rivers, tides, and winds is required just to provide maintenance, let alone growth. New sources of energy will have to be found quickly; one might even say hastily. The first step is to survey the energy resources of the earth, as so admirably done by Hubbert. Taking into consideration predictions for our own future as well as the pressures to develop the presently nonindustrialized countries, the most likely "natural" candidates for energy supply are direct use of solar energy and drilled geothermal wells; tidal and indirect solar power are too marginal and diffuse. The "unnatural" candidates are familiar: nuclear fission and nuclear fusion. But the tapping of such seemingly boundless sources of energy carries with it problems other than the obvious ones of land and water usage, radiation hazards, and air and water pollution. Continued exponential growth, be it in population or energy consumption, is not possible in the long term, if only because the planet itself is finite. The human race is on the verge of becoming more than just a clever parasite of power and thief of energy. We are now capable of interfering seriously with the earth's energy cycle, our most fundamental ecological system. Locally, as in Los Angeles or New York, such interference has already taken place. After nearly two centuries of continuous expansion, we must learn to live in a world of stability, of next to no growth, for which we have as yet no viable social or economic models.

The protean nature of energy as it passes through our society is amazing. Compared to energy, matter is stubborn, stupid stuff. Once molded into a specific configuration, matter admits to change only with an accompanying orchestration of energy—and the more profound the change sought, the more intense the accompanying energetics. Scarcely less stubborn, the medieval alchemists sought to change matter from one form to another, to transmute base and common metals into rare, precious ones. Meanwhile, all around them, technicians were performing the far more miraculous alchemy of energy. For the quality of energy is not constrained; and though it be conserved in quantity, it is readily transformed in appearance. The harnessing of energy to supply practical power proceeds by transferring it from one useless form to another, while extracting some useful work. The dynamics of the transmutation form the technical basis for thermodynamics; indeed, the science was developed largely for the purpose of analyzing how much work could be extracted during the transformation of the chemical energy of fuel into heat energy.

The cycles of transformation and the delineation of our energy ecology are the subject of "The Flow of Energy in an Industrial Society," by Earl Cook. To establish a basis for comparison, Cook has selected six stages in human development and listed both the total per capita energy consumption and the distribution of energy for each stage. The biological energy consumption of a single person is roughly equivalent to that of a one-hundred-watt light bulb: about two thousand kilocalories (kcal.) per day. With the arrival of technics and the consequent dependence on energy, the consumption of energy advanced with civilization until, in 1875, per capita consumption in the United States had risen to seventy thousand kcal. per day. In the technological society of 1970, the per capita daily energy consumption had risen to two hundred and thirty thousand kcal. (over one hundred times the amount needed to sustain life), and the multiplicative factor promises to double again by the end of this century. The difference in consumption in only the last century is enormous. Per capita energy consumption for transportation alone in the United States in 1970 nearly equalled the entire per capita energy consumption of industrialized England in 1875.

Such figures have far more than just historical import. Many countries have yet to reach the per capita energy consumption of what Cook calls "Industrial Man," and an appreciable fraction of the earth's population is still struggling to reach the advanced agricultural level. As a result, the United States, with less than 6 percent of the world's population, consumes over a third of its energy production. To put this in perspective, if worldwide zero population growth were suddenly achieved tomorrow, the total consumption of energy would still have to increase sixfold just to bring the rest of the world up to the *present* U.S. level. When we use a realistic—but very conservative—extrapolation of our own growth of energy use, and the U.N. "low" estimates for the world population, this estimate is increased by a further factor of four. This means that an overall increase in energy consumption by at least a factor of twenty would be the minimum requirement for a stable world at parity with the future U.S. level. World energy consumption is now about 6×10^{13} thermal kilowatt-hours annually. Hubbert estimates out total presently recoverable fossil fuel reserves as about 60×10^{15} thermal kilowatt-hours; a stable world at the U.S. level, with these estimates, would totally consume all our readily-accessible fossil-fuel reserves in a half-century. Unless rapid advances in technology create a breakthrough that allows the fossil reserves in oil shales and tar sands to be used, we face the prospect of a potential worldwide economic war for present supplies. Fossil-fuel use is becoming a zero-sum game, in which the economic advancement of the less-developed nations will threaten the well-being of the developed ones.

The correlation shown by Cook between gross national product and overall energy consumption demonstrates that increased energy usage is the only way to achieve prosperity on our present terms. The "overdeveloped" countries have paid a stiff price for adherence to economic growth as the barometer of economic health. Simply put, the United States is far and away the greatest "have" in the world and is steadily increasing the gap between itself and the "have-nots" who make up the bulk of the world population. For India, with a population almost triple that of our own and a per capita fossil-fuel consumption barely 1 percent of ours, the achievement of the American Dream would require a three-hundredfold increase in overall energy consumption. The prospect of competing against such countries for fossil-fuel reserves in the next century is frightening. Yet there are many who still adhere to the "growth-is-health" model born of the Industrial Revolution, even though this model ties us to an ever-expanding spiral of energy use and sets ever higher goals for the rest of the world. The development of nuclear power, or solar power, or some other as yet unspecified source, will not ameliorate this problem if we continue to widen the gap. This is the real "energy crisis": not fuel shortages in New York or gasoline rationing in California, but the dilemma into which that small fraction of humanity who have achieved the status of "technological man" have placed themselves. Either we learn to do more with less, to reduce or at least stabilize our energy use, or we gird ourselves to perpetuate by force the inequitable distribution of the world's energy resources. To maintain the present dichotomy of "have" and "have not" will require not energy, but *force majeure*.

Where is the energy going, and what is it used for that we require so much more than other societies? Cook's breakdown of U.S. energy consumption gives some answers. We use ten thousand kilocalories per day per capita for food alone—more than the *total* per capita energy consumption of Greece or Portugal. We use sixty-three thousand kcal. per day for transportation, more than the *total* per capita consumption of the USSR, France, the Scandinavian countries, or Germany—hardly underdeveloped by any standard. The 3 percent of our energy used for air conditioning alone amounts to more than the per capita consumption of India or Ghana. Our energy consumption pattern is indecently wasteful. Food is shipped thousands of miles to processing centers where it is cooked, frozen, and shipped back as TV dinners. The suburbs spread mile upon mile, housing white-collar residents who commute daily into the city center, while the blue-collar residents of the city must often travel out to the suburban factories. We are the most energy-intensive society on earth. A great part of that energy is simply squandered, wasted not only by the nature of the products consumed, but also by an economic and social structure that counts human comfort for much and energy consumption for little. As long as energy is the cheapest commodity available, and a premium is placed on human convenience, the use of energy will be maximized in almost every instance.

In the search for new energy sources, science and technology have not been idle. Given a built-in value system that admits no technical regression, the greater part of the research and development has been directed toward making sure that energy supply will always meet demand. Recent concern over nuclear plants has restimulated efforts at exploiting the "natural" sources of energy (solar, geothermal, or tidal), but most of the effort has gone into the development of the "unnatural" nuclear sources (fission and fusion). Fission power is already economically competitive, largely because most of the initial research and developmental costs were written off by the weapons program, the federal government, and the Atomic Energy Commission. Large numbers of fission plants are already completed, and far more are planned. Fusion power, in many ways more attractive, remains hypothetical pending the first successful controlled-fusion experiment.

Fission power, the subject of "the Arrival of Nuclear Power" by John F. Hogerton, is one of the few examples of a sword being successfully beaten into a plowshare (albeit a Damoclean plowshare that is only marginally less dangerous than the original sword). Hogerton refers throughout his article to "nuclear power," but it is specifically fission power that he discusses—commercial use of the energy released by the fission of uranium or plutonium in a controlled chain reaction. We could hardly find a better example of the acceleration of history in the new scientific-technological-industrial state than in Hogerton's division of the history of fission power into five "historical" periods—none of which spans more than eight years! In 1968, when the article was written, it had been twenty-five years since Enrico Fermi's pile of graphite blocks produced the first self-sustaining chain reaction, and just twenty years since the Atomic Energy Commission decided to promote peaceful uses of atomic power. This period, seen by many of fission's ardent boosters as a "delay" in deployment, has been one of on-line debugging and the acquisition of experience. Each plant was essentially handmade. But the initial debug phase is now over, and full-scale fission plants are ready to be mass-constructed as commercial propositions, not just experiments. The nuclear breakthrough, aided by objections to air pollution from fossil-fueled plants, was about to avalanche in the early 1960s. At that time it was anticipated that at least half of the newly installed capacity would be nuclear by 1970. However, the promotion of atomic power as a safe, clean, trouble-free source of energy ran into an unexpected stumbling block in the mid-1960s as a result of the growing public concern over environmental quality. The great size and lower efficiency of the fission plants create massive heat-disposal problems; the battle over thermal pollution, combined with a general fear of radioactivity, caused several years' delay in plant installation. This delay became a crucial factor in determining the deployment of fission plants, for, during those few years, it became widely understood that the earth's resources of natural fissionable material was also finite. Hubbert estimates that there will be an acute shortage of high-grade fissionable material by the end of the century, unless major new discoveries are made. In order to maintain a fission-based energy supply, the plants will have to shift from natural uranium to either plutonium-239 or uranium-233, both of which can be produced from the more abundant supplies of uranium-238 and thorium-232 by the absorption of neutrons in the neutron-rich environment of fast breeder reactors, a process described in Glenn T. Seaborg and Justin L. Bloom's article "Fast Breeder Reactors."

The Faustian bargain of the fission reactors involves radioactive wastes. Concern over these wastes has produced the most serious opposition to nuclear power. The problem of radioactive waste disposal was simply swept under the rug at first. Hogerston does not mention it at all. It was not the problem of the energy companies; their problem was cost and public concern over air pollution. Disposal was left to the AEC, which has been severely criticized for its handling of it.

Despite the inadequacy of current disposal practices, it is true that methods can be developed for storing these wastes safely. But we shall have to place our descendants in pawn for thousands of years to watch over the stored wastes and care for them. We, and they, shall have to watch over the plants even more carefully. The prospect of the seizure of a plant, its fuel, or any of the wastes by an antisocial or avaricious group, or during revolution or social upheaval, whether for the purpose of extortion, hostage, or ransom, is so frightening that stability and control are likely to become *the* overriding requisite for any government—whatever the cost to personal freedom or political reform. The breeder reactor adds a new dimension to the problem, since a breeder economy will ultimately involve the shipment, not only of wastes, but also of hundreds of tons of plutonium-239 to fuel the plants each year. Pu-239 is one of the most toxic and long-lived of radioactive isotopes. It is also a prime ingredient of atomic weapons. As John Holdren and Philip Hérrera have pointed out in *Energy* (Sierra Club, 1971):

> Even now the AEC seems unable to keep track of the plutonium it handles with an accuracy of better than about 1 percent (*Science*, April 9, 1971). If 1 percent of the plutonium the AEC estimates will be circulating in the year 2000 went astray, it would be enough to manufacture more than 1,000 Nagasaki-type bombs.

Obviously the loss or theft of only one one-hundredth of one percent of this plutonium would be cause for great—and justified—alarm if it became known.

The alternative source of nuclear power is thermonuclear fusion—the energy source of the stars and the hydrogen bomb. The fuel resources for fusion are such that supply will never be the limiting factor. Just 2 percent of the deuterium in the world's ocean could supply over one million times the energy of our entire primordial supply of fossil fuels. Fusion power is also safer, since, unlike fission, the reaction is inherently self-extinguishing if it gets out of control. In their article "The Prospects of Fusion Power," William C. Gough and Bernard J. Eastlund state that there is a growing conviction among fusion scientists that the eventual achievement of a controlled thermonuclear reaction is not blocked by the laws of nature. Such a reaction has not yet been achieved but this is "slow" progress only by the manic standards of the technological age. Fusion is, after all, a very new technology, and the developmental costs of a fusion reactor have not been written off as part of the nuclear weapons program. Yet the squeeze is definitely on, and fusion power must be shown to be at least possible within the next two decades if it is to have an appreciable impact on the U.S. energy supply by the year 2000. Plant construction times are long; utility planning times even longer; and an irrevocable commitment to breeder reactors is just over the horizon.

Fusion, of course, is not without its own problems. Tritium is also a radiological hazard, and an accident involving tritium loss would be extremely serious. But the dangers are so much smaller than those of breeder type fission reactors that—should a successful fusion reaction be achieved—there would be tremendous incentive to abandon fission immediately in favor of a crash program for fusion plants. Trying to balance the funding, however, is still difficult. The seven hundred million dollar estimate (in 1973) for an experimental four-hundred-thousand-kilowatt, liquid-metal fast breeder plant at Oak Ridge is one and one-half times the total sum spent on fusion research since its inception. The justification for this imbalance was the usual one: breeder reactors are known to work. Fusion must also compete for research funds with solar and geothermal energy, which require no scientific breakthroughs; these sources may become feasible with only incremental advances in science and technology and an immense quantity of engineering.

There is also a danger that the availability, or postulated availability, of nearly unlimited power will be taken as a mandate for encouraging unlimited consumption. Among the potential adverse side effects is the possibility that such an "energy revolution" might trigger another cycle of population growth. The human population has shown a great surge in growth at each of the major societal revolutions: cultural, agricultural, and industrial. Such growth has now slowed, partially because of the cultural effects of the soft revolution, but also at least partially because of the acknowledged shortage of world resources. A restoration of faith in the infinite manipulability of nature may in this sense be counterproductive. It would also encourage the United States to continue its headlong expansion, further increasing the gap between it and the less-developed nations and adding fuel to the race to catch up. It is far from impossible for these countries to make at least a good race of it. Chauncy Starr has pointed out the case of Puerto Rico, which (though admittedly atypical) increased its energy consumption by a factor of twenty between 1940 and 1968, achieving an annual growth rate in excess of 10 percent and a doubling time of less than seven years. The exploitation of oil shales or the deployment of nuclear power may facilitate this process in other, larger countries. The potential consequences for the energy cycle of the earth and the processes of life are the subject of the last article in this section; "Human Energy Production as a Process in the Biosphere" by S. Fred Singer. The burning of fossil fuels releases enormous quantities of air pollutants that are immediately noticeable, and even larger amounts of more subtly intrusive substances such as carbon dioxide (CO_2) that, although less noticeable, are possibly more dangerous. Increasing the pollutant dump rate by a multiplicative factor (as would be the case were other countries to industrialize to our level with fossil fuels) is not a pleasant prospect for those of us who resent being perpetual guinea pigs. Neither is it comforting to know that residents of Tokyo or Los Angeles will feel right at home in the smog of Katmandu or Accra.

Singer is perfectly correct in pointing out that such pollution is, in principle, susceptible to control. Stack scrubbers, catalytic converters, precipitators, and perhaps even water carbonators to wash out the CO_2 are all possible. They will, of course, extract a heavy price in energy consumption, as do most technological fixes, and therefore cause the net efficiency of the plant to decrease. As with the disposal of radioactive wastes and the transport of plutonium, such problems are at least hypothetically soluble. But there is one limit that remains inviolable—one problem for which there can be no fix, technological or conceptual. That limit is fundamental and thermodynamic: virtually all the energy we use must eventually end up as heat. The present world consumption of energy, 6×10^{12} watts, is twice the energy of the tides, one-seventh the energy that goes into photosynthesis, and about one twenty-thousandth of the total solar energy input. The predicted minimum factor of twenty would raise human energy consumption to roughly one one-thousandth of the total solar input at the surface, and about three times the energy going to photosynthesis. The heat inserted into the biosphere by human energy consumption is neither negligible nor avoidable; heat is the inevitable, nonrecyclable, nonreturnable waste product of technological society. As energy is the ultimate resource, heat is the ultimate waste; it is to this subject that we will turn in Part V.

21

The Energy Resources
of the Earth

by M. King Hubbert
September 1971

*They are solar energy (current and stored), the tides,
the earth's heat, fission fuels and possibly fusion fuels.
From the standpoint of human history the epoch of the
fossil fuels will be quite brief*

Energy flows constantly into and out of the earth's surface environment. As a result the material constituents of the earth's surface are in a state of continuous or intermittent circulation. The source of the energy is preponderantly solar radiation, supplemented by small amounts of heat from the earth's interior and of tidal energy from the gravitational system of the earth, the moon and the sun. The materials of the earth's surface consist of the 92 naturally occurring chemical elements, all but a few of which behave in accordance with the principles of the conservation of matter and of nontransmutability as formulated in classical chemistry. A few of the elements or their isotopes, with abundances of only a few parts per million, are an exception to these principles in being radioactive. The exception is crucial in that it is the key to an additional large source of energy.

A small part of the matter at the earth's surface is embodied in living organisms: plants and animals. The leaves of the plants capture a small fraction of the incident solar radiation and store it chemically by the mechanism of photosynthesis. This store becomes the energy supply essential for the existence of the plant and animal kingdoms. Biologically stored energy is released by oxidation at a rate approximately equal to the rate of storage. Over millions of

years, however, a minute fraction of the vegetable and animal matter is buried under conditions of incomplete oxidation and decay, thereby giving rise to the fossil fuels that provide most of the energy for industrialized societies.

It is difficult for people living now, who have become accustomed to the steady exponential growth in the consumption of energy from the fossil fuels, to realize how transitory the fossil-fuel epoch will eventually prove to be when it is viewed over a longer span of human history. The situation can better be seen in the perspective of some 10,000 years, half before the present and half afterward. On such a scale the complete cycle of the exploitation of the world's fossil fuels will be seen to encompass perhaps 1,300 years, with the principal segment of the cycle (defined as the period during which all but the first 10 percent and the last 10 percent of the fuels are extracted and burned) covering only about 300 years.

What, then, will provide industrial energy in the future on a scale at least as large as the present one? The answer lies in man's growing ability to exploit other sources of energy, chiefly nuclear at present but perhaps eventually the much larger source of solar energy. With this ability the energy resources now at hand are sufficient to sustain an industrial operation of the present magnitude for another millennium or longer. Moreover, with such resources of energy the limits to the growth of industrial activity are no longer set by a scarcity of energy but rather by the space and material limitations of a finite earth together with the principles of ecology. According to these principles both biological and industrial activities tend to increase exponentially with time, but the resources of the entire earth are not sufficient to sustain such an increase of

any single component for more than a few tens of successive doublings.

Let us consider in greater detail the flow of energy through the earth's surface environment [*see illustration on next two pages*]. The inward flow of energy has three main sources: (1) the intercepted solar radiation; (2) thermal energy, which is conveyed to the surface of the earth from the warmer interior by the conduction of heat and by convection in hot springs and volcanoes, and (3) tidal energy, derived from the combined kinetic and potential energy of the earth-moon-sun system. It is possible in various ways to estimate approximately how large the input is from each source.

In the case of solar radiation the influx is expressed in terms of the solar constant, which is defined as the mean rate of flow of solar energy across a unit of area that is perpendicular to the radiation and outside the earth's atmosphere at the mean distance of the earth from the sun. Measurements made on the earth and in spacecraft give a mean value for the solar constant of 1.395 kilowatts per square meter, with a variation of about 2 percent. The total solar radiation intercepted by the earth's diametric plane of 1.275×10^{14} square meters is therefore 1.73×10^{17} watts.

The influx of heat by conduction from the earth's interior has been determined from measurements of the geothermal gradient (the increase of temperature with depth) and the thermal conductivity of the rocks involved. From thousands of such measurements, both on land and on the ocean beds, the average rate of flow of heat from the interior of the earth has been found to be about .063 watt per square meter. For the earth's surface area of 510×10^{12} square meters the total heat flow amounts to

RESOURCE EXPLORATION is beginning to be aided by airborne side-looking radar pictures such as the one on the opposite page made by the Aero Service Corporation and the Goodyear Aerospace Corporation. The technique has advantage of "seeing" through cloud cover and vegetation. This picture, which was made in southern Venezuela, extends 70 miles from left to right.

some 32 × 10¹² watts. The rate of heat convection by hot springs and volcanoes is estimated to be only about 1 percent of the rate of conduction, or about .3 × 10¹² watts.

The energy from tidal sources has been estimated at 3 × 10¹² watts. When all three sources of energy are expressed in the common unit of 10¹² watts, the total power influx into the earth's surface environment is found to be 173,035 × 10¹² watts. Solar radiation accounts for 99.98 percent of it. Another way of stating the sun's contribution to the energy budget of the earth is to note that at 173,000 × 10¹² watts it amounts to 5,000 times the energy input from all other sources combined.

About 30 percent of the incident solar energy (52,000 × 10¹² watts) is directly reflected and scattered back into space as short-wavelength radiation. Another 47 percent (81,000 × 10¹² watts) is absorbed by the atmosphere, the land surface and the oceans and converted directly into heat at the ambient surface temperature. Another 23 percent (40,000 × 10¹² watts) is consumed in the evaporation, convection, precipitation and surface runoff of water in the hydrologic cycle. A small fraction, about 370 × 10¹² watts, drives the atmospheric and oceanic convections and circulations and the ocean waves and is eventually dissipated into heat by friction. Finally, an even smaller fraction—about

40 × 10¹² watts—is captured by the chlorophyll of plant leaves, where it becomes the essential energy supply of the photosynthetic process and eventually of the plant and animal kingdoms.

Photosynthesis fixes carbon in the leaf and stores solar energy in the form of carbohydrate. It also liberates oxygen and, with the decay or consumption of the leaf, dissipates energy. At any given time, averaged over a year or more, the balance between these processes is almost perfect. A minute fraction of the organic matter produced, however, is deposited in peat bogs or other oxygen-deficient environments under conditions that prevent complete decay and loss of energy.

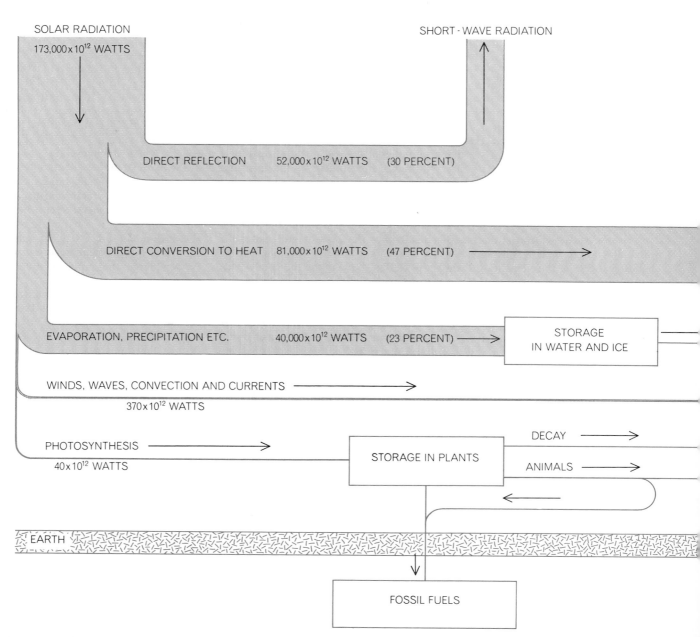

FLOW OF ENERGY to and from the earth is depicted by means of bands and lines that suggest by their width the contribution of each item to the earth's energy budget. The principal inputs are

solar radiation, tidal energy and the energy from nuclear, thermal and gravitational sources. More than 99 percent of the input is solar radiation. The apportionment of incoming solar radiation is

Little of the organic material produced before the Cambrian period, which began about 600 million years ago, has been preserved. During the past 600 million years, however, some of the organic materials that did not immediately decay have been buried under a great thickness of sedimentary sands, muds and limes. These are the fossil fuels: coal, oil shale, petroleum and natural gas, which are rich in energy stored up chemically from the sunshine of the past 600 million years. The process is still continuing, but probably at about the same rate as in the past; the accumulation during the next million years will probably be a six-hundredth of the amount built up thus far.

Industrialization has of course withdrawn the deposits in this energy bank with increasing rapidity. In the case of coal, for example, the world's consumption during the past 110 years has been about 19 times greater than it was during the preceding seven centuries. The increasing magnitude of the rate of withdrawal can also be seen in the fact that the amount of coal produced and consumed since 1940 is approximately equal to the total consumption up to that time. The cumulative production from 1860 through 1970 was about 133 billion metric tons. The amount produced before 1860 was about seven million metric tons.

Petroleum and related products were not extracted in significant amounts before 1880. Since then production has increased at a nearly constant exponential rate. During the 80-year period from 1890 through 1970 the average rate of increase has been 6.94 percent per year, with a doubling period of 10 years. The cumulative production until the end of 1969 amounted to 227 billion (227×10^9) barrels, or 9.5 trillion U.S. gallons. Once again the period that encompasses most of the production is notably brief. The 102 years from 1857 to 1959 were required to produce the first half of the cumulative production; only the 10-year period from 1959 to 1969 was required for the second half.

Examining the relative energy contributions of coal and crude oil by comparing the heats of combustion of the respective fuels (in units of 10^{12} kilowatt-hours), one finds that until after 1900 the contribution from oil was barely significant compared with the contribution from coal. Since 1900 the contribution from oil has risen much faster than that from coal. By 1968 oil represented about 60 percent of the total. If the energy from natural gas and natural-gas liquids had been included, the contribution from petroleum would have been about 70 percent. In the U.S. alone 73 percent of the total energy produced from fossil fuels in 1968 was from petroleum and 27 percent from coal.

Broadly speaking, it can be said that the world's consumption of energy for industrial purposes is now doubling approximately once per decade. When confronted with a rate of growth of such magnitude, one can hardly fail to wonder how long it can be kept up. In the case of the fossil fuels a reasonably definite answer can be obtained. Their human exploitation consists of their being withdrawn from an essentially fixed initial supply. During their use as sources of energy they are destroyed. The complete cycle of exploitation of a fossil fuel must therefore have the following characteristics. Beginning at zero, the rate of production tends initially to increase exponentially. Then, as difficulties of discovery and extraction increase, the production rate slows in its growth, passes one maximum or more and, as the resource is progressively depleted, declines eventually to zero.

If known past and prospective future rates of production are combined with a reasonable estimate of the amount of a fuel initially present, one can calculate the probable length of time that the fuel can be exploited. In the case of coal reasonably good estimates of the

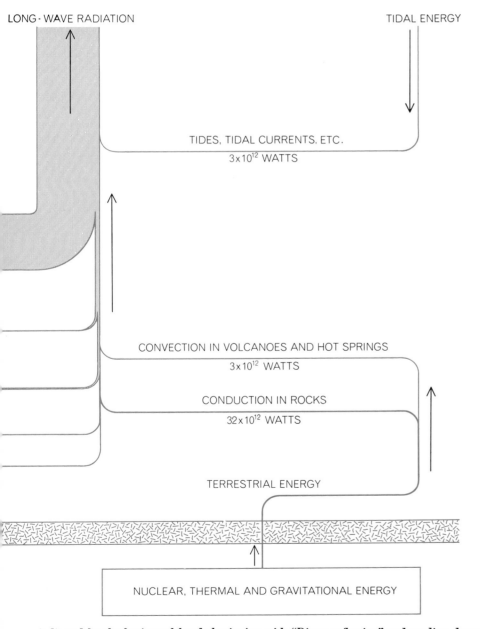

LONG-WAVE RADIATION

TIDAL ENERGY

TIDES, TIDAL CURRENTS, ETC.
3×10^{12} WATTS

CONVECTION IN VOLCANOES AND HOT SPRINGS
3×10^{12} WATTS

CONDUCTION IN ROCKS
32×10^{12} WATTS

TERRESTRIAL ENERGY

NUCLEAR, THERMAL AND GRAVITATIONAL ENERGY

indicated by the horizontal bands beginning with "Direct reflection" and reading downward. The smallest portion goes to photosynthesis. Dead plants and animals buried in the earth give rise to fossil fuels, containing stored solar energy from millions of years past.

amount present in given regions can be made on the basis of geological mapping and a few widely spaced drill holes, inasmuch as coal is found in stratified beds or seams that are continuous over extensive areas. Such studies have been made in all the coal-bearing areas of the world.

The most recent compilation of the present information on the world's initial coal resources was made by Paul Averitt of the U.S. Geological Survey. His figures [see illustration below] represent minable coal, which is defined as 50 percent of the coal actually present. Included is coal in beds as thin as 14 inches (36 centimeters) and extending to depths of 4,000 feet (1.2 kilometers) or, in a few cases, 6,000 feet (1.8 kilometers).

Taking Averitt's estimate of an initial supply of 7.6 trillion metric tons and assuming that the present production rate of three billion metric tons per year does not double more than three times, one can expect that the peak in the rate of production will be reached sometime between 2100 and 2150. Disregarding the long time required to produce the first 10 percent and the last 10 percent, the length of time required to produce the middle 80 percent will be roughly

the 300-year period from 2000 to 2300.

Estimating the amount of oil and gas that will ultimately be discovered and produced in a given area is considerably more hazardous than estimating for coal. The reason is that these fluids occur in restricted volumes of space and limited areas in sedimentary basins at all depths from a few hundred meters to more than eight kilometers. Nonetheless, the estimates for a given region improve as exploration and production proceed. In addition it is possible to make rough estimates for relatively undeveloped areas on the basis of geological comparisons between them and well-developed regions.

The most highly developed oil-producing region in the world is the coterminous area of the U.S.: the 48 states exclusive of Alaska and Hawaii. This area has until now led the world in petroleum development, and the U.S. is still the leading producer. For this region a large mass of data has been accumulated and a number of different methods of analysis have been developed that give fairly consistent estimates of the degree of advancement of petroleum exploration and of the amounts of oil and gas that may eventually be produced.

One such method is based on the principle that only a finite number of oil or gas fields existed initially in a given region. As exploration proceeds the shallowest and most evident fields are usually discovered first and the deeper and more obscure ones later. With each discovery the number of undiscovered fields decreases by one. The undiscovered fields are also likely to be deeper, more widely spaced and better concealed. Hence the amount of exploratory activity required to discover a fixed quantity of oil or gas steadily increases or, conversely, the average amount of oil or gas discovered for a fixed amount of exploratory activity steadily decreases.

Most new fields are discovered by what the industry calls "new-field wildcat wells," meaning wells drilled in new territory that is not in the immediate vicinity of known fields. In the U.S. statistics have been kept annually since 1945 on the number of new-field wildcat wells required to make one significant discovery of oil or gas ("significant" being defined as one million barrels of oil or an equivalent amount of gas). The discoveries for a given year are evaluated only after six years of subsequent development. In 1945 it required 26

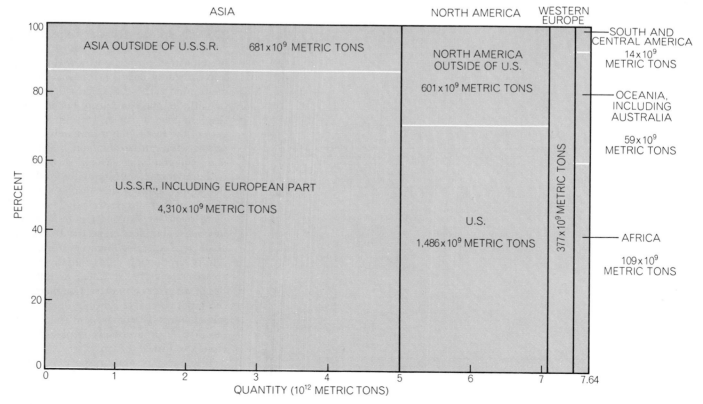

COAL RESOURCES of the world are indicated on the basis of data compiled by Paul Averitt of the U.S. Geological Survey. The figures represent the total initial resources of minable coal, which is defined as 50 percent of the coal actually present. The horizontal scale gives the total supply. Each vertical block shows the apportionment of the supply in a continent. From the first block, for example, one can ascertain that Asia has some 5 × 10¹² metric tons of minable coal, of which about 86 percent is in the U.S.S.R.

new-field wildcat wells to make a significant discovery; by 1963 the number had increased to 65.

Another way of illuminating the problem is to consider the amount of oil discovered per foot of exploratory drilling. From 1860 to 1920, when oil was fairly easy to find, the ratio was 194 barrels per foot. From 1920 to 1928 the ratio declined to 167 barrels per foot. Between 1928 and 1938, partly because of the discovery of the large East Texas oil field and partly because of new exploratory techniques, the ratio rose to its maximum of 276 barrels per foot. Since then it has fallen sharply to a nearly constant rate of about 35 barrels per foot. Yet the period of this decline coincided with the time of the most intensive research and development in petroleum exploration and production in the history of the industry.

The cumulative discoveries in the 48 states up to 1965 amounted to 136 billion barrels. From this record of drilling and discovery it can be estimated that the ultimate total discoveries in the coterminous U.S. and the adjacent continental shelves will be about 165 billion barrels. The discoveries up to 1965 therefore represent about 82 percent of the prospective ultimate total. Making

due allowance for the range of uncertainty in estimates of future discovery, it still appears that at least 75 percent of the ultimate amount of oil to be produced in this area will be obtained from fields that had already been discovered by 1965.

For natural gas in the 48 states the present rate of discovery, averaged over a decade, is about 6,500 cubic feet per barrel of oil. Assuming the same ratio for the estimated ultimate amount of 165 billion barrels of crude oil, the ultimate amount of natural gas would be about 1,075 trillion cubic feet. Combining the estimates for oil and gas with the trends of production makes it possible to estimate how long these energy resources will last. In the case of oil the period of peak production appears to be the present. The time span required to produce the middle 80 percent of the ultimate cumulative production is approximately the 65-year period from 1934 to 1999—less than the span of a human lifetime. For natural gas the peak of production will probably be reached between 1975 and 1980.

The discoveries of petroleum in Alaska modify the picture somewhat. In particular the field at Prudhoe Bay appears likely by present estimates to contain

about 10 billion barrels, making it twice as large as the East Texas field, which was the largest in the U.S. previously. Only a rough estimate can be made of the eventual discoveries of petroleum in Alaska. Such a speculative estimate would be from 30 to 50 billion barrels. One must bear in mind, however, that 30 billion barrels is less than a 10-year supply for the U.S. at the present rate of consumption. Hence it appears likely that the principal effect of the oil from Alaska will be to retard the rate of decline of total U.S. production rather than to postpone the date of its peak.

Estimates of ultimate world production of oil range from 1,350 billion barrels to 2,100 billion barrels. For the higher figure the peak in the rate of world production would be reached about the year 2000. The period of consumption of the middle 80 percent will probably be some 58 to 64 years, depending on whether the lower or the higher estimate is used [see bottom illustration on page 271].

A substantial but still finite amount of oil can be extracted from tar sands and oil shales, where production has barely begun. The largest tar-sand deposits are in northern Alberta; they have total recoverable reserves of about 300 billion

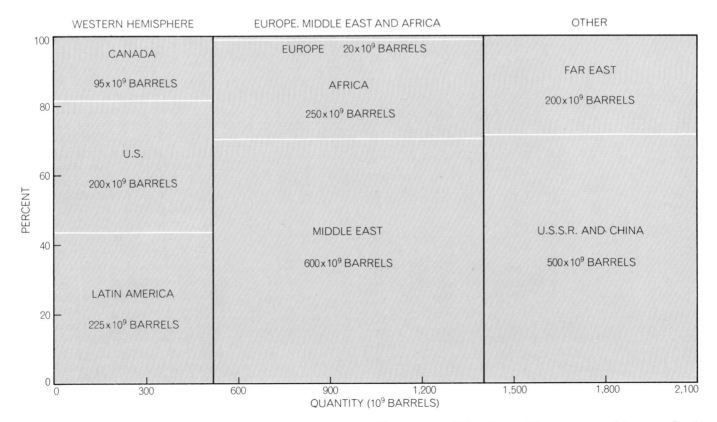

PETROLEUM RESOURCES of the world are depicted in an arrangement that can be read in the same way as the diagram of coal supplies on the opposite page. The figures for petroleum are derived from estimates made in 1967 by W. P. Ryman of the Standard Oil Company of New Jersey. They represent ultimate crude-oil production, including oil from offshore areas, and consist of oil already produced, proved and probable reserves, and future discoveries. Estimates as low as 1,350 × 10⁹ barrels have also been made.

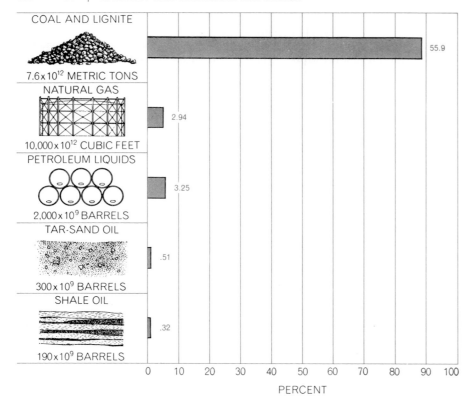

ENERGY CONTENT of the world's initial supply of recoverable fossil fuels is given in units of 10^{15} thermal kilowatt-hours (*color*). Coal and lignite, for example, contain 55.9×10^{15} thermal kilowatt-hours of energy and represent 88.8 percent of the recoverable energy.

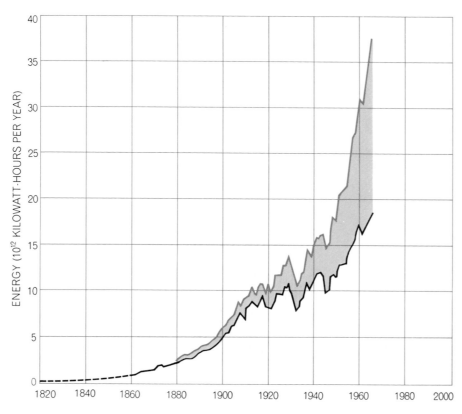

ENERGY CONTRIBUTION of coal (*black*) and coal plus oil (*color*) is portrayed in terms of their heat of combustion. Before 1900 the energy contribution from oil was barely significant. Since then the contribution from oil (*shaded area*) has risen much more rapidly than that from coal. By 1968 oil represented about 60 percent of the total. If the energy from natural gas were included, petroleum would account for about 70 percent of the total.

barrels. A world summary of oil shales by Donald C. Duncan and Vernon E. Swanson of the U.S. Geological Survey indicated a total of about 3,100 billion barrels in shales containing from 10 to 100 gallons per ton, of which 190 billion barrels were considered to be recoverable under 1965 conditions.

Since the fossil fuels will inevitably be exhausted, probably within a few centuries, the question arises of what other sources of energy can be tapped to supply the power requirements of a moderately industrialized world after the fossil fuels are gone. Five forms of energy appear to be possibilities: solar energy used directly, solar energy used indirectly, tidal energy, geothermal energy and nuclear energy.

Until now the direct use of solar power has been on a small scale for such purposes as heating water and generating electricity for spacecraft by means of photovoltaic cells. Much more substantial installations will be needed if solar power is to replace the fossil fuels on an industrial scale. The need would be for solar power plants in units of, say, 1,000 megawatts. Moreover, because solar radiation is intermittent at a fixed location on the earth, provision must also be made for large-scale storage of energy in order to smooth out the daily variation.

The most favorable sites for developing solar power are desert areas not more than 35 degrees north or south of the Equator. Such areas are to be found in the southwestern U.S., the region extending from the Sahara across the Arabian Peninsula to the Persian Gulf, the Atacama Desert in northern Chile and central Australia. These areas receive some 3,000 to 4,000 hours of sunshine per year, and the amount of solar energy incident on a horizontal surface ranges from 300 to 650 calories per square centimeter per day. (Three hundred calories, the winter minimum, amounts when averaged over 24 hours to a mean power density of 145 watts per square meter.)

Three schemes for collecting and converting this energy in a 1,000-megawatt plant can be considered. The first involves the use of flat plates of photovoltaic cells having an efficiency of about 10 percent. A second possibility is a recent proposal by Aden B. Meinel and Marjorie P. Meinel of the University of Arizona for utilizing the hothouse effect by means of selective coatings on pipes carrying a molten mixture of sodium and potassium raised by solar energy to a temperature of 540 degrees Celsius. By

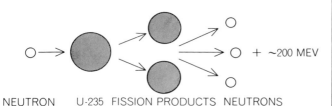

NEUTRON   U-235 FISSION PRODUCTS NEUTRONS

DEUTERIUM DEUTERIUM HELIUM 3   NEUTRON

DEUTERIUM   TRITIUM 6   HELIUM 4   NEUTRON

FISSION AND FUSION REACTIONS hold the promise of serving as sources of energy when fossil fuels are depleted. Present nuclear-power plants burn uranium 235 as a fuel. Breeder reactors now under development will be able to use surplus neutrons from the fission of uranium 235 (*left*) to create other nuclear fuels: plutonium 239 and uranium 233. Two promising fusion reactions, deuterium-deuterium and deuterium-tritium, are at right. The energy released by the various reactions is shown in million electron volts.

means of a heat exchanger this heat is stored at a constant temperature in an insulated chamber filled with a mixture of sodium and potassium chlorides that has enough heat capacity for at least one day's collection. Heat extracted from this chamber operates a conventional steam-electric power plant. The computed efficiency for this proposal is said to be about 30 percent.

A third system has been proposed by Alvin F. Hildebrandt and Gregory M. Haas of the University of Houston. It entails reflecting the radiation reaching a square-mile area into a solar furnace and boiler at the top of a 1,500-foot tower. Heat from the boiler at a temperature of 2,000 degrees Kelvin would be converted into electric power by a magnetohydrodynamic conversion. An energy-storage system based on the hydrolysis of water is also proposed. An overall efficiency of about 20 percent is estimated.

Over the range of efficiencies from 10 to 30 percent the amount of thermal power that would have to be collected for a 1,000-megawatt plant would range from 10,000 to 3,300 thermal megawatts. Accordingly the collecting areas for the three schemes would be 70, 35 and 23 square kilometers respectively. With the least of the three efficiencies the area required for an electric-power capacity of 350,000 megawatts—the approximate capacity of the U.S. in 1970—would be 24,500 square kilometers, which is somewhat less than a tenth of the area of Arizona.

The physical knowledge and technological resources needed to use solar energy on such a scale are now available. The technological difficulties of doing so, however, should not be minimized.

Using solar power indirectly means relying on the wind, which appears impractical on a large scale, or on the streamflow part of the hydrologic cycle. At first glance the use of streamflow appears promising, because the world's total water-power capacity in suitable sites is about three trillion watts, which

approximates the present use of energy in industry. Only 8.5 percent of the water power is developed at present, however, and the three regions with the greatest potential—Africa, South America and Southeast Asia—are the least developed industrially. Economic problems therefore stand in the way of extensive development of additional water power.

Tidal power is obtained from the filling and emptying of a bay or an estuary that can be closed by a dam. The enclosed basin is allowed to fill and empty only during brief periods at high and low tides in order to develop as much power as possible. A number of promising sites exist; their potential capacities range from two megawatts to 20,000 megawatts each. The total potential tidal power, however, amounts to about 64 billion watts, which is only 2 percent of the world's potential water power. Only one full-scale tidal-electric plant has been built; it is on the Rance estuary on the Channel Island coast of France. Its capacity at start-up in 1966 was 240 megawatts; an ultimate capacity of 320 megawatts is planned.

Geothermal power is obtained by extracting heat that is temporarily stored in the earth by such sources as volcanoes and the hot water filling the sands of deep sedimentary basins. Only volcanic sources are significantly exploited at present. A geothermal-power operation has been under way in the Larderello area of Italy since 1904 and now has a capacity of 370 megawatts. The two other main areas of geothermal-power production are The Geysers in northern California and Wairakei in New Zealand. Production at The Geysers began in 1960 with a 12.5-megawatt unit. By 1969 the capacity had reached 82 megawatts, and plans are to reach a total installed capacity of 400 megawatts by 1973. The Wairakei plant began operation in 1958 and now has a capacity of 290 megawatts, which is believed to be about the maximum for the site.

Donald E. White of the U.S. Geological Survey has estimated that the stored thermal energy in the world's major geothermal areas amounts to about 4×10^{20} joules. With a 25 percent conversion factor the production of electrical energy would be about 10^{20} joules, or three million megawatt-years. If this energy, which is depletable, were withdrawn over a period of 50 years, the average annual power production would be 60,000 megawatts, which is comparable to the potential tidal power.

Nuclear power must be considered under the two headings of fission and fusion. Fission involves the splitting of nuclei of heavy elements such as uranium. Fusion involves the combining of light nuclei such as deuterium. Uranium 235, which is a rare isotope (each 100,000 atoms of natural uranium include six atoms of uranium 234, 711 atoms of uranium 235 and 99,283 atoms of uranium 238), is the only atomic species capable of fissioning under relatively mild environmental conditions. If nuclear energy depended entirely on uranium 235, the nuclear-fuel epoch would be brief. By breeding, however, wherein by absorbing neutrons in a nuclear reactor uranium 238 is transformed into fissionable plutonium 239 or thorium 232 becomes fissionable uranium 233, it is possible to create more nuclear fuel than is consumed. With breeding the entire supply of natural uranium and thorium would thus become available as fuel for fission reactors.

Most of the reactors now operating or planned in the rapidly growing nuclear-power industry in the U.S. depend essentially on uranium 235. The U.S. Atomic Energy Commission has estimated that the uranium requirement to meet the projected growth rate from 1970 to 1980 is 206,000 short tons of uranium oxide (U_3O_8). A report recently issued by the European Nuclear Energy Agency and the International Atomic Energy Agency projects requirements of 430,000 short tons of uranium

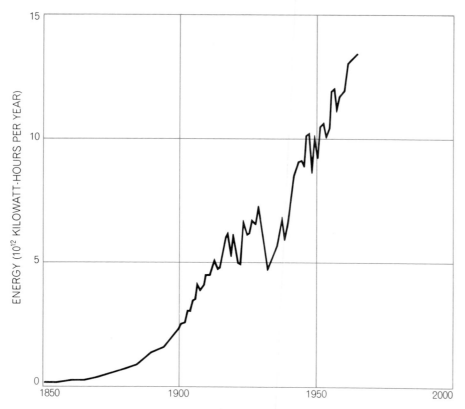

U.S. PRODUCTION OF ENERGY from coal, from petroleum and related sources, from water power and from nuclear reactors is charted for 120 years. The petroleum increment includes natural gas and associated liquids. The dip at center reflects impact of Depression.

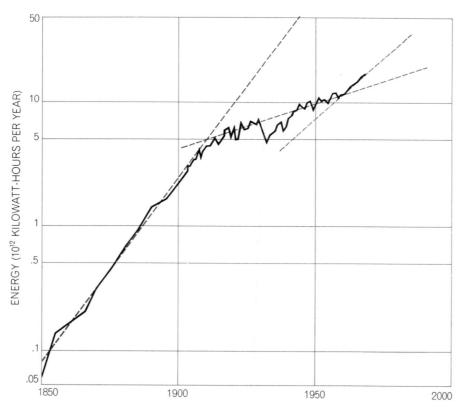

RATE OF GROWTH of U.S. energy production is shown by plotting on a semilogarithmic scale the data represented in the illustration at the top of the page. Broken lines show that the rise had three distinct periods. In the first the growth rate was 6.91 percent per year and the doubling period was 10 years; in the second the rate was 1.77 percent and the doubling period was 39 years; in the third the rate was 4.25 percent with doubling in 16.3 years.

oxide for the non-Communist nations during the same period.

Against these requirements the AEC estimates that the quantity of uranium oxide producible at $8 per pound from present reserves in the U.S. is 243,000 tons, and the world reserves at $10 per pound or less are estimated in the other report at 840,000 tons. The same report estimates that to meet future requirements additional reserves of more than a million short tons will have to be discovered and developed by 1985.

Although new discoveries of uranium will doubtless continue to be made (a large one was recently reported in northeastern Australia), all present evidence indicates that without a transition to breeder reactors an acute shortage of low-cost ores is likely to develop before the end of the century. Hence an intensive effort to develop large-scale breeder reactors for power production is in progress. If it succeeds, the situation with regard to fuel supply will be drastically altered.

This prospect results from the fact that with the breeder reactor the amount of energy obtainable from one gram of uranium 238 amounts to 8.1×10^{10} joules of heat. That is equal to the heat of combustion of 2.7 metric tons of coal or 13.7 barrels (1.9 metric tons) of crude oil. Disregarding the rather limited supplies of high-grade uranium ore that are available, let us consider the much more abundant low-grade ores. One example will indicate the possibilities.

The Chattanooga black shale (of Devonian age) crops out along the western edge of the Appalachian Mountains in eastern Tennessee and underlies at minable depths most of Tennessee, Kentucky, Ohio, Indiana and Illinois. In its outcrop area in eastern Tennessee this shale contains a layer about five meters thick that has a uranium content of about 60 grams per metric ton. That amount of uranium is equivalent to about 162 metric tons of bituminous coal or 822 barrels of crude oil. With the density of the rock some 2.5 metric tons per cubic meter, a vertical column of rock five meters long and one square meter in cross section would contain 12.5 tons of rock and 750 grams of uranium. The energy content of the shale per square meter of surface area would therefore be equivalent to about 2,000 tons of coal or 10,000 barrels of oil. Allowing for a 50 percent loss in mining and extracting the uranium, we are still left with the equivalent of 1,000 tons of coal or 5,000 barrels of oil per square meter.

Taking Averitt's estimate of 1.5 trillion metric tons for the initial minable coal in the U.S. and a round figure of 250 billion barrels for the petroleum liquids, we find that the nuclear energy in an area of about 1,500 square kilometers of Chattanooga shale would equal the energy in the initial minable coal; 50 square kilometers would hold the energy equivalent of the petroleum liquids. Adding natural gas and oil shales, an area of roughly 2,000 square kilometers of Chattanooga shale would be equivalent to the initial supply of all the fossil fuels in the U.S. The area is about 2 percent of the area of Tennessee

and a very small fraction of the total area underlain by the shale. Many other low-grade deposits of comparable magnitude exist. Hence by means of the breeder reactor the energy potentially available from the fissioning of uranium and thorium is at least a few orders of magnitude greater than that from all the fossil fuels combined.

David J. Rose of the AEC, reviewing recently the prospects for controlled fusion, found the deuterium-tritium reaction to be the most promising. Deuterium is abundant (one atom to each

6,700 atoms of hydrogen), and the energy cost of separating it would be almost negligible compared with the amount of energy released by fusion. Tritium, on the other hand, exists only in tiny amounts in nature. Larger amounts must be made from lithium 6 and lithium 7 by nuclear bombardment. The limiting isotope is lithium 6, which has an abundance of only 7.4 percent of natural lithium.

Considering the amount of hydrogen in the oceans, deuterium can be regarded as superabundant. It can also be extracted easily. Lithium is much less

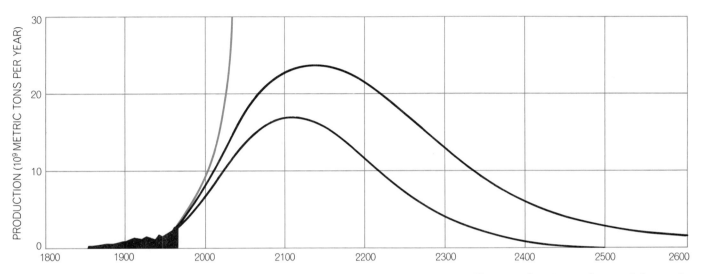

CYCLE OF WORLD COAL PRODUCTION is plotted on the basis of estimated supplies and rates of production. The top curve reflects Averitt's estimate of 7.6×10^{12} metric tons as the initial supply of minable coal; the bottom curve reflects an estimate of 4.3×10^{12} metric tons. The curve that rises to the top of the graph shows the trend if production continued to rise at the present rate of 3.56 percent per year. The amount of coal mined and burned in the century beginning in 1870 is shown by the black area at left.

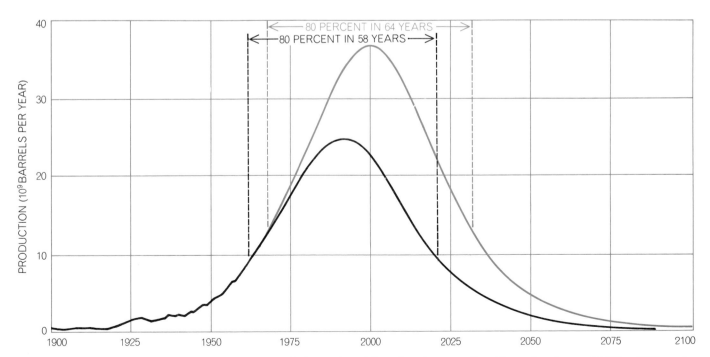

CYCLE OF WORLD OIL PRODUCTION is plotted on the basis of two estimates of the amount of oil that will ultimately be produced. The colored curve reflects Ryman's estimate of $2,100 \times 10^9$ barrels and the black curve represents an estimate of $1,350 \times 10^9$ barrels.

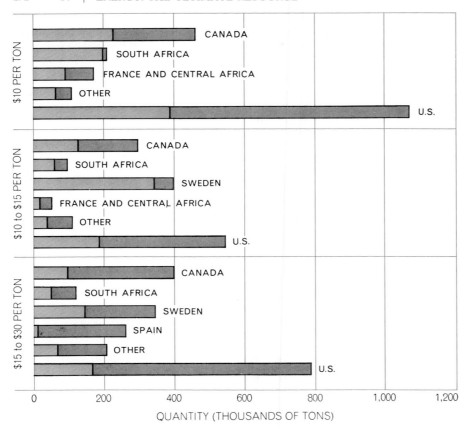

WORLD RESERVES OF URANIUM, which would be the source of nuclear power derived from atomic fission, are given in tons of uranium oxide (U_3O_8). The colored part of each bar represents reasonably assured supplies and the gray part estimated additional supplies.

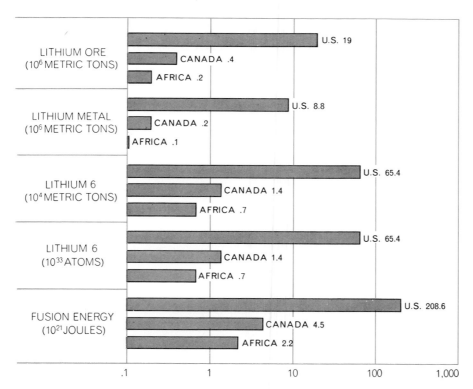

WORLD RESERVES OF LITHIUM, which would be the limiting factor in the deuterium-tritium fusion reaction, are stated in terms of lithium 6 because it is the least abundant isotope. Even with this limitation the energy obtainable from fusion through the deuterium-tritium reaction would almost equal the energy content of the world's fossil-fuel supply.

abundant. It is produced from the geologically rare igneous rocks known as pegmatites and from the salts of saline lakes. The measured, indicated and inferred lithium resources in the U.S., Canada and Africa total 9.1 million tons of elemental lithium, of which the content of lithium 6 would be 7.42 atom percent, or 67,500 metric tons. From this amount of lithium 6 the fusion energy obtainable at 3.19×10^{-12} joule per atom would be 215×10^{21} joules, which is approximately equal to the energy content of the world's fossil fuels.

As long as fusion power is dependent on the deuterium-tritium reaction, which at present appears to be somewhat the easier because it proceeds at a lower temperature, the energy obtainable from this source appears to be of about the same order of magnitude as that from fossil fuels. If fusion can be accomplished with the deuterium-deuterium reaction, the picture will be markedly changed. By this reaction the energy released per deuterium atom consumed is 7.94×10^{-13} joule. One cubic meter of water contains about 10^{25} atoms of deuterium having a mass of 34.4 grams and a potential fusion energy of 7.94×10^{12} joules. This is equivalent to the heat of combustion of 300 metric tons of coal or 1,500 barrels of crude oil. Since a cubic kilometer contains 10^9 cubic meters, the fuel equivalents of one cubic kilometer of seawater are 300 billion tons of coal or 1,500 billion barrels of crude oil. The total volume of the oceans is about 1.5 billion cubic kilometers. If enough deuterium were withdrawn to reduce the initial concentration by 1 percent, the energy released by fusion would amount to about 500,000 times the energy of the world's initial supply of fossil fuels!

Unlimited resources of energy, however, do not imply an unlimited number of power plants. It is as true of power plants or automobiles as it is of biological populations that the earth cannot sustain any physical growth for more than a few tens of successive doublings. Because of this impossibility the exponential rates of industrial and population growth that have prevailed during the past century and a half must soon cease. Although the forthcoming period of stability poses no insuperable physical or biological difficulties, it can hardly fail to force a major revision of those aspects of our current social and economic thinking that stem from the assumption that the growth rates that have characterized this temporary period can somehow be made permanent.

22

The Flow of Energy in an Industrial Society

by Earl Cook
September 1971

The U.S., with 6 percent of the world's population, uses 35 percent of the world's energy. In the long run the limiting factor in high levels of energy consumption will be the disposal of the waste heat

This article will describe the flow of energy through an industrial society: the U.S. Industrial societies are based on the use of power: the rate at which useful work is done. Power depends on energy, which is the ability to do work. A power-rich society consumes—more accurately, degrades—energy in large amounts. The success of an industrial society, the growth of its economy, the quality of the life of its people and its impact on other societies and on the total environment are determined in large part by the quantities and the kinds of energy resources it exploits and by the efficiency of its systems for converting potential energy into work and heat.

Whether by hunting, by farming or by burning fuel, man introduces himself into the natural energy cycle, converting energy from less desired forms to more desired ones: from grass to beef, from wood to heat, from coal to electricity. What characterizes the industrial societies is their enormous consumption of energy and the fact that this consumption is primarily at the expense of "capital" rather than of "income," that is, at the expense of solar energy stored in coal, oil and natural gas rather than of solar radiation, water, wind and muscle power. The advanced industrial societies, the U.S. in particular, are further characterized by their increasing dependence on electricity, a trend that has direct effects on gross energy consumption and indirect effects on environmental quality.

The familiar exponential curve of increasing energy consumption can be considered in terms of various stages of human development [see illustration on next page]. As long as man's energy consumption depended on the food he could eat, the rate of consumption was some 2,000 kilocalories per day; the do-

mestication of fire may have raised it to 4,000 kilocalories. In a primitive agricultural society with some domestic animals the rate rose to perhaps 12,000 kilocalories; more advanced farming societies may have doubled that consumption. At the height of the low-technology industrial revolution, say between 1850 and 1870, per capita daily consumption reached 70,000 kilocalories in England, Germany and the U.S. The succeeding high-technology revolution was brought about by the central electric-power station and the automobile, which enable the average person to apply power in his home and on the road. Beginning shortly before 1900, per capita energy consumption in the U.S. rose at an increasing rate to the 1970 figure: about 230,000 kilocalories per day, or about 65×10^{15} British thermal units (B.t.u.) per year for the country as a whole. Today the industrial regions, with 30 percent of the world's people, consume 80 percent of the world's energy. The U.S., with 6 percent of the people, consumes 35 percent of the energy.

In the early stages of its development in western Europe industrial society based its power technology on income sources of energy, but the explosive growth of the past century and a half has been fed by the fossil fuels, which are not renewable on any time scale meaningful to man. Modern industrial society is totally dependent on high rates of consumption of natural gas, petroleum and coal. These nonrenewable fossil-fuel resources currently provide 96 percent of the gross energy input into the U.S. economy [see top illustration on page 275]. Nuclear power, which in 1970 accounted for only .3 percent of the total energy input, is also (with present reactor technology) based on a capital source of energy: uranium 235. The energy of falling water, converted to hy-

dropower, is the only income source of energy that now makes any significant contribution to the U.S. economy, and its proportional role seems to be declining from a peak reached in 1950.

Since 1945 coal's share of the U.S. energy input has declined sharply, while both natural gas and petroleum have increased their share. The shift is reflected in import figures. Net imports of petroleum and petroleum products doubled between 1960 and 1970 and now constitute almost 30 percent of gross consumption. In 1960 there were no imports of natural gas; last year natural-gas imports (by pipeline from Canada and as liquefied gas carried in cryogenic tankers) accounted for almost 4 percent of gross consumption and were increasing.

The reasons for the shift to oil and gas are not hard to find. The conversion of railroads to diesel engines represented a large substitution of petroleum for coal. The rapid growth, beginning during World War II, of the national network of high-pressure gas-transmission lines greatly extended the availability of natural gas. The explosion of the U.S. automobile population, which grew twice as fast as the human population in the decade 1960–1970, and the expansion of the nation's fleet of jet aircraft account for much of the increase in petroleum consumption. In recent years the demand for cleaner air has led to the substitution of natural gas or low-sulfur residual fuel oil for high-sulfur coal in many central power plants.

An examination of energy inputs by sector of the U.S. economy rather than by source reveals that much of the recent increase has been going into household, commercial and transportation applications rather than industrial ones [see bottom illustration on page

275]. What is most striking is the growth of the electricity sector. In 1970 almost 10 percent of the country's useful work was done by electricity. That is not the whole story. When the flow of energy from resources to end uses is charted for 1970 [*see illustration on pages 276 and 277*], it is seen that producing that much electricity accounted for 26 percent of the gross consumption of energy, because of inefficiencies in generation and transmission. If electricity's portion of end-use consumption rises to about 25 percent by the year 2000, as is expected, then its generation will account for between 43 and 53 percent of the country's gross energy consumption. At that point an amount of energy equal to about half of the useful work done in the U.S. will be in the form of waste heat from power stations!

All energy conversions are more or less inefficient, of course, as the flow diagram makes clear. In the case of electricity there are losses at the power plant, in transmission and at the point of application of power; in the case of fuels consumed in end uses the loss comes at the point of use. The 1970 U.S. gross consumption of 64.6×10^{15} B.t.u. of energy (or 16.3×10^{15} kilocalories, or 19×10^{12} kilowatt-hours) ends up as 32.8×10^{15} B.t.u. of useful work and 31.8×10^{15} B.t.u. of waste heat, amounting to an overall efficiency of about 51 percent.

The flow diagram shows the pathways of the energy that drives machines, provides heat for manufacturing processes and heats, cools and lights the country. It does not represent the total energy budget because it includes neither food nor vegetable fiber, both of which bring solar energy into the economy through photosynthesis. Nor does it include environmental space heating by solar radiation, which makes life on the earth possible and would be by far the largest component of a total energy budget for any area and any society.

The minute fraction of the solar flux that is trapped and stored in plants provides each American with some 10,000 kilocalories per day of gross food production and about the same amount in the form of nonfood vegetable fiber. The fiber currently contributes little to the energy supply. The food, however, fuels man. Gross food-plant consumption might therefore be considered another component of gross energy consumption; it would add about 3×10^{15} B.t.u. to the input side of the energy-flow scheme. Of the 10,000 kilocalories per capita per day of gross production, handling and processing waste 15 percent. Of the remaining 8,500 kilocalories, some 6,300 go to feed animals that produce about 900 kilocalories of meat and 2,200 go into the human diet as plant materials, for a final food supply of about 3,100 kilocalories per person. Thus from field to table the efficiency of the food-energy system is 31 percent, close to the efficiency of a central power station. The similarity is not fortuitous; in both systems there is a large and unavoidable loss in the conversion of en-

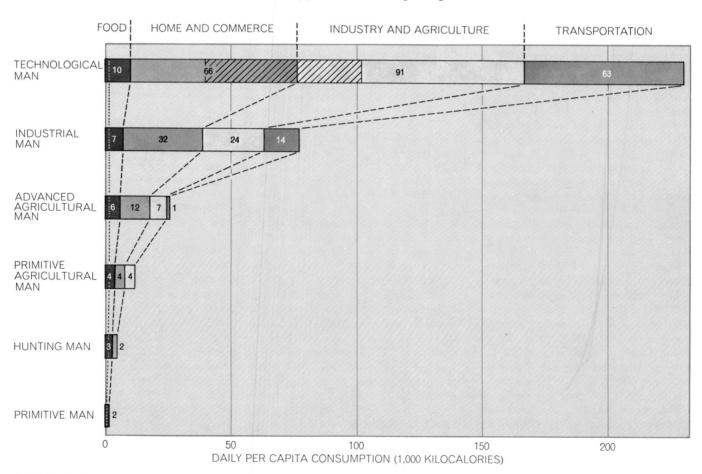

DAILY CONSUMPTION of energy per capita was calculated by the author for six stages in human development (and with an accuracy that decreases with antiquity). Primitive man (East Africa about 1,000,000 years ago) without the use of fire had only the energy of the food he ate. Hunting man (Europe about 100,000 years ago) had more food and also burned wood for heat and cooking. Primitive agricultural man (Fertile Crescent in 5000 B.C.) was growing crops and had gained animal energy. Advanced agricultural man (northwestern Europe in A.D. 1400) had some coal for heating, some water power and wind power and animal transport. Industrial man (in England in 1875) had the steam engine. In 1970 technological man (in the U.S.) consumed 230,000 kilocalories per day, much of it in form of electricity (*hatched area*). Food is divided into plant foods (*far left*) and animal foods (or foods fed to animals).

ergy from a less desired form to a more desired one.

Let us consider recent changes in U.S. energy flow in more detail by seeing how the rates of increase in various sectors compare. Not only has energy consumption for electric-power generation been growing faster than the other sectors but also its growth rate has been increasing: from 7 percent per year in 1961–1965 to 8.6 percent per year in 1965–1969 to 9.25 percent last year [*see top illustration on page 278*]. The energy consumed in industry and commerce and in homes has increased at a fairly steady rate for a decade, but the energy demand of transportation has risen more sharply since 1966. All in all, energy consumption has been increasing lately at a rate of 5 percent per year, or four times faster than the increase in the U.S. population. Meanwhile the growth of the gross national product has tended to fall off, paralleling the rise in energy sectors other than fast-growing transportation and electricity. The result is a change in the ratio of total energy consumption to G.N.P. [*see bottom illustration on page 278*]. The ratio had been in a long general decline since 1920 (with brief reversals) but since 1967 it has risen more steeply each year. In 1970 the U.S. consumed more energy for each dollar of goods and services than at any time since 1951.

Electricity accounts for much of this decrease in economic efficiency, for several reasons. For one thing, we are substituting electricity, with a thermal efficiency of perhaps 32 percent, for many direct fuel uses with efficiencies ranging from 60 to 90 percent. Moreover, the fastest-growing segment of end-use consumption has been electric air conditioning. From 1967 to 1970 consumption for air conditioning grew at the remarkable rate of 20 percent per year; it accounted for almost 16 percent of the total increase in electric-power generation from 1969 to 1970, with little or no multiplier effect on the G.N.P.

Let us take a look at this matter of efficiency in still another way: in terms of useful work done as a percentage of gross energy input. The "useful-work equivalent," or overall technical efficiency, is seen to be the product of the conversion efficiency (if there is an intermediate conversion step) and the application efficiency of the machine or device that does the work [*see bottom illustration on page 279*]. Clearly there is a wide range of technical efficiencies in energy systems, depending on the conversion devices. It is often said that

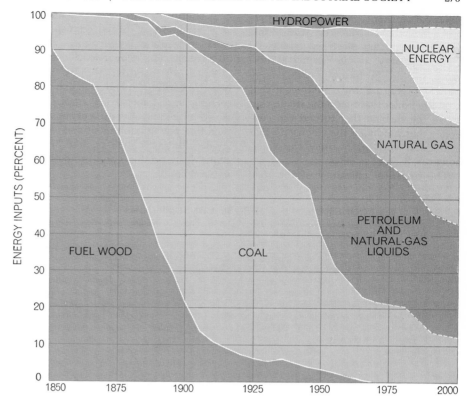

FOSSIL FUELS now account for nearly all the energy input into the U.S. economy. Coal's contribution has decreased since World War II; that of natural gas has increased most in that period. Nuclear energy should contribute a substantial percent within the next 20 years.

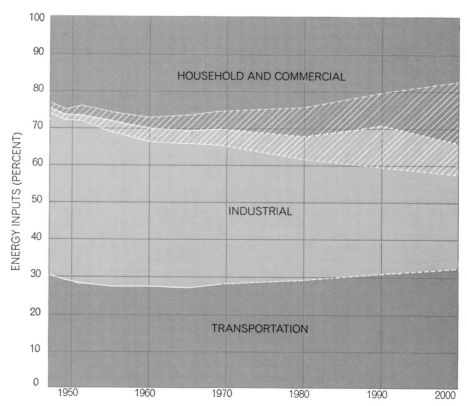

USEFUL WORK is distributed among the various end-use sectors of the U.S. economy as shown. The trend has been for industry's share to decrease, with household and commercial uses (including air conditioning) and transportation growing. Electricity accounts for an ever larger share of the work (*hatched area*). U.S. Bureau of Mines figures in this chart include nonenergy uses of fossil fuels, which constitute about 7 percent of total energy inputs.

electrical resistance heating is 100 percent efficient, and indeed it is in terms, say, of converting electrical energy to thermal energy at the domestic hot-water heater. In terms of the energy content of the natural gas or coal that fired the boiler that made the steam that drove the turbine that turned the generator that produced the electricity that heated the wires that warmed the water, however, it is not so efficient.

The technical efficiency of the total U.S. energy system, from potential energy at points of initial conversion to work at points of application, is about 50 percent. The economic efficiency of the system is considerably less. That is because work is expended in extracting, refining and transporting fuels, in the construction and operation of conversion facilities, power equipment and electricity-distribution networks, and in handling waste products and protecting the environment.

An industrial society requires not only a large supply of energy but also a high use of energy per capita, and the society's economy and standard of living are shaped by interrelations among resources, population, the efficiency of conversion processes and the particular applications of power. The effect of these interrelations is illustrated by a comparison of per capita energy consumption and per capita output for a number of countries [*see illustration on*

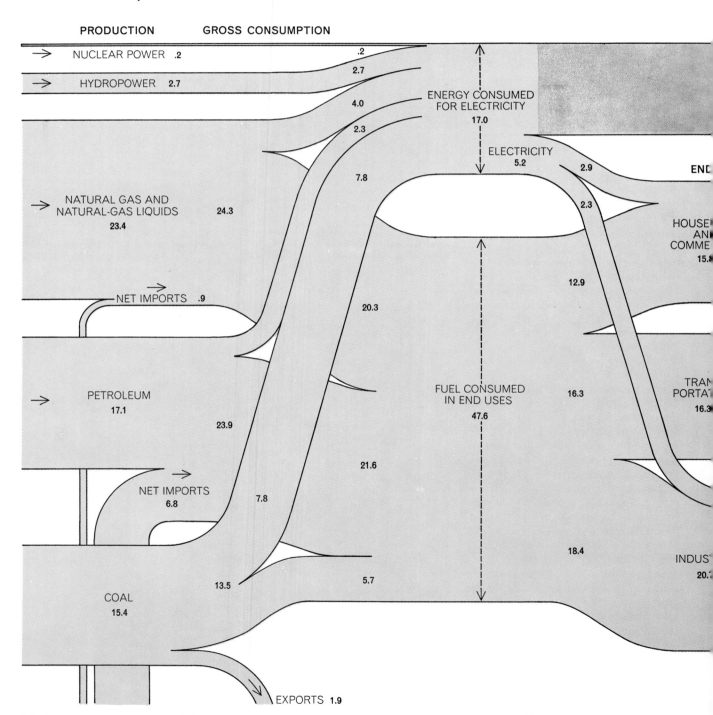

PRODUCTION   GROSS CONSUMPTION

FLOW OF ENERGY through the U.S. system in 1970 is traced from production of energy commodities (*left*) to the ultimate conversion of energy into work for various industrial end products and waste heat (*right*). Total consumption of energy in 1970 was 64.6 × 10¹⁵ British thermal units. (Adding nonenergy uses of fossil fuels, primarily for petrochemicals, would raise the total to 68.8 × 10¹⁵ B.t.u.) The overall efficiency of the system was about 51 percent. Some of the fossil-fuel energy is consumed directly and

page 280]. As one might expect, there is a strong general correlation between the two measures, but it is far from being a one-to-one correlation. Some countries (the U.S.S.R. and the Republic of South Africa, for example) have a high energy consumption with respect to G.N.P.; other countries (such as Sweden and New Zealand) have a high output with relatively less energy consumption. Such differences reflect contrasting combinations of energy-intensive heavy in-

dustry and light consumer-oriented and service industries (characteristic of different stages of economic development) as well as differences in the efficiency of energy use. For example, countries that still rely on coal for a large part of their energy requirement have higher energy inputs per unit of production than those that use mainly petroleum and natural gas.

A look at trends from the U.S. past is also instructive. Between 1800 and 1880

total energy consumption in the U.S. lagged behind the population increase, which means that per capita energy consumption actually declined somewhat. On the other hand, the American standard of living increased during this period because the energy supply in 1880 (largely in the form of coal) was being used much more efficiently than the energy supply in 1800 (largely in the form of wood). From 1900 to 1920 there was a tremendous surge in the use of energy by Americans but not a parallel increase in the standard of living. The ratio of energy consumption to G.N.P. increased 50 percent during these two decades because electric power, inherently less efficient, began being substituted for the direct use of fuels; because the automobile, at best 25 percent efficient, proliferated (from 8,000 in 1900 to 8,132,000 in 1920), and because mining and manufacturing, which are energy-intensive, grew at very high rates during this period.

Then there began a long period during which increases in the efficiency of energy conversion and utilization fulfilled about two-thirds of the total increase in demand, so that the ratio of energy consumption to G.N.P. fell to about 60 percent of its 1920 peak although per capita energy consumption continued to increase. During this period (1920–1965) the efficiency of electric-power generation and transmission almost trebled, mining and manufacturing grew at much lower rates and the services sector of the economy, which is not energy-intensive, increased in importance.

"Power corrupts" was written of man's control over other men but it applies also to his control of energy resources. The more power an industrial society disposes of, the more it wants. The more power we use, the more we shape our cities and mold our economic and social institutions to be dependent on the application of power and the consumption of energy. We could not now make any major move toward a lower per capita energy consumption without severe economic dislocation, and certainly the struggle of people in less developed regions toward somewhat similar energy-consumption levels cannot be thwarted without prolonging mass human suffering. Yet there is going to have to be some leveling off in the energy demands of industrial societies. Countries such as the U.S. have already come up against constraints dictated by the availability of resources and by damage to the environment. Another article in this

some is converted to generate electricity. The efficiency of electrical generation and transmission is taken to be about 31 percent, based on the ratio of utility electricity purchased in 1970 to the gross energy input for generation in that year. Efficiency of direct fuel use in transportation is taken as 25 percent, of fuel use in other applications as 75 percent.

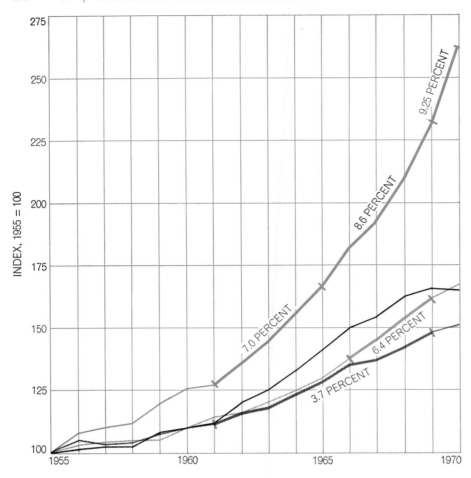

INCREASE IN CONSUMPTION of energy for electricity generation (*dark color*), transportation (*light color*) and other applications (*gray*) and of the gross national product (*black*) are compared. Annual growth rates for certain periods are shown beside heavy segments of curves. Consumption of electricity has a high growth rate and is increasing.

RATIO OF ENERGY CONSUMPTION to gross national product has varied over the years. It tends to be low when the G.N.P. is large and energy is being used efficiently, as was the case during World War II. The ratio has been rising steadily since 1965. Reasons include the increase in the use of air conditioning and the lack of advance in generating efficiency.

issue considers the question of resource availability [see the article "The Energy Resources of the Earth," by M. King Hubbert, beginning on page 263]. Here I shall simply point out some of the decisions the U.S. faces in coping with diminishing supplies, and specifically with our increasing reliance on foreign sources of petroleum and petroleum products. In the short run the advantages of reasonable self-sufficiency must be weighed against the economic and environmental costs of developing oil reserves in Alaska and off the coast of California and the Gulf states. Later on such self-sufficiency may be attainable only through the production of oil from oil shale and from coal. In the long run the danger of dependence on dwindling fossil fuels—whatever they may be —must be balanced against the research and development costs of a major effort to shape a new energy system that is neither dependent on limited resources nor hard on the environment.

The environmental constraint may be more insistent than the constraint of resource availability. The present flow of energy through U.S. society leaves waste rock and acid water at coal mines; spilled oil from offshore wells and tankers; waste gases and particles from power plants, furnaces and automobiles; radioactive wastes of various kinds from nuclear-fuel processing plants and reactors. All along the line waste heat is developed, particularly at the power plants.

Yet for at least the next 50 years we shall be making use of dirty fuels: coal and petroleum. We can improve coal-combustion technology, we can build power plants at the mine mouth (so that the air of Appalachia is polluted instead of the air of New York City), we can make clean oil and gas from coal and oil from shale, and sow grass on the mountains of waste. As nuclear power plants proliferate we can put them underground, or far from the cities they serve if we are willing to pay the cost in transmission losses. With adequate foresight, caution and research we may even be able to handle the radioactive-waste problem without "undue" risk.

There are, however, definite limits to such improvements. The automobile engine and its present fuel simply cannot be cleaned up sufficiently to make it an acceptable urban citizen. It seems clear that the internal-combustion engine will be banned from the central city by the year 2000; it should probably be banned right now. Because our cities are shaped for automobiles, not for mass transit, we shall have to develop battery-powered

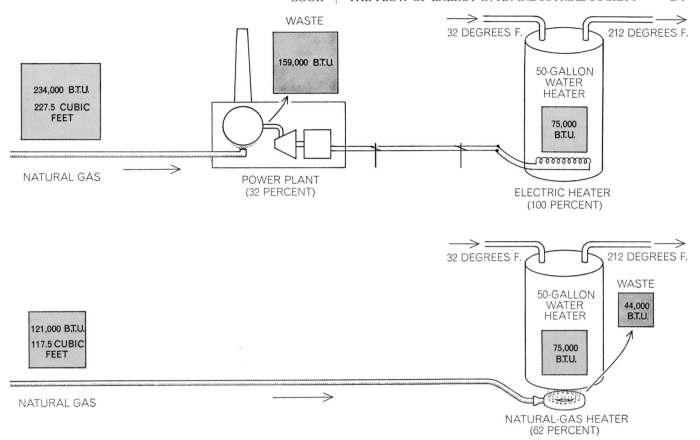

EFFICIENCIES OF HEATING WATER with natural gas indirectly by generating electricity for use in resistance heating (*top*) and directly (*bottom*) are contrasted. In each case the end result is enough heat to warm 50 gallons of water from 32 degrees Fahrenheit to 212 degrees. Electrical method requires substantially more gas even though efficiency at electric heater is nearly 100 percent.

or flywheel-powered cars and taxis for inner-city transport. The 1970 census for the first time showed more metropolitan citizens living in suburbs than in the central city; it also showed a record high in automobiles per capita, with the greatest concentration in the suburbs. It seems reasonable to visualize the suburban two-car garage of the future with one car a recharger for "downtown" and the other, still gasoline-powered, for suburban and cross-country driving.

Of course, some of the improvement in urban air quality bought by excluding the internal-combustion engine must be paid for by increased pollution from the power plant that supplies the electricity for the nightly recharging of the downtown vehicles. It need not, however, be paid for by an increased draft on the primary energy source; this is one substitution in which electricity need not decrease the technical efficiency of the system. The introduction of heat pumps for space heating and cooling would be another. In fact, the overall efficiency should be somewhat improved and the environmental impact, given adequate attention to the siting, design and operation of the substituting power plant,

	PRIMARY ENERGY INPUT (UNITS)	SECONDARY ENERGY OUTPUT (UNITS)	APPLICATION EFFICIENCY (PERCENT)	TECHNICAL EFFICIENCY (PERCENT)
AUTOMOBILE				
INTERNAL-COMBUSTION ENGINE	100		25	25
FLYWHEEL DRIVE CHARGED BY ELECTRICITY	100	32	100	32
SPACE HEATING				
BY DIRECT FUEL USE	100		75	75
BY ELECTRICAL RESISTANCE	100	32	100	32
SMELTING OF STEEL				
WITH COKE	100	94	94	70
WITH ELECTRICITY	100	32	32	32

TECHNICAL EFFICIENCY is the product of conversion efficiency at an intermediate step (if there is one) and application efficiency at the device that does the work. Losses due to friction and heat are ignored in the flywheel-drive automobile data. Coke retains only about 66 percent of the energy of coal, but the energy recovered from the by-products raises the energy conservation to 94 percent.

should be greatly alleviated.

If technology can extend resource availability and keep environmental deterioration within acceptable limits in most respects, the specific environmental problem of waste heat may become the overriding one of the energy system by the turn of the century.

The cooling water required by power plants already constitutes 10 percent of the total U.S. streamflow. The figure will increase sharply as more nuclear plants start up, since present designs of nuclear plants require 50 percent more cooling

HEAT DISCHARGE from a power plant on the Connecticut River at Middletown, Conn., is shown in this infrared scanning radiograph. The power plant is at upper left, its structures outlined by their heat radiation. The luminous cloud running along the left bank of the river is warm water discharged from the cooling system of the plant. The vertical oblong object at top left center is an oil tanker. The luminous spot astern is the infrared glow of its engine room. The dark streak between the tanker and the warm-water region is a breakwater. The irregular line running down the middle of the picture is an artifact of the infrared scanning system. The picture was made by HRB-Singer, Inc., for U.S. Geological Survey.

water than fossil-fueled plants of equal size do. The water is heated 15 degrees Fahrenheit or more as it flows through the plant. For ecological reasons such an increase in water released to a river, lake

or ocean bay is unacceptable, at least for large quantities of effluent, and most large plants are now being built with cooling ponds or towers from which much of the heat of the water is dissi-

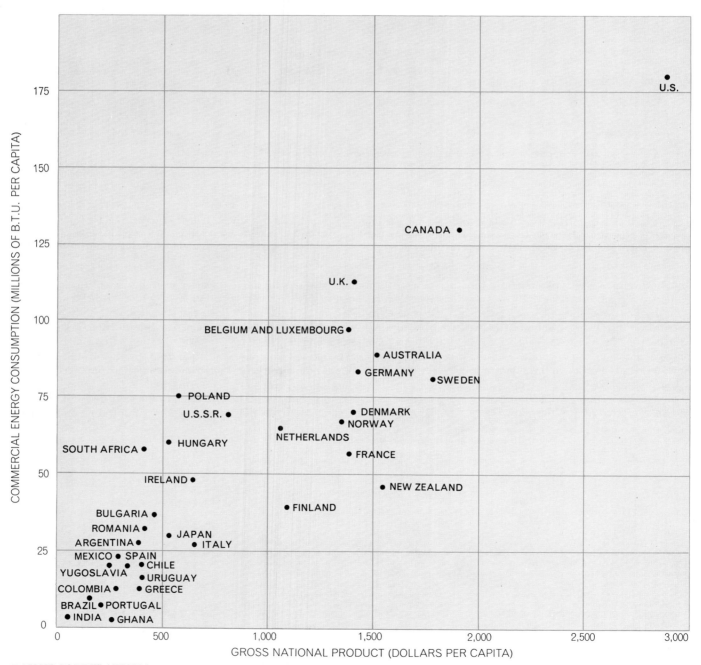

ROUGH CORRELATION between per capita consumption of energy and gross national product is seen when the two are plotted together; in general, high per capita energy consumption is a prerequisite for high output of goods and services. If the position plotted for the U.S. is considered to establish an arbitrary "line," some countries fall above or below that line. This appears to be related to a country's economic level, its emphasis on heavy industry or on services and its efficiency in converting energy into work.

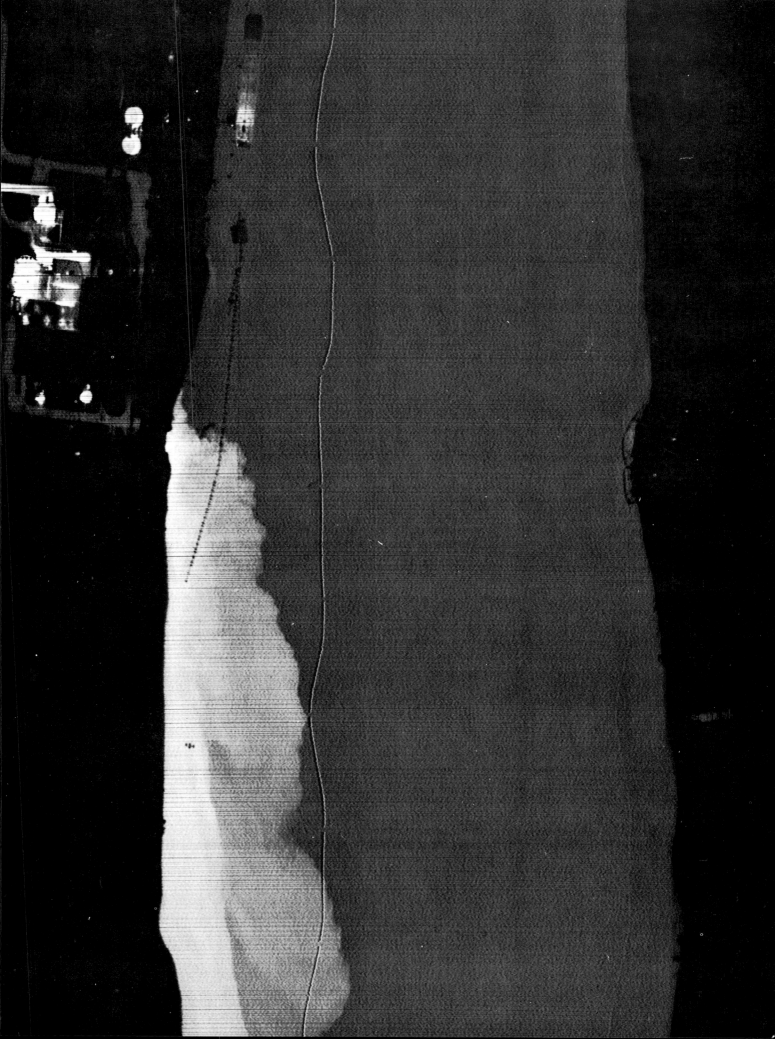

pated to the atmosphere before the water is discharged or recycled through the plant. Although the atmosphere is a more capacious sink for waste heat than any body of water, even this disposal mechanism obviously has its environmental limits.

Many suggestions have been made for putting the waste heat from power plants to work: for irrigation or aquaculture, to provide ice-free shipping lanes or for space heating. (The waste heat from power generation today would be more than enough to heat every home in the U.S.!) Unfortunately the quantities of water involved, the relatively low temperature of the coolant water and the distances between power plants and areas of potential use are serious deterrents to the utilization of waste heat. Plants can be designed, however, for both power production and space heating. Such a plant has been in operation in Berlin for a number of years and has proved to be more efficient than a combination of separate systems for power production and space heating. The Berlin plant is not simply a conserver of waste heat but an exercise in fuel economy; its power capacity was reduced in order to raise the temperature of the heated water above that of normal cooling water.

With present and foreseeable technology there is not much hope of decreasing the amount of heat rejected to streams or the atmosphere (or both) from central steam-generating power plants. Two systems of producing power without steam generation offer some long-range hope of alleviating the waste-heat problem. One is the fuel cell; the other is the fusion reactor combined with a system for converting the energy released directly into electricity [see the article

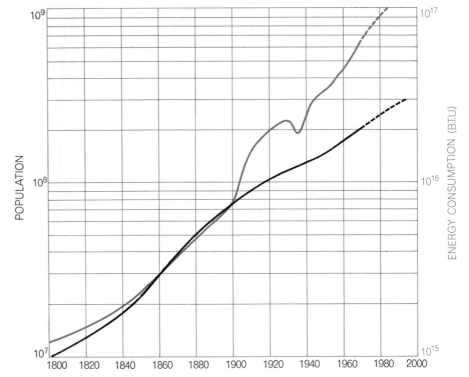

U.S. ENERGY-CONSUMPTION GROWTH (*curve in color*) has outpaced the growth in population (*black*) since 1900, except during the energy cutback of the depression years.

"The Conversion of Energy," by Claude M. Summers, beginning on page 349]. In the fuel cell the energy contained in hydrocarbons or hydrogen is released by a controlled oxidation process that produces electricity directly with an efficiency of about 60 percent. A practical fusion reactor with a direct-conversion system is not likely to appear in this century.

Major changes in power technology will be required to reduce pollution and manage wastes, to improve the efficiency of the system and to remove the resource-availability constraint. Making

the changes will call for hard political decisions. Energy needs will have to be weighed against environmental and social costs; a decision to set a pollution standard or to ban the internal-combustion engine or to finance nuclear-power development can have major economic and political effects. Democratic societies are not noted for their ability to take the long view in making decisions. Yet indefinite growth in energy consumption, as in human population, is simply not possible.

The Arrival of Nuclear Power

by John F. Hogerton
February 1968

Electric power obtained from nuclear fission has made a decisive market breakthrough and now accounts for nearly half of all the new power-generating capacity being ordered by U.S. utilities

Nuclear power, like the boy next door, seems to have grown up overnight. That it has indeed come of age is incontrovertible. For two years running it has accounted for nearly half of all the new power-generating capacity ordered by U.S. utilities. More than 50 large nuclear generating units, representing a financial commitment of $7 billion, are now under construction or awaiting construction permits. Their combined capacity exceeds 40 million kilowatts, which is as much electrical plant as the U.S. had in service going into World War II.

That maturity came quickly is also incontrovertible. The first truly large-scale nuclear unit—a 428,000-kilowatt installation at San Onofre, Calif.—was licensed for construction as recently as February 24, 1964, and announcements of commercial nuclear power projects did not begin to gain momentum until the fall of 1965; yet by the summer of 1966 nuclear power had drawn abreast of fossil power in the utility marketplace. It is safe to say that no one, not even the most optimistic reactor manufacturer, expected so rapid or decisive a market breakthrough.

This does not mean, however, that nuclear power was spared adolescence. There were many times when its progress seemed erratic and painfully slow, when friends and neighbors (particularly in the financial community) wondered if the boy next door would ever grow up. This review of the course of nuclear power development in the U.S. will explore the reasons for the uncertainty and help to place what has happened recently in better perspective. With this end in mind I have divided the history of the field into five distinct periods, which can be thought of as parts of a play, consisting of a Prologue (1942–1947), Act I (1948–1953), Act II (1954–1962), Act III (1963–1967) and an Epilogue (the present).

Strictly speaking, the years from 1942 to 1947 predate the development of nuclear power; during this period the atomic bomb held the stage almost to the exclusion of other pursuits. There are several reasons, however, for including those years in this account.

For one, 1942 was the year of the first nuclear reactor, or "pile" as it was then called. The array of uranium and graphite bricks assembled by Enrico Fermi, Walter H. Zinn and their co-workers in the famous squash court below the University of Chicago's Stagg Field was a primitive device by today's standards and had the limited objective of demonstrating that a chain reaction of fissioning atomic nuclei could be started, regulated and stopped. The pile operated at room temperature and essentially zero power. The few watts of heat that it generated were dissipated to the surrounding air. There was no reactor coolant.

The reactors built at Hanford, Wash., in 1943 and 1944 to supply plutonium for the Manhattan project ran at a temperature that can be characterized as lukewarm. They were the first water-cooled reactors. The water, drawn from the Columbia River, passed once through the reactor core and was then discharged into a storage basin; there was no recirculation.

During the Manhattan project some experience was obtained with recirculating coolant systems in small reactors built for experimental purposes, but again the temperatures were low. Design concepts for reactors capable of operating at temperatures high enough for power applications were studied as time allowed, and some development work had been started by the time the project ended.

When the civilian Atomic Energy Commission assumed its stewardship of the national atomic energy program in 1947, it was faced with the problem of deciding what avenues of power-reactor development to pursue. After taking stock of its inheritance of low-temperature reactor technology, the AEC tabled the fragmentary development work then in progress until plans could be made for a more systematic effort. By 1948 the AEC was ready to initiate a program and formed a Division of Reactor Development to carry it out. With that action the Prologue ended.

During the period from 1948 to 1953 the foundations of a diversified power-reactor technology were laid, naval-propulsion reactors were developed and, toward the end, the development of reactors for central-station electric power generation emerged as a priority objective.

The AEC's strategy of reactor development was to launch an attack along several main fronts. One front involved the pursuit of fundamental knowledge: the determination of basic physical constants; investigations of the properties of various fuels, coolants and other reactor materials; heat-transfer studies and the like. As part of this work, considerable effort was devoted to the design of supporting facilities for research and materials-testing.

A second front, and the one to which the most money was allocated, was the development of reactors for propelling submarines and aircraft. During that period two different prototype submarine reactors were successfully built, one cooled with pressurized water and the other with liquid sodium, and the submarine *Nautilus* (employing the former system) was commissioned. The attack on the intrinsically more difficult problem of nuclear aircraft propulsion did not fare as well, partly because of the

handicap of shifting military requirements and priorities, and this project was later abandoned.

The third front of the AEC reactor-development program was research on reactors for ultimate use in central-station electric power generation. A variety of design concepts were studied and experimental facilities were constructed to probe the relative merit of several of the more promising ideas. Included in the latter category were the pressurized-water and boiling-water reactor systems commonly used by utilities today; a sodium-cooled, graphite-moderated system; a system cooled and moderated with an organic chemical, and a system employing a circulating fluid fuel.

During this period civilian as well as military power-reactor technology had a security classification; access to it was restricted to those with a "need to know." Only the industrial concerns that served the AEC as contractors were in a position to follow the results of the reactor experiments, and in all but a few cases their access to information was compartmentalized. To most of industry, including the electric utilities, nuclear power was a black box.

As time went on interest in what the black box contained began to mount. The first opportunity many companies had to peer inside came in 1952, when qualified industry teams were given limited access to classified data to enable them to evaluate the economic outlook of nuclear power. There was a considerable amount of such study in 1952 and 1953. The consensus of the study teams was that if a military value were placed on the by-product plutonium, nuclear power plants could be expected to produce electricity at a cost of roughly seven mills per kilowatt-hour. The cost in conventional thermal power stations ranged at that time from five to 10 mills per kilowatt-hour and averaged about seven; thus, on paper at least, nuclear power held promise of immediate application. There was the difficulty, however, that the atomic energy statute then in force prohibited private ownership of power reactors and thereby precluded private initiative in nuclear power development.

In the spring of 1953 the late Gordon E. Dean, the chairman of the AEC, went before the Joint Committee on Atomic Energy of Congress and urged that the

INTERIOR VIEW of one of the country's first nuclear power stations shows the refueling floor of the reactor installation. It is from this floor that the fuel assemblies are lowered into the core of the reactor. The reactor housing has an inside diameter of 160 feet and is surrounded by a concrete radiation shield several feet thick. The plant, which is at Indian Point, N.Y., is owned and operated

development of nuclear power for central-station use be recognized as an important national objective and that the law be changed to allow private initiative. Later that year President Eisenhower, speaking before the United Nations, called for international cooperation in the development of nuclear power and other peaceful uses of atomic energy and for the creation of a system of international controls, backed up by inspection safeguards, to protect against misuse of the materials and facilities involved. An important factor in the President's proposed "Atoms for Peace" program was the hope that it might act to improve the climate for negotiating an effective system of control over nuclear weapons.

Thus the thrust for an accelerated nuclear power program was not sparked by any specific technical development, nor did it develop solely from the buildup of economic pressure within the U.S. power industry. Considerations of national prestige and foreign policy were of central importance. Before 1953 drew to a close the AEC began to implement the new policy of acceleration. A major first step was authorization of the construc-

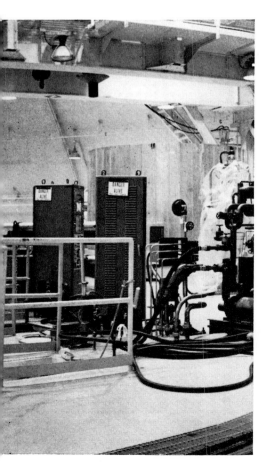

by the Consolidated Edison Co. of New York, Inc. The reactor system was supplied by the Babcock & Wilcox Company.

tion of a "full scale" (60,000-kilowatt) demonstration plant at Shippingport, Pa.

During the period from 1954 to 1962 half a dozen prototype reactors and several fair-sized demonstration plants were placed in operation. The transition from a Government program to a competitive enterprise was made, and along the way optimism about the short-term commercial prospects of the new industry took a severe buffeting.

At the start of the period many utilities and manufacturers were highly optimistic. At hearings on proposed private-ownership legislation held by the Joint Committee on Atomic Energy in the spring of 1954 there was a give-us-the-ball-and-we-will-run-with-it tenor to much of the testimony submitted by industry spokesmen. Industry was promptly given the ball, not in the sense of the Government's handing it over and retiring from the field but in the sense that a private-ownership law was promptly passed that enabled—indeed encouraged—industry to proceed with nuclear power projects. The law reserved to the Government the ownership of nuclear fuel but provided for supplies to be made available to industry under a lease arrangement. The AEC was charged with regulatory responsibility and given sole authority to grant licenses for the construction and operation of nuclear power facilities and for the possession of nuclear fuel.

The enactment of the new law triggered a good deal of behind-the-scenes industrial activity, but firm plans for actual power projects were slow in developing. Nearly seven months elapsed between the time the law became effective (August 31, 1954) and the date of the first industry application for a license to build a nuclear power plant (March 22, 1955). It would have been unrealistic, of course, to have expected an immediate flurry of license applications. Planning nuclear power projects takes time; moreover, in the fall of 1954 civilian power-reactor technology was still classified, which was doubtless a delaying factor. Yet there is little question but that industry's initial response was disappointing to a Joint Committee and an AEC bent on accelerating the national nuclear power effort.

By winter the AEC decided there was need to prime the industry pump. Accordingly, in January, 1955, it announced the "first round" of a Power Demonstration Reactor Program designed to stimulate demonstration projects. To utilities prepared to finance the construction of full-scale plants the AEC offered re-

search-and-development assistance and waiver of fuel inventory charges for the first five years of plant operation. Three projects were undertaken on this basis. In addition, two other projects were undertaken on a wholly privately financed basis outside the framework of the Power Demonstration Reactor Program. It is worth stopping the clock a moment to examine a little more closely the situation in which the utility industry found itself at that time.

By early 1955 there had been time and opportunity for industry to probe deeper into the status of nuclear power development, and it was quite clear that the first nuclear plants would not be economic. The word "acceleration" had taken on a new meaning: it meant pressing forward with nuclear power development at a faster pace and on a larger scale than the immediate economic outlook warranted. Since it was clear that the Government was committed to a policy of acceleration, the investor-owned segment of the electric-utility industry, representing about 80 percent of the total electric industry, faced a difficult choice—to build or not to build. To build full-scale noneconomic nuclear plants promised to be a costly way of speeding the development of a technology for which only a few farsighted utility executives could see a pressing need. On the other hand, not to build meant running the risk of having the Government do the job by itself. In a collective sense what the utility companies did at that critical time was to follow a middle course: they committed themselves to building a sufficient number of plants and took enough other initiatives to maintain their position with respect to the Government in an accelerated effort. In most cases the financial risk was shared by a number of companies through the device of joint projects; in one case, however, a single company (the Consolidated Edison Co. of New York, Inc.) undertook a pioneer project entirely on its own. That the collective interests of the utility industry were served was attributable to the leadership of a small number of individuals, all of whom believed strongly in the future of nuclear power and some of whom were doubtless also strongly motivated by power-policy considerations.

In subsequent months and years the AEC extended its Power Demonstration Reactor Program. The "second round" of the program, invitations for which were issued late in 1955, was designed to encourage the construction of small-scale prototypes of novel reactor systems and also to facilitate the participation of small publicly owned power entities such as rural cooperatives and municipal power authorities. To the latter end the AEC undertook to finance the reactor portion of the plant, thereby limiting the financial risk incurred by the utility participant. Five projects were proposed on this basis, but in the course of evaluation the list was reduced and only three second-round plants were actually built.

The "third round" came in 1957 and resulted in the construction of four small-scale plants, each of which broke some new technical ground either in overall design concept or in a key feature of its system. The next and, at this writing, last "round" of the Power Demonstration Reactor Program came five years later in 1962. Before describing this round the stage should be set by relating the events that led up to it.

Thus far in this account of the 1954–1962 period the focus has been on decision-making. The actual accomplishments of the period merit at least equal

time, but in the interests of brevity they will be compressed into a brief summary. During 1955 and 1956 civilian power-reactor technology was comprehensively declassified by the U.S., acting in concert with the United Kingdom and Canada. Late in 1956 the first power-reactor system designed expressly for electric-power generation—an experimental boiling-water unit at the Argonne National Laboratory– went into operation. A year later the 60,000-kilowatt Shippingport pressurized-water plant was completed. In 1960 two of the pioneer utility plants —one a 180,000-kilowatt boiling-water unit in Illinois (Dresden) and the other a 140,000-kilowatt pressurized-water unit in Massachusetts (Yankee)—began commercial service.

As operating experience was acquired with these and other early-vintage nuclear units, a favorable pattern began to emerge. Most of the debugging problems stemmed from auxiliary equipment; the reactors themselves gave surprisingly little trouble and, once they had been checked out for routine operation, gen-

erally performed with a high degree of dependability. The plants proved to be easy to operate in a utility system, under both steady-state and fluctuating-load conditions. Reactor designs were found to be conservative; in several cases it proved possible to operate the plants at power levels substantially higher than their nominal ratings. In the wash of operating experience many technical uncertainties associated with the different reactor concepts were removed and the basic validity of nuclear power was, from the technical standpoint, categorically established.

The same could not be said from the economic standpoint. Although one of the pioneer plants (Yankee) was brought in under budget, most of them cost more to build than had been anticipated, and in at least one case the final construction cost was nearly double the original estimate. The lowest generating costs achieved in these plants in their initial years of operation were 50 percent higher than the figure of seven mills per kilowatt-hour postulated by the early gener-

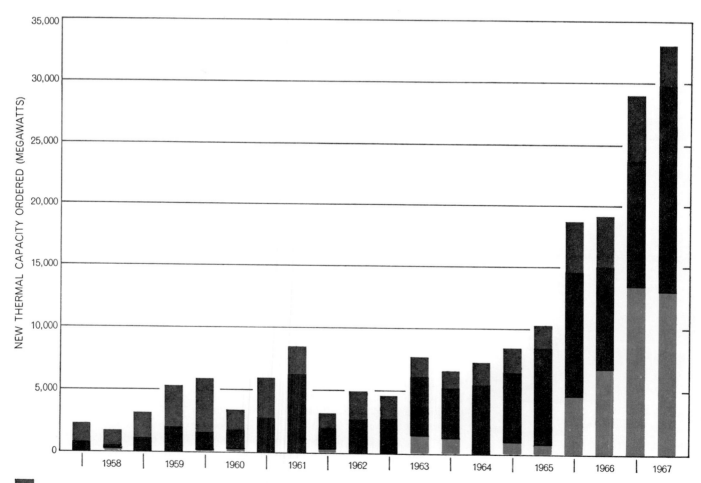

FOSSIL UNITS (<300 MEGAWATTS)

FOSSIL UNITS (⩾300 MEGAWATTS)

NUCLEAR UNITS

RECENT BREAKTHROUGH of nuclear power shows up strikingly in this chart, which breaks down the new thermal-generating capacity ordered by U.S. utilities during the past 10 years into nuclear units and fossil units (coal, oil or gas). The chart is based on the semi-annual reports issued in October and April of each year by the Edison Electric Institute.

alized studies. Moreover, the costs for conventional power generation were improving. Savings achieved by building larger, more efficient generating units had been sufficient to offset rising costs of labor and materials, and the cost of coal—the principal utility fuel and nuclear power's chief competitive target—had reversed a prior trend and was beginning to come down. No longer would a generating cost of seven mills per kilowatt-hour ensure nuclear power a large market. (Lest the wrong impression be given, it should be added that the builders of the pioneer plants did not expect to achieve anything like seven-mill power. Yankee's target, for example, was in the 11-to-13-mill range. It should also be noted that through successive core-design improvements and increases in power output, several of these plants have been able to achieve substantial reductions in generating cost. Taking Yankee again as an example, the cost has recently been running slightly under eight mills per kilowatt-hour, which in high-cost-fuel areas such as New England is not too far out of line with the cost of power today in conventional power stations of comparable size and vintage.)

Of course, the power costs achieved by the nuclear plants placed in service at the start of the 1960's were not a true indicator of the state of the art at that time. Considerable technical progress had been made since they were designed. Moreover, the experience gained in operating the "first generation" plants would bear fruit only when successor plants were designed and built. From the vantage point of the reactor manufacturers it appeared that a "second generation" of nuclear plants, larger in scale and improved in design, would prove to be economically competitive in those parts of the country where the cost of coal or other fossil fuels ran above the national average. Some utility leaders shared this belief; the majority, however, needed to be convinced.

Late in 1960 a West Coast utility expressed interest in building a large-scale pressurized-water plant in southern California and approached the AEC for assistance in what was later to become the San Onofre project. Early in 1961 another West Coast utility announced plans for a 310,000-kilowatt boiling-water plant to be located at Bodega Bay on the northern California coast and to be built without any financial assistance from the Government. The announcement of the Bodega Bay project was accompanied by cost estimates indicating that the plant would be economic under the utility's particular circumstances, the

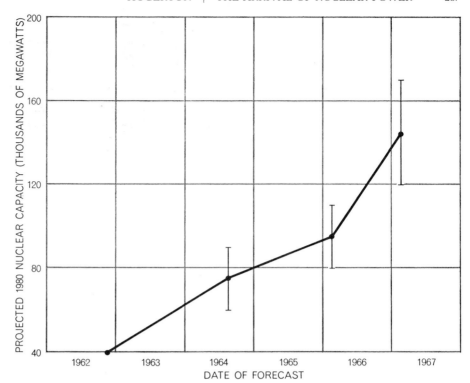

FORECASTS of the nuclear power generating capacity the U.S. will have in 1980, made periodically by the Atomic Energy Commission, have increased sharply in recent years.

central one being that fossil fuel costs in the area were already very high and were expected to rise.

In 1962, at the request of President Kennedy, the AEC made a comprehensive study of the outlook for nuclear power. The resulting report, issued in December, 1962, concluded that nuclear power had arrived at the threshold of commercial application and had a bright future. It forecast that 40 million kilowatts of nuclear generating capacity would be in service by 1980, and that by the end of the century nuclear power would account for half of the national electrical output and for essentially all subsequent power plant construction. In the light of this projected growth pattern, the report took stock of the country's nuclear fuel resources and stressed the long-range importance of using them efficiently. It went on to define three goals for the national nuclear power effort: the immediate construction of some commercial-scale plants employing reactors of already proved technology; the development of reactors with improved fuel-utilization characteristics (called "advanced converters"), and, as a long-range objective, the development of practical "breeder" reactors, which produce more fissionable material than they consume.

At the time the report to the President was being written the two California projects were enmeshed in site prob-

lems; it was not certain that either would go forward (indeed, the Bodega Bay project was subsequently canceled), and no other utility projects were in the offing. Concluding that there was need once again to prime the industry pump, the AEC announced a new round of its Power Demonstration Reactor Program aimed at stimulating the construction of commercial-scale plants of proved technology. The financial incentives offered included design as well as research and development assistance, plus waiver of initial fuel-inventory charges. At the end of the year the AEC was awaiting industry's response when, as a completely independent proposition, it received an application (later withdrawn) for a permit to build a million-kilowatt plant in metropolitan New York on a 1970-completion schedule. This was a remarkable development. Not only was the proposed plant three times larger than any previously considered but also the proposed location in the center of a densely populated area was without precedent and posed a major test both of AEC site policy and community acceptance. The proposal was several strides ahead of its time; it served, however, as a welcome affirmation of industry's confidence in nuclear power.

The period from 1963 to 1967 witnessed the breakthrough of nuclear power into the commercial market and, as something of an anticlimax, the start-up of the country's first truly large-scale nuclear plants.

Events moved swiftly in 1963. Early in the year two pressurized-water projects in the range of 400,000 to 500,000 kilowatts were undertaken within the framework of the AEC's Power Demonstration Reactor Program—one in New England (Connecticut Yankee) and the other in California (Malibu). Soon thereafter the long-pending San Onofre project finally advanced to the point of filing for a construction permit. And by the end of the year two more projects, both boiling-water plants with capacities of between 500,000 and 600,000 kilowatts,

had been undertaken on a straight commercial basis—one in New York (Nine Mile Point) and the other in New Jersey (Oyster Creek). A detailed report on the economics of the latter project, released in February, 1964, by the sponsoring utility (the Jersey Central Power & Light Company), gave the first intimation that a breakthrough might be in prospect.

Jersey Central had made an analysis of the comparative costs of power generation for three alternatives open to them: (1) building the proposed nuclear plant, (2) building a coal-fired plant of the same capacity at the same site and (3) long-distance transmission of electricity from a coal-fired plant at a "mine mouth" lo-

cation. The nuclear-cost estimates were based on a fixed-price "turnkey" bid submitted by the General Electric Company; those for the coal alternatives were based on engineering studies and reflected the best coal price offering the utility had been able to obtain from its suppliers. In the months preceding the Oyster Creek evaluation, Jersey Central had been contracting for coal at a delivered price of 30 to 31 cents per million B.T.U.'s of energy content. Under the pressure of nuclear competition coal was offered for the Oyster Creek site at a price of 26 cents. It was a remarkably good offer, but it turned out to be not good enough.

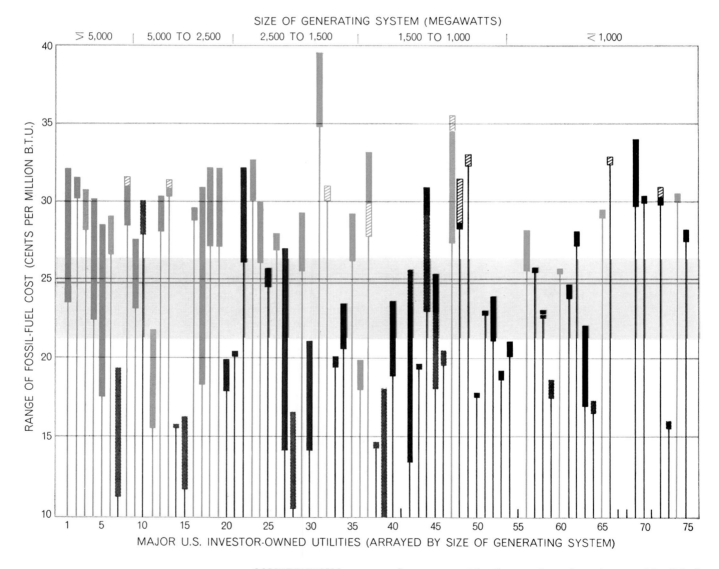

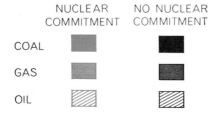

COMPETITION between nuclear power and fossil power depends on the cost of fossil fuel in different localities. In this chart vertical bars indicate the range of fossil-fuel costs in the systems of the 75 largest investor-owned electric utilities in the U.S. (Gas and oil costs have been adjusted to show the equivalent coal cost.) The utilities that have made one or more commitments for a large nuclear plant are indicated in color. The solid colored horizontal line represents the average coal cost for 1966 (24.7 cents per million B.T.U.'s). The light colored horizontal band covers the range of coal prices within which nuclear power is today competitive. The missing entries correspond either to utilities with thermal plants smaller than 300 megawatts or to utilities that are 100 percent hydroelectric.

The findings from the utility's cost analysis were, first, that if one assumed operation of the nuclear plant at its nominal capacity (515,000 kilowatts), coal would have had to have been available at the Oyster Creek site at a price slightly below 26 cents to be competitive; second, that if the nuclear plant were to achieve the higher output of which it was believed capable and for which its turbine generator had been designed (620,000 kilowatts), the price would have had to have been below 20 cents. To place these figures in perspective, the average price paid by U.S. utilities for coal in 1964 was a fraction above 26 cents, and only those utilities operating in or close to coal-mining areas enjoyed coal prices as low as 20 cents. It should perhaps be added that the Oyster Creek report placed the cost of nuclear power generation not at seven mills per kilowatt-hour, not at six mills, but at four to five mills, depending on the level of output achieved. The message, whether in terms of cents per million B.T.U.'s or mills per kilowatt-hour, was clear: If the Oyster Creek findings were at all indicative, nuclear power had arrived. The question was, how indicative was Oyster Creek?

This question quickly became a burning issue. Jersey Central spokesmen emphasized that their analysis was specific to their circumstances and might not apply under other circumstances. The company sponsoring the Nine Mile Point project, which was a parallel effort involving the same reactor manufacturer and virtually the same reactor design but a different site and different contract arrangements, did not publish a comparable cost analysis but did release some general cost estimates that indicated a much more conservative assessment.

Philip Sporn, past president of the country's largest coal-burning utility (the American Electric Power Company, Inc.) and long regarded as one of the country's ablest power men, was asked by the Joint Committee on Atomic Energy to appraise the economic position of nuclear power based on the information contained in the Oyster Creek report. His judgment was that there was insufficient margin for profit in the turnkey contract and that for this and other reasons the estimates of power cost were probably an overly optimistic indicator of the immediate outlook for nuclear power. Allowing for differences in utility circumstances, his estimate of break-even coal prices in an investor-owned system was 25 to 29 cents per million B.T.U.'s. Essentially this estimate meant that he too saw immediate market

FIRST LARGE-SCALE NUCLEAR PLANT, located at San Onofre, Calif., was completed in 1967. A pressurized-water installation capable of generating 428,000 kilowatts of electric power, it was built by the Westinghouse Electric Corporation and is operated jointly by the Southern California Edison Co. and the San Diego Gas & Electric Co. The dome-shaped building, which houses the reactor, is characteristic of a pressurized-water installation.

opportunities for nuclear power but thought they would be limited for the time being to the higher-cost-fuel areas.

Thus in the spring of 1964 there were among experts substantial differences of opinion as to the general level of costs that a utility contemplating a nuclear project might expect to encounter in building and operating a plant. The nuclear-cost picture was further complicated by differences in what the published cost estimates covered and by differences among utilities.

Even if by some magic there had been unanimity on the subject of nuclear costs, there still would have been ample room for differences of opinion on the market outlook. One reason was that important changes were taking place in coal costs, notably in the cost of transportation. In many parts of the country the expense of transporting coal from the mine to the power plant accounts for from a third to half of the utility's fuel cost, or, to put it another way, for roughly a fifth of the total cost of power generation. Most utility coal has traditionally moved by rail. Over the past decade there has been a substantial reduction in rail transportation costs, thanks to three successive innovations in rail practice. The first came in the mid-1950's when the railroads received sanction from the Interstate Commerce Commission to offer incentive rates to utilities. The second came in 1959 when the first trainload rates became effective. (Prior to that time freight rates had been based on carload quantities.) The third and by far the most important innovation came

in 1962 with the introduction of the "unit train," that is, a train that shuttles constantly back and forth between the mine and the power station, thereby achieving optimum equipment utilization. By 1964 the use of unit trains was spreading rapidly and substantial reductions in rate tariffs were being made on this basis. As with all new developments, it was difficult to foresee how far the trend might be carried and how it might affect the competitive balance between coal and nuclear power.

There were still other basic factors in the 1964 decision-making equation. The power industry was beginning to move rapidly toward interconnection, one of the benefits of which is that it often enables utilities to build larger, more economic power plants than they could manage on a single-system basis. It is characteristic of nuclear power that its competitive position improves as the size of plant increases. At the same time the larger the plant is, the more severe are the economic consequences of an unscheduled shutdown, and so a premium is placed in large installations on dependability of service. By 1964 there was conclusive evidence that utilities were experiencing a higher-than-expected rate of forced shutdown in the operation of their newest, largest and most sophisticated conventional power plants. On the other hand, nuclear power was to most utilities a completely new technology and its dependability in commercial service was a factor they had to evaluate most carefully.

Finally, in the back of every utility executive's mind were the indications, which by 1964 were beginning to become prevalent, of increased public interest in the environmental aspects of the power industry. There were then, as there are now, two diametrically opposed public attitudes regarding nuclear power. Some people resisted nuclear power because of reservations about its safety. Other people, particularly those concerned about the health implications of air pollution, favored nuclear power because of its cleanliness. At that time neither point of view commanded a large following, which meant that the "average citizen" had not yet taken a position. This left open what promised to become the most powerful factor in the utility equation.

The same month the Oyster Creek report was released the San Onofre project received its construction permit, ending a seven-year dry spell during which the only construction licenses issued in the U.S. were for experimental or prototype installations of 75,000-kilowatt capacity or less. Fortunately major plants of American design had been built overseas or were at an advanced stage of construction, and so there was not the experience gap that this statement might imply. Another notable event in 1964 was the enactment by Congress of pri-

vate-ownership legislation covering nuclear fuel. Under the new law private ownership of fuel supplies for nuclear power will be permitted at the start of 1969 and will be mandatory by mid-1973.

The leading reactor manufacturers (notably General Electric and the Westinghouse Electric Corporation) had started up their marketing machinery in 1963 and by the end of 1964 it was running in high gear. Over the 18-month period following the Oyster Creek announcement, however, only one firm sale was made. With that one exception (the Commonwealth Edison Company) utilities were not yet buying nuclear power. This doubtless meant in part that they had not yet wholly accepted the Oyster Creek message. There was another, more tangible explanation, which was that coal interests were making a valiant forestalling effort. Again and again, in situations where utilities were known to be seriously considering a nuclear project, substantial cuts were made in coal prices, mainly through the granting of favorable unit-train rates for coal delivery. In a number of cases the price concessions extended to other coal-burning plants on the same system, thereby benefiting the utility's overall power-gener-

ating economy. This practice reached a point, one suspects, where some utilities with at best a marginal interest in nuclear power went through the procedure of getting and evaluating bids from reactor manufacturers principally for the leverage this gave them in their coal negotiations.

In the late summer and early fall of 1965 the situation started to change, and by the summer of 1966 it had altered completely. Not only did utilities begin to place orders for nuclear units but also some ordered them two at a time. As was brought out at the start of this article, nuclear power rapidly built up its market position to the point where during 1966 and 1967 it accounted for nearly half of all the new power-generating capacity ordered by U.S. utilities. In the process, unit sizes leapfrogged from 500,000 to 800,000 to 1.1 million kilowatts and there was a progression of major design improvements, all of which took place before the forerunner of the 500,000-kilowatt class (San Onofre) was completed. Apart from one commitment for a 300,000-kilowatt demonstration plant of the high-temperature gas-cooled type, all the market traffic was in pressurized-water and boiling-water plants.

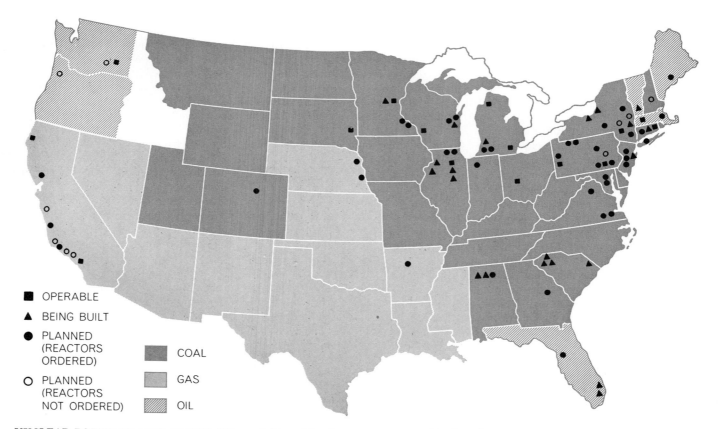

- ■ OPERABLE
- ▲ BEING BUILT
- ● PLANNED (REACTORS ORDERED)
- ○ PLANNED (REACTORS NOT ORDERED)

- ▓ COAL
- ░ GAS
- ▨ OIL

NUCLEAR POWER PLANTS IN THE U.S. are indicated by the black symbols; the map also shows by shading the principal fuel burned in conventional thermal-power plants in each state. (Idaho depends almost entirely on hydroelectric power). The plants represent a combined nuclear capacity of 59,778,300 kilowatts, which breaks down into: operable, 2,810,100 kilowatts; being built, 14,657,400 kilowatts; planned (reactors ordered), 32,210,800 kilowatts; planned (reactors not ordered), 10,100,000 kilowatts.

AERIAL VIEW of Millstone Point #1, a nuclear power station of Northeast Utilities, Inc., in Waterford, Conn., was made in September, 1967, when the plant was 35 percent complete. The reactor will be housed in the large steel dry well at right, which in turn will be completely enclosed by a rectangular building. The superstructure at left will contain the turbines. The 600,000-kilowatt boiling-water unit is scheduled for completion by the summer of 1969. The reactor system is being supplied by the General Electric Company.

Initially the sales were divided, in approximately equal measure, between General Electric and Westinghouse, which had long dominated the manufacture of turbine generators and related electrical equipment for the power industry and saw in reactor manufacture an opportunity to move in on the steam-equipment side of the industry and, perhaps even more important, to establish a position in the potentially lucrative business of supplying replacement fuel cores. Later two more companies established themselves in the water-cooled-plant market, namely the Babcock & Wilcox Company and Combustion Engineering, Inc., the two leading suppliers of conventional steam-generating systems.

The nuclear breakthrough is too recent a phenomenon to allow a definitive diagnosis, but it seems reasonably clear that there were two main factors at work. One was a divergence in marketing tactics. It cannot be documented, but there is little doubt that coal interests hardened their pricing policy sometime in the fall or early winter of 1965. With a record volume of business in hand and a record number of coal-burning power plants under construction, they may have decided to let nuclear power have its day in court, hoping all the while that it would fail to live up to its advance notices. Moreover, nuclear power was then threatening only the higher-cost-coal areas and, with the national market doubling every 10 years, the coal industry could write off these areas and still look forward to excellent growth prospects. Nuclear interests, on the other hand, kept their marketing drive going at peak intensity long after the breakthrough occurred, partly in order to achieve deeper penetration into coal territory and partly because of the stimulus of competition among reactor manufacturers. It was not until the latter part of 1966 that nuclear-equipment prices began to seek higher ground and not until 1967 that they increased substantially, and by then it was too late for coal to stem the tide of nuclear sales.

The second factor in the nuclear breakthrough was intangible and can only be described as a growing conviction among utilities that nuclear power would be the way of the future. This was particularly true of the utilities that operated coal-fired or oil-fired power stations in urban or highly industrialized areas. They could see almost daily signs of public concern about air pollution, and in not a few instances they were being confronted with new local ordinances setting stringent limits on the emission of sulfur dioxide and other stack gases. Low-sulfur fuels quickly commanded premium prices, and if the short-term outlook was poor, the long-term outlook was worse. The nation's coal resources, vast in the aggregate, were beginning to look less than vast as utilities sought to locate deposits that were cheap to mine, favorably situated and large enough to ensure a long-term supply for a large power station. It was obvious that they would look much smaller if the search had to be restricted to deposits of low sulfur content. The alternative of removing sulfur from run-of-the-mine coal, either at the mine or in the operation of the power plant, promised to present a serious cost problem. The only other alternative (apart from building nuclear plants) was to think in terms of long-distance transmission of electricity from plants located at the mines, which would usually place them at a distance from population centers. This of course had been done on a limited scale but, barring a major breakthrough in transmission technology, would be expensive as a general practice; moreover, the problems that could be anticipated in obtaining right-of-way for a multiplicity of cross-country transmission lines were formidable. In addition to all this, coal mining was itself encountering increasing public criticism on environmental issues, such as stream pollution and land despoliation, and cost increases were already beginning to be experienced on this account.

In short, utilities began to see the handwriting on the wall for coal. At the same time (and doubtless as a kind of mirror image) they began to see positive signs of growing public sentiment in favor of nuclear power.

If these were indeed the thoughts that were running through the minds of utility men, they could easily have swung marginal decisions in favor of nuclear power, particularly in cases where a company had not yet made a nuclear commitment. In effect they would have placed the burden of proof not on the new technology, where one would normally have expected it to be put, but on the old one. And even if a utility president took his staff's nuclear cost estimates with a grain of salt, he could still justify going ahead with a nuclear project on the grounds that it would give his organization essential training and experience in what seemed destined to be the coming technology.

As nuclear sales gained momentum, projections of the growth of nuclear power climbed rapidly. By mid-1967 most forecasters were predicting that there would be somewhere in the neighborhood of 150 million kilowatts of nuclear power capacity in service in the U.S. by 1980, an amount nearly four times greater than the AEC had forecast four years earlier. This translates into the expectation that, on a capacity basis, nuclear power will account for at least 50 percent of the new-plant market over the next decade and for an even higher share of base-load (as distinct from peaking) additions. By 1980 nuclear

plants would then account for about 30 percent of the country's total installed capacity and for as much as 35 to 40 percent of the gross electrical output.

This outlook is all the more remarkable when one recalls that as recently as the summer of 1967 nuclear power accounted for less than half of 1 percent of the electrical output. The amount of nuclear capacity then in service barely exceeded a million kilowatts, and the largest units were operating at power levels in the neighborhood of 200,000 kilowatts. (This statement does not take into account a large plutonium-producing reactor at Hanford, which produces electricity as a by-product.) San Onofre, the first of the truly large-scale units, did not begin preliminary operation until June 14 and did not reach full power until December. The second large-scale unit (Connecticut Yankee) followed

quickly and also reached full power in December. The next two in line (Oyster Creek and Nine Mile Point) are scheduled to begin service in 1968, and three others are scheduled for completion in 1969. It will be 1970, however, before a really large block of nuclear capacity is added to the national power grid. By then, but not much before, enough operating experience will have been accumulated to make possible a meaningful assessment of the performance of the first graduates of the 500,000-kilowatt class. In the meantime the size of the U.S. commitment to nuclear power will doubtless continue to grow, and it would not be surprising if the first member of the 1.5-million-kilowatt class will have matriculated.

Central to this account of the coming of age of nuclear power have been the pressurized-water and boiling-water reactors, which outpaced other entrants in the power-reactor sweepstakes and today dominate the commercial market. This situation might well be expected to continue indefinitely if it were not for two limitations inherent in the use of ordinary water as a reactor coolant. One is that it is impractical to achieve high enough operating temperatures inside a straightforward water-cooled system to produce high-pressure superheated steam. Power plants employing these systems are thus obliged to operate with low-quality steam and as a result are not as efficient in converting heat to electricity as the more modern conventional power plants. This means that a proportionately greater amount of waste heat is discharged into the plant's environment—in other words, into the river, lake or ocean that supplies the water used to cool the turbine condenser. Today's nuclear power plants typically discharge 30 percent more waste heat per kilowatt-hour of electricity generated than conventional plants. This could conceivably become a serious handicap. As the scale of power generation increases, utilities are finding it increasingly difficult to find suitable sites for new generating stations, and more often than not the limiting factor is the large flow of cooling water required. Of late, concern expressed by conservationists and others about the possibly adverse ecological effects of the warm-water discharge from large-scale power operations (the "thermal pollution" issue) has added a new dimension to the problem of finding suitable sites for thermal power plants. The use of cooling towers offers a way around the problem but adds to the power costs and has other disadvantages.

REACTOR VESSELS for two large-scale commercial boiling-water installations are shown undergoing final checks at the factory of the manufacturer, Combustion Engineering, Inc. Each vessel weighs more than 650 tons. The one at left is for the Niagara Mohawk Power Corporation's Nine Mile Point plant in New York; the one at right is for the Jersey Central Power & Light Company's Oyster Creek plant in New Jersey. In boiling-water reactor systems the control rods enter the core of the reactor through the holes in the bottom of the vessel.

The other limitation of pressurized-water and boiling-water reactors is that they are not efficient utilizers of nuclear fuel. Uranium 235, which is the only readily fissionable material found in nature, represents a very small fraction of the huge energy potential of the world's nuclear fuel resources. The rest is associated with the "fertile" materials, which are not themselves readily fissionable but which can be converted into fissionable form by irradiation with neutrons. For example, the fertile isotope uranium 238, which accounts for more than 99 percent of the uranium found in nature, is converted (by capturing a neutron) into fissionable plutonium 239. Water-cooled reactor systems, which are fueled with "slightly enriched" uranium (2 or 3 percent uranium 235), obtain part of their power output from plutonium formed during operation, but most of it comes from uranium 235 supplied in the reactor fuel.

The reason why greater advantage is not taken of the predominant fertile component of the fuel is that water has a pronounced tendency to absorb neutrons, and in passing through the reactor core it soaks up enough of these precious particles to prevent the system from achieving a high ratio of conversion of fertile material to fissionable material. Pressurized-water and boiling-water plants manage in spite of this handicap to achieve very low fuel costs and will continue to do so as long as reasonably priced supplies of uranium are available, which will probably be the case for at least another decade. In the long run, however, and in the interest of conserving energy resources, the nuclear power industry will need reactors that are more efficient in fuel utilization. This translates into a need to strive for better neutron economy, and since the efficiency with which heat is converted into electricity also affects fuel utilization, it makes it desirable to achieve higher operating temperatures as well.

Over the years several "advanced converters" have been carried quite far down the road of development, but at this writing only one stands in a position to challenge the water-cooled systems in the utility marketplace. That is the high-temperature gas-cooled reactor, a development of Gulf General Atomic, Inc., a subsidiary of the Gulf Oil Corporation. A 40,000-kilowatt prototype reactor of this type (Peach Bottom) was built under the "second round" of the AEC's Power Demonstration Reactor Program, and a 300,000-kilowatt demonstration plant (Fort St. Vrain) is scheduled for con-

PART OF CORE-SUPPORT STRUCTURE of a large pressurized-water reactor consists of control-rod guide tubes and the upper core grid plate. The part is made by Westinghouse.

struction. This system combines the attributes of good neutron economy and high temperature; however, until bids based on this approach have been submitted for large-scale nuclear power projects, its competitive position cannot be gauged.

If neutron losses are kept to the very minimum, it is possible for a reactor to operate at a conversion ratio high enough to achieve a net gain of fissionable material, in which event the reactor is known as a breeder. Using breeders in nuclear power generation would ensure maximum exploitation of fertile-fuel resources as well as fissionable ones. The technical feasibility of building a reactor with a sufficiently refined neutron economy to achieve breeding was demonstrated more than a decade ago. Much development work remains to be done, however, before this can be reduced to commercial practice. The AEC is sponsoring a major breeder-reactor development effort with the objective of having prototype plants in operation by the mid-1970's and of achieving full-scale com-

mercial application by the early or mid-1980's. Reactor manufacturers hope to improve this schedule by several years.

There are two basic approaches to breeding: a fast neutron system operating on the uranium-238/plutonium-239 cycle, and a slow neutron system operating on the thorium/uranium-233 cycle. Higher breeding gains are possible in the first cycle, and most of the current development efforts follow this approach. All of the five reactor manufacturers that have been mentioned so far in this article, plus a sixth (Atomics International, a division of North American Rockwell Corp.), are active in breeder development. Some utilities (notably the Detroit Edison Company) have a long history of active support of breeder development, and of late there have been signs of heightened utility interest in this exacting field of reactor technology.

It will be evident by now that this epilogue is in reality a second prologue, and that the most important history of nuclear power remains to unfold.

Fast Breeder Reactors **24**

by Glenn T. Seaborg and Justin L. Bloom
November 1970

*Nuclear reactors that use fast neutrons to produce
more fuel than they consume are a promising
approach to producing electric power with a minimum
of strain on energy resources and the environment*

The need to generate enormous additional amounts of electric power while at the same time protecting the environment is taking form as one of the major social and technological problems that our society must resolve over the next few decades. The Federal Power Commission has estimated that during the next 30 years the American power industry will have to add some 1,600 million kilowatts of electric generating capacity to the present capacity of 300 million kilowatts. As for the environment, the extent of public concern over improving the quality of air, water and the landscape hardly needs elaboration, except for one point that is often overlooked: it will take large amounts of electrical energy to run the many kinds of purification plants that will be needed to clean up the air and water and to recycle wastes.

A related problem of equal magnitude is the rational utilization of the nation's finite reserves of coal, oil and gas. In the long term they will be far more precious as sources of organic molecules than as sources of heat. Moreover, any reduction in the consumption of organic fuels brings about a proportional reduction in air pollution from their combustion products.

Nuclear reactors of the breeder type hold great promise as the solution to these problems. Producing more nuclear fuel than they consume, they would make it feasible to utilize enormous quantities of low-grade uranium and thorium ores dispersed in the rocks of the earth as a source of low-cost energy for thousands of years. In addition, these reactors would operate without adding noxious combustion products to the air. It is in the light of these considerations that the U.S. Atomic Energy Commission, the nuclear industry and the electric utilities have mounted a large-scale effort to develop the technology whereby it will be possible to have a breeder reactor generating electric power on a commercial scale by 1984.

Nuclear breeding is achieved with the neutrons released by nuclear fission. The fissioning of each atom of a nuclear fuel, such as uranium 235, liberates an average of more than two fast (high-energy) neutrons. One of the neutrons must trigger another fission to maintain the nuclear chain reaction; some neutrons are nonproductively lost, and the remainder are available to breed new fissionable atoms, that is, to transform "fertile" isotopes of the heavy elements into fissionable isotopes. The fertile raw materials for breeder reactions are thorium 232, which is transmuted into uranium 233, and uranium 238, which is transmuted into plutonium 239 [see illustrations on next page].

We have mentioned that breeding occurs when more fissionable material is produced than is consumed. A quantitative measure of this condition is the doubling time: the time required to produce as much net additional fissionable material as was originally present in the reactor. At the end of the doubling time the reactor has produced enough fissionable material to refuel itself and to fuel another identical reactor. An efficient breeder reactor will have a doubling time in the range of from seven to 10 years.

Two different breeder systems are involved, depending on which raw material is being transmuted. The thermal breeder, employing slow neutrons, operates best on the thorium 232–uranium 233 cycle (usually called the thorium cycle). The fast breeder, employing more energetic neutrons, operates best on the uranium 238–plutonium 239 cycle (the uranium cycle). Nonproductive absorption of neutrons is less in fast reactors than it is in thermal reactors, resulting in a decrease in the doubling time.

The concept of the breeder reactor is almost as old as the idea of the nuclear chain reaction. In the early stages, soon after World War II, many types of breeder reactor were visualized. Some were thermal and some were fast. Another important differentiation involved the type of coolant employed to carry off the heat of fission and deliver it to a power-generating system. Among the coolants proposed were water and molten salts for thermal breeding and inert gas (such as helium), liquid metal (such as sodium) and steam for fast breeding.

In the U.S. and several other countries decisions were made rather early that a fast breeder reactor cooled with liquid metal was the most attractive concept to pursue. This concept is known to atomic energy workers as the LMFBR (liquid-metal-cooled fast breeder reactor). Since the greater part of breeder-reactor development is now proceeding on the basis of this concept, this article is mainly devoted to the liquid-metal fast breeder. A serious alternative effort is being pursued, chiefly by utility companies, to develop the technology of a gas-cooled fast breeder reactor using

pressurized helium as the coolant. In the U.S. two thermal-breeder-reactor concepts operating on the thorium cycle are being developed: the light-water breeder reactor at the Bettis Atomic Power Laboratory and the molten-salt reactor at the Oak Ridge National Laboratory.

In the design of a liquid-metal-cooled fast breeder reactor several features are noteworthy. The core of a fast reactor can be quite small. For economic

reasons the reactor must be operated at a much higher power density than ordinary fission reactors are. The active core volume is therefore only a few cubic meters and is roughly proportional to the power output. The power density is about .4 megawatt per liter.

To carry off the heat while maintaining the fuel at a reasonable temperature sodium must flow through the core at a rate of tens of thousands of cubic meters per hour. In order to provide channels

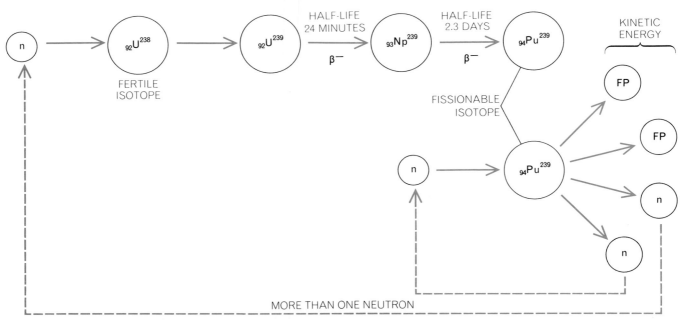

URANIUM CYCLE for breeding in a fast breeder reactor relies on fast, or highly energetic, neutrons. In the cycle an atom of fertile uranium 238 absorbs a neutron and emits a beta particle to become neptunium, which then undergoes beta decay to become fissionable plutonium 239. When an atom of plutonium 239 absorbs a neutron, it can fission, releasing energy, fission products (*FP*) and at least two neutrons. One of the neutrons is needed to continue the chain reaction, but the others are available to transform a fertile isotope into a fissionable one, thereby "breeding" fuel. Within a few years a breeder doubles its original fuel inventory.

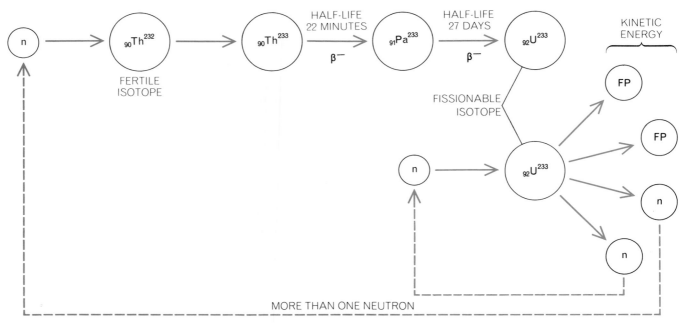

THORIUM CYCLE of breeding is similar to the uranium cycle except that it works best in a thermal breeder reactor, where it relies on thermal, or relatively slow, neutrons. Thorium 232 is the fertile isotope that becomes protactinium and then uranium 233.

for the flow of sodium the fuel is divided into thousands of slender vertical rods, which are usually referred to as pins. Each pin is sealed in stainless steel or another high-temperature alloy.

The fuel is preferably in a ceramic form such as oxide or carbide, since these ceramics are stable during long exposures to heat and radiation, have very high melting points and are relatively inert in liquid metal. The fissionable component of the fuel can be enriched uranium 235, plutonium 239 or a mixture of the two. Typically the fuel is diluted with uranium 238, so that part of the breeding takes place within the core. The uranium 238 also serves a safety function in the core, which we shall explain in more detail below. For maximum economy and performance the fuel must be able to accept neutron irradiation at many times the rate common in present nuclear reactors of commercial scale. Furthermore, the consumption of fuel between reprocessing steps is to be at least twice that of thermal reactors. The development of a fuel meeting these stringent criteria requires the testing of numerous fuel combinations in reactors and accounts for a major element of the breeder development program.

A second major feature of a fast breeder reactor is the "blanket" that surrounds the core. Much of the breeding takes place here, and so the blanket consists of uranium 238 in stainless-steel tubes. (It can be uranium that is depleted in the isotope uranium 235 as a result of enrichment procedures designed to make uranium 235 as a fuel for nuclear reactors; large stocks of such depleted uranium are now available.) Since there is a certain amount of fission in the blanket, it too must be cooled by the flow of sodium. The blanket also has an important nuclear function. Not all the neutrons entering the blanket are captured; a fairly large proportion are reflected back into the reactor core, enhancing the neutron economy there.

The sodium coolant has excellent heat-transfer characteristics. Moreover, it can be used at a fairly low pressure even though it emerges from the reactor at a temperature (above 500 degrees Celsius) that with water would give rise to high pressures. Indeed, the sodium pressure arises solely from the force required to maintain the high rate of flow through the maze of tubes in the core and the blanket. Compared with coolants such as water and gas, sodium requires low pumping power. It is not particularly corrosive to the reactor.

Sodium does, however, have certain

CORE AND BLANKET of a fast breeder reactor, Experimental Breeder Reactor II, are the heart of the breeding operation. The dark hexagonal area is the core, where fuel elements can be installed and removed by the gripper mechanism at right center. Rods clustered at left center are connected to control rods in the core. Around the core is the blanket, consisting of rods containing uranium 238, which is converted to plutonium during the breeding.

EXPERIMENTAL BREEDER REACTOR II is operated by the U.S. Atomic Energy Commission at the National Reactor Testing Station in Idaho. The primary components of the nuclear reactor are under the floor in a tank that is 26 feet high and 26 feet in diameter and contains liquid sodium, which is the coolant. Vertical assembly at right center contains mechanisms for operating the control rods and for handling fuel elements within the reactor. The reactor is used to test fuels and materials for breeder reactors of commercial scale.

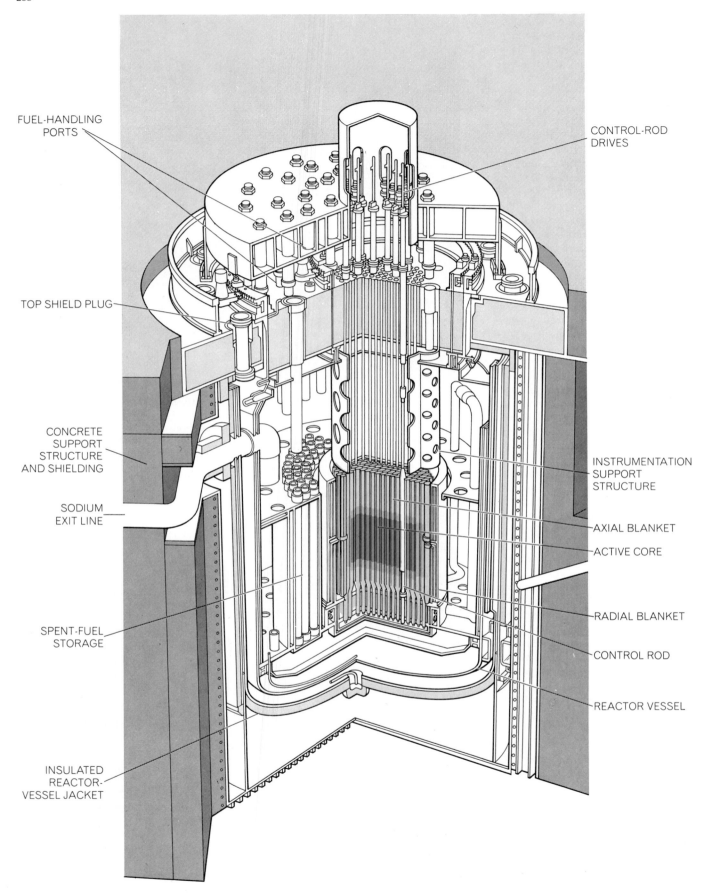

FUEL-HANDLING PORTS

CONTROL-ROD DRIVES

TOP SHIELD PLUG

CONCRETE SUPPORT STRUCTURE AND SHIELDING

SODIUM EXIT LINE

INSTRUMENTATION SUPPORT STRUCTURE

AXIAL BLANKET

ACTIVE CORE

SPENT-FUEL STORAGE

RADIAL BLANKET

CONTROL ROD

REACTOR VESSEL

INSULATED REACTOR-VESSEL JACKET

LIQUID-METAL REACTOR of the fast-breeder type is depicted on the basis of a design for a demonstration plant that would produce some 500 megawatts of electricity. A full-scale commercial plant, scheduled for operation by 1984, would be of 1,000-mega- watt capacity. This design is of the loop type, meaning that the re- actor proper, which is contained in a large tank of liquid sodium, is separated from the primary heat exchangers and the associated pumps by loops of piping through which sodium coolant flows.

disadvantages that markedly influence the design of a reactor. Since sodium is opaque, provision must be made for the maintenance and refueling of the reactor without benefit of visual observation. Sodium is of course highly reactive chemically, and it becomes intensely radioactive when it is exposed to neutrons, even though its "cross section," or neutron-absorption capacity, is relatively low. Hence the sodium must be kept out of contact with air or water, and radiation shielding must be used to protect workers who are near sodium that has been through the core and blanket of an operating reactor.

Interspersed through the core region are numerous rods with safety and control functions. They maintain the power output at the desired level and provide the means for starting and stopping the reactor. The rods are filled with neutron-absorbing material such as boron carbide or tantalum metal.

All materials have markedly lower neutron-absorption probabilities for fast neutrons than for thermal ones. The lower cross sections reduce the effectiveness of fast-reactor control rods of sizes comparable to those in thermal reactors. On the other hand, a large amount of excess fuel is present in the core of a thermal reactor to compensate for the fuel that will be consumed by fission and to counteract the poisoning effect of the fission products. (The fission products capture neutrons without yielding significant amounts of energy.) With extra fuel there must be extra controls. Fast breeder reactors require fewer control rods because their greater effectiveness in converting uranium 238 to fissionable plutonium 239 compensates for depletion of the initial fuel charge and because fast neutrons are not absorbed by fission products as much as thermal neutrons are.

During a fission reaction not all the neutrons are released at the precise instant that each nucleus disintegrates. A small proportion of the neutron population is created by the decay of fission products. One thus distinguishes delayed neutrons from the "prompt" neutrons emitted directly by the fissioning nuclei. It is the delayed neutrons that keep the chain reaction from escalating into an essentially instantaneous propagation of one generation of neutrons to the next. The fraction of delayed neutrons depends appreciably on which nucleus is fissioning. Most thermal reactors are fueled with uranium 235, whereas the fast breeder will be fueled with plutoni-

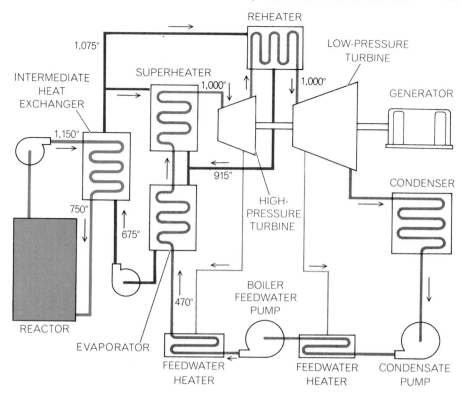

FLOW PLAN for a liquid-metal fast breeder reactor entails pumping the sodium coolant (*color*) through the reactor, where it becomes radioactive, and then to an intermediate heat exchanger, where heat is transferred to a separate stream of sodium (*dark gray*) that is not radioactive. The heat of that stream is put into a water and steam cycle (*light gray*) that is employed to generate electricity. The numerals give temperatures in degrees Fahrenheit.

um 239. The fraction of delayed neutrons produced by the fission of uranium 235 is about .0065 and by plutonium 239 fission about .003. The smaller fraction of delayed neutrons present in a fast reactor is not of major concern under normal operation. It does increase the sensitivity of the reactor to adjustments of the control rods and to other inputs that affect reactivity, such as temperature variations in the core.

Two different designs of containers for the core-and-blanket assembly and the primary heat-transfer system are under consideration: the pot type and the loop type [*see illustrations on next page*]. In the pot type a large tank filled with sodium encloses (1) the reactor vessel, (2) sodium pumps that take sodium from the pool and move it through the core and blanket and (3) intermediate heat exchangers that transfer heat from the radioactive sodium to another sodium stream. In the loop type only the reactor vessel is filled with sodium; the liquid metal is circulated by pumps through heat-exchange loops mounted outside the reactor container. The pot type has the advantage of a much greater heat capacity in the event of pump failure, but it also requires a much larger inventory of sodium.

In both the pot and the loop design the liquid-metal fast breeder reactor employs a complex heat-transfer arrangement to isolate the sodium that flows through the core from the steam-generating equipment. This is the role of the intermediate heat exchangers. They transfer heat from the radioactive sodium to nonradioactive sodium, which then flows through the steam generator. Subsidiary streams of sodium are required to superheat the steam and to reheat it from time to time as it works against the blades of the turbine.

Both the pot and the loop design require sealing of the part of the structure that is in direct contact with the radioactive core and blanket. In routine operation there would be no release of radioactive fission products to the environment. Because of the inherently low pressure of the sodium coolant, the reactor vessel and its associated piping need be designed to withstand only moderate operating stresses, in marked distinction to the pressure vessels and other primary-system components of a pressurized-water reactor, a boiling-water reactor or a gas-cooled reactor.

At present the pot design seems to be attracting the most interest. It is inherently a less complicated arrangement than the loop design. Nonetheless, it

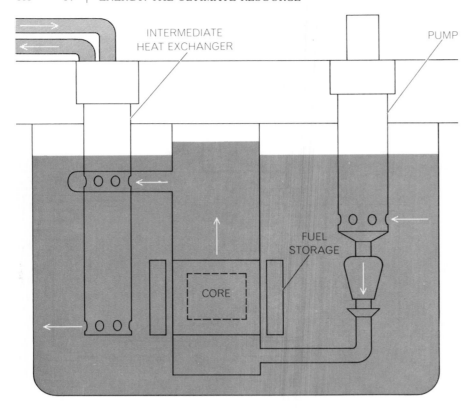

POT SYSTEM is one of two designs for containing the core-and-blanket assembly of the reactor and the primary heat-transfer system. The pot is a tank that is filled with sodium and also contains the reactor, pumps that take sodium from the pool and move it through the reactor, and intermediate heat exchanger where heat is transferred to nonradioactive sodium.

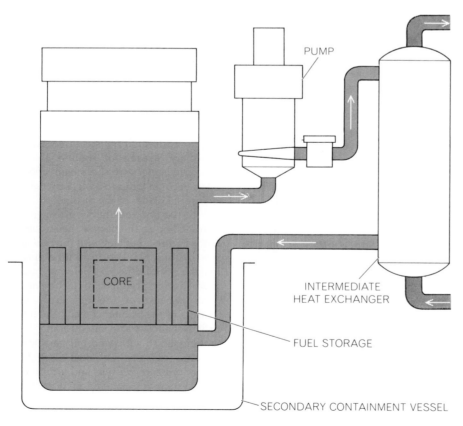

LOOP SYSTEM has most of its heat-exchange apparatus outside the reactor. Only the reactor vessel is filled with sodium, which is circulated by pumps through heat-exchange loops mounted outside the reactor vessel. In the present state of breeder-reactor technology both of the designs in the schematic illustrations on this page are being pursued.

does present certain problems, notably in gaining access to the reactor for maintenance.

The gas-cooled fast breeder reactor is receiving attention (comparatively modest so far) as a parallel and complementary concept to the liquid-metal fast breeder. Gas-cooled thermal reactors are already in operation, and a gas-cooled fast breeder would not represent a large step in terms of coolant technology. The design and testing of the fuel for a gas-cooled fast breeder have much in common with the work on fuel for the liquid-metal fast breeder.

The essential difference between the two fast breeders is that the gas-cooled one uses helium gas at a pressure of from 70 to 100 atmospheres rather than molten sodium to transport the heat from the reactor core to the steam generators. Since the gas does not become radioactive and cannot react chemically with the water in the steam generator, there is no need for an intermediate heat exchanger. The resulting simplification of the system is a helpful offset against the need to design for a higher coolant pressure with gas.

The use of helium as a coolant has other special advantages for a fast breeder reactor. Helium does not interact with the fast neutrons in the reactor core, resulting in both simplified control of the reactor and enhanced breeding of new fissionable fuel from fertile material. In addition helium is transparent and chemically inert, providing visibility during refueling and maintenance operations, a simpler engineering design and freedom from corrosion problems.

In a gas-cooled fast breeder the reactor core, helium circulators and steam generators are all contained in a prestressed-concrete reactor pressure vessel. These major components and their arrangement are almost the same as in a thermal gas-cooled reactor.

The development of a gas-cooled fast breeder reactor could result in substantial additional savings beyond those that would be achieved by liquid-metal fast breeders. Neutrons are moderated, or slowed, less in helium than they are in sodium. Hence the doubling time is short. It is also possible to foresee the development of a gas-cooled fast breeder with a direct power cycle wherein the gas coolant flows from the reactor directly to a gas turbine that drives the electrical generator. Such a cycle should help to reduce the capital cost of fast breeder reactors.

Three major reactors will carry the burden of the Atomic Energy Commission's program to develop a liquid-metal fast breeder reactor. Two of them are already in operation: the Experimental Breeder Reactor II (EBR-II) and the Zero Power Plutonium Reactor (ZPPR).

EBR-II is a fast-neutron test reactor operated by the Argonne National Laboratory at the commission's National Reactor Testing Station in Idaho. This reactor, which as of July 1 had a cumulative record of more than 35,000 megawatt-days of operation, is the focal point of the program of testing fuels and irradiating materials for the liquid-metal fast breeder reactor. At present almost 800 experimental fuel pins and more than 100 capsules containing hundreds of structural, control-rod and shield-material specimens are being irradiated in the reactor. EBR-II achieved its design power of 62.5 megawatts (thermal) last year.

The Argonne National Laboratory is also operating the ZPPR, which went into operation last year. (Zero power in this context means that the reactor does not generate a significant amount of heat.) It is the nation's largest zero-power fast reactor and the only one in the world that is big enough and has a large enough inventory of plutonium (at least 3,000 kilograms) to allow full-scale mock-ups of the plutonium fuel arrangements that will be used in the large commercial breeders envisioned for the 1980's and beyond. The reactor will provide important information on the behavior of neutrons in breeder-reactor cores.

The third reactor is now being designed on the basis of data obtained from EBR-II, the ZPPR and smaller facilities. Called the Fast Flux Test Facility, it will operate at a very high neutron flux (defined as the number of neutrons passing through a square centimeter of area per second) to produce the radiation effects on fuel and structural materials that will take place in a commercial breeder reactor. The reactor, which will cost about $100 million, will operate at a power level of 400 megawatts (thermal) with no conversion to electric power. It will be built at the Atomic Energy Commission's site in Richland, Wash.; construction should start next year and full

power should be achieved by the middle of the decade.

Following the lessons learned in the development of thermal reactors, the commission has taken the first steps toward construction of one or more liquid-metal fast-breeder demonstration plants. The cost will be shared by the Government and industry. The first such plant, with a capacity of from 300 to 500 megawatts (electric), will accumulate valuable operating experience with both the reactor and the power-conversion equipment. Such a plant will not compete economically with existing nuclear or conventional plants because of its relatively small size and early stage of development. The full-scale liquid-metal fast breeder reactor of the 1980's will be rated at 1,000 megawatts (electric) or more.

Much consideration is being given to safety in the fast breeder development program. The waste products of fission are the elements in the middle of the periodic table that represent the split nuclei of the fuel atoms. Many isotopes

| | NAME | COUNTRY | POWER | | INITIAL OPERATION | TYPE (POT OR LOOP) |
			MEGAWATTS (THERMAL)	MEGAWATTS (ELECTRICAL)		
OPERATING	BR-5	U.S.S.R.	5	—	1959	LOOP
	DFR	U.K.	60	15	1959	LOOP
	EBR-II	U.S.	62.5	20	1964	POT
	FERMI	U.S.	200	66	1963	LOOP
	RAPSODIE	FRANCE	40	—	1967	LOOP
	SEFOR	U.S.	20	—	1969	LOOP
	BOR-60	U.S.S.R.	60	12	1970	LOOP
UNDER CONSTRUCTION	BN-350	U.S.S.R.	1,000	150	1971	LOOP
	PFR	U.K.	600	250	1972	POT
	PHENIX	FRANCE	600	250	1973	POT
	BN-600	U.S.S.R.	1,500	600	1973/75	POT
	FFTF	U.S.	400	—	1974	LOOP
PLANNED	KNK-II	WEST GERMANY	58	20	1972	LOOP
	JEFR	JAPAN	100	—	1973	LOOP
	PEC	ITALY	140	—	1975	MODIFIED POT
	SNR	WEST GERMANY	730	300	1975	LOOP
	DEMO #1	U.S.	750-1,250	300-500	1976	NOT DECIDED
	JPFR	JAPAN	750	300	1976	LOOP
DECOMMISSIONED	CLEMENTINE	U.S.	.025	—	1946	LOOP
	EBR-I	U.S.	1	.2	1951	LOOP
	BR-2	U.S.S.R.	.1	—	1956	LOOP
	LAMPRE-I	U.S.	1	—	1961	LOOP

LIQUID-METAL FAST REACTORS built or planned are summarized. Those that produce electricity have far less capacity than the 1,000-megawatt commercial fast-breeder plant that the development program of the U.S. seeks to have in operation by 1984.

of these elements are radioactive. The permanent control of the fission products has been recognized as essential from the early stages of reactor development and is routinely achieved in the fuel cycle. In addition to making fission products, fast breeders will also contain large amounts of plutonium, which in certain forms is radiologically toxic. The standard procedure in both thermal and fast reactors is to ensure the confinement of all potentially hazardous substances under all foreseeable conditions, including earthquakes.

Perhaps the most significant safety feature of commercial nuclear reactors is that they are self-regulating, that is, they are designed to compensate inherently for any incident that could lead to an un-

intentional increase of power output. In water reactors the compensation is usually achieved through the decrease in reactivity caused by the decrease in the density of water as its temperature increases. In a fast reactor the change in density of the coolant with temperature may lead in the opposite direction.

Compensation is provided in a fast breeder by the Doppler effect, which results from the increase in the rate at which neutrons are absorbed by uranium 238 as the temperature of the fuel in the core rises. Since a sudden power increase will necessarily be accompanied by increased fuel temperature, there will be increased neutron absorption and a consequent tendency toward reduction of power. A small sodium-cooled fast reac-

tor has been built in Arkansas with private funds to measure this effect under conditions analogous to those in a large power reactor. It is called the Southwest Experimental Fast Oxide Reactor.

The fact that a decrease in coolant density or coolant content can result in an increase in reactivity leads to other safety considerations for sodium-cooled fast reactors. For example, if one postulates a bubble of gas or another void whereby an area of the core might overheat without detection, some of the fuel pins in the area would be expected to fail. Further disturbances of flow might ensue. A continued sequence of such events would not necessarily result in an automatic shutdown of the reactor. Thus

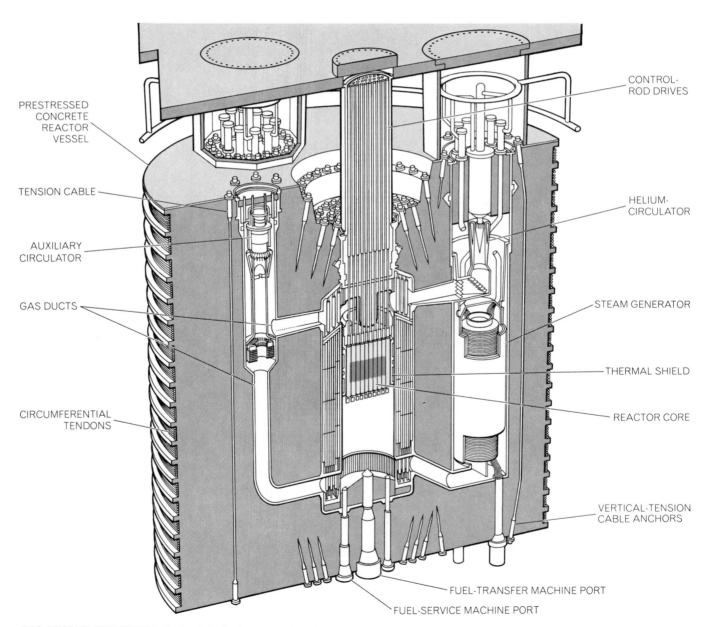

GAS-COOLED REACTOR is depicted in the form it might take for a demonstration breeder plant with a capacity of 300 megawatts of electric power. The chief difference between such a reactor and a liquid-metal one is that the coolant here is helium at high pressure instead of liquid sodium at low pressure. Because of the pressure the reactor is contained within a prestressed-concrete vessel.

it is necessary to preclude by design the propagation of fuel-pin failures.

This control can be achieved by a number of techniques. The addition of a moderator such as beryllium oxide increases the magnitude of the neutron Doppler effect. A change in the ratio of coolant to fuel can reduce the void effect. Other methods include distributing the fuel in such a way that the potential reactivity of coolant voids is decreased by increasing the number of neutrons leaking from the core.

In a gas-cooled fast reactor there is no problem with voids because a bubble cannot form in gas. There is, however, another condition to be guarded against: a sudden loss of coolant pressure resulting from an event such as rupture of the pressure vessel. This possibility is minimized by the use of the prestressed-concrete type of reactor vessel.

Having made sure that perturbations in normal operating conditions do not escalate, one looks to the possibility of other problems. One is a loss of cooling by mechanical blockage. Such accidents have occurred, but they will become less likely as engineering experience is gained. In this type of accident any significant release of fission products is precluded by providing several layers of structural containment for the entire reactor system.

Another possibility is an increase in power to the point where heat is being generated faster than it can be carried away by the coolant. Such an accident took place in EBR-I some years ago. Here again the answer has been found in improved design. Even beyond this, structural containment sufficient for any foreseeable accident will be provided.

Much consideration is also being given to environmental factors in the design of fast breeder reactors. Because fast breeder reactors will operate at far higher temperatures than are encountered in contemporary water reactors, they will have greater thermodynamic efficiency. Today's water reactors operate at an overall efficiency of about 32 percent, meaning that 32 percent of the thermal energy produced is converted to electrical energy. Modern fossil-fueled plants operate at about 39 percent efficiency. Hence water reactors add more waste heat to the environment per unit of electrical energy produced than fossil-fueled plants do. Fast breeder reactors will probably attain efficiencies equal to that of the most modern fossil-fueled plant, thereby reducing the nuclear waste-heat problem.

The release of radioactivity to the air from fast breeders will be near zero.

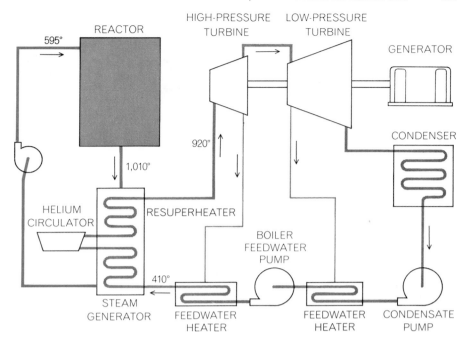

PATTERN OF FLOW in a gas-cooled fast breeder reactor entails the transfer of the heat from the helium coolant (*color*) to a water-steam cycle (*gray*). The system operates without an intermediate heat-exchange cycle. Numerals give temperatures in degrees Fahrenheit.

Even the small amounts of radioactive fission-product gases (primarily krypton 85 and tritium) now released under controlled conditions from water-moderated reactors will be eliminated because the necessity of hermetically sealing the core area will give an inherently effective method for collecting and disposing of the gases, which can then be rendered harmless. Moreover, since the coolant in a fast breeder is kept in a closed system, and since the water used to generate steam is never exposed to neutrons, there should be no formation of radioactivity in aqueous effluents from the plant.

The economic potential of fast breeder reactors lies mainly, but not entirely, in the fact that they would conserve resources of nuclear fuel. Over the next 50 years the use of breeders as planned can be expected to reduce by 1.2 million tons the amount of uranium that would be consumed without breeders. That is the energy equivalent of about three billion tons of coal.

The development of a breeder economy also appears to offer a direct dollar gain of large proportions. Studies have indicated that the cost of research and development of the liquid-metal fast breeder will be more than $2 billion through the year 2020 for the Atomic Energy Commission alone, with large industrial expenditures added. If the first commercial breeder is introduced by 1984 as planned, however, reductions in the cost of electrical energy thereafter

(to 2020) are estimated at $200 billion in 1970 dollars.

The present cost of producing electricity in the U.S. ranges from five to 10 mills per kilowatt-hour delivered to the transmission system, depending on the type, age and location of the plant. This range covers most plants, although there are a few outside of either extreme. The liquid-metal fast breeder reactor is predicted to produce power at a saving of from .5 to one mill per kilowatt-hour. Large breeder-reactor systems that eventually bring the cost of electricity down by as much as two mills per kilowatt-hour will make it possible to extract, use and reuse resources in ways that cannot be afforded today. It will be possible to tap substantial resources in the oceans and on land and to use land not now habitable or productive.

Indeed, we believe breeders will result in a transition to the massive use of nuclear energy in a new economic and technological framework. The transition may be slow, and it will require the introduction of a series of innovations in the technologies of industry, agriculture and transportation. The innovations will include large-scale, dual-purpose desalting plants; electromechanization of farms and of means of transportation; electrification of the metal and chemical industries, and more effective means for utilizing wastes. The key to these possibilities is abundant low-cost electrical energy, and the route to that is by way of the breeder reactor.

25

The Prospects of Fusion Power

by William C. Gough and Bernard J. Eastlund
February 1971

*Recent advances in the performance of several
experimental plasma containers have brought the
fusion-power option very close to the "break even"
level of scientific feasibility*

The achievement of a practical fusion-power reactor would have a profound impact on almost every aspect of human society. In the past few years considerable progress has been made toward that goal. Perhaps the most revealing indication of the significance of this progress is the extent to which the emphasis in recent discussions and meetings involving workers in the field has tended to shift from the question of purely scientific feasibility to a consideration of the technological, economic and social aspects of the power-generation problem. The purpose of this article is to examine the probable effects of the recent advances on the immediate and long-term prospects of the fusion-power program, with particular reference to mankind's future energy needs.

The Role of Energy

The role of energy in determining the economic well-being of a society is often inadequately understood. In terms of *total* energy the main energy source for any society is the sun, which through the cycle of photosynthesis produces the food that is the basic fuel for sustaining the population of that society. The efficiency with which the sun's energy can be put to use, however, is determined by a feedback loop in which auxiliary energy sources form a critical link [*see illustration, page 306*]. The auxiliary energy (derived mainly from fossil fuels, water power and nuclear-fission fuels) "opens the gate" to the efficient use of the sun's energy by helping to produce fertilizers, pesticides, improved seeds, farm machinery and so on. The result is that the food yield (in terms of energy content) produced per unit area of land in a year goes up by orders of magnitude. This auxiliary energy input, when it is transformed into food energy, enables large populations to live in cities and develop new ways to multiply the efficiency of the feedback loop. If a society is to raise its standard of living by increasing the efficiency of its agricultural feedback loop, clearly it must expand its auxiliary energy sources.

The dilemma here is that the economically less developed countries of the world cannot *all* industrialize on the model of the more developed countries, for the simple reason that the latter countries, which contain only a small fraction of the world's population, currently maintain their high standard of living by consuming a disproportionately large share of the world's available supply of auxiliary energy. Just as there is a direct, almost linear, relation between a nation's use of auxiliary energy and its standard of living, so also there is a similar relation between energy consumption and the amount of raw material the nation uses and the amount of waste material it produces. Thus the more developed countries consume a correspondingly oversized share of the world's reserves of material resources and also account for most of the world's environmental pollution.

In order to achieve a more equitable and stable balance between the standards of living in the more developed countries and those in the less developed countries, only two alternatives exist. The more developed countries could reduce their consumption of auxiliary energy (thereby lowering their standard of living as well) or they could contribute to the development of new, vastly greater sources of auxiliary energy in order to help meet the rising demands for a better standard of living on the part of the rapidly growing populations of the less developed countries.

When one projects the world's long-term energy requirements against this background, another important factor must be taken into account. There are finite limits to the world's reserves of material resources and to the ability of the earth's ecological system to absorb pollutants safely. As a consequence future societies will be forced to develop "looped," or "circular," materials economies to replace their present, inherently wasteful "linear" materials economies [*see bottom illustration, page 315*]. In such a "stationary state" system, limits on the materials inventory, and hence on the total wealth of the society, would be set by nature. Within these limits, however, the standard of living of the population would be higher if the rate of flow of materials were lower. This maximizing of the life expectancy of the materials inventory could be accomplished in two ways: increasing the durability of individual commodities and developing the technological means to recycle the limited supply of material resources.

The conclusion appears radical. Future societies must *minimize* their physical flow of production and consumption. Since a society's gross national product for the most part measures the flow of physical things, it too would be reduced.

But all nations now try to *maximize* their gross national product, and hence their rate of flow of materials! The explanation of this paradox is that in the existing linear economies the inputs for increasing production must come from the environment, which leads to depletion, while an almost equal amount of materials in the form of waste must be returned to the environment, which leads to pollution. This primary cause of pollution is augmented by the pollution that is produced by the energy sources used to drive the system.

In order to make the transition to a stationary-state world economy, the wealthier nations will have to develop the technology—and the concomitant auxiliary energy sources—necessary to operate a closed materials economy. This capability could then be transferred to the poorer nations so that they could develop to the level of the wealthier nations without exhausting the world's supply of resources or destroying the environment. Thus some of the causes of international conflict would be removed, thereby reducing the danger of nuclear war.

Of course any effort to bring about a rapid change from linear economies to looped economies will encounter the massive economic, social and political forces that sustain the present system. The question of how to distribute the stock of wealth, including leisure, within a stationary-state economy will remain. In summary, the world's requirements for energy are intimately related to the issues of population expansion, economic development, materials depletion, pollution, war and the organization of human societies.

The Energy Options

What are the available energy options for the future? To begin with there are the known finite and irreplaceable energy sources: the fossil fuels and the better-grade, or easily fissionable, nuclear fuels such as uranium 235. Estimates of the life expectancy of these sources vary, but it is generally agreed that they are being used up at a rapid rate—a rate that will moreover be accelerated by increases in both population and living standards. In addition, environmental considerations could further restrict the use of these energy sources.

Certain other known energy sources, such as water power, tidal power, geothermal power and wind power are "infinite" in the sense of being continuously replenished. The total useful *amount* of energy available from these sources, however, is insufficient to meet the needs of the future.

Direct solar radiation, resulting from the fusion reactions that take place in the core of the sun, is an abundant as well as effectively "infinite" energy source. The immediate practical obstacle to the direct use of the sun's energy as an effective auxiliary energy source is the necessity of finding some way to economically concentrate the available low energy density of solar radiation. Controlled fusion is another potentially "infinite" energy source; its energy output arises from the reduction of the total mass of a nuclear system that accompanies the merger of two light

U.S. TOKAMAK, a toroidal plasma-confinement machine used in fusion research, was recently put into operation at the Plasma Physics Laboratory of Princeton University. Until about a year ago this machine, formerly known as the Model-C stellarator and now called the Model ST tokamak, had been the largest of the stellarator class of experimental plasma containers developed primarily at the Princeton laboratory. The decision to convert it to the closely related tokamak design followed the 1969 announcement by the Russian fusion-research group of some important new results obtained from their Model T-3 machine, the most advanced of the tokamak class of plasma containers developed mainly at the I. V. Kurchatov Institute of Atomic Energy near Moscow. In large part because of the cooperative nature of the world fusion-research program, this conversion was accomplished quickly and the Model ST has already produced results comparable to those obtained by the Russians. Several other tokamak-type machines are being built in this country.

RUSSIAN STELLARATOR is now the largest representative of this class of experimental plasma containers in the world. It is named the Uragan (or "hurricane") stellarator and is located at the Physico-Technical Institute of the Academy of Sciences of the Ukrainian S.S.R. at Kharkov. In both photographs on this page the large circular structures surrounding and almost completely obscuring the toroidal plasma chambers are the primary magnet coils. The main difference between the tokamak design and the stellarator design is that in a tokamak a secondary plasma-stabilizing magnetic field is generated by an electric current flowing axially through the plasma itself, whereas in a stellarator this secondary magnetic field is set up by external helical coils situated just inside the primary coils and hence not visible.

nuclei. The most likely fuel for a fusion-power energy source is deuterium, an abundant heavy isotope of hydrogen easily separated from seawater.

In addition to these two primary "infinite" energy sources, secondary "infinite" energy sources could be made by using neutrons to transmute less useful elements into other elements capable of being used effectively as fuels. Thus for fission systems the vast reserves of uranium 238 could be converted by neutron bombardment into easily fissionable plutonium 239; similarly, thorium 232 could be converted into uranium 233. For fusion systems lithium could be converted into tritium, another heavy isotope of hydrogen with a comparatively low resistance to entering a fusion reaction and a comparatively high energy output once it does.

The prime hope for extending the world's reserves of nuclear-fission fuels is the development of the neutron-rich fast breeder fission reactors [see the article "Fast Breeder Reactors," by Glenn T. Seaborg and Justin L. Bloom, beginning on page 295]. Another potential source of abundant, inexpensive neutrons is a fusion-fission hybrid system, an alternative

that will be discussed further below.

Fusion Energy

Nuclear fusion, the basic energy process of the stars, was first reproduced on the earth in 1932 in an experiment involving the collision of artificially accelerated deuterium nuclei. Although it was thereby shown that fusion energy could be released in this way, the use of particle accelerators to provide the nuclei with enough energy to overcome their Coulomb, or mutually repulsive, forces was never considered seriously as a practical method for power generation. The reason is that the large majority of the nuclei that collide in an accelerator scatter without reacting; thus it is impossible to produce more energy than was used to accelerate the nuclei in the first place.

The uncontrolled release of a massive amount of fusion energy was achieved in 1952 with the first thermonuclear test explosion. This test proved that fusion energy could be released on a large scale by raising the temperature of a high-density gas of charged particles (a plas-

ma) to about 50 million degrees Celsius, thereby increasing the probability that fusion reactions will take place within the gas.

Coincident with the development of the hydrogen bomb, the search for a more controlled means of releasing fusion energy was begun independently in the U.S., Britain and the U.S.S.R. Essentially this search involves looking for a practical way to maintain a comparatively low-density plasma at a temperature high enough so that the output of fusion energy derived from the plasma is greater than the input of some other kind of energy supplied to the plasma. Since no solid material can exist at the temperature range required for a useful energy output (on the order of 100 million degrees C.) the principal emphasis from the beginning has been on the use of magnetic fields to confine the plasma.

The variety of magnetic "bottles" designed for this purpose over the years can be arranged in several broad categories in order of increasing plasma density [see illustration, pages 310 and 311]. First there are the basic plasma devices. These are low-density, low-temperature systems used primarily to study the fundamental properties of plasmas. Their configuration can be either linear (open) or toroidal (closed). Linear basic-plasma devices include simple glow-discharge systems (similar in operation to ordinary fluorescent lamps) and the more sophisticated "Q-machines" ("Q" for "quiescent") found in many university plasma-physics laboratories. Toroidal representatives of this class include the "multipole" devices, developed primarily at Gulf Energy & Environmental Systems, Inc. (formerly Gulf General Atomic Inc.) and the University of Wisconsin, and the spherator, developed at the Plasma Physics Laboratory of Princeton University.

Next there are the medium-density plasma containers; these are defined as systems in which the outward pressure of the plasma is much less than the inward pressure of the magnetic field. A typical configuration in this density range is the linear magnetic bottle, which is usually "stoppered" at the ends by magnetic "mirrors": regions of somewhat greater magnetic-field strength that reflect escaping particles back into the bottle. In addition extra current-carrying structures are often used to improve the stability of the plasma. These structures were originally proposed on theoretical grounds in 1955 by Harold Grad of New York University. They were first used successfully in an experimental test in 1962 by the Russian physicist M. S. Ioffe.

ROLE OF AUXILIARY ENERGY in determining the economic well-being of a society is illustrated by these two diagrams of agricultural feedback loops. In an economically less developed country (top) the bulk of the population must be devoted to the agricultural transformation of the sun's energy into food in order to support itself at a subsistence level. In an economically more developed industrial country (bottom) auxiliary energy sources "open the gate" to the more efficient utilization of the sun's energy, making it possible for the entire population to maintain a higher standard of living and freeing many people to live in cities and develop new ways to multiply the efficiency of the feedback loop.

		LIFE EXPECTANCY OF KNOWN RESERVES (YEARS)		LIFE EXPECTANCY OF POTENTIAL RESERVES (YEARS)		LIFE EXPECTANCY OF TOTAL RESERVES (YEARS)	
		AT .17Q	AT 2.8Q	AT .17Q	AT 2.8Q	AT .17Q	AT 2.8Q
FINITE ENERGY SOURCES	FOSSIL FUELS (COAL, OIL, GAS)	132	8	2,700	165	2,832	173
	MORE ACCESSIBLE FISSION FUELS (URANIUM AT $5 TO $30 PER POUND OF U_3O_8 BURNED AT 1.5 PERCENT EFFICIENCY)	66	4	66	4	132	8
	LESS ACCESSIBLE FISSION FUELS (URANIUM AT $30 TO $500 PER POUND OF U_3O_8 BURNED AT 1.5 PERCENT EFFICIENCY)	43,000	2,600	129,000	7,800	172,000	10,400
"INFINITE" NATURAL ENERGY SOURCES	WATER POWER, TIDAL POWER, GEOTHERMAL POWER, WIND POWER	INSUFFICIENT		INSUFFICIENT		INSUFFICIENT	
	SOLAR RADIATION	10 BILLION	10 BILLION			10 BILLION	10 BILLION
	FUSION FUELS (DEUTERIUM FROM OCEAN)	45 BILLION	2.7 BILLION			45 BILLION	2.7 BILLION
"INFINITE" ARTIFICIAL ENERGY SOURCES (ELEMENTS TRANSMUTED FROM OTHER ELEMENTS BY NEUTRON BOMBARDMENT)	FISSION FUELS (PLUTONIUM 239 FROM URANIUM 238; URANIUM 233 FROM THORIUM 232)	8.8 MILLION	536,000	21 MILLION	1.3 MILLION	30 MILLION	1.8 MILLION
	FUSION FUELS (TRITIUM FROM LITHIUM) a) ON LAND b) IN OCEAN	48,000 / 120 MILLION	2,900 / 7.3 MILLION	UNKNOWN	UNKNOWN	48,000+ / 120 MILLION	2,900+ / 7.3 MILLION

WORLD ENERGY RESERVES are listed in this table in terms of their life expectancy estimated on the basis of two extreme assumptions, which were chosen so as to bracket a reasonable range of values. First, the assumption was made that the world population would remain constant at its 1968 level of 3.5 billion persons and that the energy-consumption rate of this population would remain constant at the estimated 1968 rate of .17 Q (Q is a unit of heat measurement equal to 10^{18} BTU, or British Thermal Units). Second, the assumption was made that the world population would eventually reach seven billion and that this population would consume energy at a per capita rate of 400 million BTU per year (about 20 percent higher than the present U.S. rate), giving a total world energy-consumption rate of 2.8 Q per year. (A commonly projected world energy-consumption rate for the year 2000 is one Q.) Current fission-converter reactors use only between 1 and 2 percent of the uranium's potential energy content, since the component of the ore that is burned as fuel is primarily high-grade, or easily fissionable, uranium 235. The world fission-fuel reserves were derived by multiplying the U.S. reserves times the ratio of world land area to the U.S. land area (approximately 16.2 to one). For fusion-converter reactors lithium-utilization studies show that natural lithium, a mixture of lithium 6 and lithium 7, would be superior to pure lithium 6 in a tritium-breeding reactor "blanket" and would yield an energy output of about 86.4 million BTU per gram. The figure for known world lithium reserves is based on a study carried out last year by James J. Norton of the U.S. Geological Survey. The potential reserves of lithium are unknown, since there has been no exploration program comparable to that undertaken for, say, uranium. Lithium, however, is between five and 15 times more abundant in the earth's crust than uranium. Finally, the life expectancy of the earth—and hence that of potentially useful solar radiation—is predicted to be at most 10 billion years.

The straight rods used by Ioffe in his experiment have come to be called Ioffe bars, but such stabilizing structures can assume various other shapes. For example, in one series of medium-density linear devices they resemble the seam of a baseball; accordingly these devices, developed at the Lawrence Radiation Laboratory of the University of California at Livermore, are named Baseball I and Baseball II.

Medium-density plasma containers with a toroidal geometry include the stellarators, originally developed at the Princeton Plasma Physics Laboratory, and the tokamaks, originally developed at the I. V. Kurchatov Institute of Atomic Energy near Moscow. The only essential difference between these two machines is that in a stellarator a secondary, plasma-stabilizing magnetic field is set up by external helical coils, whereas in a tokamak this field is generated by an electric current flowing through the plasma itself. The close similarity between these two designs was emphasized recently by the fact that the Princeton Model-C stellarator was rather quickly converted to a tokamak system following the recent announcement by the Russians of some important new results from their Tokamak-3 machine.

The astron concept, also developed at the Lawrence Radiation Laboratory at Livermore, is another example of a medium-density plasma container. In overall geometry it shares some characteristics of both the linear and the toroidal designs.

Higher-density plasma containers, defined as those in which the plasma pressure is comparable to the magnetic-field pressure, have also been built in both the linear and the toroidal forms. In one such class of devices, called the "theta pinch" machines, the electric current is in the theta, or azimuthal, direction (around the axis) and the resulting magnetic field is in the zeta, or axial, direction (along the axis). The Scylla and Scyllac machines at the Los Alamos Scientific Laboratory are respectively examples of a linear theta-pinch design and a toroidal theta-pinch design.

As the density of the plasma is increased further, one reaches a technological limit imposed by the inability of the materials used in the magnet coils to withstand the pressure of the magnetic field. Consequently very-high-density plasma systems are often fast-pulsed and obtain their principal confining forces from "self-generated" magnetic fields (fields set up by electric currents in the plasma itself), from electrostatic fields or from inertial pressures. In this very-high-density category are the "zeta pinch" machines, devices in which the electric current is in the zeta direction and the resulting magnetic field is in the theta direction. An example of this type of configuration is the Columba device at Los Alamos.

Other very-high-density, fast-pulsed systems include the "strong focus" designs, in which a stream of plasma in a cylindrical, coaxial pipe is heated rapidly by shock waves as it is brought to a sharp focus by self-generated magnetic forces, and laser designs, in which a pellet of fuel is ionized instantaneously by a pulse from a high-power laser, producing an "inertially confined" plasma. Still another confinement scheme that has been investigated in this general density range includes an electrostatic device in which the plasma is confined by inertial forces generated by concentric spherical electrodes.

The Fusion-Power Balance

What are the fundamental requirements for a meaningful release of fusion energy in a reactor? First, the plasma must be hot enough for the production of fusion energy to exceed the energy loss due to bremsstrahlung radiation (radiation resulting from near-collisions between electrons and nuclei in the plasma). The temperature at which this transition occurs is called the ignition temperature. For a fuel cycle based on fusion reactions between deuterium and tritium nuclei the ignition temperature is about 40 million degrees C. Second, the plasma must be confined long enough to release a significant net output of energy. Third, the energy must be recovered in a useful form.

In the first years of the controlled-fusion research program one of the major goals was to achieve the ignition temperature in a fairly dense laboratory plasma. Steady progress was made toward this goal, culminating in 1963, when the ignition temperature (for a deuterium-tritium fuel mixture) was reached in one of the Scylla devices at Los Alamos. This test, which was performed in a pure deuterium plasma to avoid the generation of excessive neutron flux, resulted in the release of fusion energy: about a thousandth of a joule per pulse, or 370 watts of fusion power during the three-microsecond duration of the pulse. If the test had been performed using a deuterium-tritium mixture, it would have released approximately a half-joule of fusion energy per pulse, or 180,000 watts of fusion power.

Today a large number of different devices have either achieved the deuterium-tritium ignition temperature or are very close to it [see bottom illustration on opposite page]. The main difficulties encountered in reaching this goal were comparatively straightforward energy-loss processes involving impurity atoms that entered the plasma from the walls of the container. A large research effort in the areas of vacuum and surface technology was a major factor in surmounting the ignition-temperature barrier.

The problem of confining a plasma long enough to release a significant net amount of energy has proved to be even more difficult than the problem of achieving the ignition temperature. Extremely rapid energy-loss processes—known collectively as "anomalous diffusion" processes—appeared to prevent the attainment of adequate confinement times. Plasma instabilities were the primary cause of this rapid plasma leakage [see "The Leakage Problem in Fusion Reactors," by Francis F. Chen; SCIENTIFIC AMERICAN, July, 1967]. Within the past few years, however, several large containment devices have reduced these instabilities to such a low amplitude that other more subtle effects, such as convective plasma losses and magnetic-field imperfections, can be studied. As a result it has been shown that there is no basic law of physics (such as an instability-initiated anomalous plasma loss) that prevents plasma confinement for times long enough to release significant net fusion energy. In fact, "classical," or ideal, plasma confinement has been achieved in several machines; this is the best confinement possible and yields a plasma-loss rate much lower than that required for a fusion reactor.

The twin achievements of ignition temperature and adequate confinement time, it should be noted, have taken place in quite different machines, each

FUSION REACTIONS regarded as potentially useful in full-scale fusion reactors are represented in this partial list. The two possible deuterium-deuterium reactions occur with equal probability. The deuterium-tritium fuel cycle has been considered particularly attractive because this mixture has the lowest ignition temperature known (about 40 million degrees Celsius). Other fuel cycles, including many not shown in this list, have been attracting increased attention lately, since certain plasma-confinement schemes actually operate better at higher temperatures and offer the advantage of direct conversion to electricity. The energy released by each reaction is given at right in millions of electron volts (MeV).

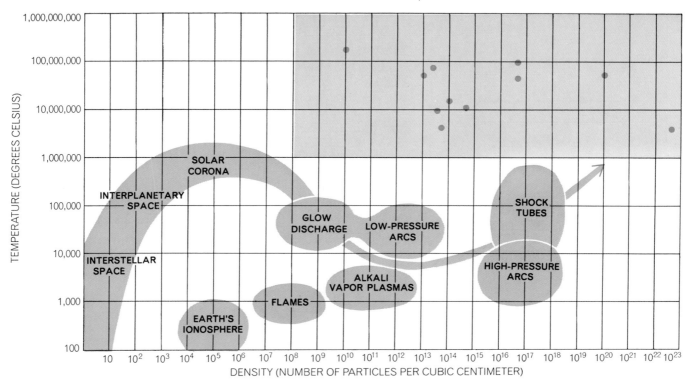

INDUSTRIALLY UNEXPLORED RANGE of plasma temperatures and densities has already been made available by the fusion-power research program. These experimental plasmas (*colored dots*), which range in temperature from 500,000 to a billion degrees C. and in density from 10^9 to 10^{22} ions per cubic centimeter, are compared here with various other industrial and natural plasmas.

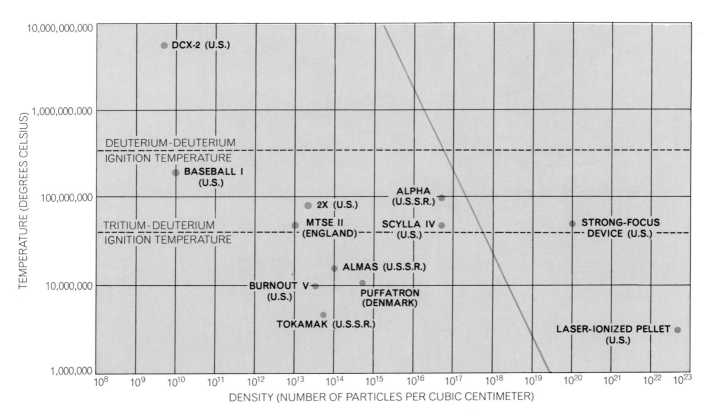

PLASMA EXPERIMENTS that have achieved temperatures near or above the fusion ignition temperatures of a deuterium-tritium fuel (*bottom horizontal line*) and a deuterium-deuterium fuel (*top horizontal line*) are identified by the name of the experimental device and the country in which the experiment took place in this enlargement of the upper right-hand section of the illustration at top. The diagonal colored line represents the limit beyond which the materials used to construct the magnet coils can no longer withstand the magnetic-field pressure required to confine the plasma (assumed to be 300,000 gauss in this case). Beyond this limit only fast-pulsed systems (in which the magnetic fields are generated by intense currents inside the plasma itself) or systems operating on entirely different principles (such as laser-produced, inertially confined plasmas) are possible. The record of six billion degrees C. was achieved with the aid of a high-energy ion-injection system associated with DCX-2 device at the Oak Ridge National Laboratory.

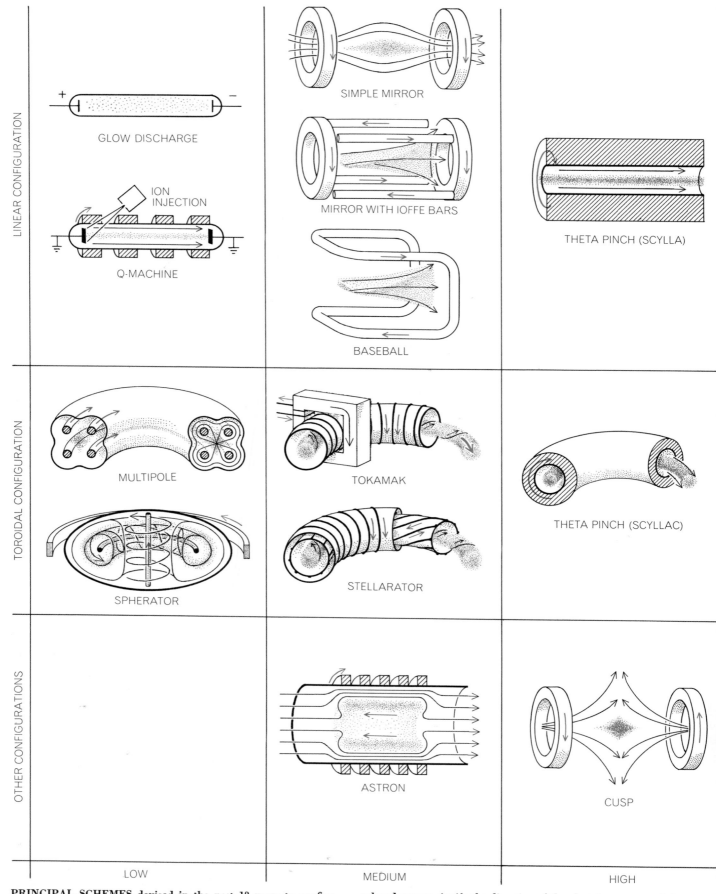

LINEAR CONFIGURATION

GLOW DISCHARGE

ION INJECTION

Q-MACHINE

SIMPLE MIRROR

MIRROR WITH IOFFE BARS

BASEBALL

THETA PINCH (SCYLLA)

TOROIDAL CONFIGURATION

MULTIPOLE

SPHERATOR

TOKAMAK

STELLARATOR

THETA PINCH (SCYLLAC)

OTHER CONFIGURATIONS

ASTRON

CUSP

LOW

MEDIUM

HIGH

PRINCIPAL SCHEMES devised in the past 18 years to confine plasmas for fusion research are arranged in the illustration on these two pages in order of increasing plasma density (*left to right*) and overall geometry (*top to bottom*). Only a few examples are depicted in each category. In every case the plasma is in color, the colored arrows signify the direction of the electric current and the black arrows denote the direction of the resultant magnetic field. Various structural details have been omitted for clarity. For each example shown there are a large number of variations either already in existence or in the conceptual stage. Furthermore, the

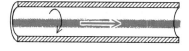

ZETA PINCH

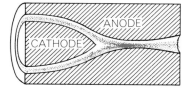

PLASMA FOCUS

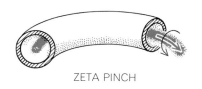

ZETA PINCH

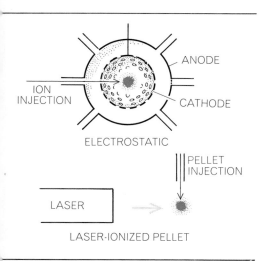

ELECTROSTATIC

LASER-IONIZED PELLET

VERY HIGH

fact that an example is given in one category does not necessarily mean that that configuration is not applicable to some other category; there are, for instance, toroidal Q-machines and medium-density cusp designs.

specially designed to maximize the conditions for reaching one goal or the other. How does one compare the performances of these machines in order to gauge how near one is to the combined conditions needed to operate a practical fusion-power reactor? The basic criterion for determining the length of time a plasma must be confined at a given density and temperature to produce a "break even" point in the power balance was laid down in 1957 by the British physicist J. D. Lawson. Combining data on the physics of fusion reactions with some estimates of the efficiency of energy recovery from a hypothetical fusion reactor, Lawson derived a factor, which he called R, that denoted the ratio of energy output to energy input needed to compensate for all possible plasma losses. Lawson's criterion is still in general use as a convenient yardstick for measuring the extent to which losses must be controlled in order to make possible the construction of a fusion reactor. Although more recent calculations consider many other physical constraints in order to arrive at the break-even power balance, these criteria still give values very close to those derived by Lawson.

For a deuterium-tritium fuel mixture Lawson found that at temperatures higher than the ignition temperature the product of density and confinement time must be equal to 10^{14} seconds per cubic centimeter in order to achieve the break-even condition. This criterion defines a surface in three-dimensional space, the coordinates being the logarithmic values of density, temperature and confinement time [see illustration on page 313]. The goal of a break-even release of energy will have been achieved when the set of conditions for a given machine reaches this surface. It should be emphasized that the exact location and shape of the surface is a function of both the fuel cycle used and the recovery efficiency of the hypothetical reactor system. Fuels other than the deuterium-tritium mixture would increase the temperature needed to achieve a break-even power balance.

The extraordinary progress made recently by various groups in learning how to raise the combination of density, temperature and confinement time to a set of values approaching this break-even surface can be appreciated by referring to the illustration of the Lawson-criterion surface. The several plasma systems shown range in density from about 10^9 ions per cubic centimeter to 5×10^{22} ions per cubic centimeter. (Below a density of about 10^{11} ions per cubic centimeter the power density would be so low that it would require an impractically

large reactor.) The particular density range chosen for investigation in each case is a function of the scientific preferences of the investigators concerning the best route to fusion power and the available technology (magnets, power supplies, lasers and so forth). Thus there are various trajectories to the break-even surface being followed through the three-dimensional "parameter space" of the illustration. Closing the gap between where each trajectory is now and the break-even surface depends in some cases (for example the tokamak devices) on obtaining a better understanding of the physical principles required to develop reliable scaling rules, whereas in other cases (such as the linear theta-pinch devices) all that may be required is an economic solution to the engineering problem of building a large enough system.

Fusion-Reactor Designs

How would a full-scale fusion reactor operate? In the first place fusion reactors, like fission reactors, could be run on a variety of fuels. The nature of the fuel used in the core of a fusion reactor would, however, have a decisive effect on the method used to recover the fusion energy and the uses to which the recovered energy might be put. Most research on reactor technology has centered on the use of a deuterium-tritium mixture as a fuel. The reason is that the mixture has the lowest ignition temperature, and hence the lowest rate of energy loss by radiation, of any possible fusion fuel. Nonetheless, other combinations of light nuclei have been considered for many years as potential fusion fuels. Prominent among these are reactions involving a deuterium nucleus and a helium-3 nucleus and reactions involving a single proton (a hydrogen nucleus) and a lithium-6 nucleus. Because containment based on the magnetic-mirror concept actually operates better at higher temperatures, a number of other fuels have been attracting increased attention [see illustration on page 308].

Depending on the fuel used, a fusion reactor could release its energy in several ways. For example, neutrons, which are produced at various rates by different fusion reactions, can cross magnetic fields and penetrate matter quite easily. A reactor based on, say, a deuterium-tritium fuel cycle would release approximately 80 percent of its energy in the form of highly energetic neutrons. Such a reactor could be made to produce electricity by absorbing the neutron energy in a liquid-lithium shield, circulating the

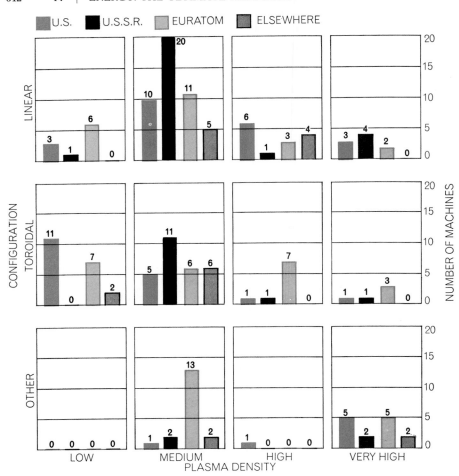

INVENTORY of the number of machines now operating throughout the world in each of the broad categories represented in the illustration on the preceding two pages is given in this table. The total number in each category is broken down into subtotals for the U.S., the U.S.S.R., the European Atomic Energy Community, or Euratom, countries (Belgium, France, Germany, Italy, Luxembourg and the Netherlands) and the rest of the world (principally Japan, Sweden and Australia). Britain, although not officially a member of Euratom, is included in the Euratom subtotal. The figures are drawn mainly from a recent survey compiled by Amasa S. Bishop and published by the International Atomic Energy Commission. The U.S. fusion-power program currently represents about a fifth of the world total.

plates such a system could theoretically be made to operate at a conversion efficiency of 90 percent.

J. Rand McNally, Jr., of the Oak Ridge National Laboratory has suggested that a long sequence of fusion reactions similar to those that power the stars could be reproduced in a fusion reactor. The data necessary to evaluate fuel cycles operating in this manner, however, do not exist at present.

The characteristics of a full-scale fusion reactor would depend not only on the fuel cycle but also on the particular plasma-confinement configuration and density range chosen. Thus it is probable that there eventually will exist a number of different forms of fusion reactor. For example, medium-density magnetic-mirror reactors and very-high-density laser-ignited reactors could be expected to operate at power levels as low as between five and 50 megawatts, which could make them potentially useful for fusion-propulsion schemes.

For central-station power generation the medium-density reactors would most likely operate on a deuterium-tritium fuel cycle in order to take advantage of the mixture's low ignition temperature. Because of the high neutron output associated with this fuel, a heat-cycle conversion system would be appropriate.

A reactor of this type would operate most efficiently with a power output in the billion-watt range. Before such a reactor can be built, it will be necessary to prove that the plasma will remain stable as present devices are scaled to reactor sizes and temperatures. Problems likely to be encountered in this effort involve the long-term equilibrium of the plasma, the interaction of the plasma with the walls of the container and the necessity of pumping large quantities of liquid lithium across the magnetic field.

Medium-density linear reactors would be better suited for fuel cycles that yield a major part of their energy output in the form of charged particles, since this approach would allow the direct recovery of the kinetic energy of these reaction products through schemes such as Post's. Such fuel cycles could be based on a deuterium-deuterium reaction, a deuterium-helium reaction or a proton-lithium reaction. A system operating on this principle could be made to produce direct-current electricity at a potential of about 400 kilovolts, which would be ideal for long-distance cryogenic (supercooled) power transmission.

Although the break-even conditions would be lowered in this case (because of the high energy-conversion efficien-

liquid lithium to a heat exchanger and there heating water to produce steam and so drive a conventional steam-generator electric power plant [*see top illustration on page 314*].

This general approach could also lead to an attractive new technique for converting the world's reserves of uranium 238 and thorium 232 to suitable fuels for fission reactors—the fusion-fission hybrid system mentioned above. By employing the abundance of inexpensive, energetic neutrons produced by the deuterium-tritium fuel cycle to synthesize fissionable heavy nuclei, a fusion reactor could act as a new type of breeder reactor. This could have the effect of lowering the break-even surface defined by Lawson's criterion, bringing the fusion-breeding scheme actually closer to feasibility than the generation of electricity solely by fusion reactions. Cheap fuel might thus be made for existing fission

reactors in systems that could be inherently safe. A "neutron-rich" economy created by fusion reactors would have other potential benefits. For example, it has been suggested that large quantities of neutrons could be useful for "burning" various fission products, thereby alleviating the problem of disposing of radioactive wastes.

Fuel cycles that release most of their energy in the form of charged particles offer still other avenues for the recovery of fusion energy. For example, Richard F. Post of the Lawrence Radiation Laboratory at Livermore has proposed a direct energy-conversion scheme in which the energetic charged particles produced in a fusion-reactor core are slowed directly by an electrostatic field set up by an array of large electrically charged plates [*see bottom illustration on page 314*]. By a judicious arrangement of the voltages applied to the

cy), it still remains to be shown that existing experiments can be scaled to large sizes and higher temperatures. Some major technological obstacles that need to be overcome include the construction of large atomic-beam injectors and extremely strong magnetic mirrors.

For reactors operating on the basis of any of the higher-density schemes, such as the theta-pinch machines or the fast-pulsed systems, major technological hurdles include the development of efficient energy-storage and energy-transfer techniques and problems related to heating techniques such as lasers.

In addition to generating electric power and possibly serving in a propulsion system, fusion reactors are potentially useful for other applications. For example, fusion research has already made available plasmas that range in temperature from 500,000 to a billion degrees C. and in density from 10^9 to 10^{22} ions per cubic centimeter. Almost all industrial processes that use plasmas fall outside this range [*see top illustration on page 309*]. In order to suggest how this industrially unexplored range might be exploited, we recently put forward the concept of the "fusion torch." The gen-

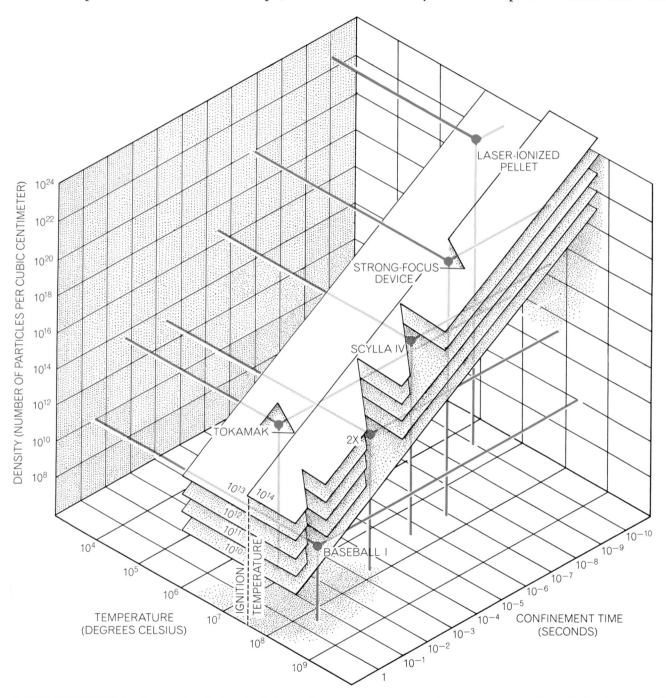

BASIC CRITERION for determining the length of time a plasma must be confined at a given density and temperature to achieve a "break even" point in the fusion-power balance is represented in this three-dimensional graph. The graph is based on a method of analysis devised in 1957 by the British physicist J. D. Lawson. For a deuterium-tritium fuel mixture in the temperature range from 40 million degrees C. to 500 million degrees C., Lawson found that the product of density and confinement time must be close to 10^{14} seconds per cubic centimeter to achieve the break-even condition (based on an assumed energy-conversion efficiency of 33 percent). This criterion corresponds to the top layer in the stack of planes in the illustration. The lower planes, which correspond to successively smaller values of density times confinement time, are included in order to give some idea of the positions of the best confirmed results from several experimental devices with respect to the combination of parameters needed to operate a full-scale fusion reactor.

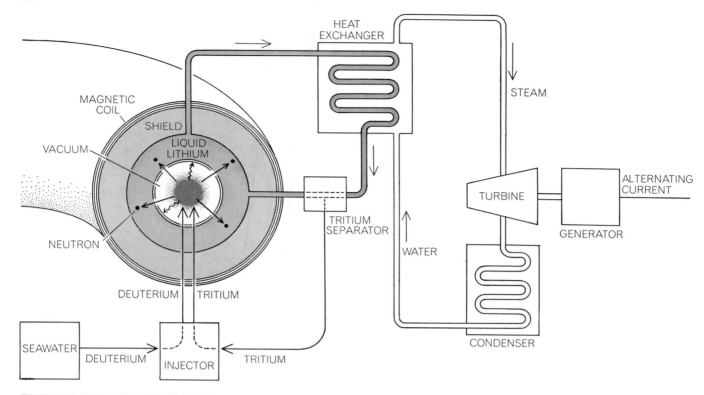

THERMAL ENERGY CONVERSION would be most effective in a fusion reactor based on a deuterium-tritium fuel cycle, since such a fuel would release approximately 80 percent of its energy in the form of highly energetic neutrons. The reactor could produce elec-tricity by absorbing the neutron energy in a liquid-lithium shield, circulating the liquid lithium to a heat exchanger and there heating water to produce steam and thus drive a conventional steam-gener-ator plant. The reactor core could be either linear or toroidal.

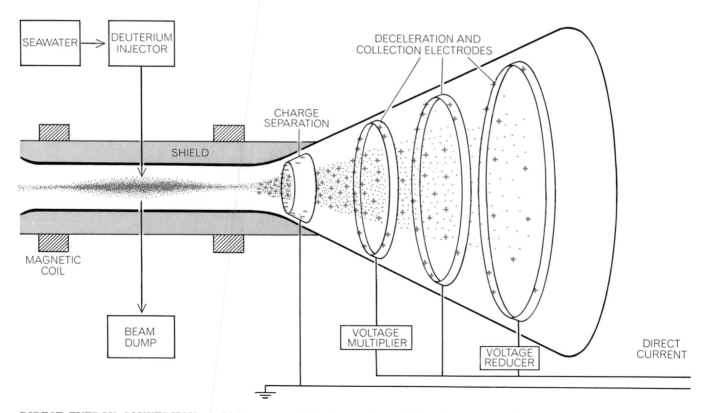

DIRECT ENERGY CONVERSION would be more suitable for fusion fuel cycles that release most of their energy in the form of charged particles. In this novel direct energy-conversion scheme, first proposed by Richard F. Post of the Lawrence Radiation Lab-oratory of the University of California at Livermore, the energetic charged particles (primarily electrons, protons and alpha particles) produced in the core of a linear fusion reactor would be released through diverging magnetic fields at the ends of the magnetic bot-tle, lowering the density of the plasma by a factor of as much as a million. A large electrically grounded collector plate would then be used to remove only the electrons. The positive reaction prod-ucts (at energies in the vicinity of 400 kilovolts) would finally be collected on a series of high-voltage electrodes, resulting in a direct transfer of the kinetic energy of the particles to an external circuit.

eral idea here is to use these ultrahigh-density plasmas, possibly directly from the exhaust of a fusion reactor, to vaporize, dissociate and ionize any solid or liquid material [*see top illustration at right*]. The potential uses of such a fusion-torch capability are intriguing. For one thing, an operational fusion torch in its ultimate form could be used to reduce all kinds of wastes to their constituent atoms for separation, thereby closing the materials loop and making technologically possible a stationary-state economy. On a shorter term the fusion torch offers the possibility of processing mineral ores or producing portable liquid fuels by means of a high-temperature plasma system.

The fusion-torch concept could also be useful in transforming the kinetic energy of a plasma into ultraviolet radiation or X rays by the injection of trace amounts of heavy atoms into the plasma. The large quantity of radiative energy generated in this way could then be used for various purposes, including bulk heating, the desalting of seawater, the production of hydrogen or new chemical-processing techniques. Because such new industrial processes would make use of energy in the form of plasmas rather than in the form of, say, chemical solvents, they would be far less likely to pollute the environment. Although the various fusion-torch possibilities are largely untested and many aspects may turn out to be impractical, the concept is intended to stimulate new ideas for the industrial use of the ultrahigh-temperature plasmas that have already been developed in the fusion program as well as those plasmas that would be produced in large quantities by future fusion reactors.

Environmental Considerations

The environmental advantages of fusion power can be broken down into two categories: those advantages that are inherent in all fusion systems and those that are dependent on particular fuel cycles and reactor designs. Among the inherent advantages, one of the most important is the fact that the use of fusion fuel requires no burning of the world's oxygen or hydrocarbon resources and hence releases no carbon dioxide or other combustion products to the atmosphere. This advantage is shared with nuclear-fission plants.

Another advantage of fusion power is that no radioactive wastes are produced as the result of the fuel cycles contemplated. The principal reaction products would be neutrons, nonradioactive heli-

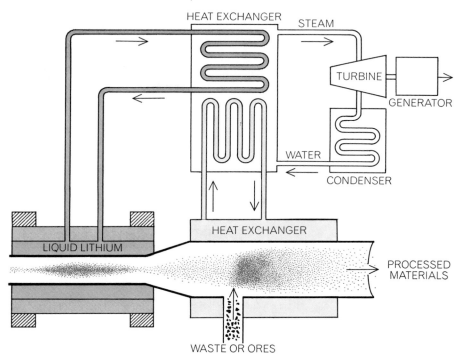

POTENTIAL NONPOWER USE of fusion energy is represented by the concept of the "fusion torch," which was put forward recently by the authors as a suggestion intended to stimulate new ideas for the industrial exploitation of the ultrahigh-temperature plasmas already made available by the fusion-research program as well as those that would be produced by fusion reactors. The general idea is to use some of the energy from these plasmas to vaporize, dissociate and ionize any solid or liquid material. In its ultimate form the fusion torch could be used to reduce any kind of waste to its constituent atoms for separation.

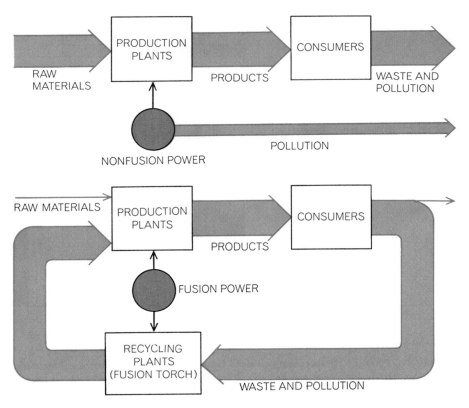

CLOSED MATERIALS ECONOMY could be achieved with the aid of the fusion-torch concept illustrated at the top of this page. In contrast to present systems, which are based on inherently wasteful linear materials economies (*top*), such a stationary-state system would be able to recycle the limited supply of material resources (*bottom*), thus alleviating most of the environmental pollution associated with present methods of energy utilization.

um and hydrogen nuclei, and radioactive tritium nuclei. It is true that tritium emits low-energy ionizing radiation in the form of beta particles (electrons), but since tritium is also a fusion fuel, it could be returned to the system to be burned. This situation is strongly contrasted with that in nuclear fission, which by its very nature must produce a multitude of highly radioactive waste elements.

Fusion reactors are also inherently incapable of a "runaway" accident. There is no "critical mass" required for fusion. In fact, the fusioning plasma is so tenuous (even in the "high density" machines) that there is never enough fuel present at any one time to support a nuclear excursion. This situation is also in contrast to nuclear-fission reactors, which must contain a critical mass of fissionable material and hence an extremely large amount of potential nuclear energy.

Among the system-dependent environmental advantages of fusion power must be counted the fact that the only radioactive fusion fuel considered so far is tritium. The amount of tritium present in a fusion reactor can range from near zero for a proton-lithium fuel cycle to a maximum for a deuterium-tritium cycle, where a "blanket" for the production of tritium must be included. Tritium, however, is one of the least toxic of the radioactive isotopes, whereas the fission fuel plutonium is one of the most toxic radioactive materials known.

The most serious radiological prob-

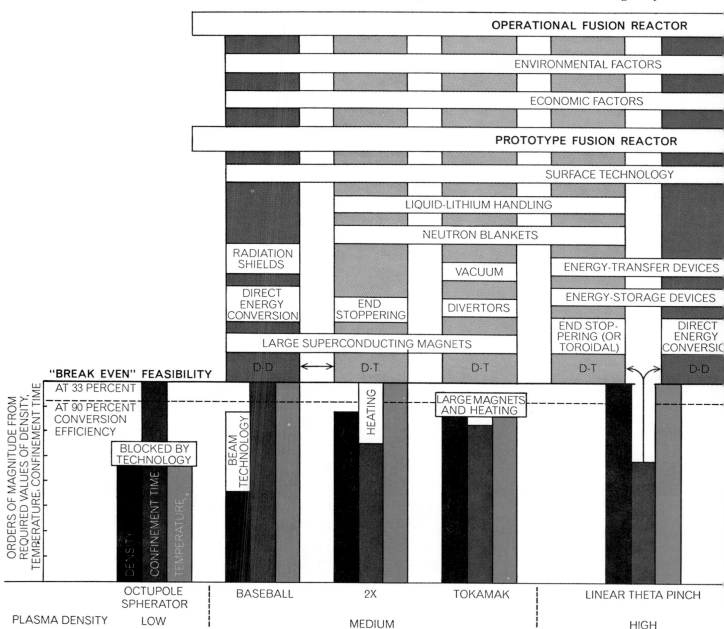

REMAINING PROBLEMS that must be solved before the goal of useful, economic fusion power can be achieved are depicted schematically in this illustration. The major experimental routes to the goal are ordered according to plasma density. Various experimental devices are represented by bars indicating the best combination of plasma density, confinement time and temperature achieved by each device; the logarithmic scale at lower left gauges how far each of these essential parameters is from the values needed to attain break-even feasibility. Technological problems that must be solved in each case are labeled. The achievement of a prototype reactor will be a function not only of plasma technology but also of the fuel cycle and the method of energy conversion chosen. Thus medium-density magnetic-mirror devices could be built to operate with either a deuterium-tritium (D-T) fuel mixture or a deuterium-deuterium (D-D) fuel mixture; the arrows signify these alternatives. If a D-D cycle is chosen, then direct energy conversion is possible, and once the converters are developed very few obstacles would remain to delay the construction of a prototype reactor. If, on the other hand, a D-T cycle is chosen, then conventional thermal energy conversion would be needed, and the listed technological

lems for fusion would exist in a reactor burning and producing tritium. A representative rate of tritium consumption for a 2,000-megawatt deuterium-tritium thermal plant would be about 260 grams per day. Tritium "holdup" in the blanket and other elements of the tritium loop would dictate the tritium inventory. Holdup is estimated to be about 1,000 grams in a 2,000-megawatt plant. If necessary, the doubling time of breeding tritium could be less than two months in order to meet the needs of an expanding

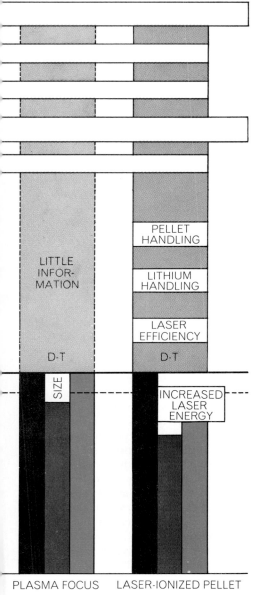

PLASMA FOCUS   LASER-IONIZED PELLET

VERY HIGH

hurdles would have to be overcome. Other systems operating on the D-T cycle would have to climb past similar hurdles. The high-density linear theta-pinch device could take either the thermal-conversion path or the direct-conversion path (*arrows*). The final step, from a prototype to an operational reactor, would proceed through a region in which economic and environmental considerations can be expected to be paramount.

economy. The amount of tritium produced by the plant is controllable, however, and need not exceed the fuel requirements of the plant.

Careful design to prevent the leakage of tritium fuel from a deuterium-tritium reactor is mandatory. Engineering studies that take into account economic considerations indicate that the leakage rate can be reduced to .0001 percent per day. The conclusion is that even for an all-deuterium-tritium fusion economy the genetic dose rate from worldwide tritium distribution would be negligible.

In fact, for a given total power output the tritium inventory for an all-deuterium-tritium fusion economy (including both the inventory within the plant and that dispersed in the biosphere) would be between one and 100 times what it would be for an all-fission economy. It is true that tritium would be produced in a deuterium-tritium fusion reactor at a rate of from 1,000 to 100,000 times faster than in various types of fission reactor. Since tritium is burned as a fuel, however, it has an effective half-life of only about three days rather than the normal 12 years.

A technology-dependent but possibly serious limitation on deuterium-tritium fusion plants could be the release of tritium to the local environment. The level would be quite low but the long-term consequences from tritium emission to the environment in the vicinity of a deuterium-tritium reactor needs to be explored. In general the biological-hazard potential of the tritium fuel inventory in a deuterium-tritium reactor is lower by a factor of about a million than that of the volatile isotope iodine 131 contained in a fission reactor. Of course there is no expectation in either case that such a release would occur.

The radioactivity induced in the surrounding structures by a fusion reactor is dependent on both the fuel cycle and the engineering design of the plant. This radioactivity could range from zero for a fuel cycle that produces no neutrons up to very high values for a deuterium-tritium cycle if the engineering design is such that the type and amount of structural materials could become highly activated under neutron bombardment. Cooling for "after heat" will be required for systems that have intense induced radioactivity. Even if the cooling system should fail, however, there could be no nuclear excursion that would disperse the radioactivity outside the plant.

Other system-dependent environmental advantages of fusion power include safety in the event of sabotage or natural disaster, reduced potential for the diversion of weapons-grade materials and low

waste heat. In fact, the potential exists for fusion systems to essentially eliminate the problem of thermal pollution by going to charged-particle fuel cycles that result in direct energy conversion. Finally, there is the advantage of the materials-recycling potential of the fusion-torch concept.

The Timetable to Fusion Power

The construction and operation of a power-producing controlled-fusion reactor will be the end product of a chain of events that is already to a certain extent discernible. For controlled fusion, however, there can never be an instant equivalent to the one that demonstrated the "feasibility" of a fission-power reactor (the Stagg Field experiment of Enrico Fermi in 1942). To reach the plasma conditions required for a net release of fusion power it is necessary to first develop many new technologies. In this context the term "scientific feasibility" cannot be precisely defined. To some investigators it means simply the achievement of the basic plasma conditions necessary to reach the break-even surface in the illustration on page 313. To others it represents reaching the same surface—but with a system that can be enlarged to a full-scale, economic power plant. To a few it represents the attainment of a full understanding of all the phenomena involved.

Although these differing interpretations of what is needed to give confidence in our ability to construct a fusion reactor may be somewhat confusing, each interpretation nevertheless contains a modicum of truth. To depict the complexity of this drive toward the goal of fusion power we have prepared the highly schematic illustration on the opposite page. The goal is to achieve useful, economic fusion power. The major routes to the goal are ordered in the illustration according to plasma density. Various individual experiments have climbed past various obstacles to reach positions close to the break-even level. In fact, in some instances two of the three essential parameters (density, temperature and confinement time) have already been achieved. The ignition temperature has been achieved in a number of cases. The rest of the climb to the break-even level in some cases involves a better understanding of the physics of the plasma-confinement system, but in others it may involve only engineering problems. Indeed, the location of the break-even level is a function of the technology used. Direct energy conversion, for example, would lower this level.

The next portion of the climb, the

construction of a prototype reactor, will be a function of the route taken to scientific feasibility. For instance, if a deuterium-tritium mixture is the fuel, this would require the development of components such as lithium blankets, large superconducting magnets, radiation-resistant vacuum liners, fueling techniques and heat-transfer technology. If fuels that release most of their energy in the form of charged particles are considered, however, then in the case of mirror reactors direct-conversion equipment may be part of a device used to demonstrate break-even feasibility. The step from that device to a prototype reactor could then be very short because the conversion equipment would be already developed. Other devices would face similar problems of differing magnitude in prototype construction. The final step from a prototype to an operational reactor would proceed through a much more nebulous region in which economic and environmental considerations would influence the comparative desirability of different power plants.

At present the main factor limiting the rate of progress toward fusion power is financial. The annual operating and equipment expenditures for the U.S.

fusion program, when one uses the consumer price indexes to adjust these dollars for inflation, has remained fairly constant for the past eight years [see illustration below]. The total amount spent on the program since its inception is the cost equivalent of a single Apollo moon shot. The annual funding rate of about $30 million per year is the equivalent of 15 cents per person per year in the U.S.

The road to fusion power is a cumulative one in that successive advances can be built on earlier advances. At present the U.S. has a fairly broad program of investigations approaching the break-even surface for net energy release. It is essential that larger (and thus more expensive) devices be built if the goal of the break-even surface is to be reached. The surface should be broken through in a number of places so that the relative advantages of the possible routes beyond that surface to an eventual fusion-power reactor can be assessed.

Clearly the timetable to fusion power is difficult to predict. If the level of effort on fusion research remains constant or decreases slightly, the requirement for larger devices and advanced engineering will automatically cause a premature

narrowing of the density range under investigation. This increases the risk of reaching the goal in a given time scale. To put it another way, it extends one's estimate of the probable time scale. If the level of fusion research expands sufficiently to maintain a fairly broad program across the entire density range, the probability of success increases and the probable time scale decreases. If fusion power is pursued as a "national objective," expanded programs could be carried out across the entire density range accompanied by parallel strong programs of research on the remaining engineering and materials problems to determine as quickly as possible the best routes to practical fusion-power systems. Therefore, depending on one's underlying assumptions on the level of effort and the difficulties ahead, the time it would take to produce a large prototype reactor could range from as much as 50 years to as little as 10 years.

There is at least one case in which the fusion break-even surface could be reached without making any new scientific advances and without developing any new technologies. This "brute force" approach, which might not be the optimum route to an eventual power reactor, would involve simply extending the length of the existing theta-pinch linear devices. It has been estimated that to reach the break-even surface by this method such a system would have to be about 2,000 feet long—less than a fifth of the length of the Stanford Linear Accelerator. This one fusion device, however, would cost an order of magnitude more than any experimental fusion device built to date. Even though a simple scaling of this type would introduce no new problems in plasma physics, one could not exclude the possibility of unexpected difficulties arising solely from the extended length of the system.

The length of such a device could be shortened by as much as 90 percent by installing magnetic mirrors at the ends, by increasing the diameter of the plasma or by making the system toroidal, but these steps would introduce new physical conditions. The system could also be shortened by the use of a direct energy-conversion approach, but this would introduce an unproved technology. At present a significant portion of the fusion-power program is concentrating on developing the new physics and technology that would reduce the cost of such break-even experiments. This continuing effort is sustained by the growing conviction that the eventual attainment of a practical fusion-power reactor is not blocked by the laws of nature.

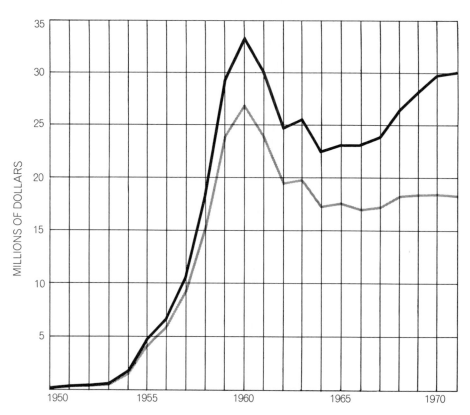

FINANCIAL SUPPORT is currently the main factor limiting the rate of progress toward the goal of fusion power. The solid curve shows the annual operating and equipment expenditures for the U.S. fusion program. The gray curve shows these expenditures adjusted for inflation. The adjustment shows that fusion research has been funded by the Atomic Energy Commission at an essentially constant rate for the past eight years. Smaller research programs have been funded by both private industry and other Government agencies.

Human Energy Production as a Process in the Biosphere

by S. Fred Singer
September 1970

In releasing the energy stored in fossil and nuclear fuels man accelerates slow cycles of nature. The waste products of power generation then interact with the fast cycles of the biosphere

The earth in general and the biosphere in particular have grand-scale pathways of energy metobolism. For example, solar energy falls on the earth, green plants utilize a tiny fraction of it to manufacture energy-rich compounds and some of these compounds are stored in the earth's crust as what we have come to call fossil fuels. The primary fission fuel uranium and the potential fusion fuel deuterium were originally "cooked" in the interior of stars. In releasing the energy of these chemical and nuclear fuels man is in effect racing the slow cycles of nature, with inevitable effects on the cycles themselves.

Before 1800 the power available to human societies was limited to solar energy that had only recently been radiated to the earth. The most direct form of such power was human or animal power; the energy came from the metabolism of food, which is to say from the biological oxidation of compounds storing solar energy. The burning of wood or oils of animal or vegetable origin to provide light and heat also represented the conversion of recently stored solar energy. By the same token the use of moving air or falling water to drive mills or pumps constituted the use of recently arrived solar energy. Among the other limitations of such power sources was the fact that they could not be readily transported and that their energy could not be transmitted any considerable distance.

This picture has of course changed completely since 1800, and it has assumed significant new dimensions in the past two decades with the advent of nuclear power. The most striking measure of these changes is the increased per capita consumption of energy in the developed countries. Indeed, the correlation between a nation's per capita use of energy and its level of economic development is almost linear [*see illustration on page 321*]. The minimum per capita consumption of energy is what is required in food for a man to stay alive, namely about 2,000 kilocalories per day or 100 watts (thermal). Today the per capita use of energy in the U.S. is 10,000 watts, and the figure is rising by some 2.5 percent per year.

Hand in hand with the advance in the rate of energy consumption has gone the introduction of the new sources of energy: fossil and nuclear fuels. In contrast to the sources used before 1800, fossil and nuclear fuels represent energy that reached the earth millions and even some billions of years ago. Except occasionally for political reasons, it matters little where the new fuels are found; they can be transported readily, and the energy produced from them can be transmitted over great distances.

On first consideration it might seem that fossil and nuclear fuels are fundamentally different, in that the energy of one is released by oxidation, or burning, and the energy of the other is released by fission or fusion. In a deeper sense, however, the two kinds of fuel are closely related. Fossil fuels store the radiant energy originally produced by nuclear reactions in the interior of the sun. Nuclear fuels store energy produced by another set of nuclear reactions in the interior of certain stars. When such stars exploded, they showered into space the elements that had been synthesized within them. These elements then went into the formation of younger stars such as the sun, together with its family of planets.

The production of fossil fuels is based on the carbon cycle. In the process of photosynthesis plants use radiant energy from the sun to convert carbon dioxide and water into carbohydrates, at the same time releasing oxygen into the atmosphere. When the plant materials decompose or are eaten by animals, the process is reversed: oxygen is used to convert carbohydrates into energy plus carbon dioxide and water.

The amount of carbon dioxide involved in photosynthesis annually is about 110 billion tons, or roughly 5 percent of the carbon dioxide in the atmosphere. The consumption of carbon dioxide through photosynthesis is matched to one part in 10,000 by the annual release of carbon dioxide to the atmosphere through oxidation. Under normal conditions the amounts of carbon dioxide and oxygen in the atmosphere remain approximately in equilibrium from year to year.

There are, however, small long-term imbalances in the carbon cycle, and it is owing to them that the fossil fuels being exploited today all derive from plants and animals that lived long ago. Over a span of geologic history extending back into the Cambrian period of some 500 million years ago, a small fraction of these organisms have been buried in sediments or mud under conditions that prevented complete oxidation. Various chemical changes have transformed them into fossil fuels: coal, oil, natural gas, lignite, tar and asphalt. Although the same geological processes are still operative, they function over vast periods of time, and so the amount of new fossil fuel that is likely to be produced during the next few thousand years is inconsequential. Therefore one can assume that the existing fossil fuels constitute a nonrenewable resource.

Coal has been burned for some eight centuries, but it was consumed in negligible amounts until early in the 19th century. Since the middle of that century the rise in the consumption of coal has been spectacular: in 1870 the world production rate of coal was about 250 million metric tons per year, whereas this year it will be about 2.8 billion tons. The rate of increase, however, is lower now than it was at the beginning of the period, having declined from an average of 4.4 percent per year to 3.6 percent,

largely because of the rapid increase in the fraction of total industrial energy contributed by oil and gas. In the U.S. that fraction rose from 7.9 percent in 1900 to 67.9 percent in 1965, whereas the contribution of coal declined from 89 percent to 27.9 percent.

World production of crude oil was negligible as recently as 1890; now it is close to 12 billion barrels per year. The rise in the rate of production has been nearly 7 percent per year, so that the amount of oil extracted has doubled every 10 years. As yet there is no sign of a deceleration in this rate.

Nonetheless, the finiteness of the earth's fossil fuel supplies gives rise to the question of how long they will last. M. King Hubbert of the U.S. Geological

Survey has estimated that the earth's coal supply can serve as a major source of industrial energy for another two or three centuries. His estimate for petroleum is 70 to 80 years. However much these periods may be stretched by unforeseen discoveries and improved technology, the end of the fossil fuel era will inevitably come. From the perspective of that time—perhaps the 23rd century—the period of exploitation of fossil fuels will be seen as only a brief episode in the span of human history.

This year the U.S. will consume some 685,000 million million B.T.U. of energy, most of it derived from fossil fuels. (One short ton of coal has a thermal value of 25.8 million B.T.U. The thermal

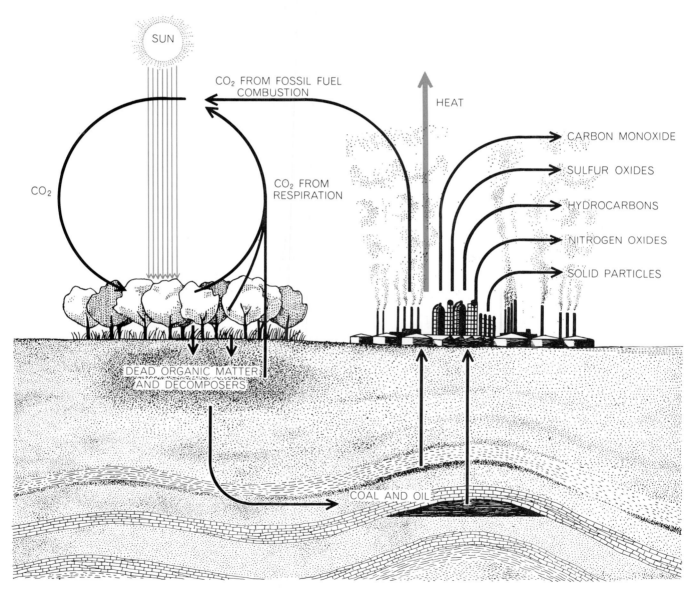

ENERGY CYCLE involved in the combustion of fossil fuels begins with solar energy employed in photosynthesis millions of years ago. A small fraction of the plants is buried under conditions that prevent complete oxidation. The material undergoes chemical changes that transform it into coal, oil and other fuels. When they are burned to release their stored energy, only part of the energy goes into useful work. Much of the energy is returned to the atmosphere as heat, together with such by-products of combustion as carbon dioxide and water vapor. Other emissions in fossil fuel combustion are listed at right in the relative order of their volume.

value of one barrel of oil is 5.8 million B.T.U.) Industry takes more than 35 percent of the total energy consumption. About a third of industry's share is in the form of electricity, which, as of 1960, was generated roughly 50 percent from coal, 20 percent from water power, 20 percent from natural gas and 10 percent from oil.

The nation's homes use almost as much energy as industry does. A major consumer is space heating, which for the average home requires as much energy as the average family car: about 70 million B.T.U. per year, or the equivalent of 900 gallons of oil. The other domestic uses are for cooking, heating water, lighting and air conditioning.

Transportation accounts for 20 percent of the annual energy consumption, mainly in the form of gasoline for automobiles. Another 10 percent goes for commercial consumption in stores, offices, hotels, apartment houses and the like. Agriculture probably consumes no more than 1 percent of all the energy, chiefly for the operation of tractors and for running irrigation and drainage equipment.

Looking at the use of fossil fuels from another viewpoint, one finds that most of the coal goes into the generation of electricity. Oil and natural gas tend to be used directly, either for heating purposes or to provide motive power for vehicles. Fossil fuels are also used as the raw materials for the petrochemical industry. Notwithstanding that industry's rapid growth, however, it still accounts for less than 2 percent of the annual consumption of fossil fuels.

Clearly the production of energy from fossil fuels on the scale typical of a modern industrial nation represents an enormous amount of combustion, with attendant effects on the biosphere. By far the greatest effect is the emission of carbon dioxide. Combustion also injects a number of pollutants into the air. In the U.S. the five most common air pollutants, listed in the order of their annual tonnage, are carbon monoxide, sulfur oxides, hydrocarbons, nitrogen oxides and solid particles. The major sources are automobiles, industry, electric power plants, space heating and refuse disposal. The burning of fossil fuels also produces effects on water: chemical effects when the air pollutants are washed down by water and thermal effects arising from the dispersal of waste heat from thermal power plants.

Carbon dioxide is the only combustion product whose increase has been documented on a worldwide basis. The

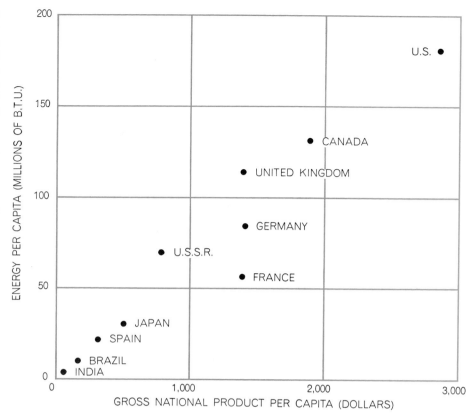

CLOSE RELATION between a nation's consumption of energy and its gross national product is depicted on the basis of a study made by the Office of Science and Technology in 1961. Most of the nations covered beyond the 10 shown would be in the lower left-hand rectangle.

injection of large quantities of carbon dioxide into the atmosphere in the past few decades has been extremely sudden in relation to important natural time scales. For example, although the surface of the sea can adjust to changes in the level of carbon dioxide in the atmosphere in about five years, the deeper layers require some hundreds or thousands of years to adjust. If the oceans were perfectly mixed at all times, carbon dioxide added to the atmosphere would distribute itself about five-sixths in the water and about one-sixth in the air. In actuality the distribution is about equal.

It appears that between 1860 and the present the concentration of carbon dioxide in the atmosphere has increased from about 290 parts per million to about 320 parts per million, an increase of more than 10 percent. Precise measurements by Charles D. Keeling of the Scripps Institution of Oceanography have established that the carbon dioxide content increased by six parts per million between 1958 and 1968. Reasonable projections indicate an increase of 25 percent (over 1970) to about 400 parts per million by the turn of the century and to between 500 and 540 parts per million by 2020.

The most widely discussed matter related to these increases is the possibility that they will lead to a worldwide rise in temperature. The molecule of carbon dioxide has strong absorption bands, particularly in the infrared region of the spectrum at wavelengths of from 12 to 18 microns. This is the spectral region where most of the thermal energy radiating from the earth into space is concentrated. By increasing the absorption of this radiation and by reradiating it at a lower temperature corresponding to the temperature of the upper atmosphere the carbon dioxide reduces the amount of heat energy lost by the earth to outer space. The phenomenon has been called the "greenhouse effect," although the analogy is inexact because a real greenhouse achieves its results less from the fact that the glass blocks reradiation in the infrared than from the fact that it cuts down the convective transfer of heat.

The possibility that additional carbon dioxide from the burning of fossil fuels could produce a worldwide increase in temperature seems to have been raised initially by the American geologist P. C. Chamberlain in 1899. In 1956 Gilbert N. Plass calculated that a doubling of

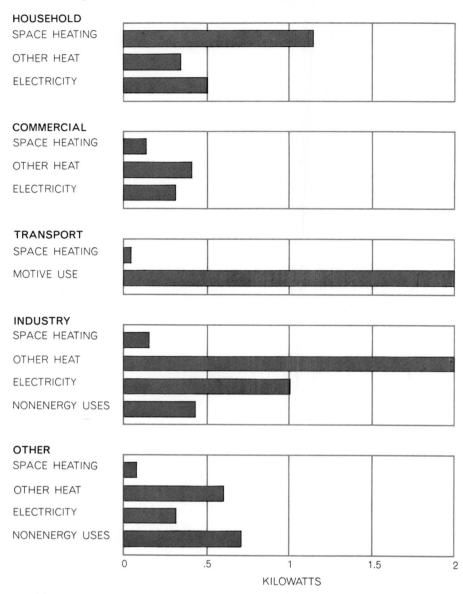

USE OF ENERGY in the U.S. is expressed in terms of thermal kilowatts per capita in 1967 in five major categories. All together the consumption averages 10,000 watts per person, which is 100 times the food-intake level of 100 watts that is barely exceeded in many nations.

the carbon dioxide content of the atmosphere would result in a rise of 6.5 degrees Fahrenheit at the earth's surface. In 1963 Fritz Möller calculated that a 25 percent increase in atmospheric carbon dioxide would increase the average temperature by one to seven degrees F., depending on the effects of water vapor in the atmosphere. The most extensive calculations have been made by Syukuro Manabe and R. T. Wetherald, who estimate that a rise in atmospheric carbon dioxide from 300 to 600 parts per million would increase the average surface temperature by 4.25 degrees, assuming average cloudiness, and by 5.25 degrees, assuming no clouds.

Unfortunately the problem is more complicated than these calculations im-ply. An increase of temperature at the surface of the earth and in the lower levels of the atmosphere not only increases evaporation but also changes cloudiness. Changes of cloudiness alter the albedo, or average reflecting power, of the earth. The normal average albedo is about 30 percent, meaning that 30 percent of the sunlight reaching the earth is immediately reflected back into space. Changes in cloudiness, therefore, can have a pronounced effect on the atmospheric temperature and on climate.

The situation is further complicated by atmospheric turbidity. J. Murray Mitchell, Jr., of the Environmental Science Services Administration has determined that atmospheric temperatures rose generally between 1860 and 1940. Between 1940 and 1960, although warming occurred in northern Europe and North America, there was a slight lowering of temperature for the world as a whole. Mitchell finds that a cooling trend has set in; he believes it is owing partly to the dust of volcanic eruptions and partly to such human activities as agricultural burning in the Tropics. (In the future the condensation trails left by jet airplanes may contribute to this problem.)

In sum, the fact that the carbon dioxide content of the atmosphere has increased is firmly established by reliable measurements. The effect of the increase on climate is uncertain, partly because no good worldwide measurements of radiation are available and partly because of the counteractive effects of changes in cloudiness and in the turbidity of the atmosphere. An exciting technological possibility is the use of a weather satellite to keep track of the energy radiated back into space by the earth. The data would provide a basis for the first reliable and standardized measurement of the "global radiation climate."

In any event, the higher levels of carbon dioxide may not persist for long. For one thing, the oceans, which contain 60 times as much carbon dioxide as the atmosphere does, will begin to absorb the excess as the mixing of the intermediate and deeper levels of water proceeds. For another, the increased atmospheric content of carbon dioxide will stimulate a more rapid growth of plants—a phenomenon that has been utilized in greenhouses. It is true that the carbon dioxide thus removed from the atmosphere will be returned when the plants decay. Forests, however, account for about two-thirds of the photosynthesis taking place on land (and therefore for nearly half of the world total), and since forests are long-lived, they tend to spread over a long period of time the return of carbon dioxide to the atmosphere.

The five major air pollutants resulting from the combustion of fossil fuels also interact with the biosphere in various ways, not all of them clearly understood. One tends to think of pollutants as harmful, but the situation is not that simple, as becomes apparent in a consideration of the pollutants and their known effects.

Carbon monoxide appears to be almost entirely a man-made pollutant. The only significant source known is the imperfect combustion of fossil fuels, resulting in incomplete oxidation of the

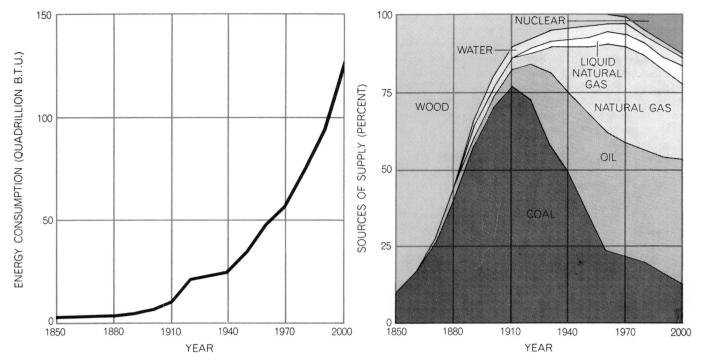

CHANGING SOURCES of energy in the U.S. since 1850 are com-pared (*right*) with total consumption (*left*) over the same period. At right one can see that in 1850 fuel wood was the source of 90 percent of the energy and coal accounted for 10 percent. By 2000 it is foreseen that coal will be back to almost 10 percent and that other sources will be oil, natural gas, liquid natural gas, hydro-electric power, fuel wood and nuclear energy. The estimates were made by Hans H. Landsberg of Resources for the Future, Inc.

carbon. Although carbon monoxide is emitted in large amounts, it does not seem to accumulate in the atmosphere. The mechanism of removal is not known, but it is probably a biological sink, such as soil bacteria.

Sulfur, which occurs as an impurity in fossil fuels, is among the most trou-blesome of the air pollutants. Although there are natural sources of sulfur di-oxides, such as volcanic gases, more than 80 percent is estimated to come from the combustion of fuels that contain sulfur. The sulfur dioxide may form sulfuric acid, which often becomes associated with atmospheric aerosols, or it may re-act further to form ammonium sulfate. A typical lifetime in the atmosphere is about a week.

When the sulfur products are removed from the atmosphere by precipitation, they increase the acidity of the rainfall. Values of pH of about 4 have been found in the Netherlands and Sweden, probably because of the extensive indus-trial activity in western Europe. As a re-sult small lakes and rivers have begun to show increased acidity that endangers the stability of their ecosystems. Certain aquatic animals, such as salmon, cannot survive if the pH falls below 5.5.

Nothing is known about the global effects of sulfur emission, but they are believed to be small. In any case most of the sulfur ends up in the oceans. It is possible, however, that sulfur com-pounds are accumulating in a layer of sulfate particles in the stratosphere. The layer's mechanism of formation, its ef-fects and its relation to man-made emis-sions are not clear. The fine particles of the layer could have an effect on radia-tion from the upper atmosphere, thereby affecting mean global temperatures.

Hydrocarbons are emitted naturally into the atmosphere from forests and

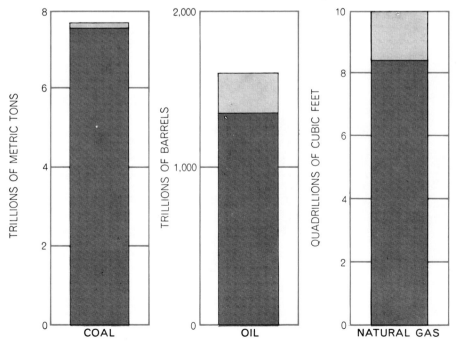

FOSSIL FUEL SUPPLIES remaining in the world are indicated by a scheme wherein the entire gray bar represents original resources, light gray portion shows how much has been extracted and dark gray area shows what remains. Figures reflect estimates by M. King Hubbert of the U.S. Geological Survey and could be changed by unforeseen discoveries.

vegetation and in the form of methane from the bacterial decomposition of organic matter. Human activities account for only about 15 percent of the emissions, but these contributions are concentrated in urban areas. The main contributor is the processing and combustion of petroleum, particularly gasoline for the internal-combustion engine.

The reactions of hydrocarbons with nitrogen oxides in the presence of ultraviolet radiation produce the photochemical smog that appears so often over Los Angeles and other cities. The biological effects of several of the products of the reactions, including ozone and complex organic molecules, can be quite severe. Some of the products are thought to be carcinogenic. Ozone has highly detrimental effects on vegetation, but fortunately they are localized. As yet no regional or worldwide effects have been discovered.

Hydrocarbon pollutants in the form of oil spills are well known to have drastic ecological effects. The spill in the Santa Barbara Channel last year, which involved some 10,000 tons, and the *Torrey Canyon* spill in 1967, involving about 100,000 tons, produced intense local concentrations of oil, which is toxic to many marine organisms. Besides these well-publicized events there is a yearly worldwide spillage from various oil operations that adds up to about a million tons, even though most of the individual

spills are small. There are also natural oil seeps of unknown magnitude. Added to all of these is the dumping of waste motor oil; in the U.S. alone about a million tons of such oil is discarded annually. Up to the present time no worldwide effects of these various oil spills are detectable. It can therefore be assumed that bacteria degrade the oil rapidly.

Nitrogen oxides occur naturally in the atmosphere as nitrous oxide (N_2O), nitric oxide (NO) and nitrogen dioxide (NO_2). Nitrous oxide is the most plentiful at .25 part per million and is relatively inert. Nitrogen dioxide is a strong absorber of ultraviolet radiation and triggers photochemical reactions that produce smog. In combination with water it can form nitric acid.

The production of nitrogen oxides in combustion is highly sensitive to temperature. It is particularly likely to result from the explosive combustion taking place in the internal-combustion engine. If this engine is ever replaced by an external-combustion engine that operates at a steady and relatively low temperature rather than at high peaks, the emission of nitrogen oxides will be greatly reduced.

Solid particles are injected into the lower atmosphere from a number of sources, with the combustion of fossil fuels making a major contribution. The technology of pollution control is adequate for limiting such emissions. If it

is applied, solid particles will become insignificant pollutants.

Although the fossil fuels still predominate as sources of power, the introduction of nuclear fuels into the generation of power is changing both the scale of energy conversion and the effects of that conversion on the biosphere. Nuclear energy can be considered as a heat source differing from coal or oil, but once the energy has been released in the form of heat it is used in the same way as heat from other sources. Therefore the problem of waste heat is the same. The pollution characteristics of nuclear energy, however, differ from those of the fossil fuels, being radioactive rather than chemical.

Two processes are of concern: the fission of heavy nuclei such as uranium and the fusion of light nuclei such as deuterium. The fission reaction has to start with uranium 235, because that is the only naturally occurring isotope that is fissioned by the capture of slow neutrons. On fissioning the uranium 235 supplies the neutrons needed to carry out other reactions.

Each fission event of uranium 235 releases some 200 million electron volts of energy. One gram of uranium 235 therefore corresponds to 81,900 million joules, an energy equivalent of 2.7 metric tons of coal or 13.7 barrels of crude oil. A nuclear power plant producing

SOURCES OF WASTE HEAT are evident in a thermal infrared image, made at an altitude of 2,000 feet, of an industrial concentration along the Detroit River in Detroit. The whiter an object is, the hotter it was when the image was made. The complex at left

1,000 electrical megawatts with a thermal efficiency of 33 percent would consume about three kilograms of uranium 235 per day.

A nuclear "burner" uses up large amounts of uranium 235, which is in short supply since it has an abundance of only .7 percent of the uranium in natural ore. If reactor development proceeds as foreseen by the Atomic Energy Commission, inexpensive reserves of uranium (costing less than $10 per pound) would be used up within about 15 years and medium-priced fuel (up to $30 per pound) would be used up by the year 2000. Hence there has been concern that present reactors will deplete these supplies of uranium before converter and breeder reactors are developed to make fissionable plutonium 239 and uranium 233. Either of these isotopes can be used as a catalyst to burn uranium 238 or thorium 232, which are relatively abundant. Thorium and uranium together have an abundance of about 15 parts per million in the earth's crust, representing therefore a source of energy millions of times larger than all known reserves of fossil fuel.

The possibility of generating energy by nuclear fusion is more remote. Of the two processes being considered—the deuterium-deuterium reaction and the deuterium-tritium reaction—the latter is somewhat easier because it proceeds at a lower temperature. In it lithium 6 is the basic fuel, because it is needed to make tritium by nuclear bombardment. The amount of energy available in this way is limited by the abundance of lithium 6 in the earth's crust, namely about two parts per million. The deuterium-deuterium reaction, on the other hand, would represent a practically inexhaustible source of energy, since one part in 5,000 of the hydrogen in the oceans is deuterium.

One must hope, then, that breeder reactors and perhaps fusion reactors will be developed commercially before the supplies of fossil fuel and uranium 235 are exhausted. With inexhaustible (but not cheap) supplies of nuclear energy, automobiles may run on artificially produced ammonia or methane; coal and oil shale will be used as the basis for chemicals, and electricity generated in large breeder or fusion reactors will be used for such purposes as the manufacture of ammonia and methane, the reduction of ores and the production of fertilizers.

It is difficult at this stage to predict the effects of large-scale use of nuclear energy on the biosphere. One must make certain assumptions about the disposal of radioactive wastes. A reasonable assumption is that they will be rendered harmless by techniques whereby long-lived radioactive isotopes are made into solids and buried. (They are potentially dangerous now because of the technique of storing them as liquids in underground tanks.) Short-lived radioactive wastes can presumably be stored safely until they decay.

For both nuclear energy and for processes involving fossil fuels the major problem and the major impact of human energy production is the dissipation of waste heat. The heat has direct effects on the biosphere and could have indirect effects on climate. It is useful to distinguish between local problems of thermal pollution, meaning the problems that arise in the immediate vicinity of a power plant, and the global problem of thermal balance created by the transformation of steadily rising amounts of energy.

The efficiency of a power plant is determined by the laws of thermodynamics. No matter what the fuel is, one tries to create high-temperature steam for driving the turbines and to condense the steam at the lowest possible temperature. Water is the only practical medium for carrying the heat away. Hence more than 80 percent of the cooling water used by U.S. industry is accounted for by electric power plants. For every kilowatt-hour of energy produced about 6,000 B.T.U. in heat must be dissipated from a fossil fuel plant and about 10,000 B.T.U. from a contemporary nuclear plant.

In the U.S., where the consumption of power has been doubling every eight to 10 years, the increase in the number and size of electric power plants is putting a severe strain on the supply of cooling water. By 1980 about half of the normal runoff of fresh water will be needed for this purpose. Even though some 95 percent of the water thus used is returned to the stream, it is not the same: its increased temperature has a number of harmful effects. Higher temperatures decrease the amount of dissolved oxygen and therefore the capacity of the stream to assimilate organic wastes. Bacterial decomposition is accelerated, further depressing the oxygen level. The reduction of oxygen decreases the viability of aquatic organisms while at the same time the higher temperature raises their metabolic rate and therefore their need for oxygen.

In the face of stringent requirements being laid down by the states and the Department of the Interior, power companies are installing devices that cool water before it is returned to the stream. The devices include cooling ponds, spray ponds and cooling towers. They function by evaporating some of the cooling

center, identifiable by a distinctly warm effluent entering the river, is a power plant. Group of hot buildings at right is a steel mill. Cool land area at bottom is part of Grosse Ile.

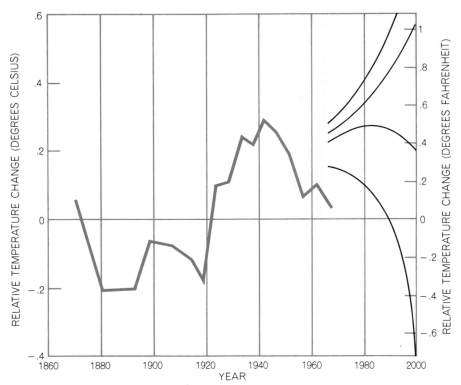

TEMPERATURE TREND in Northern Hemisphere is portrayed as observed (*color*) and as predicted under various conditions (*black*). The top black curve assumes an effect from carbon dioxide only; the other black curves also take account of dust. Second and third curves assume doubling of atmospheric dust in 20 and 10 years respectively; bottom curve, doubling in 10 years with twice the thermal effect thought most probable. Chart is based on work of J. Murray Mitchell, Jr., of the Environmental Science Services Administration.

surrounding countryside affects the ecology and biospheric activity in metropolitan areas in numerous ways. For example, the release of heat in a relatively small local area causes a change in the convective pattern of the atmosphere. The addition of large amounts of particulate matter from industry, space heating and refuse disposal provides nuclei for the condensation of clouds. A study in the state of Washington showed an increase of approximately 30 percent in average precipitation over long periods of time as a result of air pollution from pulp and paper mills.

The worldwide consumption of energy can be estimated from the fact that the U.S. accounts for about a third of this consumption. The U.S. consumption of 685,000 million million B.T.U. per year is equivalent to 2.2 million megawatts. World consumption is therefore some 6.6 million megawatts. Put another way, the present situation is that the per capita consumption of energy in the U.S. of 10,000 watts compares with somewhat more than 100 watts (barely above the food-intake level) in most of the rest of the world.

Projections for the future depend on the assumptions made. If one assumes that in 50 years the rest of the world will reach the present U.S. level of energy consumption and that the population will be 10 billion, the total man-made energy would be 110 million megawatts. The energy would of course be distributed in a patchy manner reflecting the location of population centers and the distributing effects of the atmosphere and the oceans.

That figure is numerically small compared with the amount of solar energy the earth radiates back into space. Over the entire earth the annual heat loss is about 120,000 million megawatts, or more than 1,000 times the energy that

water, so that the excess heat is dissipated into the atmosphere rather than into the stream.

This strategy of spreading waste heat has to be reexamined as the scale of the problem increases. It is already apparent that the "heat islands" characteristic of metropolitan areas have definite meteorological effects—not necessarily all bad. The fact that a city is warmer than the

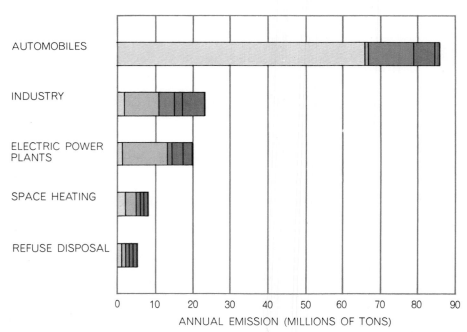

SOURCES OF EMISSIONS from combustion are ranked. Five parts of each bar represent (*from left*) carbon monoxide, sulfur oxides, hydrocarbons, nitrogen oxides and particles.

WASTE HEAT that is an inevitable accompaniment of the human use of energy is evident in the thermal infrared image of New York on the opposite page. The thermogram was made with a Barnes thermograph that depicts emissions of energy on a color scale ranging from black for the coolest objects through green, yellow and red to red-purple for the hottest ones. Some of the emissions represent solar energy stored in the walls of buildings, but a large fraction is waste heat from the human consumption of energy. The rectangular elements of the image result from digitized output of thermograph. Empire State Building is at center.

would be dissipated by human activity if the level of energy consumption projected for 2020 were reached. It would be incautious to assume, however, that the heat put into the biosphere as a result of human energy consumption can be neglected because it is so much smaller than the solar input. The atmospheric engine is subtle in its operation and delicate in its adjustments. Extra inputs of energy in particular places can have significant and far-reaching consequences.

V

HEAT: THE ULTIMATE WASTE

With the advent of nuclear power, energy can now be created as well as exploited. The second law of thermodynamics tells us that such energy can only flow from a more to a less useful form, that it must all eventually end up as waste heat. As long as human use of energy merely diverted some of the energy flow of the earth without adding to it, the heat was not a severe problem. The burning of the accumulated fossil-fuel energy reserves in so short a time has increased local heat loads far above the natural level, and the predicted shift to the less efficient nuclear power plants will raise them even more sharply. The next five articles discuss the actual and potential consequences of the addition of such huge quantities of thermal waste to the earth's carefully balanced energy budget.

V HEAT: THE ULTIMATE WASTE

INTRODUCTION

Chaos *Umpire sits,*
And by decision more imbroils the fray
By which he reigns: next him high Arbiter
Chance *governs all. Into this wild Abyss,*
The Womb of nature and perhaps her Grave,
Of neither Sea, nor Shore, nor Air, nor Fire,
But all these in this pregnant causes mixt
Confus'dly, and which thus must ever fight,
His dark materials to create more Worlds,
Into this wild Abyss the wary fiend
Stood on the brink of Hell and look'd a while,
Pondering his voyage . . .

John Milton
Paradise Lost

The importance to human culture of the Promethean gift of fire was well appreciated from earliest times. With or without tools, the control of the forces of nature was made possible by the domestication of fire and the consequent harnessing of energy. Fire was used for warmth, which made possible permanent settlements in forbidding climates. It was also used to predigest food and thus enormously expand available food supplies, which made the abandonment of a purely hunting culture possible. The cooking and baking of cereal grains was a necessary prerequisite for the agricultural revolution. Fire supplied heat to cook with, to bake with, to make pottery, to refine ores and smelt metals, and, beginning in the eighteenth century, to be harnessed for the production of motion by the heat-engines of the Industrial Revolution. Heat and fire also had their dangerous aspects, but until recently only uncontrolled fire was considered foe rather than friend. In the past thirty years, however, we have unleased a potential source of fire and heat that might cause Prometheus to reconsider. Not only through the sun-bright flame of nuclear weapons, but also through the quiet and slow burning of nuclear fuels, the vast quantity of heat generated by industrialized societies threatens to upset the delicate equilibrium of more than one biocycle, and possibly even the equilibrium of the total energy cycle of the earth itself.

The recognition that heat is a form of energy is less than 150 years old. Before that time, fire was considered to be an elemental substance, as in the division of all matter into Earth, Air, Fire, and Water, a theory advanced by Empedocles in the fifth century B.C. In postulating the existence of the four elements as underlying the sensory qualities of the physical world, Empedocles was the first to distinguish between passive matter and force (love and strife), which moved the cosmos. These four elements were taken to be primordial; ordinary matter was considered to be a mixture of the elements. Although the appearance of matter might be altered, the quantity of the elemental substances was held to be conserved. According to the interpretation of Aristotle, the four elements were characterized by pairs of the four basic qualities: heat, cold, dryness, and moisture. Matter was transmuted from one form to the other by altering the mix of the four elements, which could be accomplished by the transference of one or more of the four qualities.

Aristotle's formulation of the structure of matter continued to be used, to a greater or lesser extent, for two millennia—up to the birth of modern chemistry in the seventeenth century—and provided the rational basis for the study of alchemy. The view of heat as being a manifestation of some tangible substance was not materially challenged until the opening years of the scientific revolution. Galileo, Robert Boyle, Robert Hooke, Christiaan Huygens, and Francis Bacon, among others, believed heat to be a motion of small particles, although it was not clear whether these were particles of the matter itself or of a separate fluid. In any case, their views seem colored more by an attempt to fit all of nature into the newly mechanized world picture than by available scientific evidence. The mechanical theory of heat in this period was no better grounded than the contemporary mechanical theory of light. *Phlogiston*, the "matter of fire" was introduced by J. J. Becher in the mid-seventeenth century; but it was soon evicted out the front door in 1787 by Antoine Laurent Lavoisier, while the equally material *caloric*, the "substance of heat," was let in by the back door. This left unaltered the general view of heat as a property that a body possessed in some fixed quantity. Heat was first demonstrated to be a form of energy and not of matter in the cannon-boring experiment of Benjamin Thompson, Count Rumford, which was published in 1798. By showing that the frictional heat generated when boring a cannon was produced continuously, Rumford demonstrated that no finite amount of "heat fluid" was contained in either the cannon or the bore. Although this caused him to return to the old theory of heat as a vibratory motion of particles, he did not connect the quantity of heat produced to the work done by his horses in operating his boring machinery.

The impetus for the development of thermodynamics in the nineteenth century was acquired not so much from science as from the needs of technology. For, as Stanley W. Angrist describes in his survey "Perpetual Motion Machines," magicians were not alone in attempting to obtain something for nothing from nature. The mechanical analog of enrichment through enchantment was the perpetual motion machine—a device thought to be capable of generating enough power to operate itself with no further energy input once started. By ingenious design, it was to have enough power left over to do useful work as well as keep going. Some of the more notorious of these machines—such as the recycling water mill and the overbalanced wheel—are catalogued by Angrist. These could not be dismissed by purely physical arguments before the nineteenth century; until it was proven that heat was itself a form of energy, no energy conservation law could be convincingly demonstrated, nor could it be proven that such a machine could not be built. Even the Paris Academy's famous decision in 1775 that it would no longer receive papers on squaring the circle, trisecting the angle, duplicating the cube, or perpetual motion could be attacked as being arbitrary, high-handed, and motivated more by internal pressures than by the scientific knowledge of the time.

The first law of thermodynamics, the law of conservation of energy, was initially advanced by Julius Robert Mayer in 1842 as the rule of the indestructability of force. This was followed in 1843 by James Prescott Joule's measurements of the mechanical equivalent of heat, which showed that a fixed amount of work done generated a fixed quantity of heat. The quantification of the relation between heat and work was subsequently formalized into the first law in 1847 by Hermann von Helmholtz, whose analysis began with an axiomatic denial of the possibility of perpetual motion machines. The work of Mayer, Joule, and Helmholtz was not well received at first, and it was not until 1851 that the first law became widely accepted. This acceptance was due largely to the work of William Thomson, Lord Kelvin, who also introduced the word *energy* to describe the common physical property of heat and work that was both interchanged and conserved. It is the first law that proves that a perpetual motion machine of the "first kind"—such as Robert Fludd's water mill—is impossible.

There is another class of impossible machines that does not violate the first law, yet these machines are still based upon a form of perpetual motion— for example, devices that transfer heat from a colder to a warmer object without needing any external power or engines such as John Gamgee's that absorb ambient heat and convert it to work. The former would spontaneously turn lukewarm water into ice and steam. The latter would enable a ship to ply the oceans without fuel by extracting heat from the water and using the energy to turn the engine. A mechanical theory of heat does not in itself preclude such machines, for mechanisms are reversible. If heat were simply mechanical, then a clever enough technician should be able to build a device which would cause Rumford's cannon to rotate when heated. Machines such as these are perpetual motion machines of the "second kind": engines that do not violate conservation of energy and are not forbidden on strictly mechanical grounds. Yet, once started, they would continue to produce useful work at no cost perpetually. Clearly, a second law is needed.

The second law of thermodynamics, conceptually far subtler than the first, was actually established a few years earlier. In 1824 the brilliant young French engineer Nicolas Léonard Sadi Carnot published his one work, "Reflections on the Motive Power of Fire, and on Machines Fitted to Develop that Power." Until that time, heat engines had been constructed on a purely *ad hoc* basis. Unlike the case of purely mechanical engines, such as water wheels or windmills, there was no analytical physical theory by which the development of heat engines could be directed, nor was there any way to estimate their maximum efficiency. Carnot sought to convert an almost entirely empirical technology into a scientific one by establishing an analytic method for gauging the efficiency of such engines. He was guided by the general principles that perpetual motion was impossible and that the most efficient machine was a reversible one; and his principle methodological contribution was to abstract the operation of a heat engine from its structure, to generalize the treatment to an idealized machine whose efficiency under any given set of conditions could not be exceeded. To perform the analysis, he invented the method of the closed cycle; the cycle he devised to maximize efficiency, the Carnot cycle, remains the most efficient cycle possible for any engine operating under a given set of conditions. His book, which was almost unknown, was read by Émile Clapeyron, who used Carnot's ideas but gave them a more correct, mathematical, and general form. Clapeyron's paper, first published in 1834, was not widely read until its translation into German and republication in 1843. The first law had already been advanced at that time, and the value of Clapeyron's work and Carnot's method were quickly recognized by scientists working in the field. Using the cyclic engine as a model, the first law can be stated: *It is impossible to construct an engine which, operating in a cycle, produces more energy than is contained in the heat and work put into it.* Perpetual motion of the first kind is prohibited.

Both Kelvin and Rudolph Clausius applied themselves to extending Clapeyron's work into a second law as theoretically rigorous as the first. Clausius (who states in his paper of 1850 that he *still* has been unable to obtain a copy of Carnot's book and knows of it only through Clapeyron) introduced the word *entropy* in 1852 to describe that quantity, related to heat capacity, that is conserved in a reversible process. Kelvin and Clausius, by disentangling entropy flow from heat flow, both established the second law. Clausius's statement of this law was: *It is impossible to build an engine which, operating in a cycle, has no effect other than to extract heat from a cooler body and deliver it to a warmer.* Kelvin arrived at another version: *It is impossible to build a machine which, operating in a cycle, has no effect other than the extraction of heat from a body and the delivery of an equal amount of work.* Clausius's statement requires that a refrigerator have an engine to drive it; Kelvin's requires that an engine that delivers useful work have both a heat source and a heat sink. Both can be shown to be equivalent forms of the second law, which demonstrates the impossibility of constructing a perpetual motion machine of the second kind. Unlike energy, which must be conserved, entropy tends to increase (or at best stay constant) in any process. The total amount of entropy in the universe can only increase. The notion of irreversibility was thus introduced into physics.

The first and second laws—which are of such fundamental importance that they should be struck in stone above the door of every school, laboratory, or business that deals with energy in any form—lacked only an absolute scale for relating entropy to energy. Kelvin provided this by showing that there is an absolute zero of temperature at which the heat capacity and the entropy of any substance are at an absolute minimum, so that the total entropy change of a body is as exactly calculable as its total energy. Eventually this came to be known as the third law of thermodynamics: *The entropy of the body tends to a minimum at absolute zero.* In a popular student paraphrase, the three laws can be stated thus:

1. You can't win.
2. In the long run, you can't break even.
3. You can't get out of the game.

The final step in the program of thermodynamics—the mechanical theory of heat—was to connect energy and entropy to the motion of atoms and molecules. Through the work of James Clerk Maxwell, Ludwig Boltzmann, and Josiah Willard Gibbs, the heat energy of a body was shown to be contained in the kinetic energy of its molecules. This motion also contains the entropy, not as an inherent property, but as a measure of the randomness of molecular motion. The first law is a law of kinetics and, therefore, of perfection: the conservation of energy is a consequence of mechanics. The second law is a law of statistics. It quantifies imperfection and subjects chaos to the rule of scientific law. Entropy measures the quantity of disorder in a system.° When a falling rock hits the ground, its energy of motion is converted to heat, to random motions of atoms. The second law simply states that, once randomized this way, it is extremely improbable that all the atoms will rearrange their motions again in such a way as to restore the original condition and cause the rock to fly upward. The second law predicts a statistical evolution toward disorder but can only make accurate predictions for complex systems, on the average, or in the long run. Entropy can be locally decreased—as, for instance, in the evolution of life or the manufacture of beer cans—but only at the expense of increasing it somewhere else by at least an equal amount. The second law also determines which way heat flows. Since the energy of a body decreases more rapidly than its entropy does as its temperature is lowered, energy tends to flow toward lower temperatures to increase its disorder. Therefore, a heat engine ultimately produces waste heat in such a manner as to decrease the order of the universe; and the universe itself will suffer an ultimate "heat death" when its energy is spread out as thinly as possible and ceases to flow at all.

In "The Conversion of Energy," Claude M. Summers analyzes the thermal efficiency of a number of extant and hypothetical technologies for the "al-

---

°See "What is Heat?" by Freeman J. Dyson, *Scientific American*, September 1954.

chemy of energy," the conversion of the energy supplied by nature into the many forms in which it is consumed. His use of the term *efficiency* is, unfortunately, confusing. Thermal efficiency, in the technical, scientific sense of the second law, is the ratio of work output to energy input. In this sense, all heat output is waste. An electric space heater, which converts 100 percent of the electrical work from the power company's lines into heat is a totally inefficient machine. In the usage adopted by Summers, however, it is a perfectly "efficient" machine in the sense that it performs its assigned task with no waste. Certainly no engineer would improve the "efficiency" of a space heater by arranging for it to jump up and down to perform work, but for other mechanisms this is not so obvious. An incandescent light bulb has a technical efficiency of only 5 percent for converting electricity to light, but if the other 95 percent that goes to heat is used for part of the space-heating of a large building, the practical efficiency of the system (for converting electricity) may be as high as 90 percent or more. However, the efficiency of converting fuel energy to electricity is only about 35 percent, so it would be better to burn the fuel directly for heat. Efficiency must be carefully defined.

Purely mechanical energy carries almost no entropy; it is of extremely high technical quality because it is almost completely usable. Heat energy at very high temperatures is still of fairly high quality; since its entropy is relatively low, it is largely usable. Heat energy at low temperatures is of very poor quality; due to its high entropy only a small portion of it is recoverable and convertible to usable work. For this reason, energy converted to heat is referred to as energy degraded. Concentrated energy is high-quality energy—energy convertible at small cost. Diffuse energy, waste heat, is not only of low technical quality but is also virtually unusable for any commercial or social purpose. It is thermal garbage, and a major concern of environmentalists. Entropy-free mechanical energy can in principle be interconverted freely. The efficiency of the water wheel or dynamo is limited only by friction, the final mediator of motion, and by technical ingenuity in eliminating losses. The burning of fuel to produce heat has a technical efficiency of zero, but it has a practical efficiency limited only by problems of containment and insulation. It is in the critical transition from heat to work that the restrictions of the second law become important. Until recently, the Carnot limit for the efficiency of heat engines was far beyond the reach of technique. Even today, only the steam turbine has come close to the maximum possible efficiency and at great engineering cost. Most of our other popular machines remain far below these limits. We have too much inefficiency built into our system. Houses are often underinsulated, and buildings overlit. Fuel burned to generate electricity for space heating dissipates more than half of its energy to the plant environment. Automobiles are inefficient, overpowered, and overweight, so that vast quantities of fuel are needed to accelerate them (all of which ends up as heat, just as if the gasoline were spilled on the ground and burned). Such technical clumsiness markedly increases the waste heat dumped into the environment, and it would be good policy to minimize it whether there were an energy shortage or not. But, as Summers shows, energy conservation is only a short-term solution to our supply problem unless the growth of energy consumption is limited. The fourfold increase in the efficiency with which fossil fuels have been used since 1900 has purchased only fifty years of grace, by current estimates. A further increase by a factor of two, which is probably the best we could possibly do in light of the second law, would postpone the crisis by perhaps twenty years more.

More and more of the energy we consume is in the form of electricity, and the generation of electricity produces waste heat in quantities far beyond our ability to direct it to any useful purpose. The cheapest way to get rid of this heat has been to dump it into water, and the consequences of using the rivers, lakes, and oceans as a thermal "garbage can" have proved to be severe, as discussed by John R. Clark in his article "Thermal Pollution and Aquatic Life." Aquatic ecosystems are far more fragile than commonly believed, and the oceans are far more limited as reservoirs of food than was formerly thought to be the case. By comparison with the land, the deep ocean is largely a sparse pasture grazed by migratory schools of fish. Only in shallower water or where water and land are in intimate contact (such as, rivers, lakes, bays, tidal estuaries, and continental shelves) do rich populations consistently develop. But it is just these waters that are in demand for dumping or cooling purposes. Clark estimates that as much as a third of the fresh water runoff in the United States will be required for plant cooling by the year 2000. In California, where free runoff is scarce, rising energy demands have raised the spectre of a coastal nuclear plant in every bay and estuary.

Fish and shellfish populations are the most vulnerable part of a delicate aquatic ecosystem. Even a small increase in water temperature can have a drastic effect. It is not just a matter of massive fish kills, although these are bound to become more common as the number of power plants increases. A temperature that is a few degrees too high upsets biological regulatory mechanisms, particularly if the rise is abrupt, and is capable of completely altering the species mix in fish populations. Similar upsets due to other causes have occurred before, and not just from overfishing or pollution. The introduction of the lamprey into the four lower Great Lakes by the opening of the St. Lawrence Seaway resulted in almost total destruction of the commercially valuable and abundant lake trout and whitefish populations. The introduction of new species, depletion of breeding stocks, and chemical pollution are all known to be dangerous and are beginning to be carefully regulated and guarded against. At the same time, we have been allowing thermal pollution to practice precarious experiments in the same waters.

As Clark points out, there is no technological fix for waste heat. It must exist; it must be unloaded into the local environment. There are, nevertheless, some alternate methods for the distribution of the heat. The load could be shifted from the more fragile portions of the ecosystem onto possibly more durable ones. Artificial lakes may serve in some places. Long conduits leading the heated water far out to sea and down past the rich life zone may be a solution for others. In many situations, the most practicable solution is the cooling tower, either wet or dry, which takes the heat load off the immediate surface of the earth and places it into the atmosphere. This method is discussed by Riley D. Woodson in his article "Cooling Towers." Several types of these towers are presently used in places such as Great Britain and Europe where there is a shortage of freshwater runoff. They constitute an adequate technical solution for the moment; but as heat production continues to increase, even cooling towers may adversely affect the surrounding environment. Due to the high comparative cost of dry cooling towers, where the heat is ultimately carried away by the air (as in the radiator of an automobile), virtually all of the towers presently being built are of the "wet" type, where the heat is carried off by the evaporation of water. Such towers can produce a local fogging effect, although proper design to maximize condensation can do much to minimize it. Nevertheless, even wet towers are more desirable on environmental grounds than a once-through system. Their water supply is nearly a closed cycle, and only a relatively small amount of water needs to be added for replenishment.

Technical solutions such as cooling towers do work at some times and in some places. They are expensive, however, and capital investment is still the dominant factor in making these decisions. It seems overly optimistic to assume that, if excessive heating of natural waters becomes apparent, it can be quickly stopped. Present experience indicates otherwise: whales continue to be fished to extinction; automobile engine size continues to be unregulated; and, under the pressure of a shortage of oil, the Alaska pipeline was approved.

Even a clear and present danger is often not enough to win the case for an environmental issue when it conflicts with an economic or political one. To plan a conversion of a plant, arrange for its shutdown, and then install cooling towers will be very expensive; three million dollars is Woodson's estimate for the cost of the cheapest kind of tower for an eight-hundred-thousand-kilowatt plant. There will be a corresponding reluctance to implement the change and a correspondingly persuasive argument that the threat has been exaggerated.

The environmental problems of waste heat would have been taken even less seriously if it were not for the advanced urbanization of technical societies and the resulting concentration of a large fraction of their total energy consumption into a very small area. The resulting thermal imbalance can appreciably modify local climates, as discussed in "The Climate of Cities," by William P. Lowry. There are a number of reasons for the mean temperature of a city being several degrees higher than the surrounding countryside, such as differences in surface, shape, drainage, and air flow. A large part of the effect is nevertheless directly attributable to the heat generated through the consumption of energy for transportation, lighting, heating, and cooling. The direct effects of the heat dumped into the city's microclimate are further enhanced by interaction with air pollutants. The net effect is a rise in the average temperature and in the incidence of fog. Since temperature differences can exceed ten degrees at times, the effect has become known as the "heat island"; and the data of J. Murray Mitchell, which indicate that cities are cooler on Sundays, clearly show that human energy consumption is largely responsible. In many of the larger cities, the total energy consumption (all of which ends up as heat) has become an appreciable fraction of the total solar input. According to Donald R. F. Harleman (*Technology Review*, December 1971), the total energy consumption in Manhattan is six times the net average input from solar radiation; in Moscow it is three times the solar input; and in the entire ten-thousand-square-kilometer Los Angeles basin, it is about 7 percent of solar input. Extrapolations differ, particularly since the increased incidence of fog and decreased sunshine tend to compensate slightly. But by some estimates there will be some megalopolitan areas where the thermal output over a huge area will equal the net solar input by the year 2000. Such cities will be heated as if there were a second sun in the sky.

Since heating consumes a large share of the energy, waste heat is partially useful in midwinter. But, in midsummer, a thermal output of even 25 percent of the solar input can be of great consequence to the human population. The effects go beyond the increased discomfort that is already commonly known and noted. Many homes and businesses and most large office buildings are air-conditioned. The heat output from the air-conditioners adds considerably to the total thermal burden in midsummer; and the hotter it is, the more waste heat they put out. (One of the restrictions of the second law is that, the greater the temperature difference a refrigeration device tries to maintain, the greater the work that needs to be put in, and the greater the waste heat it will put out.) If present trends toward increased consumption continue, the feeding back of the waste heat of the city into its own cooling machines will cause the consumption of electricity to rise more steeply than the number of installations. Such a positive feedback system always has an instability point, at which the results of feeding the output back to the input defeat all efforts of the system to regulate itself. In terms of the cities, this means possible thermal runaway, where the combination of heat output and the drop in electrical voltage as the load increases causes the cooling systems to draw more current. This, in turn, increases the thermal dump. Sooner or later the system will collide with the limit on deliverable electricity, and the power will go out. It can be argued that whatever adverse effects energy consumption may have on the cities is relevant only as an index of comfort, since the presence of the city has already destroyed the natural environment. It can be further argued that climate control through central heating and cooling could be brought back from the brink of runaway by better design (such as keeping individual air conditioners away from each other so they do not have to cope with each other's waste heat) and increased electrical capacity. But Lowry's article points out that there already exist areas of the world where the local climate has been altered through the direct intervention of thermal pollution. If the cities generated all of their own power locally, the problem would be all theirs, and the outcry against excessive use of energy would be quickly forthcoming. As it is, about half of the thermal and climatological costs will be borne by the far more fragile ecosystems of the nonurban areas surrounding the generating plants.

Conventional economics lies at the heart of the matter. Conventional cost-benefit analysis includes no built-in incentive to encourage the power companies to go to the enormous expense of diverting their waste heat. Cooling towers were installed in Europe not for environmental reasons but because they were economically more attractive. Given the must higher labor costs in the United States and the comparative abundance of rivers, lakes, and accessible ocean sites, towers are unlikely to become the least expensive cooling method here for many years to come. Woodson's chart tells some of the story. Only the mechanical-draft wet tower comes close to being economically competitive for cooling the hypothetical eight-hundred-thousand-kilowatt plant, but the listed cost is only capital investment. The fans to run the tower will use some of the electricity, raising delivered costs and slightly increasing the local heat load. Owing to its lower operating costs, the cheapest alternative solution is probably the natural draft wet tower, which costs only about 4.6 million dollars. Yet, if the cost of repairing and maintaining the environment and the price of leaving the natural water supplies untouched could be weighted into the cost-benefit, dry towers would be the best bargain in many (if not all) installations, despite their high initial investment. There is simply no way to include environmental costs as yet. The passage of the National Environmental Protection Act (NEPA) has done much to alleviate this situation, but under the menace of an impending fuel shortage as of fall, 1973, many of the key provisions of NEPA were suspended or bypassed. This was possible primarily because NEPA is still seen as being an external restriction on industry, not as an integral part of the economics of energy production. Unfortunately, solutions like those of the NEPA are the only way to deal with environmental protection at the moment. There is no way to quantify the price of the extinction of a species or relate it to the quantity of heat delivered at a cooling outlet or the temperature rise of the water.

The 1973-74 crisis, although it caused an abrogation of the NEPA, may turn out to be constructive in the long run if it causes a reexamination of our patterns of energy consumption. As fossil-fuel supplies run out, and we turn to newer and potentially inexhaustible sources of energy, it would be wise to consider the possible consequences of the heat load we impose upon the earth by our use of energy. The burning of fossil fuels has released in decades the accumulated energy of nearly a billion years of sunlight, measurably adding to the heat burden. The less efficient nuclear power plants and the attendant shift from direct fuel consumption to electrical power will cause the waste heat to grow faster than the net consumption of usable energy. Whether the energy industry shifts to cooling towers and artificial lakes or goes over to geothermal or solar energy, consumption must still be closely regulated. There will still be problems with local heating due to the concentration of consumption in urban areas and to the waste heat of the generation systems. The threat to local environments and local climates will still remain, even if the overall dangers to the total energy cycle are removed through the use of solar power or the strict regulation of overall consumption. At present levels of use, local heating effects on urban climates are already noticeable. Further increases may cause effects which are wider in scope—continental or even global. As both Singer and Lowry point out in their articles, the stability of atmospheric climate cycles is unknown, and local dumping of large quantities of heat in amounts comparable to or exceeding solar input may have profound and widespread consequences. But the ultimate ceiling on heat production comes from the energy balance of the planet itself. Earth absorbs energy from the sun and radiates it away as heat. In order to maintain a constant temperature, the energy radiated out must equal the net input. If extra heat is generated within the system, the earth must raise its temperature slightly to increase its radiation and balance its energy budget in accordance with the first law. A heat production equal to 1 percent of the net solar input will cause a mean temperature rise of about one degree Fahrenheit. The global effects of an appreciable mean temperature rise are unknown, but the figure of one degree has frequently been taken as the maximum tolerable limit above which no guarantees against major climatological effects can be given. The actual threshold may be much lower.

Each year the United States alone spends some tens of billions of dollars for military defense. Less than 1 percent of that sum is allocated to defending our physical environment from ourselves. Conventional wisdom, conventional politics, and conventional economic theory were all developed during the centuries of scientific and technical expansion, when the resources of the earth seemed limitless. With the present growing sense of the finiteness of this singular and beautiful planet, the costs of repairing and maintaining it are properly coming to be included in the price of goods and services. Waste heat, however, cannot be scrubbed, catalyzed, recycled, buried, ignored, or eliminated. A highly industrialized nation such as ours is committed to paying a high cost in energy for the benefits conferred, and a high cost in heat for the energy. The wasteful consumption of energy can be regulated, and new sources of power that do not add to the overall heat load can be developed. Given adequate funding and federal support, pilot plants could be produced in a relatively short time. But time is definitely short; we are on the threshold of being irrevocably committed to an energy economy based on thermal generation of electricity by fast breeder reactors. Within a few years the proliferation of such plants will present us with a *fait accompli*. Before we surrender ourselves and our children to a future inextricably linked to the thermal inefficiency and dangers of the widespread use of fission power, we should at least do all in *our* power to buy time. We should at least keep alive the possibility of a choice.

Perpetual Motion Machines

by Stanley W. Angrist
January 1968

Over the past 400 years numerous inventors have proposed marvelous ways of getting something for nothing. All these proposals have foundered on either the first or the second law of thermodynamics

The interwoven tapestry of history sometimes displays odd relationships. Who would think, for example, that two medical men would be leading figures in the history of efforts to make a perpetual motion machine? One of them, the 17th-century English physician Robert Fludd, is usually mentioned as one of the first to propose a perpetual motion machine to do useful work. The other, the 19th-century German physician Julius Robert Mayer, was among those who established as a law of nature the conservation of energy, which dooms proposals such as Fludd's.

The notion of getting something for nothing that underlies all speculations about perpetual motion is as old as Archimedes and may be a good deal older. In classical times, however, there was a tendency to depend on supernatural power sources. A more down-to-earth approach to the subject grew out of economic considerations as the first labor-saving machines, in particular water mills, spread across Europe. Originally used to grind flour, water mills evolved rapidly in later Roman times. Although they were never especially popular in the Mediterranean area, quite the opposite was the case in western Europe. By A.D. 400 water-driven flour mills and sawmills were common in France. Twenty years after the Norman Conquest some 5,600 water mills were operating in 3,000 English communities, and before the end of the 14th century in England waterpower had been harnessed not only to grind flour and saw wood but also to tan leather, to full woolens and to grind pigments for paint. Soon almost every English manor that was situated on a stream—roughly a third of all the manors in the Domesday Book—had its own mill. Elsewhere floating mills were anchored in rivers and tidal mills stood in estuaries.

Villagers and townspeople who had no access to running water naturally sought alternative sources of power. One result was the windmill, a thoroughly practical invention. A less practical result was a series of proposals for closed-cycle water mills such as the one that Fludd put forward in 1618. The proposal must have seemed sensible enough at the time. If the water that turns a mill wheel could be collected from the race at the foot of the wheel and somehow put back into the reservoir above the wheel, the need for a source of running water would disappear. Centuries of experience had shown that mill wheels could turn big grindstones or raise heavy hammers. Why couldn't the wheel also drive a pump that would recycle the mill's water supply? In Fludd's day there was little reason to deny the possibility.

The same was true half a century later, when John Wilkins, Bishop of Chester and an early official of the Royal Society, put forward his views on the subject. In the 1670's Wilkins envisioned three natural power sources that might be harnessed to provide perpetual motion. These, in his words, were "Chymical Extractions," "Magnetical Virtues" and "the Natural Affection of Gravity."

Wilkins' third power source embraces the entire family of overbalanced wheels; that is, wheels that turn because they are perpetually heavier on one side than the other. He specifically mentioned only one formula for chemical extraction; its underlying concept may have arisen from a misunderstood observation of the ceaseless motion of small particles visible in a fluid that we know as Brownian movement. Wilkins also designed, but almost certainly never tried to build, a machine to utilize magnetic attraction. At no point, however, did he suggest a way of obtaining useful work out of the proposed perpetual motions.

As can be judged by Wilkins' leading role in the scientific community, speculation on perpetual motion machines was not yet considered a crackpot activity. Robert Boyle recounted in detail his examination of a fluid, compounded of bituminous oils and similar ingredients, that an engineer of his acquaintance had prepared as a charge for fire bombs. The engineer had mixed the ingredients over a fire and was surprised to find that days after the pot had been left to cool the fluid in it still swirled about. Keeping the pot in his laboratory for a time, Boyle observed that the oilier constituents of the fluid continued to stream, alternately spreading across the surface and then sinking out of sight. Again he made no proposal for harnessing the motion.

How was the tolerant attitude of early scientists toward perpetual motion transformed into today's skepticism? Clearly we now have far more theoretical knowledge and can make much more refined devices such as bearings, linkages and heat exchangers. Cannot this combination of talents close the apparently tiny gap between the designs of earlier times and the construction of actual working models? The answer, of course, is an emphatic no. For a perpetual motion machine to function, whatever its design, would require that it violate either the first or the second law of thermodynamics.

The first law of thermodynamics—the principle of energy conservation that Mayer helped to formulate—can be stated in various ways. One way of putting it says that a fixed amount of mechanical work always gives rise to the equivalent amount of heat. Thus energy can be converted from work into heat, but it can neither be created nor destroyed. There are more complex formulations of the first law but all eventually arrive at the same conclusion: The total energy of the universe is constant.

Even before Mayer, pioneer studies of heat phenomena by the Scottish chemist Joseph Black and the American-born Count Rumford had helped to clear the way for deeper understanding. Black established the vital distinction between heat (as a quantity of something) and temperature (as an index of heat's intensity). The interrelationship of heat, energy and temperature is a complex one that can be explained by an analogy. After rain falls into a lake it is no longer rain but simply water; after heat is transferred to a body (because of a temperature difference between the cool body and its warm surroundings) it is no longer heat but simply energy. If the lake has no outlets, the rain raises the water level; if the body cannot get rid of energy, the heat transfer adds to its total energy and thereby raises its index of heat—its temperature.

In Black's time variations in temperature and energy were attributed to the presence or absence of the intangible fluid called caloric. Rumford, in turn, struck a deathblow to the concept of caloric with his experiments in a Bavarian cannon foundry. Bringing water to a boil solely with the heat generated by the boring of a cannon barrel, he concluded that the heat was due to friction. This was the first demonstration of the connection between heat and work, but it was soon confirmed by Humphry Davy's experiment in which the rubbing together of two pieces of ice was shown to produce heat. It was a number of years, however, before the equivalence of work and heat was determined with any precision.

This brings us up to Mayer. In 1840, when he was 27, Mayer sailed from Rotterdam as ship's physician on the schooner *Java*, bound for the East Indies. Although it is doubtful that he knew anything about Black's work or Rumford's, he had brought along Antoine Laurent Lavoisier's treatise on chemistry, and he soon became fascinated by Lavoisier's suggestion that animal heat is generated by the slow internal combustion of food.

When the *Java* reached the East Indies, 28 of its crew were ill with fever. The treatment for fever in those days was to bleed the patient, and when Mayer did so, he observed that the crewmen's venous blood was bright red rather than the normal dark red—almost as red as arterial blood. Now, one of La-

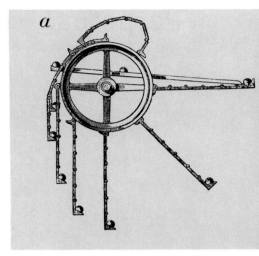

OVERBALANCED WHEELS have been the most common prime movers of perpetual motion machines. Just as the water's weight overbalances a mill wheel and makes it turn, so various means of apparently adding weight to one side of a wheel were expected

voisier's comments was that, when the body is in warm surroundings, less internal combustion is required to keep it warm than when it is in cold ones. In support of this view he and others pointed to variations in the color of venous blood. Mayer concluded that his pa-

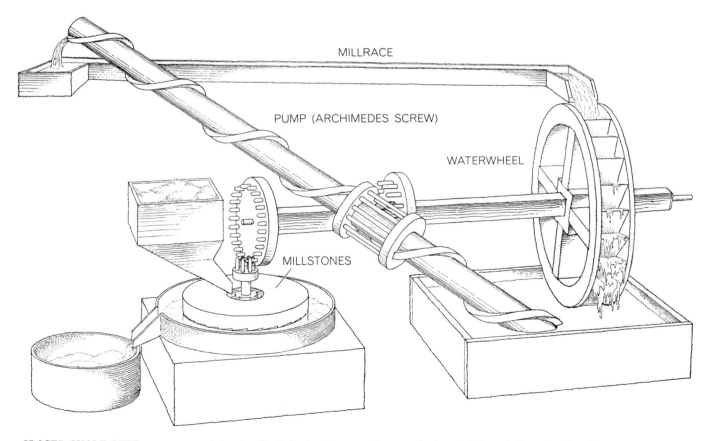

CLOSED-CYCLE MILL was proposed by the English physician Robert Fludd in 1618 as a source of perpetual power in areas that lacked streams. The fact that such devices could not work because they required a violation of the principle of energy conservation, formally known as the first law of thermodynamics, was not recognized by the scientific community until two centuries after Fludd.

b *c* *d*

to move the four machines shown above. The first device (*a*) was expected to turn when jointed arms, with weights that rolled to their ends, were extended on one side; actually the wheel remains exactly balanced, whether or not the arms are extended. A much more complex wheel (*b*) was designed with the same objective. Like *a*, however, it is actually in balance in spite of its shifting weights. A pair of buoys within a water-filled drum were expected to move weights that would overbalance the next device (*c*). Finally (*d*), a starkly simple design reflects the inventor's conviction that his overbalanced wheel rim would spin between two rollers in spite of its lack of any support. These engravings and four on the following pages appeared in early issues of SCIENTIFIC AMERICAN.

tients' venous blood looked like arterial blood because, like arterial blood, it had a high content of oxygen. It seemed that in the tropical East Indies the crewmen's bodies did not consume as much oxygen as they did in cooler latitudes.

At this point Mayer went a step beyond Lavoisier to conjecture that the body heat evolved by the metabolism of food should be exactly balanced by a combination of two opposing factors. These were, first, the heat lost by the body to its surroundings and, second, the work the body performed. Mayer was soon saying that heat and work are merely different manifestations of energy (which he called "force"), and that the two manifestations are equivalent.

The young physician was not able to obtain experimental proof of his conjecture; he lacked both money and laboratory facilities. He did, however, analyze data collected by other investigators on the specific heat of air, and he managed to calculate a numerical relation between heat and units of mechanical work. In effect he had determined the mechanical equivalent of heat. He offered an account of his work to the foremost scientific journal of his day, *Annalen der Physik und Chemie,* but it was refused. In 1842 a revised account appeared in another journal, and Mayer's version of the first law of thermodynamics was formally put forward. "Once in existence," he wrote, "force cannot be annihilated; it can only change its form."

James Prescott Joule, the son of a prosperous English brewer, was born four years later than Mayer. Joule stud-

ied chemistry in Manchester with John Dalton, but soon he developed an enthusiasm for experiments in electricity and electromagnetism, a field in which he was largely self-taught. In the early 1840's he carefully measured the amount of work required to raise the temperature of a pound of water from 60 degrees Fahrenheit to 61 degrees. Joule announced his result in 1843: the amount of mechanical energy required was 838 foot-pounds. In later years he refined this figure to 772 foot-pounds, a value remarkably close to today's standard (778.16 foot-pounds).

Joule had thus quantified the relation between work and heat that Mayer had propounded. Four more years were to elapse, however, before a third young investigator, Hermann von Helmholtz, convinced the international scientific community that the first law was a valid generalization. In 1847, when he was 26, Helmholtz presented his formulation of the first law before the Physical Society of Berlin in a paper titled "On the Conservation of Force." He began his analysis by declaring that perpetual motion machines were axiomatically impossible. In physics, as in mathematics, axioms are distinct from theorems. A theorem is a conclusion that is logically deduced from an axiom. An axiom does not require logical proof. The validity of a physical axiom can be based instead on repeated observations of nature. Thus Helmholtz did not need to prove his axiom; it was enough to point out that no one had yet built a successful perpetual motion machine. Helmholtz observed further that he was not alone in his view. Nicolas

Léonard Sadi Carnot, an early student of the theoretical basis for the steam engine, had started with a similar axiom and had reached a number of significant conclusions concerning the dynamics of heat. As we shall see, Carnot's work, particularly his 1824 study "Reflections on the Motive Power of Heat," forms the basis of the second law of thermodynamics.

Proceeding from his axiom, Helmholtz next showed that the failure of perpetual motion machines led logically to the conclusion that energy is always conserved. He went on to demonstrate that both heat (regarded as small-scale motion) and work (regarded as large-scale motion) were forms of energy and that what was conserved was the total of the two forms rather than either heat or work taken separately. Helmholtz showed that the findings of Joule's experiments were in general agreement with calculations of the kind made by Mayer. Like Mayer, Helmholtz submitted his paper to *Annalen der Physik und Chemie,* and it too was refused.

I have given this brief history of the first law because it is the law that most would-be inventors of perpetual motion machines attempt to evade. Their expectation is that more energy can be wrung out of some device incorporating falling or turning bodies than is required to restore the device to its original state. Curiously one of the most persistent proposals is Fludd's closed-cycle water mill. As late as 1871 an American patent attorney noted with some asperity that inventors submitted one or another variation on Fludd's mill to him every year, inquiring whether the concept was patentable. Over the years, however, devices that depended for their power on overbalanced wheels gradually abandoned running water in favor of ingenious weight-shifting systems.

Many inventors have preferred power sources more sophisticated than the overbalanced wheel. Both early and late they have turned to magnets, at first natural magnets and then electrically powered ones. Bishop Wilkins' design for a magnetic device depended on a lodestone, which was to be strong enough to pull an iron ball up a ramp. Just before the ball had climbed all the way up to the lodestone, it would drop through a hole and roll back down a curved second ramp. The ball would then pass through a door and reach the first ramp again, where it would resume its upward journey. It is easy enough to find the flaw in Wilkins' proposal today: any lodestone strong enough to pull the ball up the ramp would be too strong to let it fall back to its starting point.

A 19th-century device solved a similar problem by incorporating an electromagnet that was alternately turned on and off. When the circuit to the magnet was closed, the magnet's attraction was supposed to pull a connecting rod that acted through a crank to impart rotary motion to a disk. The spinning of the disk between two brushes was then expected to generate enough electricity to energize the magnet. Once the machine was started by hand the inventor expected it to run forever, or at least until the contact points on the switches wore out. As so often happens in the design of perpetual motion machines, the inventor had made no allowance for the energy lost to friction and, in this case, to electrical resistance as well.

It is scarcely surprising that the chimera of perpetual motion has attracted not only savants and optimists but also rascals. One of the many outright frauds in the history of perpetual motion machines was perhaps the most elegant overbalanced wheel ever built. It was the work of a skilled Connecticut machinist, E. P. Willis. A large gear wheel, mounted at an angle to the horizontal and fitted with a complex system of weights, purportedly drove a smaller hollow flywheel. After the machine had attracted much attention in New Haven, where Willis charged admission for viewing it, he moved it to New York in 1856. There the same attorney who was to comment on the perpetual rediscovery of Fludd's water mill went to see it. The exhibitors, he noted, were careful

FRAUDULENT MACHINE that purported to demonstrate perpetual motion was built by a Connecticut machinist in the 1850's. Ostensibly each pair of rod-linked weights that rested atop the tilted wheel (*right*) was shifted in position as the wheel turned, so that the uphill weight extended beyond the wheel's perimeter. The resulting imbalance was said to be sufficient to keep the wheel turning and to drive a flywheel (*left*). Actually compressed air passed through a strut (*A, far left*), turning both of the wheels.

not to claim that Willis had achieved perpetual motion; rather, they challenged any visitor to provide another explanation for the machine's motion. Although a glass case kept viewers from inspecting the machine closely, the attorney noted that there was a suspiciously nonfunctional strut below the edge of the hollow flywheel. Evidently a steady flow of compressed air, undetectable outside the glass case, kept the flywheel turning. Thus it was actually the flywheel that drove the overbalanced wheel, rather than the reverse.

The Willis fraud, Fludd's water mill and all similar devices are based on the assumption that the first law of thermodynamics can be violated. Some perpetual motion machines, however, do not violate the first law; neither friction nor electrical resistance is a significant problem in their design. They are nonetheless impossibilities because they attempt instead to circumvent the second law of thermodynamics.

The foundation of the second law was laid down by the observations of Carnot, and the law was first fully formulated by the German physicist Rudolf Clausius. The first law, as we have seen, demonstrates that a fixed amount of mechanical work can always be converted into the equivalent amount of heat. But the most casual observation of a heat engine in operation—for example a steam engine—makes it plain that the reverse of the first law's axiom is not precisely true: a fixed amount of heat cannot be completely converted into the same amount of work. When heat is transformed into work, some of the initial energy is unavoidably wasted. In the case of a real steam engine operating in the real world, some of the wasted energy goes to overcoming friction, some is lost through warming the engine and the surrounding atmosphere, some through leakage and some through other avenues of dissipation.

Carnot wanted to find out whether improved design could eliminate all steam engine losses. He created in his imagination an ideal engine; it was leakproof, completely insulated and frictionless. He then ran the imaginary engine through a full operating "cycle" (a concept, by the way, that Carnot was the first to develop). In one ideal cycle water is heated until it vaporizes into steam, and the pressure of the steam forces the engine's piston to move; the cycle is completed when the expanded steam cools and condenses into water again, allowing the piston to return to

PERPETUAL MOTION powered by "Magnetical Virtues" was to be achieved by a steel bullet as it rolled up and down a pair of ramps according to a design proposed by the Bishop of Chester in the 1670's. The lodestone placed on the top of the pedestal was expected to draw the bullet up the straight ramp, whereupon it would fall through a hole and roll back to its starting position. The bishop did not propose harnessing the device to obtain power.

PERPETUAL MOTION powered by electricity was often favored by 19th-century inventors. In this design the attraction of an electromagnet worked through a crank to turn a wheel; the wheel's rotation was then supposed to generate enough electricity to work the magnet. As usual the inventor neglected to allow for the losses from friction and resistance.

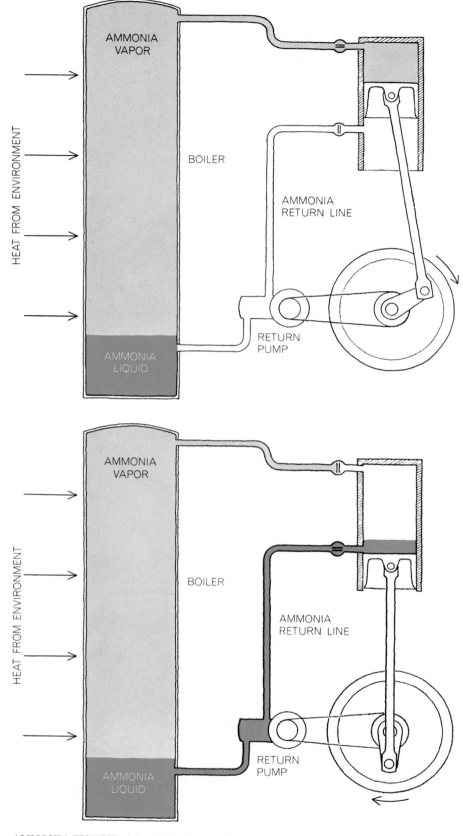

AMMONIA ENGINE of the 1880's, designed by John Gamgee, was based on the expectation that free power would be produced because heat transferred from the surroundings would turn the ammonia from liquid to gas. The gas pressure is enough to drive a piston (*top*). When the gas then expanded in the cylinder (*bottom*), the inventor expected that it would condense spontaneously and return to the boiler as a liquid to repeat the cycle. He did not anticipate the need to refrigerate the return side of his engine in order to convert the ammonia gas to liquid. The energy needed to do this, of course, is more than the engine produces.

its starting position. Thinking through the steps in this ideal cycle, Carnot realized that a complete conversion of heat into work was impossible; an unavoidable loss of thermal energy occurred in the process of cooling and condensation.

The language Carnot used to state his conclusions is strange to our ears because, like others in his day, he talked about heat in terms of caloric. What he had to say was nonetheless the earliest statement of the second law. The transformation of heat into motive power, Carnot wrote, "is fixed solely by the temperature of the bodies between which is effected...the transfer of the caloric." This is to say that, in order to do work, heat must "run downhill" as water does and, just as with water, the farther it runs downhill, the greater the amount of work it does. This is the concept we express today by saying that heat must be transferred from a higher temperature to a lower one to do work.

Building on Carnot, Clausius applied the word "entropy" (from the Greek for "turning") to the index used to measure the amount of heat that is unavoidably lost. The modern formulation of the second law that says that entropy always increases arises from the earlier realization that heat is a downhill flow. Because the supply of energy in the universe is a constant that cannot be increased or decreased, and because at the same time the downhill flow of heat is accompanied by inevitable losses, a time will inevitably come when the entire universe will be at the same temperature. With no more hills of heat and therefore, in Carnot's terms, no further transfers of caloric, there can be no work. This inevitable end, sometimes called the "heat death" of the universe, concerns us here because perpetual motion machines that attempt to violate the second law are expected to achieve a localized halt in the inevitable increase of entropy and produce a decrease of entropy instead.

The fact that, on the average, entropy continually increases does not, of course, rule out the possibility that occasional local decreases of entropy can take place. It is only that the odds against such an event are extraordinarily long. The bed of a river could suddenly cool, yielding its energy to the running water, and this energy could be applied in some way to make the water run uphill. But riverbeds do not cool and water does not run uphill. A similar loan of thermal energy from the river's environment could al-

"GENERATOR" AND "MOTOR" of a supposed perpetual motion device were exhibited in Philadelphia for more than a decade late in the 19th century. The inventor, John E. W. Keely, contended that the generator (*left*) turned tap water into high-pressure "etheric vapor" when "vibratory energy" was applied. After Keely's death the totally fraudulent device was found to run on compressed air.

low the water to dissociate spontaneously into hydrogen and oxygen. But the water does not dissociate spontaneously. Furthermore, an old man on the riverbank, watching the water flow by, could grow younger rather than older, but he doesn't. Rivers continue to flow downhill, H_2O remains water and man inevitably ages. The chemist Henry A. Bent has calculated the odds against a local reversal of entropy, specifically the probability that one calorie of thermal energy could be converted completely into work. His result can be expressed in terms of a familiar statistical example: the probability that a group of monkeys hitting typewriter keys at random could produce the works of Shakespeare. According to Bent's calculation, the likelihood of such a calorie conversion is about the same as the probability that the monkeys would produce Shakespeare's works 15 quadrillion times in succession without error.

It is against these odds that the would-be inventor of a perpetual motion heat engine must struggle. One such inventor was John Gamgee, who was active in Washington, D.C., during the 1880's. He developed a heat engine that he called the zeromotor because its normal operating temperature was zero degrees centigrade. The zeromotor was not unlike an ordinary steam engine except that the working fluid was ammonia rather than water. Liquid ammonia vaporizes into a gas at a low temperature, and at zero degrees C. the gas exerts a pressure of four atmospheres. Gamgee reasoned that the transfer of heat from the environment, rather than the energy supplied by the combustion of fuel, would be enough to transform the ammonia working fluid from a liquid to a gas. He reasoned further that the ammonia gas, on driving the piston back and expanding, would cool, condense and drain into a reservoir, whereupon the cycle could begin again [*see illustration on opposite page*].

Anyone with the slightest knowledge of Carnot's cycle, let alone the second law of thermodynamics, could scarcely take such an idea seriously, yet Gamgee and his supporters were undoubtedly sincere. They had either incorrectly calculated or failed to calculate the zeromotor's temperature requirements. The heat transfer from the environment was indeed sufficient to convert ammonia from a liquid to a gas, but this advantage is nullified in the system as a whole by the cooling of the gas on expansion. Starting at zero degrees C. and a pressure of four atmospheres, the temperature of the gas has fallen to −33 degrees by the time its volume has quadrupled. If the gas is to condense into a liquid, both the condenser and the reservoir must be at a temperature lower than −33 degrees. Gamgee had not provided for this cooling, and if he had, the cooling process would of course have required more energy than the zeromotor could produce.

One of Gamgee's principal supporters was B. F. Isherwood, Chief Engineer of the U.S. Navy. In March, 1881, Isherwood reported favorably on the zeromotor to the Secretary of the Navy, in spite of the fact that scholars had pointed out that the engine fatally violated the second law. Official Washington came close to embracing the inventor. The Secretary of the Navy was not the only high official who inspected a model of the zeromotor with interest; so did other Cabinet members and President Garfield himself. Isherwood's gullibility may be hard to understand, but not his interest. This was an era when in order to keep the U.S. fleet at sea it was necessary to maintain a complicated and expensive network of coaling stations abroad. If the Gamgee engine had worked, coaling stations could have been forgotten and all the energy the Navy would have needed to power its fleet could have been provided by the thermal energy contained in the seawater in which the ships floated.

The surprisingly wide acceptance of proposals such as Gamgee's can be explained, of course, by general ignorance of known principles. As early as 1775 the French Academy of Sciences passed a resolution refusing to entertain any future communications concerning perpetual motion. The U.S. Patent Office has long declined to examine applications for patents covering perpetual motion machines unless the applicant furnishes a working model or "other demonstration...of the operativeness of the invention," a ruling that has produced much hostile correspondence but no working models. In spite of such official opposition public sophistication regarding the possibility of building perpetual

PERPETUAL WATERFALL, one of many "impossible objects" conceived by the contemporary Dutch artist Maurits C. Escher, seems to drive a mill wheel endlessly. Mills that produced enough power to recirculate the water needed to drive them were among the first perpetual motion machines proposed in Europe (*see bottom illustration on page 340*). Their designers did not realize that, because of the energy losses due to friction, no mill is capable of pumping all its water supply back to the uphill starting position.

motion machines was slow to develop.

Perhaps the most ingenious, and certainly the longest-lived, swindle involving a supposed perpetual motion machine began in 1875, when John E. W. Keely unveiled a combined "generator" and engine at his home in Philadelphia. There was nothing unusual about Keely's engine, which was a variation on the conventional steam engine. Keely's generator, however, was extraordinary. It was an elaborate combination of metal globes, tubes, petcocks, nozzles, valves and gauges, but its operation was deceptively simple. Keely would blow into a nozzle for half a minute and then pour five gallons of tap water into the generator through the same nozzle. After turning various petcocks and valves he would show onlookers a pressure gauge indicating that the generator was full of a mysterious "vapor" with a pressure of 10,000 pounds per square inch. "People have no idea of the power in water," Keely would say. "A bucket of water has enough of this vapor to produce a power sufficient to move the world out of its course."

Keely and his associates formed the Keely Motor Company, capitalized at $1 million. They raised much of the money from gullible New York businessmen. As the years passed, although no engines other than the first one were ever built, Keely's showmanship became more polished. By 1881 he had begun to attribute the production of vapor to "vibratory energy," and he would "vivify" the vapor during demonstrations with a giant tuning fork. By 1884 he had so mastered what he now called the "etheric vapor" or the "interatomic ether" that he demonstrated a new device: a cannon, complete with a "vibrator" near the breech, that was capable of propelling a ball 500 yards with a muzzle velocity of 500 feet per second.

Keely died in 1898. The son of one of his major backers promptly rented his Philadelphia house and explored the premises in the company of reputable witnesses, seeking evidence of fraud. Under the floor of the house the searchers found a three-ton metal tank that had evidently served as a reservoir for compressed air. In the walls were found quantities of brass tubing, and a false ceiling suggested the means by which Keely and his associates had conducted the compressed air to his generator. Whatever other laws he may have broken in his long career, Keely had left the first and second laws of thermodynamics inviolate.

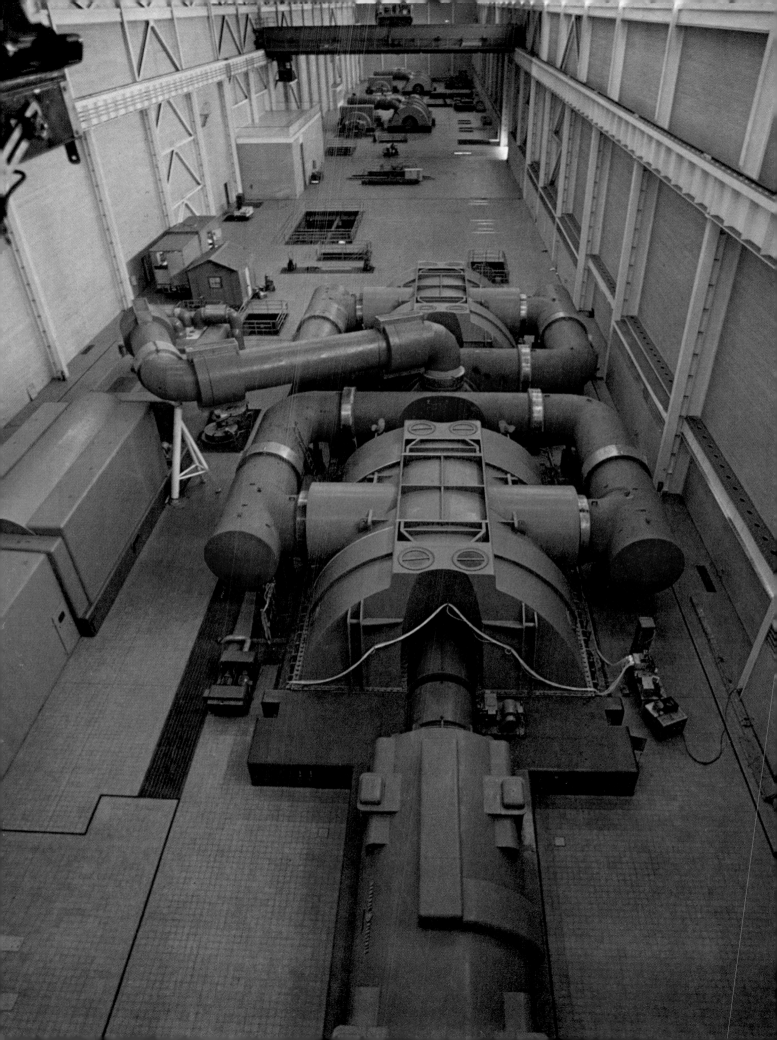

The Conversion of Energy

by Claude M. Summers
September 1971

The efficiency of home furnaces, steam turbines, automobile engines and light bulbs helps in fixing the demand for energy. A major need is a kind of energy source that does not add to the earth's heat load

A modern industrial society can be viewed as a complex machine for degrading high-quality energy into waste heat while extracting the energy needed for creating an enormous catalogue of goods and services. Last year the U.S. achieved a gross national product of just over $1,000 billion with the help of 69×10^{15} British thermal units of energy, of which 95.9 percent was provided by fossil fuels, 3.8 percent by falling water and .3 percent by the fission of uranium 235. The consumption of 340 million B.t.u. per capita was equivalent to the energy contained in about 13 tons of coal or, to use a commodity now more familiar, 2,700 gallons of gasoline. One can estimate very roughly that between 1900 and 1970 the efficiency with which fuels were consumed for all purposes increased by a factor of four. Without this increase the U.S. economy of 1971 would already be consuming energy at the rate projected for the year 2025 or thereabouts.

Because of steadily increasing efficiency in the conversion of energy to useful heat, light and work, the G.N.P. between 1890 and 1960 was enabled to grow at an average annual rate of 3.25 percent while fuel consumption increased at an annual rate of only 2.7 per-cent. It now appears, however, that this favorable ratio no longer holds. Since 1967 annual increases in fuel consumption have risen faster than the G.N.P., indicating that gains in fuel economy are becoming hard to achieve and that new goods and services are requiring a larger energy input, dollar for dollar, than those of the past. If one considers only the predicted increase in the use of nuclear fuels for generating electricity, it is apparent that an important fraction of the fuel consumed in the 1980's and 1990's will be converted to a useful form at lower efficiency than fossil fuels are today. The reason is that present nuclear plants convert only about 30 percent of the energy in the fuel to electricity compared with about 40 percent for the best fossil-fuel plants.

It is understandable that engineers should strive to raise the efficiency with which fuel energy is converted to other and more useful forms. For industry increased efficiency means lower production costs; for the consumer it means lower prices; for everyone it means reduced pollution of air and water. Electric utilities have long known that by lowering the price of energy for bulk users they can encourage consumption. The recent campaign of the utility in-dustry to "save a watt" marks a profound reversal in business philosophy. The difficulty of finding acceptable new sites for power plants underscores the need not only for frugality of use but also for efficiency of use. Having said this, one must emphasize that even large improvements in efficiency can have only a modest effect in extending the life of the earth's supply of fossil and nuclear fuels. I shall develop the point more fully later in this article.

The efficiency with which energy contained in any fuel is converted to useful form varies widely, depending on the method of conversion and the end use desired. When wood or coal is burned in an open fireplace, less than 20 percent of the energy is radiated into the room; the rest escapes up the chimney. A well-designed home furnace, on the other hand, can capture up to 75 percent of the energy in the fuel and make it available for space heating. The average efficiency of the conversion of fossil fuels for space heating is now probably between 50 and 55 percent, or nearly triple what it was at the turn of the century. In 1900 more than half of all the fuel consumed in the U.S. was used for space heating; today less than a third is so used.

The most dramatic increase in fuel-conversion efficiency in this century has been achieved by the electric-power industry. In 1900 less than 5 percent of the energy in the fuel was converted to electricity. Today the average efficiency is around 33 percent. The increase has been achieved largely by increasing the temperature of the steam entering the turbines that turn the electric generators and by building larger generating units [*see illustration on opposite page*]. In 1910 the typical inlet temperature was 500 degrees Fahrenheit; today the latest

STEAM-DRIVEN TURBOGENERATOR at Paradise power plant of the Tennessee Valley Authority near Paradise, Ky., has a capacity of 1,150 megawatts. When placed in operation in February, 1970, it was the largest unit in the world. The turbine, built by the General Electric Company, is a cross-compound design in which steam first enters a high-pressure turbine, below the angled pipe at the left, then flows through the angled pipe to pass in sequence through an intermediate-pressure turbine (*blue casing at rear*) and then through a low-pressure turbine (*blue casing in foreground*). The high-pressure turbine is connected to one generator (*gray housing at left*) and the other two turbine sections to a second generator of the same capacity (*gray housing in foreground*). The entire unit is driven by eight million pounds of steam per hour, which enters the high-pressure turbine at 3,650 pounds per square inch and 1,003 degrees Fahrenheit. The daily coal consumption is 10,572 tons, enough to fill 210 railroad coal cars. The unit has a net thermal efficiency of 39.3 percent. Two smaller turbogenerators, each rated at 704 megawatts, are visible in the background.

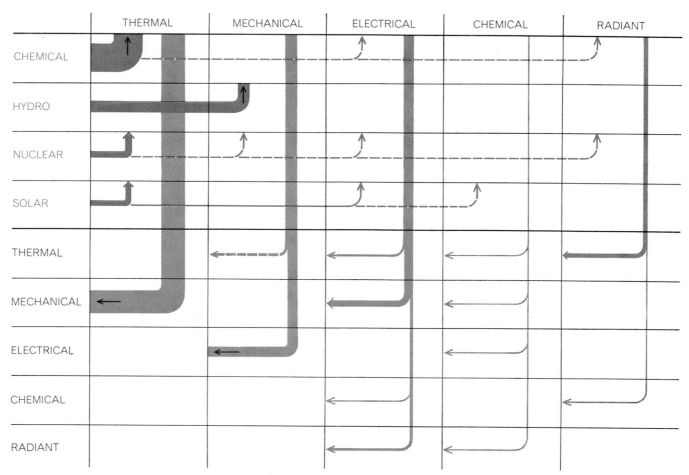

CONVERSION PATHWAYS link many of the familiar forms of
energy. The four forms shown in color are either important
sources of power today or, in the case of solar energy, potentially
important. The broken lines indicate rare, incidental or theoretically useful conversions. The gray lines follow the destiny of intermediate forms of energy. Except for the thermal energy used for
space heating, most is converted to mechanical energy. Mechanical
energy is used directly for transportation (see illustration below)
and for generating electricity. Electrical energy in turn is used for
lighting, heating and mechanical work. As a secondary form, chemical energy is found in dry cells and storage batteries. The radiant energy produced by electric lamps ends up chiefly as heat.

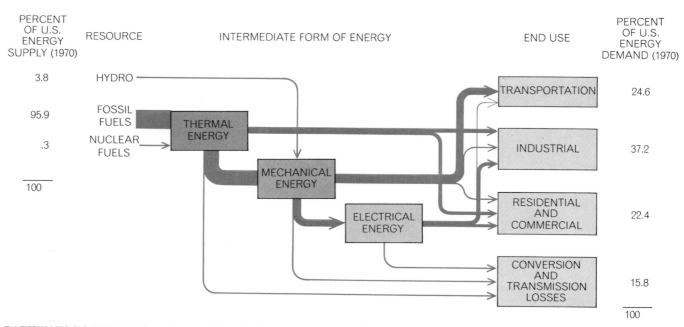

PATHWAYS TO END USES are depicted for the three principal
sources of energy. The most direct and most efficient conversion
is from falling water to mechanical energy to electrical energy. The
energy locked in fossil and nuclear fuels must first be released in
the form of thermal energy before it can be converted to mechanical energy and then, if it is desired, to electric power. Conversion
and transmission losses include various nonenergy uses of fossil
fuels, such as the manufacture of lubricants and the conversion
of coal to coke. The biggest loss, however, arises from the generation of electric power at an average efficiency of 32.5 percent.

units take steam superheated to 1,000 degrees. The method of computing the maximum theoretical efficiency of a steam turbine or other heat engine was enunciated by Nicolas Léonard Sadi Carnot in 1824. The maximum achievable efficiency is expressed by the fraction $(T_1 - T_2)/T_1$, where T_1 is the absolute temperature of the working fluid entering the heat engine and T_2 is the temperature of the fluid leaving the engine. These temperatures are usually expressed in degrees Kelvin, equal to degrees Celsius plus 273, which is the difference between absolute zero and zero degrees C. In a modern steam turbine T_1 is typically 811 degrees K. (1,000 degrees Fahrenheit) and T_2 degrees K. (100 degrees F.). Therefore according to Carnot's equation the maximum theoretical efficiency is about 60 percent. Because the inherent properties of a steam cycle do not allow the heat to be introduced at a constant upper temperature, the maximum theoretical efficiency is not 60 percent but more like 53 percent. Modern steam turbines achieve about 89 percent of that value, or 47 percent net.

To obtain the overall efficiency of a steam power plant this value must be multiplied by the efficiencies of the other energy converters in the chain from fuel to electricity. Modern boilers can convert about 88 percent of the chemical energy in the fuel into heat. Generators can convert up to 99 percent of the mechanical energy produced by the steam turbine into electricity. Thus the overall efficiency is 88 × 47 (for the turbine) × 99, or about 41 percent.

Nuclear power plants operate at lower efficiency because present nuclear reactors cannot be run as hot as boilers burning fossil fuel. The temperature of the steam produced by a boiling-water reactor is around 350 degrees C., which means that the T_1 in the Carnot equation is 623 degrees K. For the complete cycle from fuel to electricity the efficiency of a nuclear power plant drops to about 30 percent. This means that some 70 percent of the energy in the fuel used by a nuclear plant appears as waste heat, which is released either into an adjacent body of water or, if cooling towers are used, into the surrounding air. For a fossil-fuel plant the heat wasted in this way amounts to about 60 percent of the energy in the fuel.

The actual heat load placed on the water or air is much greater, however, than the difference between 60 and 70 percent suggests. For plants with the same kilowatt rating, a nuclear plant produces about 50 percent more waste

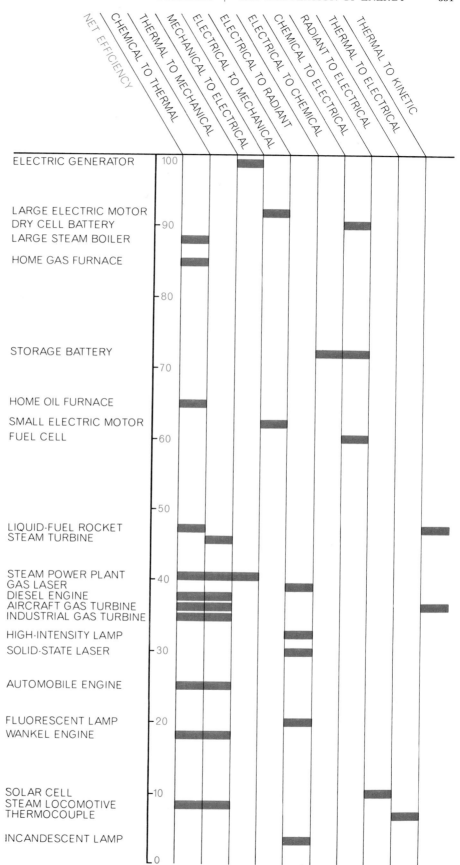

EFFICIENCY OF ENERGY CONVERTERS runs from less than 5 percent for the ordinary incandescent lamp to 99 percent for large electric generators. The efficiencies shown are approximately the best values attainable with present technology. The figure of 47 percent indicated for the liquid-fuel rocket is computed for the liquid-hydrogen engines used in the Saturn moon vehicle. The efficiencies for fluorescent and incandescent lamps assume that the maximum attainable efficiency for an acceptable white light is about 400 lumens per watt rather than the theoretical value of 220 lumens per watt for a perfectly "flat" white light.

heat than a fossil-fuel plant. The reason is that a nuclear plant must "burn" about a third more fuel than a fossil-fuel plant to produce a kilowatt-hour of electricity and then wastes 70 percent of the larger B.t.u. input.

Of course, no law of thermodynamics decrees that the heat released by either a nuclear or a fossil-fuel plant must go to waste. The problem is to find something useful to do with large volumes of low-grade energy. Many uses have been proposed, but all run up against economic limitations. For example, the low-pressure steam discharged from a steam turbine could be used for space heating. This is done in some communities, notably in New York City, where Consolidated Edison is a large steam supplier. Many chemical plants and refineries also use low-pressure steam from turbines as process steam. It has been suggested that the heated water released by power plants might be beneficial in speeding the growth of fish and shellfish in certain localities. Nationwide, however, there seems to be no attractive use for the waste heat from the present fossil-fuel plants or for the heat that will soon be pouring from dozens of new nuclear power plants. The problem will be to limit the harm the heat can do to the environment.

From the foregoing discussion one can see that the use of electricity for home heating (a use that is still vigorously promoted by some utilities) represents an inefficient use of chemical fuel. A good oil- or gas-burning home furnace is at least twice as efficient as the average electric-generating station. In some locations, however, the annual cost of electric space heating is competitive with direct heating with fossil fuels even at the lower efficiency. Several factors account for this anomaly. The electric-power rate decreases with the added load. Electric heat is usually installed in new constructions that are well insulated. The availability of gas is limited in some locations and its cost is higher. The delivery of oil is not always dependable. As fossil fuels become scarcer, their cost will increase, and the production of electrical energy with nuclear fuels will increase. Unfortunately we must expect that a greater percentage of our fuel resources (particularly nuclear fuels) will be consumed in electric space heating in spite of the less efficient use of fuel.

The most ubiquitous of all prime movers is the piston engine. There are two in many American garages, not counting the engines in the power mower, the snowblower or the chain saw. The piston engines in the nation's more than 100 million motor vehicles have a rated capacity in excess of 17 billion horsepower, or more than 95 percent of the capacity of all prime movers (defined as engines for converting fuel to mechanical energy). Although this huge capacity is unemployed most of the time, it accounts for more than 16 percent of the fossil fuel consumed by the U.S. Transportation in all forms—including the propulsion systems of ships, locomotives and aircraft—absorbs about 25 percent of the nation's energy budget.

Automotive engineers estimate that the efficiency of the average automobile engine has risen about 10 percent over the past 50 years, from roughly 22 percent to 25 percent. In terms of miles delivered per gallon of fuel, however, there has actually been a decline. From 1920 until World War II the average automobile traveled about 13.5 miles per gallon of fuel. In the past 25 years the average has fallen gradually to about 12.2 miles per gallon. This decline is due to heavier automobiles with more powerful engines that encourage greater acceleration and higher speed. It takes about eight times more energy to push a vehicle through the air at 60 miles per hour than at 30 miles per hour. The same amount of energy used in accelerating the car's mass to 60 miles per hour must be absorbed as heat, primarily in the brakes, to stop the vehicle. Therefore most of the gain in engine efficiency is lost in the way man uses his machine. Automobile air conditioning has also played a role in reducing the miles per gallon. With the shift in consumer preference to smaller cars the figure may soon begin to climb. The efficiency of the basic piston engine, however, cannot be improved much further.

If all cars in the year 2000 operated on electric batteries charged by electricity generated in central power stations, there would be little change in the nation's overall fuel requirement. Although the initial conversion efficiency in the central station might be 35 percent compared with 25 percent in the piston engine, there would be losses in distributing the electrical energy and in the conversion of electrical energy to chemical energy (in the battery) and back to electrical energy to turn the car wheels. Present storage batteries have an overall efficiency of 70 to 75 percent, so that there is not much room for improvement. Anyone who believes we

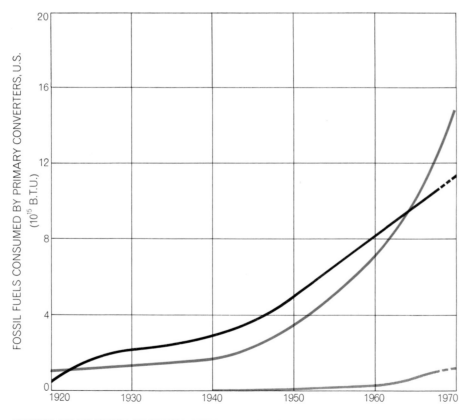

THREE OF FASTEST-GROWING ENERGY USERS are electric utilities (*color*), motor vehicles (*black*) and aircraft (*gray*). Together they now consume about 40 percent of all the energy used in the U.S. As recently as 1940 they accounted for only 18 percent of a much smaller total. The demand for aircraft fuel has more than tripled in 10 years.

would all be better off if cars were electrically powered must consider the problem of increasing the country's electric-generating capacity by about 75 percent, which is what would be required to move 100 million vehicles.

The difficulty of trying to trace savings produced by even large changes in efficiency of energy conversion is vividly demonstrated by what happened when the railroads converted from the steam locomotive (maximum thermal efficiency 10 percent) to diesel-electric locomotives (thermal efficiency about 35 percent). In 1920 the railroads used about 135 million tons of coal, which represented 16 percent of the nation's total energy demand. By 1967, according to estimates made by John Hume, an energy consultant, the railroads were providing 54 percent more transportation than in 1920 (measured by an index of "transportation output") with less than a sixth as many B.t.u. This increase in efficiency, together with the railroads' declining role in the national economy, had reduced the railroads' share of the nation's total fuel budget from 16 percent to about 1 percent. If one looks at a curve of the country's total fuel consumption from 1920 to 1967, however, the impact of this extraordinary change is scarcely visible.

Perhaps the least efficient important use for electricity is providing light. The General Electric Company estimates that lighting consumes about 24 percent of all electrical energy generated, or 6 percent of the nation's total energy budget. It is well known that the glowing filament of an ordinary 100-watt incandescent lamp produces far more heat than light. In fact, more than 95 percent of the electric input emerges as infrared radiation and less than 5 percent as visible light. Nevertheless, this is about five times more light than was provided by a 100-watt lamp in 1900. A modern fluorescent lamp converts about 20 percent of the electricity it consumes into light. These values are based on a practical upper limit of 400 lumens per watt, assuming the goal is an acceptable light of less than perfect whiteness. If white light with a totally flat spectrum is specified, the maximum theoretical output is reduced to 220 lumens per watt. If one were satisfied with light of a single wavelength at the peak sensitivity of the human eye (555 nanometers), one could theoretically get 680 lumens per watt.

General Electric estimates that fluorescent lamps now provide about 70 percent of the country's total illumination and that the balance is divided between incandescent lamps and high-

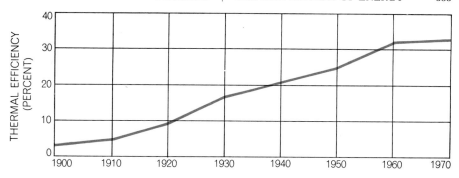

EFFICIENCY OF FUEL-BURNING POWER PLANTS in the U.S. increased nearly tenfold from 3.6 percent in 1900 to 32.5 percent last year. The increase was made possible by raising the operating temperature of steam turbines and increasing the size of generating units.

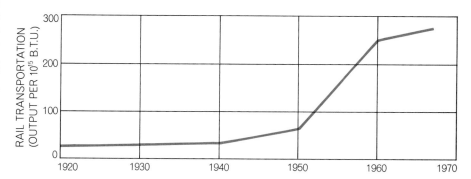

EFFICIENCY OF RAILROAD LOCOMOTIVES can be inferred from the energy needed by U.S. railroads to produce a unit of "transportation output." The big leap in the 1950's reflects the nearly complete replacement of steam locomotives by diesel-electric units.

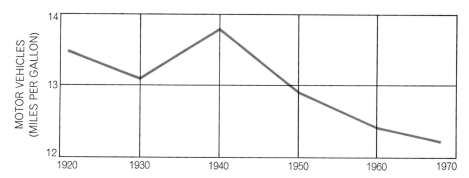

EFFICIENCY OF AUTOMOBILE ENGINES is reflected imperfectly by miles per gallon of fuel because of the increasing weight and speed of motor vehicles. Manufacturers say that the thermal efficiency of the 1920 engine was about 22 percent; today it is about 25 percent.

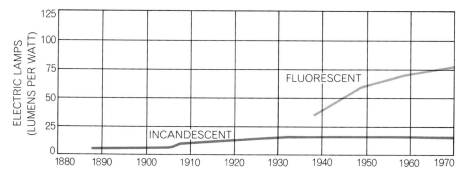

EFFICIENCY OF ELECTRIC LAMPS depends on the quality of light one regards as acceptable. Theoretical efficiency for perfectly flat white light is 220 lumens per watt. By enriching the light slightly in mid-spectrum one could obtain about 400 lumens per watt. Thus present fluorescent lamps can be said to have an efficiency of either 36 percent or 20.

intensity lamps, which have efficiencies comparable to, and in some cases higher than, fluorescent lamps. This division implies that the average efficiency of converting electricity to light is about 13 percent. To obtain an overall efficiency for converting chemical (or nu-clear) energy to visible light, one must multiply this percentage times the average efficiency of generating power (33 percent), which yields a net conversion efficiency of roughly 4 percent. Nevertheless, thanks to increased use of fluorescent and high-intensity lamps, the nation was able to triple its "consumption" of lighting between 1960 and 1970 while only doubling the consumption of electricity needed to produce it.

This brief review of changing efficiencies of energy use may provide some perspective when one tries to evaluate

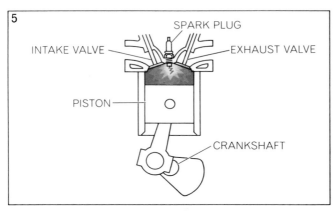

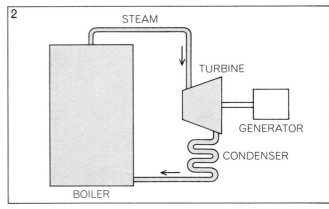

ELECTRIC-POWER GENERATING MACHINERY now in use extracts energy from falling water, fossil fuels or nuclear fuels. The hydroturbine generator (1) converts potential and kinetic energy into electric power. In a fossil-fuel steam power plant (2) a boiler produces steam; the steam turns a turbine; the turbine turns an electric generator. In a nuclear power plant (3) the fission of ura-

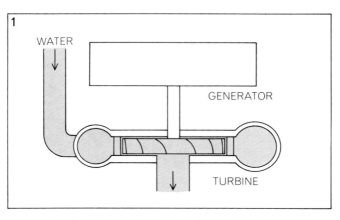

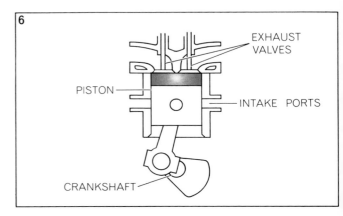

PROPULSION MACHINERY converts the energy in liquid fuels into forms of mechanical or kinetic energy useful for work and transportation. In the piston engine (5) a compressed charge of fuel and air is exploded by a spark; the expanding gases push against the piston, which is connected to a crankshaft. In a diesel engine (6) the compression alone is sufficient to ignite the charge

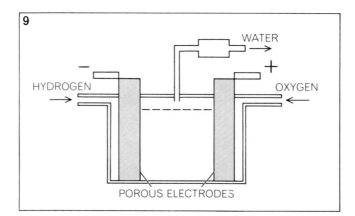

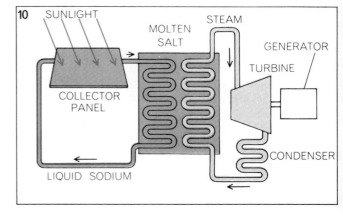

NOVEL ENERGY CONVERTERS are being designed to exploit a variety of energy sources. The fuel cell (9) converts the energy in hydrogen or liquid fuels directly into electricity. The "combustion" of the fuel takes place inside porous electrodes. In a recently proposed solar power plant (10) sunlight falls on specially coated collectors and raises the temperature of a liquid metal to 1,000 degrees F. A heat exchanger transfers the heat so collected to steam, which then turns a turbogenerator as in a conventional power plant. A salt reservoir holds enough heat to keep generating steam during the night and when the sun is hidden by clouds. In a mag-

the probable impact of novel energy-conversion systems now under development. Two devices that have received much notice are the fuel cell and the magnetohydrodynamic (MHD) generator. The former converts chemical energy directly into electricity; the latter is potentially capable of serving as a high-temperature "topping" device to be operated in series with a steam turbine and generator in producing electricity. Fuel cells have been developed that can "burn" hydrogen, hydrocarbons or alcohols with an efficiency of 50 to 60 percent. The hydrogen-oxygen fuel cells used in the Apollo space missions, built by the Pratt & Whitney division of United Aircraft, have an output of 2.3 kilowatts of direct current at 20.5 volts.

A decade ago the magnetohydrodynamic generator was being advanced as

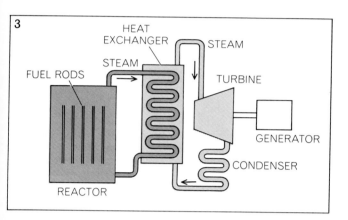

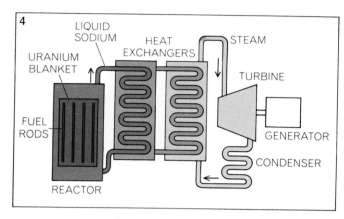

nium 235 releases the energy to make steam, which then goes through the same cycle as in a fossil-fuel power plant. Under development are nuclear breeder reactors (4) in which surplus neutrons are captured by a blanket of nonfissile atoms of uranium 238 or thorium 232, which are transformed into fissile plutonium 239 or U-233. The heat of the reactor is removed by liquid sodium.

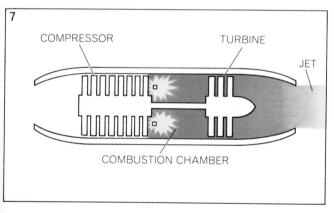

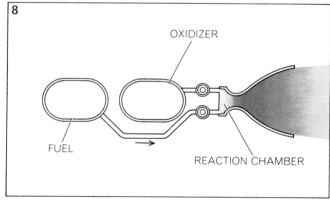

of fuel and air. In an aircraft gas turbine (7) the continuous expansion of hot gas from the combustion chamber passes through a turbine that turns a multistage air compressor. Hot gases leaving the turbine provide the kinetic energy for propulsion. A liquid-fuel rocket (8) carries an oxidizer in addition to fuel so that it is independent of an air supply. Rocket exhaust carries kinetic energy.

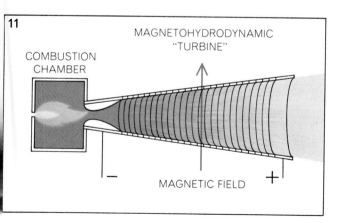

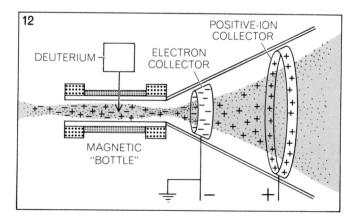

netohydrodynamic "turbine" (11) the energy contained in a hot electrically conducting gas is converted directly into electric power. A small amount of "seed" material, such as potassium carbonate, must be injected into the flame to make the hot gas a good conductor. Electricity is generated when the electrically charged particles of gas cut through the field of an external magnet. A long-range goal is a thermonuclear reactor (12) in which the nuclei of light elements fuse into heavier elements with the release of energy. High-velocity charged particles produced by a thermonuclear reaction might be trapped in such a way as to generate electricity directly.

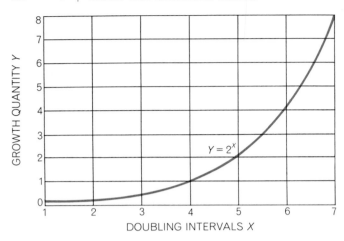

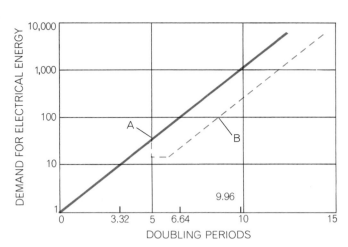

DOUBLING CURVE (*left*) rises exponentially with time. It shows how many doubling intervals are needed to produce a given multiplication of the growth quantity. Thus if electric-power demand continues to double every 10 years, the demand will increase eightfold in three doubling periods, that is, by the year 2001. When ex-ponential growth curves are plotted on a semilogarithmic scale, the result is a straight line (*right*). If electric-power consumption were cut in half at *A*, held constant for 10 years and allowed to return to the former growth rate, time needed to reach a given demand (*B*) would be extended by only two doubling periods, or 20 years.

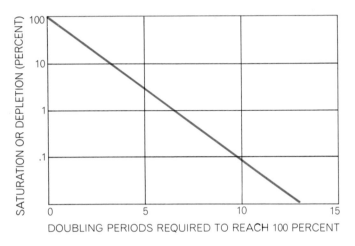

SATURATION OR DEPLETION (PERCENT)	DOUBLING PERIODS TO REACH 100 PERCENT	YEARS FROM NOW
100.0	0.0	0
10.0	3.32	33
1.0	6.64	66
.1	9.96	100
.01	13.28	133
.001	16.60	166
.0001	19.92	199
.00001	23.24	232
.000001	26.56	266
.0000001	29.88	299
.00000001	33.20	332

DEPLETION OF A RESOURCE can be read from the curve at the left. Thus if .1 percent of world's oil has now been extracted, all will be gone in just under 10 doubling periods, or 100 years if the doubling interval is 10 years. The table at the right shows that the ultimate depletion date is changed very little by large changes in the estimate of amount of resource that has been extracted to date.

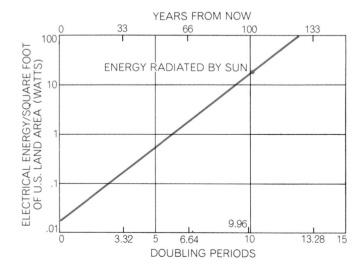

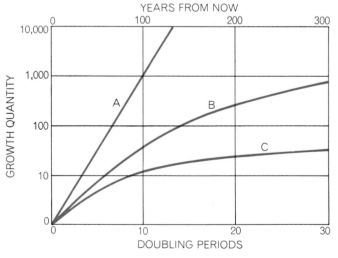

ENERGY RECEIVED FROM SUN on an average square foot of the U.S. will be equaled by production of electrical energy in roughly 100 years if the demand continues to double every 10 years.

THREE GROWTH CURVES are compared. Curve *A* is exponential. In Curve *B* each doubling period is successively increased by 20 percent. In Curve *C* the growth per doubling period is constant.

the energy converter of the future. In such a device the fuel is burned at a high temperature and the gaseous products of combustion are made electrically conducting by the injection of a "seed" material, such as potassium carbonate. The electrically conducting gas travels at high velocity through a magnetic field and in the process creates a flow of direct current [see No. 11 in illustrations on pages 354 and 355]. If the MHD technology can be developed, it should be possible to design fossil-fuel power plants with an efficiency of 45 to 50 percent. Since MHD requires very high temperatures it is not suitable for use with nuclear-fuel reactors, which produce a working fluid much cooler than one can obtain from a combustion chamber fired with fossil fuel.

If ever an energy source can be said to have arrived in the nick of time, it is nuclear energy. Twenty-two nuclear power plants are now operating in the U.S. Another 55 plants are under construction and more than 40 are on order. This year the U.S. will obtain 1.4 percent of its electrical energy from nuclear fission; it is expected that by 1980 the figure will reach 25 percent and that by 2000 it will be 50 percent.

Although a 1,000-megawatt nuclear power plant costs about 10 percent more than a fossil-fuel plant ($280 million as against $250 million), nuclear fuel is already cheaper than coal at the mine mouth. Some projections indicate that coal may double in price between now and 1980. One reason given is that new Federal safety regulations have already reduced the number of tons produced per man-day from the 20 achieved in 1969 to fewer than 15.

The utilities are entering a new period in which they will have to rethink the way in which they meet their base load, their intermediate load (which coincides with the load added roughly between 7:00 A.M. and midnight by the activity of people at home and at work) and peak load (the temperature-sensitive load, which accounts for only a few percent of the total demand). In the past utilities assigned their newest and most efficient units to the base load and called on their older and smaller units to meet the variable daily demand. In the future, however, when still newer capacity is added, the units now carrying the basic load cannot easily be relegated to intermittent duty because they are too large to be easily put on the line and taken off.

There is therefore a need for a new kind of flexible generating unit that may be best satisfied by coupling an industrial gas turbine to an electric generator and using the waste heat from the gas turbine to produce low-pressure steam for a steam turbine–generator set. Combination systems of this kind are now being offered by General Electric and the Westinghouse Electric Company. Although somewhat less efficient than the best large conventional units, the gas-turbine units can be brought up to full load in an hour and can be installed at lower cost per kilowatt. To meet brief peak demands utilities are turning to gas turbines (without waste-heat boilers that can be brought up to full load in minutes) and to pumped hydrostorage systems. In the latter systems off-peak capacity is used to pump water to an elevated reservoir from which it can be released to produce power as needed.

Westinghouse has recently estimated that U.S. utilities must build more than 1,000 gigawatts (GW, or 10^9 watts) of new capacity between 1970 and 1990, or more than three times the present installed capacity of roughly 300 GW. Of the new capacity 500 GW, or half, will be needed to handle the anticipated increase in base load and 75 percent of the 500 GW will be nuclear. More than 400 GW of new capacity will be needed to meet the growing intermediate load, and a sizable fraction of it will be provided by gas turbines. The new peaking capacity, amounting to some 170 GW, will be divided, Westinghouse believes, between gas turbines and pumped storage in the ratio of 10 to seven.

Such projections can be regarded as the conventional wisdom. Does unconventional wisdom have anything to offer that may influence power generation by 2000, if not by 1990? First of all, there are the optimists who believe prototype nuclear-fusion plants will be built in the 1980's and full-scale plants in the 1990's. In a sense, however, this is merely conventional wisdom on an accelerated time scale. Those with a genuinely unconventional approach are asking: Why do we not start developing the technology to harness energy from the sun or the wind or the tides?

Many people still remember the Passamaquoddy project of the 1930's, which is once more being discussed and which would provide 300 megawatts (less than a third the capacity of the turbogenerator shown on page 348) by exploiting tides with an average range of 18 feet in the Bay of Fundy, between Maine and Canada. A working tidal power plant of 240 megawatts has recently been placed in operation by the French government in the estuary of the Rance River, where the tides average 27 feet. How much tidal energy might the U.S. extract if all favorable bays and inlets were developed? All estimates are subject to heavy qualification, but a reasonable guess is something like 100 GW. We have just seen, however, that the utilities will have to add 10 times that much capacity just to meet the needs of 1990. One must conclude that tidal power does not qualify as a major unconventional resource.

What about the wind? A study we conducted at Oklahoma State University a few years ago showed that the average wind energy in the Oklahoma City area is about 18.5 watts per square foot of area perpendicular to the wind direction. This is roughly equivalent to the amount of solar energy that falls on a square foot of land in Oklahoma, averaging the sunlight for 24 hours a day in all seasons and under all weather conditions. A propeller-driven turbine could convert the wind's energy into electricity at an efficiency of somewhere between 60 and 80 percent. Like tidal energy and other forms of hydropower, wind power would have the great advantage of not introducing waste heat into the biosphere.

The difficulty of harnessing the wind's energy comes down to a problem of energy storage. Of all natural energy sources the wind is the most variable. One must extract the energy from the wind as it becomes available and store it if one is to have a power plant with a reasonably steady output. Unfortunately technology has not yet produced a practical storage medium. Electric storage batteries are out of the question.

One scheme that seems to offer promise is to use the variable power output of a wind generator to decompose water into hydrogen and oxygen. These would be stored under pressure and recombined in a fuel cell to generate electricity on a steady basis [see illustration on next page]. Alternatively the hydrogen could be burned in a gas turbine, which would turn a conventional generator. The Rocketdyne Division of North American Rockwell has seriously proposed that an industrial version of the hydrogen-fueled rocket engine it builds for the Saturn moon vehicle could be used to provide the blast of hot gas needed to power a gas turbine coupled to an electric generator. Rocketdyne visualizes that a water-cooled gas turbine could operate at a higher temperature than conventional fuel-burning gas turbines and achieve our overall plant efficiency of 55 percent. If the Rocketdyne concept were successful, it could use hydrogen from any source. A wind-driven hydrogen-rocket gas-turbine power plant should be unconventional enough to please the most exotic taste.

By comparison most proposals for harnessing solar energy seem tame indeed. One fairly straightforward proposal has recently been made to the Arizona Power Authority on behalf of the University of Arizona by Aden B. Meinel and Marjorie P. Meinel of the university's Optical Sciences Center. They suggest that if the sunlight falling on 14 percent of the western desert regions of the U.S. were efficiently collected, it could be converted into 1,000 GW of power, or approximately the amount of additional power needed between now and 1990. The Meinels believe it is within the reach of present technology to design collecting systems capable of storing solar energy as heat at 1,000 degrees F., which could be converted to electricity at an overall efficiency of 30 percent.

The key to the project lies in recently developed surface coatings that have high absorbance for solar radiation and low emittance in the infrared region of the spectrum. To achieve a round-the-clock power output, heat collected during daylight hours would be stored in molten salts at 1,000 degrees F. A heat exchanger would transfer the stored energy to steam at the same temperature. The thermal storage tank for a 1,000-megawatt generating plant would require a capacity of about 300,000 gallons. The Meinels propose that industry and the Government immediately undertake design and construction of a 100-megawatt demonstration plant in the vicinity of Yuma, Ariz. The collectors for such a plant would cover an area of 3.6 million square meters (slightly more than a square mile). The Meinels estimate that after the necessary development has been done a 1,000-megawatt solar power station might be built for about $1.1 billion, or about four times the present cost of a nuclear power plant. As they point out: "Solar power faces the economic problem that energy is purchased via a capital outlay rather than an operating expense." They calculate nevertheless that a plant with an operating lifetime of 40 years should produce power at an average cost of only half a cent per kilowatt hour.

A more exotic solar-power scheme has been advanced by Peter E. Glaser of Arthur D. Little, Inc. The idea is to place a lightweight panel of solar cells in a synchronous orbit 22,300 miles above the earth, where they would be exposed to sunlight 24 hours a day. Solar cells (still to be developed) would collect the radiant energy and convert it to electricity with an efficiency of 15 to 20 percent. The electricity would then be converted electronically in orbit to microwave energy with an efficiency of 85 percent, which is possible today. The microwave radiation would be at a wavelength selected to penetrate clouds with little or no loss and would be collected by a suitable antenna on the earth. Present techniques can convert microwave energy to electric power with an efficiency of about 70 percent, and 80 to 85 percent should be attainable. Glaser calculates that a 10,000-megawatt (10 GW) satellite power station, large enough to meet New York City's present power needs, would require a solar collector panel five miles square.

WIND GENERATOR

ELECTROLYSIS CELL

O₂

H₂

OXYGEN STORAGE

HYDROGEN STORAGE

RECTIFIER

O₂

FUEL CELL

H₂

H₂O

PUMP BACK CONTROL

INVERTER

60-CYCLE POWER LINE

WIND AS POWER SOURCE is attractive because it does not impose an extra heat burden on the environment, as is the case with energy extracted from fossil and nuclear fuels. Unlike hydropower and tidal power, which also represent the entrapment of solar energy, the wind is available everywhere. Unfortunately it is also capricious. To harness it effectively one must be able to store the energy captured when the wind blows and release it more or less continuously. One scheme would be to use the electricity generated by the wind to decompose water electrolytically. The stored hydrogen and oxygen could then be fed at a constant rate into a fuel cell, which would produce direct current. This would be converted into alternating current and fed into a power line. Off-peak power generated elsewhere could also be used to run the electrolysis cell whenever the wind was deficient.

The receiving antenna on the earth would have to be only slightly larger: six miles square. Since the microwave energy in the beam would be comparable to the intensity of sunlight, it would present no hazard. The system, according to Glaser, would cost about $500 per kilowatt, roughly twice the cost of a nuclear power plant, assuming that space shuttles were available for the construction of the satellite. The entire space station would weigh five million pounds, or slightly less than the Saturn moon rocket at launching.

To meet the total U.S. electric-power demand of 2,500 GW projected for the year 2000 would require 250 satellite stations of this size. Since the demand to 1990 will surely be met in other ways, however, one should perhaps think only of meeting the incremental demand for the decade 1990–2000. This could be done with about 125 power stations of the Glaser type.

The great virtue in power schemes based on using the wind or solar energy collected at the earth's surface, farfetched as they may sound today, is that they would add no heat load to the earth's biosphere; they can be called invariant energy systems. Solar energy collected in orbit would not strictly qualify as an invariant system, since much of the radiant energy intercepted at an altitude of 22,300 miles is radiation that otherwise would miss the earth. Only the fraction collected when the solar panels were in a line between the sun and the earth's disk would not add to the earth's heat load. On the other hand, solar collectors in space would put a much smaller waste-heat load on the environment than fossil-fuel or nuclear plants. Of the total energy in the microwave beam aimed at the earth all but 20 percent or less would be converted to usable electric power. When the electricity was consumed, of course, it would end up as heat.

To appreciate the long-term importance of developing invariant energy systems one must appreciate what exponential growth of any quantity implies. The doubling process is an awesome phenomenon. In any one doubling period the growth quantity—be it energy use, population or the amount of land covered by highways—increases by an amount equal to its growth during its entire past history. For example, during the next doubling period as much fossil fuel will be extracted from the earth as the total amount that has been extracted to date. During the next 10 years the U.S. will generate as much electricity as

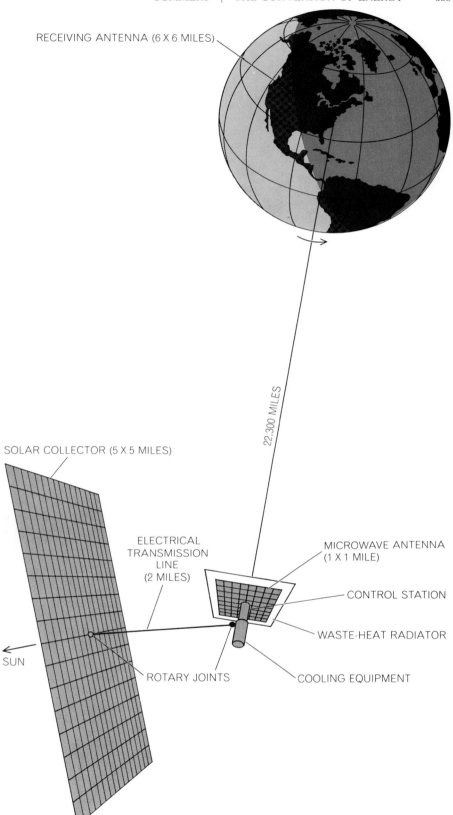

RECEIVING ANTENNA (6 X 6 MILES)

22,300 MILES

SOLAR COLLECTOR (5 X 5 MILES)

ELECTRICAL TRANSMISSION LINE (2 MILES)

MICROWAVE ANTENNA (1 X 1 MILE)

CONTROL STATION

WASTE-HEAT RADIATOR

COOLING EQUIPMENT

ROTARY JOINTS

SUN

SOLAR COLLECTOR IN STATIONARY ORBIT has been proposed by Peter E. Glaser of Arthur D. Little, Inc. Located 22,300 miles above the Equator, the station would remain fixed with respect to a receiving station on the ground. A five-by-five-mile panel would intercept about 8.5×10^7 kilowatts of radiant solar power. Solar cells operating at an efficiency of about 18 percent would convert this into 1.5×10^7 kilowatts of electric power, which would be converted into microwave radiation and beamed to the earth. There it would be reconverted into 10^7 net kilowatts of electric power, or enough for New York City. The receiving antenna would cover about six times the area needed for a coal-burning power plant of the same capacity and about 20 times the area needed for a nuclear plant.

it has generated since the beginning of the electrical era.

Such exponential growth curves are usually plotted on a semilogarithmic scale in order to provide an adequate span. By selecting appropriate values for the two axes of a semilogarithmic plot one can also obtain a curve showing the number of doubling periods to reach saturation or depletion from any known or assumed percentage position [see left half of middle illustration on page 356]. As an example, let us assume that we have now extracted .1 percent of the earth's total reserves of fossil fuels and that the rate of extraction has been doubling every 10 years. If this rate continues, we shall have extracted all of these fuels in just under 10 doubling periods, or in 100 years. We have no certain knowledge, of course, what fraction of all fossil fuels has been extracted. To be conservative let us assume that we have extracted only .01 percent rather than .1 percent. The curve shows us that in this case we shall have extracted 100 percent in 13.3 doubling periods, or 133 years. In other words, if our estimate of the fuel extracted to this moment is in error by a factor of 10, 1,000 or even 100,000, the date of total exhaustion is not long deferred. Thus if we have now depleted the earth's total supply of fossil fuel by only a millionth of 1 percent (.000001 percent), all of it will be exhausted in only 266 years at a 10-year doubling rate [see right half of middle illustration on page 356]. I should point out that the actual extraction rate varies for the different fossil fuels; a 10-year doubling rate was chosen simply for the purpose of illustration.

In estimating how many doubling periods the nation can tolerate if the demand for electricity continues to double every 10 years (the actual doubling rate), the crucial factor is probably not the supply of fuels—which is essentially limitless if fusion proves practical—but the thermal impact on the environment of converting fuel to electricity and electricity ultimately to heat. For the sake of argument let us ignore the burden of waste heat produced by fossil-fuel or nuclear power plants and consider only the heat content of the electricity actually consumed. One can imagine that by the year 2000 most of the power will be generated in huge plants located several miles offshore so that waste heat can be dumped harmlessly (for a while at least) into the surrounding ocean.

In 1970 the U.S. consumed 1,550 billion kilowatt hours of electricity. If this were degraded into heat (which it was) and distributed evenly over the total land area of the U.S. (which it was not), the energy released per square foot would be .017 watt. At the present doubling rate electric-power consumption is being multiplied by a factor of 10 every 33 years. Ninety-nine years from now, after only 10 more doubling periods, the rate of heat release will be 17 watts per square foot, or only slightly less than the 18 or 19 watts per square foot that the U.S. receives from the sun, averaged around the clock. Long before that the present pattern of power consumption must change or we must develop the technology needed for invariant energy systems.

Let us examine the consequences of altering the pattern of energy growth in what may seem to be fairly drastic ways. Consider a growth curve in which each doubling period is successively lengthened by 20 percent [see bottom illustration at right on page 356]. On an exponential growth curve it takes 3.32 doubling periods, or 33 years, to increase energy consumption by a factor of 10. On the retarded curve it would take five doubling periods, or 50 years, to reach the same tenfold increase. In other words, the retardation amounts to only 17 years. The retardation achieved for a hundredfold increase in consumption amounts to only 79 years (that is, the difference between 145 years and 66 years).

Another approach might be to cut back sharply on present consumption, hold the lower value for some period with no growth and then let growth resume at the present rate. One can easily show that if consumption of power were immediately cut in half, held at that value for 10 years and then allowed to return to the present pattern, the time required to reach a hundredfold increase in consumption would be stretched by only 20 years: from 66 to 86 years [see right half of top illustration on page 356].

For long-term effectiveness something like a constant growth curve is required, that is, a curve in which the growth increases by the same amount for each of the original doubling periods. On such a curve nearly 1,000 years would be required for electric-power generation to reach the level of the radiant energy received from the sun instead of the 100 years predicted by a 10-year doubling rate. One can be reasonably confident that the present doubling rate cannot continue for another 100 years, unless invariant energy systems supply a large part of the demand, but what such systems will look like remains hidden in the future.

29

Thermal Pollution and Aquatic Life

by John R. Clark
March 1969

The increasing use of river and lake waters for industrial cooling presents a real threat to fish and other organisms. To avoid an ecological crisis new ways must be found to get rid of waste heat

Ecologists consider temperature the primary control of life on earth, and fish, which as cold-blooded animals are unable to regulate their body temperature, are particularly sensitive to changes in the thermal environment. Each aquatic species becomes adapted to the seasonal variations in temperature of the water in which it lives, but it cannot adjust to the shock of abnormally abrupt change. For this reason there is growing concern among ecologists about the heating of aquatic habitats by man's activities. In the U.S. it appears that the use of river, lake and estuarine waters for industrial cooling purposes may become so extensive in future decades as to pose a considerable threat to fish and to aquatic life in general. Because of the potential hazard to life and to the balance of nature, the discharge of waste heat into the natural waters is coming to be called thermal pollution.

The principal contributor of this heat is the electric-power industry. In 1968 the cooling of steam condensers in generating plants accounted for about three-quarters of the total of 60,000 billion gallons of water used in the U.S. for industrial cooling. The present rate of heat discharge is not yet of great consequence except in some local situations; what has aroused ecologists is the ninefold expansion of electric-power production that is in prospect for the coming years with the increasing construction of large generating plants fueled by nuclear energy. They waste 60 percent more energy than fossil-fuel plants, and this energy is released as heat in condenser-cooling water. It is estimated that within 30 years the electric-power industry will be producing nearly two million megawatts of electricity, which will require the disposal of about 20 million billion B.T.U.'s of waste heat per day. To carry off that heat by way of natural waters would call for a flow through power plants amounting to about a third of the average daily freshwater runoff in the U.S.

The Federal Water Pollution Control

HEATED EFFLUENT from a power plant on the Connecticut River is shown in color thermograms (*see page 369*), in which different temperatures are represented by different hues. At the site (*above*) three large pipes discharged heated water that spreads across the river at slack tide (*top thermogram*) and tends to flow downriver at ebb tide (*bottom*). An infrared camera made by the Barnes Engineering Company scans the scene and measures the infrared radiation associated with the temperature at each point in the scene (350 points on each of 180 horizontal lines). Output from an infrared radiometer drives a color modulator, thus changing the color of a beam of light that is scanned across a color film in synchrony with the scanning of the scene. Here the coolest water (*black*) is at 59 degrees Fahrenheit; increasingly warm areas are shown, in three-degree steps, in blue, light blue, green, light green, yellow, orange, red and magenta. The effluent temperature was 87 degrees. A tree (*dark object*) appears in lower thermogram because the camera was moved.

NUCLEAR POWER PLANT at Haddam on the Connecticut River empties up to 370,000 gallons of coolant water a minute through a discharge canal (*bottom*) into the river. In this aerial thermogram, made by HRB-Singer, Inc., for the U.S. Geological Survey's

Administration has declared that waters above 93 degrees Fahrenheit are essentially uninhabitable for all fishes in the U.S. except certain southern species. Many U.S. rivers already reach a temperature of 90 degrees F. or more in summer through natural heating alone. Since the waste heat from a single power plant of the size planned for the future (some 1,000 megawatts) is expected to raise the temperature of a river carrying a flow of 3,000 cubic feet per second by 10 degrees, and since a number of industrial and power plants are likely to be constructed on the banks of a single river, it is obvious that many U.S. waters would become uninhabitable.

A great deal of detailed information is available on how temperature affects the life processes of animals that live in the water. Most of the effects stem from the impact of temperature on the rate of metabolism, which is speeded up by heat in accordance with the van't Hoff principle that the rate of chemical reaction increases with rising temperature. The acceleration varies considerably for particular biochemical reactions and in different temperature ranges, but gener-

ally speaking the metabolic rate doubles with each increase of 10 degrees Celsius (18 degrees F.).

Since a speedup of metabolism increases the animal's need for oxygen, the rate of respiration must rise. F. E. J. Fry of the University of Toronto, experimenting with fishes of the salmon family, found that active fish increased their oxygen consumption as much as fourfold as the temperature of the water was raised to the maximum at which they could survive. In the brown trout the rate of oxygen consumption rose steadily until the lethal temperature of 79 degrees F. was reached; in a species of lake trout, on the other hand, the rate rose to a maximum at about 60 degrees and then fell off as the lethal temperature of 77 degrees was approached. In both cases the fishes showed a marked rise in the basal rate of metabolism up to the lethal point.

The heart rate often serves as an index of metabolic or respiratory stress on the organism. Experiments with the crayfish (*Astacus*) showed that its heart rate increased from 30 beats per minute at a water temperature of 39 degrees F. to 125 beats per minute at 72 degrees and then slowed to a final 65 beats per min-

ute as the water approached 95 degrees, the lethal temperature for this crustacean. The final decrease in heartbeat is evidence of the animal's weakening under the thermal stress.

At elevated temperatures a fish's respiratory difficulties are compounded by the fact that the hemoglobin of its blood has a reduced affinity for oxygen and therefore becomes less efficient in delivering oxygen to the tissues. The combination of increased need for oxygen and reduced efficiency in obtaining it at rising temperatures can put severe stress even on fishes that ordinarily are capable of living on a meager supply of oxygen. For example, the hardy carp, which at a water temperature of 33 degrees F. can survive on an oxygen concentration as low as half a milligram per liter of water, needs a minimum of 1.5 milligrams per liter when the temperature is raised to 95 degrees. Other fishes can exist on one to two milligrams at 39 degrees but need three to four milligrams merely to survive at 65 degrees and five milligrams for normal activity.

The temperature of the water has pronounced effects on appetite, digestion and growth in fish. Tracer experiments

Water Resources Division, temperature is represented by shades of gray. The hot effluent (*white*) is at about 93 degrees F.; ambient river temperature (*dark gray*) is 77 degrees. The line across the thermogram is a time marker for a series of absolute measurements.

with young carp, in which food was labeled with color, established that they digest food four times as rapidly at 79 degrees F. as they do at 50 degrees; whereas at 50 degrees the food took 18 hours to pass through the alimentary canal, at 79 degrees it took only four and a half hours.

The effects of temperature in regulating appetite and the conversion of the food into body weight can be used by hatcheries to maximize fish production in terms of weight. The food consumption of the brown trout, for example, is highest in the temperature range between 50 and 66 degrees. Within that range, however, the fish is so active that a comparatively large proportion of its food intake goes into merely maintaining its body functions. Maximal conversion of the food into a gain in weight occurs just below and just above that temperature range. A hatchery can therefore produce the greatest poundage of trout per pound of food by keeping the water temperature at just under 50 degrees or just over 66 degrees.

It is not surprising to find that the activity, or movement, of fish depends considerably on the water temperature.

By and large aquatic animals tend to raise their swimming speed and to show more spontaneous movement as the temperature rises. In many fishes the temperature-dependent pattern of activity is rather complex. For instance, the sockeye salmon cruises twice as fast in water at 60 degrees as it does in water at 35 degrees, but above 60 degrees its speed declines. The brook trout shows somewhat more complicated behavior: it increases its spontaneous activity as the temperature rises from 40 to 48 degrees, becomes less active between 49 and 66 degrees and above 66 degrees again goes into a rising tempo of spontaneous movements up to the lethal temperature of 77 degrees. Laboratory tests show that a decrease in the trout's swimming speed potential at high temperatures affects its ability to feed. By 63 degrees trout have slowed down in pursuing minnows, and at 70 degrees they are almost incapable of catching the minnows.

That temperature plays a critical role in the reproduction of aquatic animals is well known. Some species of fish spawn during the fall, as temperatures drop; many more species, however,

spawn in the spring. The rising temperature induces a seasonal development of their gonads and then, at a critical point, triggers the female's deposit of her eggs in the water. The triggering is particularly dramatic in estuarine shellfish (oysters and clams), which spawn within a few hours after the water temperature reaches the critical level. Temperature also exerts a precise control over the time it takes a fish's eggs to hatch. For example, fertilized eggs of the Atlantic salmon will hatch in 114 days in water at 36 degrees F. but the period is shortened to 90 days at 45 degrees; herring eggs hatch in 47 days at 32 degrees and in eight days at 58 degrees; trout eggs hatch in 165 days at 37 degrees and in 32 days at 54 degrees. Excessive temperatures, however, can prevent normal development of eggs. The Oregon Fish Commission has declared that a rise of 5.4 degrees in the Columbia River could be disastrous for the eggs of the Chinook salmon.

Grace E. Pickford of Yale University has observed that "there are critical temperatures above or below which fish will not reproduce." For instance, at a temperature of 72 degrees or higher the banded sunfish fails to develop eggs. In the case of the carp, temperatures in the range of 68 to 75 degrees prevent cell division in the eggs. The possum shrimp *Neomysis,* an inhabitant of estuaries, is blocked from laying eggs if the temperature rises above 45 degrees. There is also the curious case of the tiny crustacean *Gammarus,* which at temperatures above 46 degrees produces only female offspring.

Temperature affects the longevity of fish as well as their reproduction. D'Arcy Wentworth Thompson succinctly stated this general life principle in his classic *On Growth and Form:* "As the several stages of development are accelerated by warmth, so is the duration of each and all, and of life itself, proportionately curtailed. High temperature may lead to a short but exhausting spell of rapid growth, while the slower rate manifested at a lower temperature may be the best in the end." Thompson's principle has been verified in rather precise detail by experiments with aquatic crustaceans. These have shown, for example, that *Daphnia* can live for 108 days at 46 degrees F. but its lifetime at 82 degrees is 29 days; the water flea *Moina* has a lifetime of 14 days at 55 degrees, its optimal temperature for longevity, but only five days at 91 degrees.

Other effects of temperature on life processes are known. For example, a century ago the German zoologist Karl Möbius noted that mollusks living in cold waters grew more slowly but attained larger size than their cousins living in warmer waters. This has since been found to be true of many fishes and other water animals in their natural habitats.

Fortunately fish are not entirely at the mercy of variations in the water tem-perature. By some process not yet understood they are able to acclimate themselves to a temperature shift if it is moderate and not too sudden. It has been found, for instance, that when the eggs of largemouth bass are suddenly transferred from water at 65 or 70 degrees to water at 85 degrees, 95 percent of the eggs perish, but if the eggs are acclimated by gradual raising of the temperature to 85 degrees over a period of 30 to 40 hours, 80 percent of the eggs will survive. Experiments with the possum shrimp have shown that the lethal temperature for this crustacean can be raised by as much as 24 degrees (to a high of 93 degrees) by acclimating it through a series of successively higher temperatures. As a general rule aquatic animals can acclimate to elevated temperatures faster than they can to lowered temperatures.

Allowing for maximum acclimation (which usually requires spreading the gradual rise of temperature over 20 days), the highest temperatures that most fishes of North America can tolerate range from about 77 to 97 degrees F. The direct cause of thermal death is not known in detail; various investigators have suggested that the final blow may be some effect of heat on the nervous system or the respiratory system, the coagulation of the cell protoplasm or the inactivation of enzymes.

Be that as it may, we need to be concerned not so much about the lethal temperature as about the temperatures that may be *unfavorable* to the fish. In the long run temperature levels that adversely affect the animals' metabolism, feeding, growth, reproduction and other vital functions may be as harmful to a fish population as outright heat death.

Studies of the preferred, or optimal, temperature ranges for various fishes have been made in natural waters and in the laboratory [see illustration on page 366]. For adult fish observed in nature the preferred level is about 13 degrees F. below the lethal temperature on the average; in the laboratory, where the experimental subjects used (for convenience) were very young fish, the preferred level was 9.5 degrees below the lethal temperature. Evidently young fish need warmer waters than those that have reached maturity do.

The optimal temperature for any water habitat depends not only on the preferences of individual species but also on the well-being of the system as a whole. An ecological system in dynamic balance is like a finely tuned automobile engine, and damage to any component can disable or impair the efficiency of the entire mechanism. This means that if we are to expect a good harvest of fish, the temperature conditions in the water medium must strike a favorable balance for all the components (algae and other plants, small crustaceans, bait fishes and so on) that constitute the food chain producing the harvested fish. For example, above 68 degrees estuarine eelgrass does not reproduce. Above 90 degrees there is extensive loss of bottom life in rivers.

So far there have been few recorded instances of direct kills of fish by thermal pollution in U.S. waters. One recorded kill occurred in the summer of 1968 when a large number of menhaden acclimated to temperatures in the 80's became trapped in effluent water at 93 to 95 degrees during the testing of a new power plant on the Cape Cod Canal. A very large kill of striped bass occurred

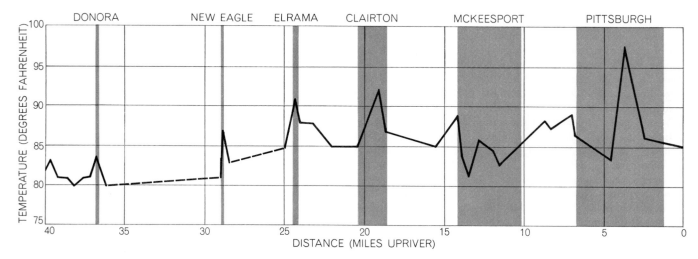

WATER TEMPERATURES can become very high, particularly in summer, along rivers with concentrated industry. The chart shows the temperature of the Monongahela River, measured in August, along a 40-mile stretch upriver from its confluence with the Ohio.

in the winter and early spring of 1963 at the nuclear power plant at Indian Point on the Hudson River. In that instance the heat discharge from the plant was only a contributing factor. The wintering, dormant fish, attracted to the warm water issuing from the plant, became trapped in its structure for water intake, and they died by the thousands from fatigue and other stresses. (Under the right conditions, of course, thermal discharges benefit fishermen by attracting fish to discharge points, where they can be caught with a hook and line.)

Although direct kills attributable to thermal pollution apparently have been rare, there are many known instances of deleterious effects on fish arising from natural summer heating in various U.S. waters. Pollution by sewage is often a contributing factor. At peak summer temperatures such waters frequently generate a great bloom of plankton that depletes the water of oxygen (by respiration while it lives and by decay after it dies). In estuaries algae proliferating in the warm water can clog the filtering apparatus of shellfish and cause their death. Jellyfish exploding into abundant growth make some estuarine waters unusable for bathing or other water sports, and the growth of bottom plants in warm waters commonly chokes shallow bays and lakes. The formation of hydrogen sulfide and other odorous substances is enhanced by summer temperatures. Along some of our coasts in summer "red tides" of dinoflagellates occasionally bloom in such profusion that they not only bother bathers but also may poison fish. And where both temperature and sewage concentrations are high a heavy toll of fish may be taken by proliferating microbes.

This wealth of evidence on the sensitivity of fish and the susceptibility of aquatic ecosystems to disruption by high temperatures explains the present concern of biologists about the impending large increase in thermal pollution. Already last fall 14 nuclear power plants, with a total capacity of 2,782 megawatts, were in operation in the U.S.; 39 more plants were under construction, and 47 others were in advanced planning stages. By the year 2000 nuclear plants are expected to be producing about 1.2 million megawatts, and the nation's total electricity output will be in the neighborhood of 1.8 million megawatts. As I have noted, the use of natural waters to cool the condensers would entail the heating of an amount of water equivalent to a third of the yearly freshwater runoff in the U.S.; during low-flow periods in summer the requirement would be 100 percent of the runoff. Obviously thermal pollution of the waters on such a scale is neither reasonable nor feasible. We must therefore look for more efficient and safer methods of dissipating the heat from power plants.

One might hope to use the heated water for some commercial purpose. Unfortunately, although dozens of schemes have been advanced, no practicable use has yet been found. Discharge water is not hot enough to heat buildings. The cost of transmission rules out piping it to farms for irrigation even if the remaining heat were enough to improve crop production. More promising is the idea of using waste heat in desalination plants to aid in the evaporation process, but this is still only an idea. There has also been talk of improving the efficiency of sewage treatment with waste heat from power plants. Sea farm-

ing may offer the best hope of someday providing a needed outlet for discharges from coastal power plants; pilot studies now in progress in Britain and the U.S. are showing better growth of fish and shellfish in heated waters than in normal waters, but no economically feasible scheme has yet emerged. It appears, therefore, that for many years ahead we shall have to dispose of waste heat to the environment.

The dissipation of heat can be facilitated in various ways by controlling the passage of the cooling water through the condensers. The prevailing practice is to pump the water (from a river, a lake or an estuary) once through the steam-condensing unit, which in a 500-megawatt plant may consist of 400 miles of one-inch copper tubing. The water emerging from the unit has been raised in temperature by an amount that varies from 10 to 30 degrees F., depending on the choice of manageable factors such as the rate of flow. This heated effluent is then discharged through a channel into the body of water from which it was taken. There the effluent, since it is warmer and consequently lighter than the receiving water, spreads in a plume over the surface and is carried off in the direction of the prevailing surface currents.

The ensuing dispersal of heat through the receiving water and into the atmosphere depends on a number of natural factors: the speed of the currents, the turbulence of the receiving water (which affects the rate of mixing of the effluent with it), the temperature difference between the water and the air, the humidity of the air and the speed and direction of the wind. The most variable and most important factor is wind: other things being equal, heat will be dissipated from

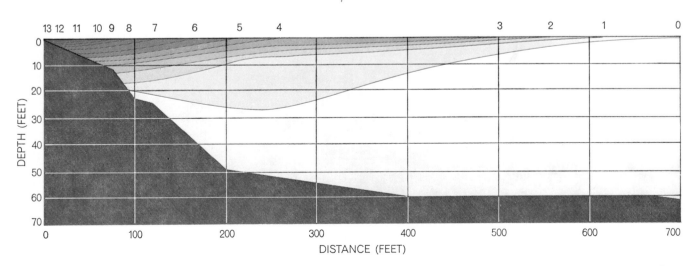

HOT-WATER "PLUME" that would result from an Indian Point nuclear power plant mixes with cooler Hudson River as shown by the one-degree contour lines. This section across the river shows temperature structure that was predicted by engineering studies.

the water to the air by convection three times faster at a wind speed of 20 miles per hour than at a wind speed of five miles.

In regulating the rate of water flow through the condenser one has a choice between opposite strategies. By using a rapid rate of flow one can spread the heat through a comparatively large volume of cooling water and thus keep down the temperature of the effluent; conversely, with a slow rate of flow one can concentrate the heat in a smaller volume of coolant. If it is advantageous to obtain good mixing of the effluent with the receiving water, the effluent can be discharged at some depth in the water rather than at the surface. The physical and ecological nature of the body of water will determine which of these strategies is best in a given situation. Where the receiving body is a swift-flowing river, rapid flow through the condenser and dispersal of the low-temperature effluent in a narrow plume over the water surface may be the most effective way to dissipate the heat into the atmosphere. In the case of a still lake it may be best to use a slow flow through the condenser so that the comparatively small volume of effluent at a high tem-

perature will be confined to a small area in the lake and still transfer its heat to the atmosphere rapidly because of the high temperature differential. And at a coastal site the best strategy may be to discharge an effluent of moderate temperature well offshore, below the ocean-water surface.

There are many waters, however, where no strategy of discharge will avail to make the water safe for aquatic life (and where manipulation of the discharge will also be insufficient to avoid dangerous thermal pollution), particularly where a number of industrial and power plants use the same body of water for cooling purposes. It therefore appears that we shall have to turn to extensive development of devices such as artificial lakes and cooling towers.

Designs for such lakes have already been drawn up and implemented for plants of moderate size. For the 1,000-megawatt power plant of the future a lake with a surface area of 1,000 to 2,000 acres would be required. (A 2,000-acre lake would be a mile wide and three miles long.) The recommended design calls for a lake only a few feet deep at one end and sloping to a depth of 50

feet at the other end. The water for cooling is drawn from about 30 feet below the surface at the deep end and is pumped through the plant at the rate of 500,000 gallons per minute, and the effluent, 20 degrees higher in temperature than the inflow, is discharged at the lake's shallow end. The size of the lake is based on a pumping rate through the plant of 2,000 acre-feet a day, so that all the water of the lake (averaging 15 feet in depth) is turned over every 15 days. Such a lake would dissipate heat to the air at a sufficient rate even in prolonged spells of unfavorable weather, such as high temperature and humidity with little wind.

Artificial cooling lakes need a steady inflow of water to replace evaporation and to prevent an excessive accumulation of dissolved material. This replenishment can be supplied by a small stream flowing into the lake. The lake itself can be built by damming a natural land basin. A lake complex constructed to serve not only for cooling but also for fishing might consist of two sections: the smaller one, in which the effluent is discharged, would be stocked with fishes tolerant to heat, and the water from this basin, having been cooled by exposure to

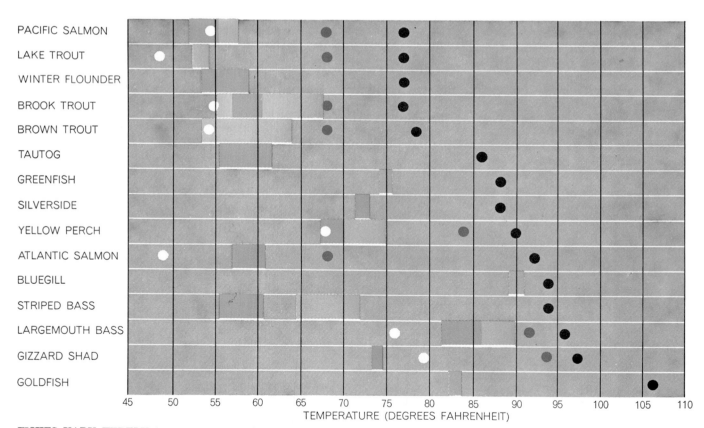

FISHES VARY WIDELY in temperature preference. Preferred temperature ranges are shown here for some species for which they have been determined in the field (*dark colored bands*) and, generally for younger fish, in the laboratory (*light color*). The chart also indicates the upper lethal limit (*black dot*) and upper limits recommended by the Federal Water Pollution Control Administration for satisfactory growth (*colored dot*) and spawning (*white dot*). Temperatures well below lethal limit can be in stress range.

the air, would then flow into a second and larger lake where other fishes could thrive.

In Britain, where streams are small, water is scarce and appreciation of aquatic life is high, the favored artificial device for getting rid of waste heat from power plants has been the use of cooling towers. One class of these towers employs the principle of removing heat by evaporation. The heated effluent is discharged into the lower part of a high tower (300 to 450 feet) with sloping sides; as the water falls in a thin film over a series of baffles it is exposed to the air rising through the tower. Or the water can be sprayed into the tower as a mist that evaporates easily and cools quickly. In either case some of the water is lost to the atmosphere; most of it collects in a basin and is pumped into the waterway or recirculated to the condensers. The removal of heat through evaporation can cool the water by about 20 degrees F.

The main drawback of the evaporation scheme is the large amount of water vapor discharged into the atmosphere. The towers for a 1,000-megawatt power plant would eject some 20,000 to 25,000 gallons of evaporated water per minute—an amount that would be the equivalent of a daily rainfall of an inch on an area of two square miles. On cold days such a discharge could condense into a thick fog and ice over the area in the vicinity of the plant. The "wet" type of cooling tower may therefore be inappropriate in cold climates. It is also ruled out where salt water is used as the coolant: the salt spray ejected from a single large power plant could destroy vegetation and otherwise foul the environment over an area of 160 square miles.

A variation of the cooling-tower system avoids these problems. In this refinement, called the "dry tower" method, the heat is transferred from the cooling water, through a heat exchanger something like an enormous automobile radiator, directly to the air without evaporation. The "dry" system, however, is two and a half to three times as expensive to build as a "wet" system. In a proposed nuclear power plant to be built in Vernon, Vt., it is estimated that the costs of operation and amortization would be $2.1 million per year for a dry system and $800,000 per year for a wet system. For the consumer the relative costs would amount respectively to 2.6 and 1 percent of the bill for electricity.

The public-utility industry, like other industries, is understandably reluctant to incur large extra expenses that add sub-

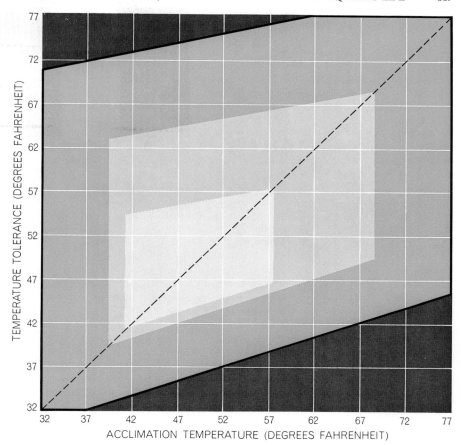

ACCLIMATION extends the temperature range in which a fish can thrive, but not indefinitely. For a young sockeye salmon acclimated as shown by the horizontal scale, spawning is inhibited outside the central (*light color*) zone and growth is poor outside the second zone (*medium color*); beyond the outer zone (*dark color*) lie the lethal temperatures.

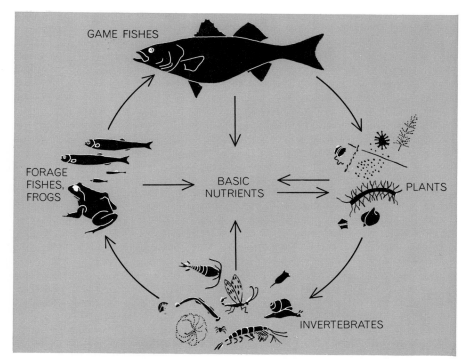

AQUATIC ECOSYSTEM is even more sensitive to temperature variation than an individual fish. A single game-fish species, for example, depends on a food chain involving smaller fishes, invertebrates, plants and dissolved nutrients. Any change in the environment that seriously affects the proliferation of any link in the chain can affect the harvest of game fish.

POWER PLANT being completed by the Tennessee Valley Authority on the Green River in Kentucky will be the world's largest coal-fueled electric plant. Its three wet, natural-draft cooling towers, each 437 feet in height and 320 feet in diameter at ground level, will each have a capacity of 282,000 gallons of water a minute, which they will be able to cool through a range of 27.5 degrees.

INDUSTRIAL PLANTS of various kinds can use towers to cool process water. These two five-cell cooling towers were built by the Marley Company for a chemical plant. They are wet, mechanical-draft towers of the cross-flow type: a fan in each stack draws air in through the louvers, across films of falling water and then up. The towers cool 120,000 gallons a minute through a 20-degree range.

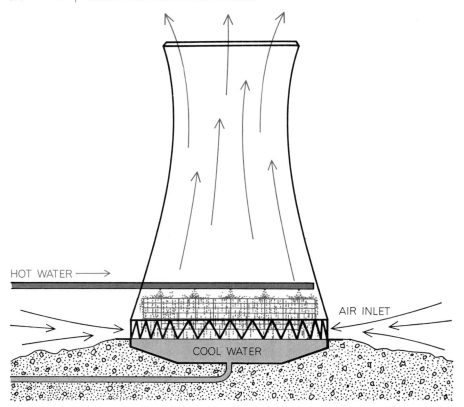

COOLING TOWER is one device that can dissipate industrial heat without dumping it directly into rivers or lakes. This is a "wet," natural-draft, counterflow tower. Hot water from the plant is exposed to air moving up through the chimney-like tower. Heat is removed by evaporation. The cooled water is emptied into a waterway or recirculated through the plant. In cold areas water vapor discharged into the atmosphere can create a heavy fog.

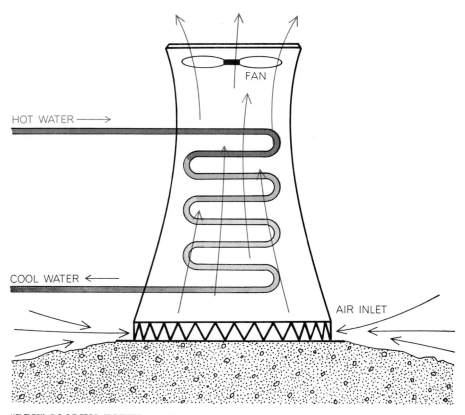

"DRY" COOLING TOWER avoids evaporation. The hot water is channeled through tubing that is exposed to an air flow, and gives up its heat to the air without evaporating. In this mechanical-draft version air is moved through the tower by a fan. Dry towers are costly.

stantially to the cost of its product and services. There is a growing recognition, however, by industry, the public and the Government of the need for protecting the environment from pollution. The Federal Water Pollution Control Administration of the Department of the Interior, with help from a national advisory committee, last year established a provisional set of guidelines for water quality that includes control of thermal pollution. These guidelines specify maximum permissible water temperatures for individual species of fish and recommend limits for the heating of natural waters for cooling purposes. For example, they suggest that discharges that would raise the water temperature should be avoided entirely in the spawning grounds of cold-water fishes and that the limit for heating any stream, in the most favorable season, should be set at five degrees F. outside a permitted mixing zone. The size of such a zone is likely to be a major point of controversy. Biologists would limit it to a few hundred feet, but in one case a power group has advocated 10 miles.

It appears that the problem of thermal pollution will receive considerable attention in the 91st Congress. Senator Edward M. Kennedy has proposed that further licensing of nuclear power-plant construction be suspended until a thorough study of potential hazards, including pollution of the environment, has been made. Senator Edmund S. Muskie's Subcommittee on Air and Water Pollution of the Senate Committee on Public Works last year held hearings on thermal pollution in many parts of the country.

Thermal pollution of course needs to be considered in the context of the many other works of man that threaten the life and richness of our natural waters: the discharges of sewage and chemical wastes, dredging, diking, filling of wetlands and other interventions that are altering the nature, form and extent of the waters. The effects of any one of these factors might be tolerable, but the cumulative and synergistic action of all of them together seems likely to impoverish our environment drastically.

Temperature, Gordon Gunter of the Gulf Coast Research Laboratory has remarked, is "the most important single factor governing the occurrence and behavior of life." Fortunately thermal pollution has not yet reached the level of producing serious general damage; moreover, unlike many other forms of pollution, any excessive heating of the waters could be stopped in short order by appropriate corrective action.

Cooling Towers

by Riley D. Woodson
May 1971

*Hot water carrying waste heat from a large industrial
installation such as a power plant is often best
handled by a tower that breaks the water up, cools it
in a cross draft of air and then recirculates it*

A modern steam-turbine electric power plant has a thermal efficiency of about 40 percent, meaning that 40 percent of the heat energy put into the system by the burning of fuel emerges from the plant as electrical energy. Most of the remaining heat has to be eliminated as waste. In the majority of such plants the heat is removed by drawing water from a stream or a lake or even a body of salt water, running it through the plant once and then returning the heated water to its source. This type of cooling is becoming less acceptable as questions arise about the adequacy of cooling-water supplies and about the biological effects of warming natural bodies of water.

What are the alternatives for cooling power plants and other industrial facilities? Air is a perfectly good coolant, and indeed a few plants are cooled by "dry" systems. Since air is a gas, however, one must move huge volumes of it and provide elaborate heat-exchange surfaces in order to cool a large plant effectively. A more practical alternative is the "wet" cooling tower, where both air and water serve as coolants. In such a tower the cooling water being circulated through a plant falls through a draft of air and heat is carried away mostly by evaporation. The rest of the water is collected at the bottom of the tower and returned to the cooling cycle.

Although the principle is simple, the tower itself is likely to be quite an elaborate structure. One reason is the large amount of heat that must be removed from many industrial plants. Another is the need to pass the water through the tower in the form of fine droplets in order to achieve maximum contact between the water and the air. As a result a typical tower of a kind now in service, incorporating high-capacity fans to assist the flow of air, is about 55 feet wide at

the bottom, 75 feet wide at the top and 325 feet long; it consists of several cooling cells, each with a draft fan as large as 28 feet in diameter powered by a 200-horsepower motor. The chimney-like natural-draft towers that are more common in Europe than in the U.S. can be 300 feet or more in diameter at the base and as high as 500 feet. If one thinks in terms of a fossil-fueled power plant with

a capacity of 800,000 kilowatts, the cost of an adequate mechanical-draft tower would be about $1.7 million and of a natural-draft tower about $4.2 million. The cost of operating the natural-draft tower would be lower, however, since such a structure acts as a chimney and does not need electric power to drive fans.

Cooling towers of much smaller scale,

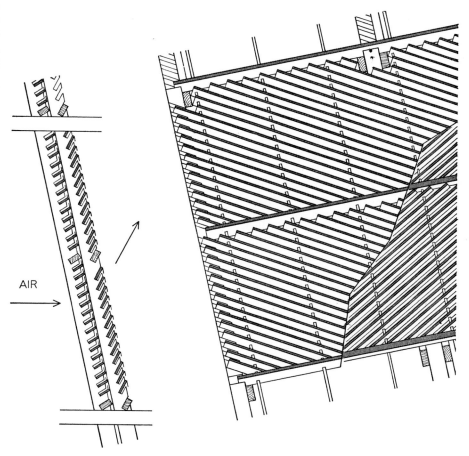

AIR

DRIFT ELIMINATOR is depicted from an end (*left*) and the exhaust side (*right*). The configuration forces outgoing air to make two abrupt turns; the resulting centrifugal force separates water droplets from the air so that moisture downwind of the tower is reduced.

such as the ones used in central-air-conditioning systems, are usually factory-built, bought from a catalogue and installed as a component of a building. The large-scale cooling towers of the kind found at generating stations, petrochemical plants and other major industrial operations are designed individually and built at the site.

The designer of a large cooling tower must deal with a variety of physical factors. He needs to know what the temperature of the water will be when it enters the tower and what temperature it should be when it leaves the tower. The difference between them defines the cooling range of the tower. In general the range is between 10 and 40 degrees Fahrenheit.

Another factor to be considered is the normal summertime temperature of the area where the tower is to be built, since the cooling problem increases with higher ambient temperature. For this calculation one uses the "wet bulb" temperature of the environment rather than the temperature recorded on a standard dry thermometer. The wet-bulb temperature is the lowest water temperature that can be realized by evaporation into the surrounding environment. It is measured by whirling a thermometer whose bulb is covered with wet cloth, and it represents the theoretical limit of cooling that a tower can achieve. To attain it would require a tower with a thermal efficiency of 100 percent, which in practice is beyond reach.

The difference between the theoretical limit and the actual cooling accomplished by a tower is called the tower's "approach to wet bulb." The approach is seldom closer to the theoretical limit than five degrees F., and more often it ranges between seven and 15 degrees. The wet-bulb temperatures for which cooling towers are designed vary from 60 to 83 degrees F. The usual criterion is that the tower's designed wet-bulb temperature is exceeded no more than 5 percent of the total summer operating time of the tower.

Since the cooling is mainly evaporative, part of the water circulating through the system is lost by evaporation. The rule of thumb is 1 percent of evaporation loss for each 10 degrees F. of cooling. There are two other sources of water loss. The first is drift, meaning water that is simply carried out as very fine droplets with the air leaving the tower. The cooling-tower industry routinely guarantees that the loss by drift will not exceed .2 percent of the water circulated. The second source of loss results from the need

to bleed off part of the water in order to get rid of dissolved solids that might form scale and interfere with the operation of the system. In the industry this process is called blowdown. As a rule it amounts to about .3 percent for each 10 degrees F. of cooling.

The sum of these losses has to be made up by adding water to the system. Although the losses are only a small percentage of the total water flow, the amount of water required for "makeup" in a large cooling tower is considerable. In a modern 680,000-kilowatt generating plant the amount of water circulating through the tower units under average summer conditions would be about 345 million gallons per day and the requirement for makeup would be about 6.5 million gallons per day.

The first industrial cooling towers, which were built early in this century, depended on the natural movement of air. Since natural wind is seldom strong enough to produce a brisk movement of air across a wide structure, such an atmospheric tower had to be long, narrow and high and built broadside to the prevailing wind. Decks were built at intervals inside the tower to break up the falling water. The towers entailed no cost for mechanical equipment, but because they were high and the water to be cooled had to be pumped to the top, the pumping costs were fairly substantial. Moreover, the amount of cooling accomplished by an atmospheric tower with a given amount of water was erratic because it depended not only on the environmental temperature but also on the speed and direction of the wind.

The modern counterpart of the atmospheric tower is the hyperbolic tower, which operates on the natural-draft principle. The term "hyperbolic" refers to the fact that the tower characteristically has the form of a hyperbola [*see illustration on these two pages*]. In such a tower air moves upward as a result of the chimney effect created by the difference in density between the warm, moist air inside the tower and the colder, denser air outside. Hyperbolic towers are best suited to regions with high humidity, such as western Europe and parts of the eastern and northwestern U.S. They are also desirable in populated areas because the height of the exhaust from such a tower helps to prevent the formation of fog along the ground.

Economic reasons as well as climatic ones account for the greater prevalence of hyperbolic towers in Europe. A hyperbolic tower is usually built of thin-shell reinforced concrete, and its construction calls for much labor, which costs less in

Europe than in the U.S. The cost of electric power, however, is higher in Europe, and so it is advantageous to be able to operate a cooling tower without a fan. In recent years conditions have changed, so that more hyperbolic towers are being built in the U.S. The reasons include improved materials and building techniques, the rising cost of land (a hyperbolic tower takes less land than other types) and the effectiveness of the hyperbolic tower in dissipating large quantities of water vapor.

Mechanical-draft towers, using fans to either force or induce the movement of air, first came into use about 40 years ago. In a forced-draft tower the fan is at the bottom and pushes the air up through the tower; in an induced-draft tower the fan is at the top and pulls the air up. The forced-draft regime creates several problems in large-scale towers, chief among which is recirculation: vapor leaving the tower at low velocity tends to reenter the tower. As a result the wet-bulb temperature of the entering

FOUR COOLING TOWERS serve the Keystone Steam Electric Station in western Pennsylvania. Each pair of towers serves one of the station's two 820,000-kilowatt genair is increased and the performance of the tower is impaired. Another problem is that the uniform distribution of air through the tower that is necessary for efficient operation is difficult to attain in large forced-draft units. For these reasons modern mechanical-draft towers of industrial size operate on the induced-draft principle.

Depending on the particular requirements for a mechanical-draft tower, the designer will arrange for either a counterflow or a crossflow movement of the air past the falling water. In a counterflow tower the water moves down and the air moves up; in a crossflow tower the water moves down and the air moves horizontally. For a number of reasons the crossflow principle is much more widely used than the counterflow principle.

A distinct advantage of the crossflow design is its low loss of draft. In a counterflow tower the air travels all the way from the bottom to the top, passing through all the water and meeting great-

er resistance from it. A significant loss of draft results. In a crossflow tower the air cuts across the path of the water and travels a shorter distance vertically, with a lower loss of draft.

Another consideration favoring crossflow towers stems from the fact that the cooling performance of a tower rises with increased height of "fill": the material in the tower that breaks up the water into fine droplets. The crossflow tower benefits particularly from increased height because the distance the air travels is independent of fill height. Therefore the distance the air travels and the height of the fill (or the distance the water travels through air) can be adjusted to minimize the loss of draft. In a counterflow tower the distance the air travels varies directly with the fill height, so that the fill height can be increased only at the expense of increasing the horsepower of the fan to compensate for the greater loss of draft through the fill. The crossflow tower is superior to the counterflow tower in handling heavy loads of water

with long cooling ranges and close approaches to the wet-bulb temperature.

In "dry" cooling towers, where air alone is the ultimate coolant, the water that holds the heat removed from the plant is circulated through a closed system of tubes exposed to air. Such a tower, operating without losses of water from evaporation and drift, is advantageous in areas where the supply of water is limited. The towers are very costly, however, and so they have been built only at a small number of generating stations with special needs. Most of the dry units are outside the U.S. and at generating stations of relatively small size (250,000 kilowatts or less).

In a wet tower the heated water entering the top of the tower for cooling is distributed evenly for its fall through the tower. In most crossflow towers the water simply pours over baffles that break up the flow into droplets. In counterflow towers, however, it is difficult to achieve uniform distribution of water by gravity,

erating units. The towers are 325 feet high, 247 feet in diameter at the base and 142 feet in diameter at the top. They are natural-draft towers of typical hyperbolic form, with most of their height functioning to help induce a strong draft of air that moves up the inside of the tower as warm water from the plant falls through the draft near the bottom of the tower. Clouds of vapor emerge from the towers at left; two towers at right were not operating when photograph was made. Plant is jointly owned by seven electric utilities.

a

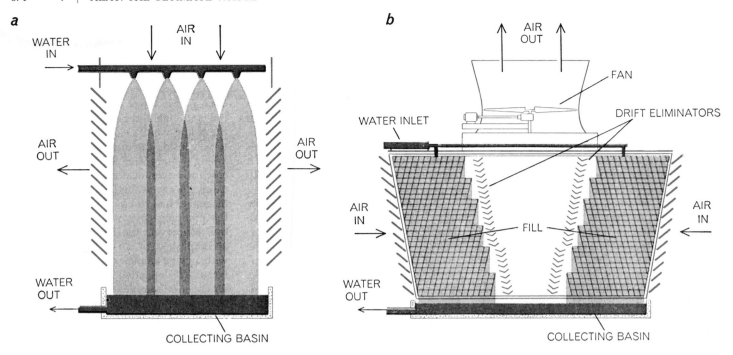

b

TYPES OF TOWER are depicted schematically. The atmospheric tower (*a*), which was one of the earliest types of cooling tower, relied on natural winds and the aspirating effect of the spray nozzles to provide cooling air. The mechanical-draft crossflow tower (*b*), which is the kind in widest use at present, has a fan at the top to draw air in through the louvers, across the path of the falling water and out of the stack. "Fill" is the material that breaks up water into fine droplets; the drift eliminators retard the movement of

and so pressure is supplied to start the water downward as a spray emerging from nozzles.

The distribution system functions in the highly corrosive oxidizing atmosphere of the tower. Accordingly the materials for the system must be chosen with care. Piping made of redwood, cast iron, galvanized steel and, more recently, plastic is installed in order to keep down maintenance costs.

Water coming out of the distribution system falls onto the first of many layers of fill. The function of fill, which occupies much of the interior of a cooling tower, is to speed up the dissipation of heat from the water by increasing the amount and duration of contact between the water and the air in the tower. Fill accomplishes this objective by promoting the formation of droplets and water

MECHANICAL-DRAFT CROSSFLOW TOWERS serve the 640,-000-kilowatt power plant of the Northeastern Station of the Public Service Company of Oklahoma. One cell of one tower is shown on the cover of this issue. In all the plant is served by two six-cell

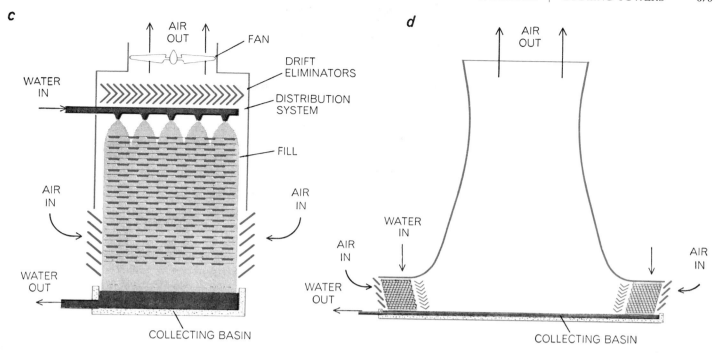

c

AIR OUT — FAN

DRIFT ELIMINATORS

WATER IN

DISTRIBUTION SYSTEM

FILL

AIR IN   AIR IN

WATER OUT

COLLECTING BASIN

d

AIR OUT

WATER IN

AIR IN

WATER OUT

AIR IN

COLLECTING BASIN

water out of the stack with the departing air. The towers shown on the cover of this issue and in the photograph at the bottom of these two pages are of this type. The mechanical-draft counterflow tower (*c*) is similar except that the air moves vertically upward through the falling water. The hyperbolic natural-draft tower (*d*) has no fan; the draft results both from the height of the structure and the difference in density between the warm, moist air inside the tower and the colder, denser air outside. Discharge is far above ground.

film and by increasing the wetted surface in contact with air. At the same time it should provide low resistance to the flow of air and should maintain a uniform distribution of water and air.

In most industrial towers fill is of the splash type, meaning that it consists of small bars or planking set horizontally [*see top illustration on page 377*]. Water splashes downward from one level of bars to the next. Good exposure of water surfaces to the passing air is thus achieved because the water is constantly broken up into fine droplets and films by splash-

ing as it falls.

Wood is the usual material for splash fill, although in recent years strips of perforated plastic have been developed for the purpose. Redwood has long been popular; lately fir has gained in use as a result of improvements that have been

towers, which are in the foreground at left, and two seven-cell towers, which are in the background at left. The generating station is at right. In normal operation the flow of water through the entire set of four cooling towers is about 523 million gallons per day.

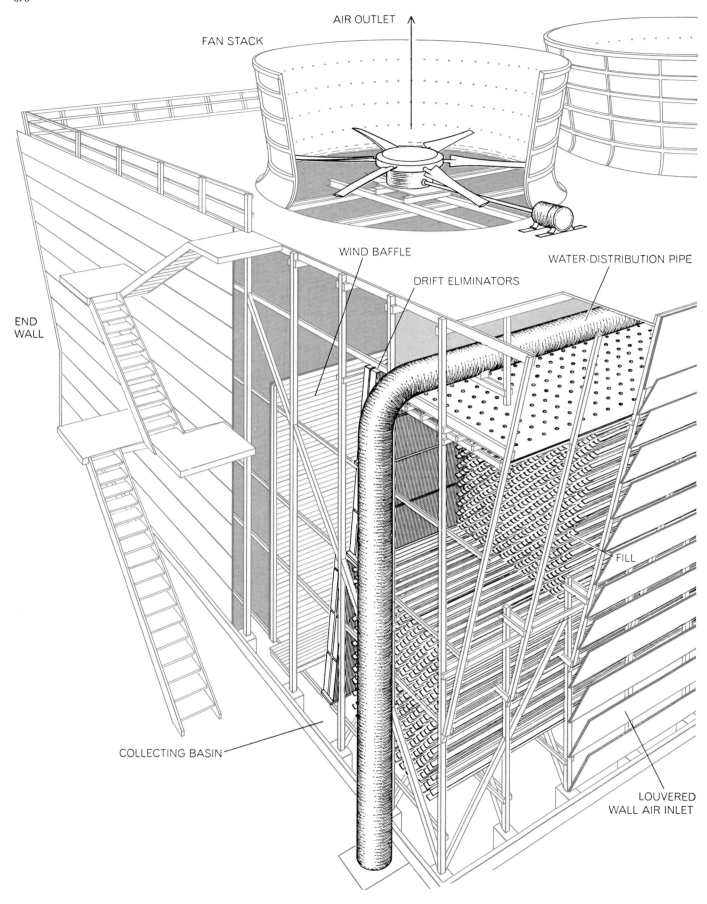

FAN STACK

AIR OUTLET

WIND BAFFLE

DRIFT ELIMINATORS

WATER-DISTRIBUTION PIPE

END
WALL

FILL

COLLECTING BASIN

LOUVERED
WALL AIR INLET

DETAILS OF CONSTRUCTION are depicted for one cell of a mechanical-draft crossflow tower. Ordinarily much of a tower is made of wood, although plastic is coming into use for the slatlike members that constitute the fill. In this design the large pipe near the top, which distributes the hot water that will fall through the tower to the collecting basin, is made of plastic reinforced with glass fiber. The tower is louvered on two sides to admit the cooling air that flows across the falling water. The two end walls are solid.

made in wood preservatives. Occasionally one finds metal or ceramic splash bars, but they are costly and so tend to be installed only in special applications.

In a tower with splash fill it is important for the splash bars to be maintained in a horizontal position. If they sag, water will tend to flow through the tower in channels and the cooling capacity of the tower will be reduced. It is also important for the tower to be kept level; otherwise water will run to the lower ends of splash bars and channeling will again result.

Another type of fill that is gaining acceptance for smaller towers is called film fill because it tends to spread the falling water into a thin film flowing over a large total surface area. Film fill is made of plastic, metal or asbestos and is formed into complex shapes such as honeycombs. It is more expensive than splash fill, but it provides more efficient cooling within a given amount of space. On the other hand, film fill is more sensitive than splash fill to irregularities in the flow of air and the distribution of water, so that the tower needs to be designed to ensure that such irregularities do not arise. The fill must be uniformly spaced and well supported.

At the end of its passage through the tower the cooled water falls into a collecting basin at the bottom of the tower. From there it is pumped back to the place where it picked up heat—for example a steam condenser in a generating station—and begins another cycle.

In addition to the fan and its motor, both of which have to be capable of long operation under rugged conditions, several other parts of a mechanical-draft tower have a bearing on the movement of air through the tower. The characteristic flared stack enclosing the fan on each cell of a tower performs several functions. It acts as a fan guard and as a means of conveying water vapor and air away from the tower. The stack also plays a role in the efficiency of the fan: the closer it can be built to the tips of the fan blades, the higher will be the efficiency of the fan (and the lower the noise of the fan's operation).

The louvers on the sides of the tower where air is admitted have the obvious function of bringing the air in efficiently and the less obvious function of preventing the loss of water. The louvers must admit air in such a way that it is distributed uniformly and undergoes a minimum drop in pressure. To prevent the loss of water they must be designed with the proper width, spacing and slope. In crossflow towers that operate under

SPLASH-BAR FILL is usually installed in large cooling towers. Laid horizontally to a considerable depth in the tower, the bars interrupt the fall of the water, causing it to break up into fine droplets. Bars are usually made of wood or, as in this case, perforated plastic.

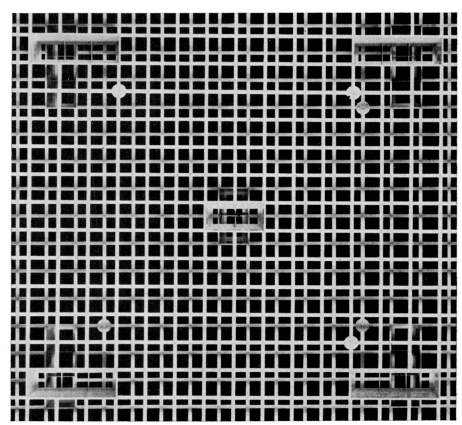

FILL OF FILM TYPE is sometimes used in smaller towers. It is made in geometric shapes that cause the falling water to form many surfaces and thus gain more exposure to air.

COOLING SYSTEM	CONDENSER AND AUXILIARIES		COOLING TOWERS AND BASIN		PUMPS AND CONDUITS		MAKEUP SYSTEM		COOLING LAKE		TOTAL INVESTMENT	
OCEAN	1.6	2.4	–	–	3.3	4.8	–	–	–	–	4.9	7.2
RIVER	1.7	2.6	–	–	3.3	4.8	–	–	–	–	5	7.4
LAKE	1.5	2.3	–	–	2.7	3.9	–	–	2.6	3.4	6.8	9.6
MECHANICAL-DRAFT WET TOWER	1.9	2.8	1.7	2.6	1.6	2.4	1.2	1.6	–	–	6.4	9.4
NATURAL-DRAFT WET TOWER	1.9	2.8	4.2	7.1	1.7	2.5	1.2	1.6	–	–	9	14
MECHANICAL-DRAFT DRY TOWER	1.1	1.7	12.2	19.3	2.1	3.4	–	–	–	–	15.4	24.4
NATURAL-DRAFT DRY TOWER	1.1	1.7	28	45	2.1	3.4	–	–	–	–	31.2	50.1

COST OF COOLING SYSTEMS is compared for 800,000-kilowatt power stations burning fossil fuel (*black*) and nuclear fuel (*color*). The figures represent millions of dollars at 1970 prices. In a dry tower the warm water circulates in pipes that are exposed to air.

winter conditions the louvers are often placed so that they slope back under the fill. If the fan is on, the falling water washes the louvers and so alleviates icing. If the fan is off, the louvers are bathed in warm water and ice usually does not form.

The problem of drift—the loss of water with the air leaving the tower—is dealt with by eliminators installed in the airstream to the fans. They are built in intricate geometric patterns that cause the rising air to shift direction abruptly at least once. The centrifugal force that results from the shift separates the drops of water from the air, depositing them on the surface of the eliminator, where they accumulate and eventually flow back into the tower. A secondary function of the eliminator is equalizing air-flow through the fill. The eliminator offers resistance to the flow, thereby producing a uniform pressure in the space between the eliminator and the fan. This equalization of pressure tends to provide a uniform flow of air through the fill.

An efficient eliminator must decrease drift by an acceptable amount without unduly resisting the air. Too much resistance would necessitate higher horsepower for the fan. The eliminator must also be able to collect the drift water and return it to the tower basin without reintroducing it into the stream of air being discharged from the tower. Finally, the eliminator functions in an environment that is both corrosive and erosive, and so it must be made of such materials as treated wood, galvanized or stainless

steel, plastic and asbestos.

To support all these structural elements, withstand moving water and air without deterioration and provide stiffness against wind a cooling tower needs a substantial framework. The design of a frame must be worked out with care, taking into account not only engineering factors but also thermal, aerodynamic and economic ones. Usually frames are built of wood, with redwood favored because of its natural resistance to decay. In recent years treated fir has also been extensively used. Occasionally a frame is made of galvanized steel.

In operating a wet cooling tower one must pay close attention to the condition of the water. The pH must be controlled; intermittent chlorination is necessary to control microorganisms, and corrosive substances must be removed. The problem of discharging blowdown water into a natural body of water can be alleviated by treatment and in some situations by using blowdown water to supply other plant processes.

An example of an installation with large amounts of heat to carry off is an 800,000-kilowatt steam-electric generating plant. If the plant is a modern fossil-fuel-burning one, the amount of heat rejected to the condenser from the turbine exhaust steam is about 50 percent of the heat input to the boiler. This is equivalent to about 1.33 kilowatts of heat rejected for each kilowatt of electric power generated. For a plant of similar size operating on nuclear fuel about 66 per-

cent of the heat input to the boiler is rejected to the condenser; this is equivalent to about two kilowatts of heat for each kilowatt of net generation.

Five options are available for cooling the steam condenser. The plant can use a once-through cooling water system or any of the four main types of cooling tower: wet (mechanical draft or natural draft) or dry (mechanical draft or natural draft). The total investment in a cooling system for an 800,000-kilowatt plant will range from about $4.9 million for a fossil-fueled plant with a once-through system using seawater to $50.1 million for a nuclear plant with a natural-draft dry tower [see *illustration above*]. These figures, which cover the condenser and its auxiliaries, the pumps and conduits and a cooling tower if there is one, are based on 1970 prices and include 16 percent for direct overhead costs. In general the capital costs of a given type of cooling system are about 50 percent higher for a nuclear plant than for one using fossil fuel. Natural-draft towers cost more than mechanical-draft towers and dry towers more than wet ones. Since investment costs are rising, the impact of fixed charges will increase the cost handicap against natural-draft towers and dry towers.

Comparing partial production costs, which include capital and fuel but not such items as labor and maintenance, one finds that there is not much difference between a once-through system and a wet cooling tower. With a wet cooling tower the partial production costs would be increased between 3 and 10 percent, which represents an additional .1 to .2 mill per kilowatt-hour. A more significant difference appears in a comparison of partial production costs between a once-through system and a dry cooling tower. Here the partial production costs would be increased between 30 and 75 percent, corresponding to an additional one mill to 1.5 mills per kilowatt-hour.

Currently under discussion is the use of saltwater cooling towers for large generating stations along seacoasts. (The few saltwater towers in existence serve relatively small plants.) The first question to be decided in the public interest is whether or not any additional coastal water should be used for cooling. According to tower manufacturers, it would be technologically feasible to operate a saltwater tower at a large generating station. The investment cost of such a tower would be about 25 percent higher than the cost of a freshwater tower, primarily because of the need for more expensive materials and for high-performance drift eliminators. Maintenance and energy costs for such a system would also be higher because of the problems involved in working with salt water.

The conversion of existing once-through systems to cooling-tower systems may be needed in special cases but probably will not be done widely. For full conversion the capital cost at an 800,000-kilowatt plant would be about $3 million. Alternatively the conversion might involve only the supplemental use of a tower at certain times during the year to keep the temperature of discharge water below a given level.

The increasing size of large generating units will require larger wet cooling towers, which are already being designed. From the typical large mechanical-draft wet tower now in use—a multiple-cell unit about 55 feet wide at the base, 75 feet wide at the top and 325 feet long, with 28-foot fans powered by 200-horsepower motors—one can expect a progression to fans of larger diameter, with tower dimensions increasing proportionately. Present towers are adequate for cooling up to 400,000 kilowatts per continuous tower unit at a fossil-fueled plant; two tower units of seven cells or more each would be needed to serve an 800,000-kilowatt generating unit. The mechanical-draft wet tower of the near future will probably be capable of serving a million-kilowatt fossil-fueled plant with a single tower structure, although the economics and environmental characteristics of such a tower have not yet been established.

This typical large generating unit could also be served by a single natural-draft wet cooling tower about 410 feet in diameter at the base by 370 feet in height. Natural-draft wet towers now available can handle a nuclear unit as large as 1.1 million kilowatts, which is equivalent in cooling requirements to a fossil-fueled plant of about 1.65 million kilowatts. The present maximum size of natural-draft wet towers—about 600 feet in diameter and 500 feet high—is not expected to increase appreciably in the near future.

As for dry towers, the configuration for application to generating stations is not expected to change much from what is already used extensively in such industrial operations as large petrochemical plants. One standard module now produced is 60 feet long, 150 feet wide and has four 28-foot fans. The fan size is expected to increase, as in wet towers, but the module size will probably stay about the same.

The use of cooling towers in the water systems that cool the condensers in generating plants can be expected in an increasing proportion of new generating plants located inland. The number of towers operating with seawater may also rise. Most of the new towers will be of the mechanical-draft wet type, because that is the most economical type for nearly all applications.

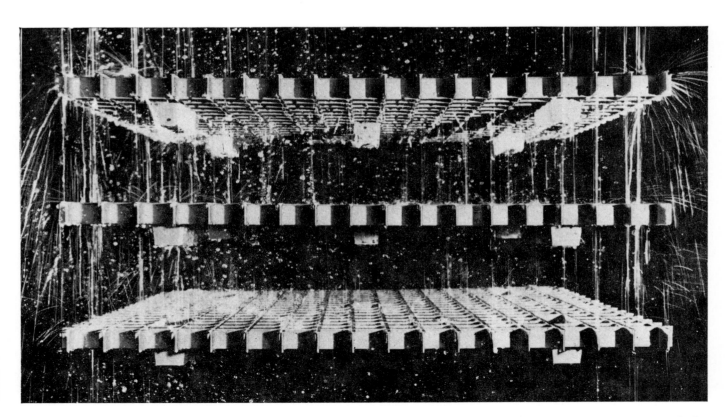

BREAKUP OF WATER by fill is shown in this special arrangement of film fill. The falling water repeatedly hits pieces of fill and breaks up into small drops and films. In this way the water presents a large surface area to the air that is being moved through the tower. Cooling takes place mainly through the evaporation of a small part of the water that is circulated through the system.

The Climate of Cities

by William P. Lowry
August 1967

The variables of climate are profoundly affected by the physical characteristics and human activities of a city. Knowledge of such effects may make it possible to predict and even to control them

It is widely recognized that cities tend to be warmer than the surrounding countryside, and one is reminded almost daily by weather forecasts such as "low tonight 75 in the city and 65 to 70 in the suburbs." Exactly what accounts for the difference? Meteorological studies designed to answer such questions have now been made in a number of cities. Much work remains to be done, but one thing is clear. Cities differ from the countryside not only in their temperature but also in all other aspects of climate.

By climate is meant the net result of several interacting variables, including temperature, the amount of water vapor in the air, the speed of the wind, the amount of solar radiation and the amount of precipitation. The fact that the variables do not usually change in the same way in a city as they do in the open country nearby can often be measured directly in differences of temperature, humidity, precipitation, fog and wind speed between a city and its environs. It is also apparent in such urban phenomena as persistent smog, the earlier blooming of flowering plants and longer periods free of frost.

The city itself is the cause of these differences. Its compact mass of buildings and pavement obviously constitutes a profound alteration of the natural landscape, and the activities of its inhabitants are a considerable source of heat. Together these factors account for five basic influences that set a city's climate apart from that of the surrounding area.

The first influence is the difference between surface materials in the city and in the countryside. The predominantly rocklike materials of the city's buildings and streets can conduct heat about three times as fast as it is conducted by wet, sandy soil. This means that the city's materials can accept more heat energy in less time, even though it takes roughly a third more energy to heat a given amount of rock, brick or concrete to a certain temperature than to heat an equal amount of soil. The temperature of soil at the warmest time of the day may be higher than that of a south-facing rock wall, but the temperature three or four inches below the surface will probably be higher in the wall. At the end of a day the rocky material will have stored more heat than an equal volume of soil.

Second, the city's structures have a far greater variety of shapes and orientations than the features of the natural landscape. The walls, roofs and streets of a city function like a maze of reflectors, absorbing some of the energy they receive and directing much of the rest to other absorbing surfaces [*see top illustration on page* 5]. In this way almost the entire surface of a city is used for accepting and storing heat, whereas in a wooded or open area the heat tends to be stored in the upper parts of plants.

Since air is heated almost entirely by contact with warmer surfaces rather than by direct radiation, a city provides a highly efficient system for using sunlight to heat large volumes of air. In addition, the city's many structures have a braking effect on the wind, thereby increasing its turbulence and reducing the amount of heat it carries away.

Third, the city is a prodigious generator of heat, particularly in winter, when heating systems are in operation. Even in summer, however, the city has many sources of heat that the countryside either lacks or has in far smaller numbers. Among them are factories, vehicles and even air conditioners, which of course must pump out hot air in order to produce their cooling effect.

Fourth, the city has distinctive ways of disposing of precipitation. If the precipitation is in the form of rain, it is quickly removed from the surface by drainpipes, gutters and sewers. If it is snow, much of it is cleared from the surface by plows and shovels, and significant amounts are carried away. In the country much precipitation remains on the surface or immediately below it; the water is thus available for evaporation, which is of course a cooling process powered by heat energy. Because there is less opportunity for evaporation in the city, the heat energy that would have gone into the process is available for heating the air.

Finally, the air in the city is different

HEAT PATTERNS in the lower Manhattan area of New York City on a summer day are shown by infrared photography. In the photographs, which were made with a Barnes thermograph, the lightest areas are the warmest and the darkest are the coolest. The view above shows the buildings at about 11:00 A.M. and the view below at about 3:30 P.M. The day was sunny but hazy; the temperature in the city during the time covered by the photographs was about 75 degrees Fahrenheit. The storage of heat by buildings affects a city's climate.

in that it carries a heavy load of solid, liquid and gaseous contaminants. About 80 percent of the solid contaminants are in the form of particles that are small enough to remain suspended for several days in still air. Although these particles collectively tend to reflect sunlight, thereby reducing the amount of heat reaching the surfaces, they also retard the outflow of heat. The gaseous contaminants, which usually have a greater total mass than the solid ones, come primarily from the incomplete combustion of fuels. One of the principal gases in many cities is sulfur dioxide; when this gas is dissolved under the appropriate meteorological conditions in cloud droplets or raindrops, it is oxidized to form dilute sulfuric acid.

Let us consider how these five influences act over a period of time on the climate of a large city. Our hypothetical city lies in an area of flat or gently rolling countryside and has no large bodies of water nearby. The day is a Sunday, so that no substantial amounts of fuel are being used for industrial purposes. It is a summer day, with clear skies and light winds.

As the sun rises it shines equally on city and country. The sunlight strikes the flat, open country at a low angle; much of it is reflected from the surface. The many vertical walls of the city, however, are almost perpendicular to the sun's rays. In spite of the fact that when the sun is low in the sky its rays are less intense because they must pass through more of the earth's atmosphere, the walls begin almost at once to absorb heat. In the country little heat is being absorbed, even in the sunlit areas.

Later in the day the rural areas begin to respond more like the city. The sun has risen high enough for its radiation to impinge on the surface more directly and with less reflection. The air outside the city begins to warm rapidly. The city has already been warming for some time, however, and so it has a large lead toward the day's maximum temperature.

The warm air in the city concentrates near the city's center of mass. Toward midmorning the air in the center begins to rise. Being warmer at each level than the air at the same level in the surrounding countryside, the city air continues to rise in a gentle stream flowing upward from the center. The air that rises must be replaced; hence a flow from the rural areas into the city begins in the layers near the ground. The air from the country must also be replaced, and gradually a slow circulation is established. Air

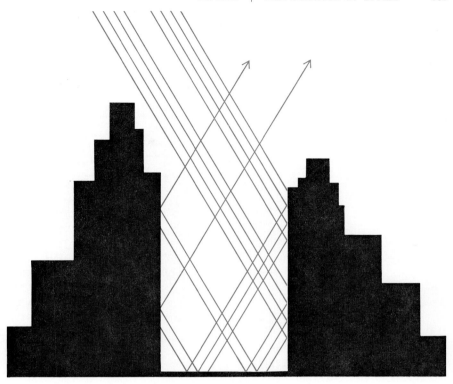

SHAPE AND ORIENTATION OF SURFACES in a city have a strong bearing on the climate. Vertical walls tend to reflect solar radiation toward the ground instead of the sky. Rocklike materials also store heat, so that the city often becomes warmer than its environs.

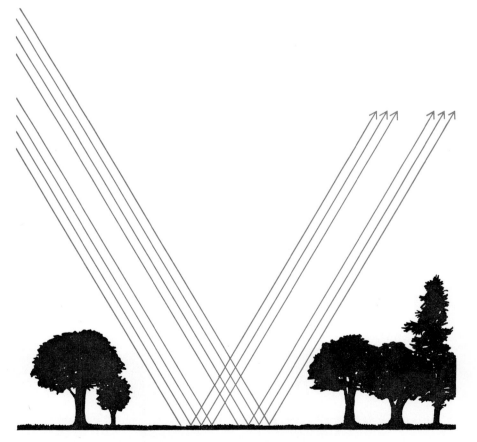

RADIATION IN COUNTRYSIDE tends to be reflected back to the sky because the countryside has fewer vertical surfaces than the city. Toward midday, however, when the sun's rays are perpendicular to the ground, city and country temperatures may be about the same.

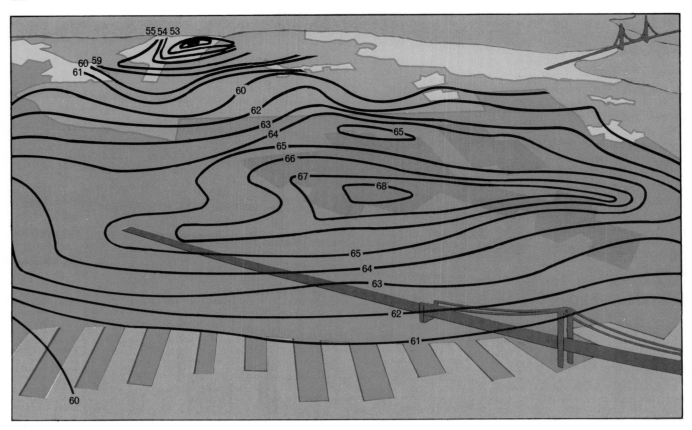

TEMPERATURE DISTRIBUTION in San Francisco on a spring evening is depicted by means of isotherms, which are lines of equal temperature. The shading ranges from the most densely built-up areas (*dark*) through less dense sections to open country (*light*).

EMISSIONS OF HEAT from an automobile with its engine idling appear in an infrared photograph made with a Barnes thermograph. Bright area below the car is pavement, which was in direct sunlight. Vehicles are a major source of heat production in a city.

moves into the center of the city in the lower layers, rises in the central core, flows outward again at a higher altitude and as it cools settles down over the open country to complete the cycle.

Near midday the sunlight strikes the open country still more directly, and the difference in temperature between city and country becomes quite small. Now the air rising over the city is not appreciably warmer than the surrounding air, so that in the early afternoon the cycle of circulation is considerably weakened. As the afternoon progresses, however, a situation similar to that of the early morning develops. The sun sinks, its rays striking the open country at a lower and lower angle; an increasing proportion of its radiation is reflected. During this time the walls in the city are still intercepting the sun's radiation quite directly. The difference in temperature between city and country begins to increase again, and the circulation of air rising over the city and sinking outside it is reinvigorated. Just before sunset the circulation is fairly strong, but it weakens again as darkness falls. At about this time one would be likely to find the temperature at a weather station outside the city (such as at an airport) lower than the temperature at the downtown weather office.

During the night the surfaces that radiate their warmth to the sky most rapidly are the streets and the rooftops. If much of the rooftop area of the city is at about the same height, there will be a strong tendency for a cool layer of air to be formed at that level. With cool air at the rooftops now lying below warmer air just above it, a rather stable stratification of air develops, and any tendency for upward movement of warm air in the spaces between buildings is inhibited.

The overall situation now is that the rural area is cooling rapidly and the city area is cooling slowly. Heat is being removed from the fields by light winds and by almost unobstructed radiation to the night sky. In the city, however, pockets of air are trapped. They cannot move upward, and they are still receiving heat from the release of energy stored in the walls of the buildings during the day. Through the night both the city and the countryside will continue to cool, but by dawn the city is still likely to be four or five degrees warmer than its surroundings.

Early Monday morning the factories in the city begin to put forth heat, smoke and gases. Automobiles, trucks and buses start to emit large quantities

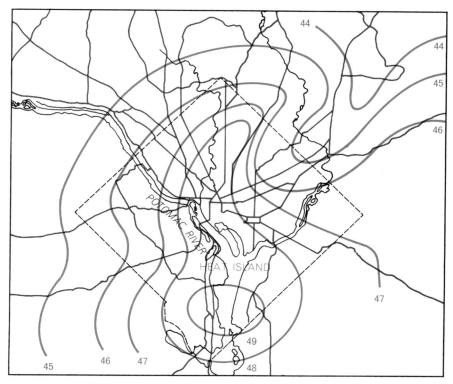

ANNUAL TEMPERATURE RECORD of Washington, D.C., and its environs gives the average of annual minimum temperatures for the period 1946–1960. The areas inside closed isotherms constitute what is known as the heat island. Here as in other cities the island is associated with the most densely built-up part of the urban complex. This map and the one below are based on data obtained by Clarence A. Woollum of the U.S. Weather Bureau.

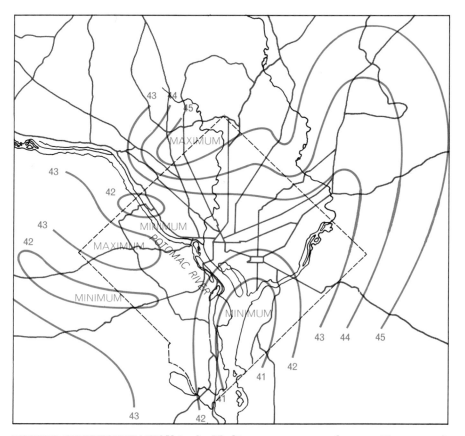

RECORD OF PRECIPITATION in the Washington area covers the same 15 years as the temperature record. Both topography and the existence of the city affect precipitation.

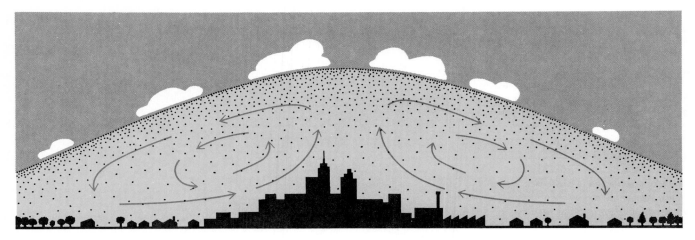

DUST DOME takes shape periodically over large cities because of the particles of dust and smoke that enter the air as a result of activities in the city. Air tends to rise over the warmer central part of the city and to settle over the cooler environs, so that a circulatory system develops. Dome is likely to persist, significantly affecting the city's climate, until a strong wind or a heavy rain carries it away.

of heat and fumes. Even stoves in kitchens constitute a source of heat that cannot be neglected. Artificial heating and air pollution thus become meteorologically significant as the day begins.

As before, the early sun starts to warm the city's walls and streets, and heat begins to accumulate in the downtown area. Today, however, there is a difference because of the heat being added to the system by the tall chimneys of factories. Ordinarily air rising to the height of the chimney tops would have had a chance to cool, but now it receives more heat at that level and will probably rise higher above the city than it did on Sunday. Moreover, the column of air now carries a freight of particles of dust and smoke. The smallest particles will fall only after they have been carried away from the rising column of air and out over the suburbs. Other particles will remain suspended over the city all day.

Over a long period of time the con-

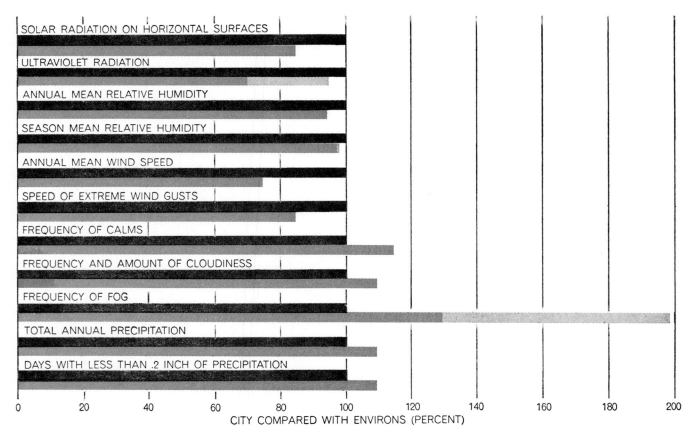

MAJOR DIFFERENCES IN CLIMATE between a city (*color*) and its environs (*gray*) are set out in terms of the percentage by which the city has more or less of each climatic variable during a year than is experienced in the countryside. For example, the city receives 5 percent less ultraviolet radiation than the countryside in summer, 30 percent less in winter; frequency of fog in city is 30 percent higher in summer and 100 percent higher in winter. Findings were made by Helmut E. Landsberg of the University of Maryland.

tinuous introduction and movement of particles creates a dome-shaped layer of haze over the city. This structure, variously called the "dust dome" and the "haze hood," has long been characteristic of large cities, although in recent years the general dirtiness of the air has made the dome harder to distinguish from its surroundings than it was several decades ago. Nonetheless, it still has a marked effect on the city's climate.

At night, as the particles in the dome cool, they can become nuclei on which the moisture in the air condenses as fog. The phenomenon occurs over cities in the middle latitudes when conditions are precisely right. The first layers of fog will usually form near the top of the dome, where the particles cool most rapidly by radiation; the blanket becomes thicker by downward growth until it reaches the ground as smog. This extra covering of water droplets over the city further retards nighttime cooling. Fog helps to perpetuate the dust dome by preventing the suspended particles from moving upward out of the system. Thus one day's contribution of solid contaminants will remain in the air over the city to be added to the next day's.

In the absence of a strong wind or a heavy rain to clear away the dust dome, the haze becomes denser each day. In winter, since less and less sunshine penetrates the dome to warm the city naturally, more and more fuel is burned to make up the difference. The combustion contributes further to the processes that build up smog. It is in this gradual but inexorable way that the smog problem has attained serious dimensions in many large cities.

In sum, a city's effect on its own climate is complex and far-reaching. Helmut E. Landsberg of the University of Maryland, who until recently was director of climatology in the U.S. Weather Bureau, has drawn up a balance sheet showing the net effect of the variables [see *bottom illustration on opposite page*]. Among other things, he has concluded that cities in the middle latitudes receive 15 percent less sunshine on horizontal surfaces than is received in surrounding rural areas and that they receive 5 percent less ultraviolet radiation in summer and 30 percent less in winter. Landsberg's figures also show that the city, compared with the countryside, has a 6 percent lower annual mean relative humidity, 10 percent more precipitation, 10 percent more cloudiness, 25 percent lower mean annual wind speed and 30

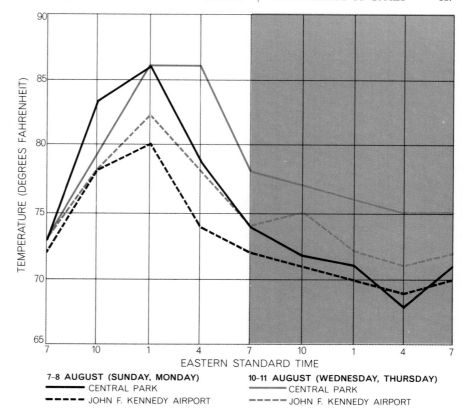

TEMPERATURE DIFFERENCES appear in readings at a weather station in New York City and one at an airport in the environs for two 24-hour periods in August, 1966. The graph begins at 7:00 A.M. for each period. Temperature differences are often less pronounced on weekends than on weekdays because fewer of a city's heat sources are operating on weekends.

percent more fog in summer and 100 percent more in winter.

T. J. Chandler, director of the London Climatic Survey, has compiled a number of records for the London area. He has found that over a period of 30 years the average maximum temperatures in the city, the suburbs and the surrounding countryside were respectively 58.3, 57.6 and 57.2 degrees and the average minimums 45.2, 43.1 and 41.8 degrees. His figures also show that over the period the city had consistently less sunshine than its environs did.

Some of these broad findings merit closer consideration. The patterns of temperature in a city can be shown on maps by drawing isotherms, or lines of equal temperature, for various times. Under a great variety of wind, cloud and sunshine conditions isotherm maps all show the highest temperatures clustered near the center of the city, with lower temperatures appearing radially toward the suburbs and the countryside. The resulting pattern of isotherms suggests the term "heat island" for the warmest area [see *top illustration on page 385*]. The term is used regularly by meteorologists to describe this major feature of a city's climate.

The heat island has been observed in many cities, some large and some small, some near water and some not, some with hills and some with none. How, then, can one be sure that the heat island, and thus the city climate itself, is really attributable to the works of man? J. Murray Mitchell, urban climatologist in the U.S. Weather Bureau, has considered the question and found three kinds of evidence that the city climate is caused by the city itself.

First, cities exhibit the heat island whether they are flat like Indianapolis or built on hills like San Francisco. Hence topography cannot explain the heat-island pattern. Second, temperature records averaged by day of the week show marked differences between Sundays and other days. Since many of the heat-creating processes distinctive to cities are inactive on Sundays, it is evident that those man-made processes account for the heat island. Finally, Mitchell has carefully examined the population and temperature records of a number of cities and found that the size of the heat island and the difference in temperature between it and surrounding areas increase as population does.

Another fact to be noted about tem-

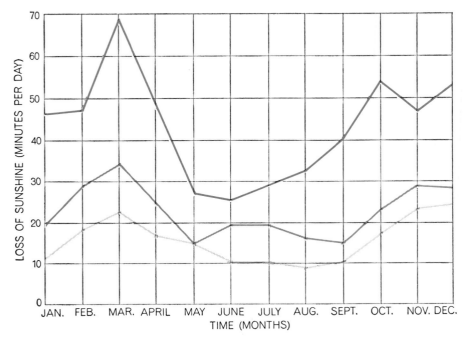

LOSS OF BRIGHT SUNSHINE in London compared with areas surrounding the city is expressed in terms of minutes per day for each month. The figures show the city's average loss during the period 1921 to 1950. London area's districts are represented by the dark line at top, the inner suburbs by the middle line and the outer suburbs by the bottom line.

peratures is that the maximum difference between city and countryside appears to be about 10 to 15 degrees Fahrenheit, regardless of the size of the city. Chandler has found this to be the case in London, which has a population of eight million; my colleagues and I have found the same in Corvallis, Ore., which has a population of about 20,000.

Chandler's figures for the loss of sunlight in London show larger losses in winter, when the sun is low, than in summer, when sunlight takes a shorter path through the atmosphere. The amount of reduction increases markedly toward the center of the city, showing both the greater depth of the dust dome and the greater density of pollutants

there. Part of the reduction of sunlight in London and other cities can be laid to the fact that a city tends to be more cloudy than its environs. Warm air rising over the center of the city provides a mechanism for the formation of clouds on many days when clouds fail to form in the country.

The frequency of fogs during the winter has to do with the greater relative reductions in sunshine during the winter months. One cannot simply say, however, that the greater frequency of fog explains the reduced total of sunshine. A feedback process is involved. Once fog forms, a weak sun has most of its energy reflected from the top of the fog layer. Little of the energy penetrates the fog to warm the city, and so the fog

tends to perpetuate itself until the climatic situation changes.

Another connection between winter and the higher frequency of fog arises from the low temperatures. After an incursion of cold arctic air the residents of the city increase their rate of fuel consumption. The higher consumption of fuel produces more particulate pollutants and more water vapor. The air above a city is usually quite stagnant following the arrival of a cold wave, and thus the stage is set for the generation of fog. Lacking ventilation, the city's atmosphere fills with smoke, dirt and water vapor. The particles of smoke and dirt act as nuclei for the condensation of the water vapor. Because the water is shared among a large number of nuclei, the air contains a large number of small water droplets. Such a size distribution of water droplets forms a persistent fog, and the fog retards warming of the city. Retarded warming prolongs the need for extra heating. Only another change of air mass will relieve the situation. This chain of events has been associated with nearly every major disaster resulting from air pollution.

Reduction of visual range by smoke alone is not regularly recorded in cities. It is recorded at airports, however, and Landsberg has been able to use data from the Detroit City Airport, which is near the center of the city, and Wayne County Airport, which is in a more rural area, to deduce something about climatic differences between a city and the nearby countryside. The records indicate that a city will have, in the course of a year, 10 times more hours in which smoke restricts visibility to a mile or less than will be experienced in rural areas.

Contrary to what one might think, this situation may be improving somewhat. Robert Beebe of the U.S. Weather Bureau recently studied records of the visual range at the major municipal airports that did not change either their location or their schedule of weather observation between 1945 and 1965. He found that the number of times when smoke reduces horizontal visibility at the airports is less now than it was in 1945. The change might be explained by efforts to control air pollution, resulting in reduced concentrations of smoke and in changes in the size and character of smoke particles.

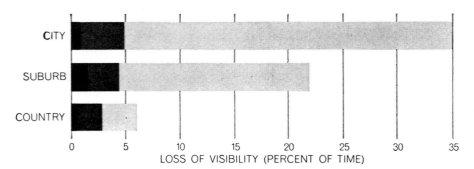

FOG IN PARIS has cut visibility more in the city than in the surrounding areas. Data are for winter and show the percent of time when visibility was reduced to between one mile and a quarter-mile by light fog (*light*), a quarter-mile to 300 feet by moderate fog (*medium*) and less than 300 feet by dense fog (*dark*). In the summer there were far fewer days of fog.

The differences in moisture and precipitation between a city and its environs are somewhat contradictory. During periods without rain the relative scarcity of water for evaporation in the city results in a reduced concentration of water vapor in the air. Expressed as relative humidity, the difference gives the city a reduction of 6 percent in the annual average of the countryside, of 2 percent in the winter average and of 8 percent in the summer.

Even though the city is somewhat drier than its environs, on the days when rain or snow falls there is likely to be more in the city than in the countryside. The difference amounts to 10 percent in a year. It builds up mostly as an accumulation of small increments on drizzly days, when not much precipitation falls anywhere in the area. On such days the updrafts over the warm city provide enough extra lift so that the clouds there produce a slightly higher amount of precipitation.

Perhaps the catalogue of differences I have cited will leave the reader thinking that the city climate offers no advantages over the country climate. Actually there are several, including lower heating bills, fewer days with snow and a longer gardening season. Landsberg has estimated that a city has about 14 percent fewer days with snow than the countryside. The season between the last freeze in the spring and the first freeze in the fall may be three or four weeks longer in the city than in the countryside.

Both the advantages and the disadvantages of city climate testify to the fact that the city's climate is distinctly different from the countryside's. Every major aspect of climate is changed, if only slightly, by an urban complex. The differences in a small city may be only occasional; in a large city every day is different climatically from what it would have been if the city were not there.

Fuller understanding of the climatic changes created by a city may make it possible to manage city growth in such a way that the effect of troublesome changes will be minimal. Perhaps the changes can even be made beneficial. Several organizations are accumulating climatological data on cities. I have already mentioned the London Climatic Survey. Similar work is in progress in the U.S. Environmental Science Services Administration, at the University of California at Los Angeles, at New York University and in the research laboratories of the Travelers Insurance Company. Meteorologists in those organizations are driving instrumented automobiles, flying instrumented aircraft and operating hundreds of ground stations to obtain weather data. Although the studies are aimed primarily at understanding the meteorological problems of air pollution, other aspects of the local modification of climate by cities will be better understood as a result.

What may be even more important is the possibility of ascertaining the potential of extensive urbanization for causing large-scale changes of climate over entire continents. The evidence is not yet substantial enough to show that urbanization does cause such changes, but it is sufficient to indicate that the possibility cannot be ignored. The acquisition of more knowledge about the climate of cities may in the long run be one of the keys to man's survival.

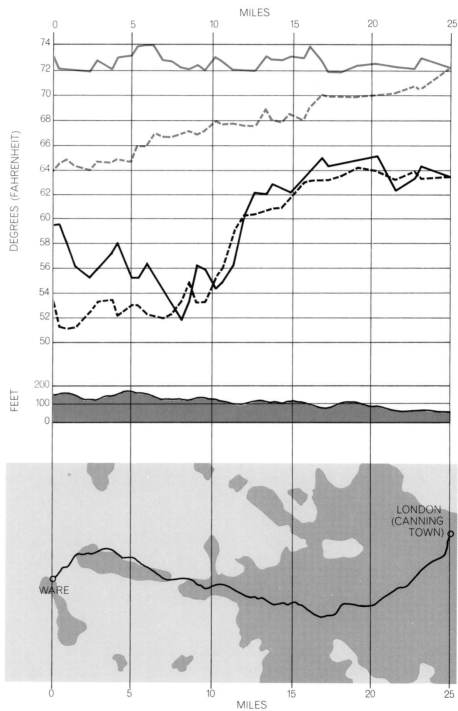

TEMPERATURE TRAVERSES between the Canning Town section of London and the community of Ware 25 miles north were made on a June day (*color*) and night (*black*) by T. J. Chandler of the London Climatic Survey. In each case he made an outbound trip (*solid line*) and an inbound one (*broken line*). Dark shading at bottom shows heavily built-up areas.

BIBLIOGRAPHIES

When the articles in this volume appeared in the SCIENTIFIC AMERICAN, they were accompanied by these bibliographies.

I TECHNICS AND EMPIRICAL TECHNOLOGY

1. Tools and Human Evolution

CEREBRAL CORTEX OF MAN. Wilder Penfield and Theodore Rasmussen. Macmillan Company, 1950.

THE EVOLUTION OF MAN, edited by Sol Tax. University of Chicago Press, 1960.

THE EVOLUTION OF MAN'S CAPACITY FOR CULTURE. Arranged by J. N. Spuhler. Wayne State University Press, 1959.

HUMAN ECOLOGY DURING THE PLEISTOCENE AND LATER TIMES IN AFRICA SOUTH OF THE SAHARA. J. Desmond Clark in Current Anthropology, Vol. I, pages 307–324; 1960.

2. The Beginnings of Wheeled Transport

ANCIENT EUROPE FROM THE BEGINNINGS OF AGRICULTURE TO CLASSICAL ANTIQUITY. Stuart Piggott. Aldine Publishing Company, 1965.

CLAY MODELS OF BRONZE AGE WAGONS AND WHEELS IN THE MIDDLE DANUBE BASIN (PLATES LXI-LXVIII). I. Bóna in Acta Archaeologica Academiae Scientiarum Hungaricae, Tomus XII, pages 83–111; Akadémiai Kiado, 1960.

THE DIFFUSION OF WHEELED VEHICLES. V. Gordon Childe in Ethnographish Archäologische Forschungen: Vol. II, edited by H. Kothe and K. H. Otto. Veb Deutscher Verlag der Wissenschaften, 1954.

THE FIRST WAGGONS AND CARTS—FROM THE TIGRIS TO THE SEVERN. V. Gordon Childe in Proceedings of the Prehistoric Society for 1951, New Series, Vol. 17, Part 2, pages 177–194; 1951.

3. Medieval Uses of Air

EILMER OF MALMESBURY, AN ELEVENTH CENTURY AVIATOR: A CASE STUDY OF TECHNOLOGICAL INNOVATION, ITS CONTEXT AND TRADITION. Lynn White, Jr., in Technology and Culture, Vol. 2, No 2, pages 97–111; Spring, 1961.

HELICOPTERS AND WHIRLIGIGS. Ladislao Reti in Raccolta Vinciana, Vol. 20, pages 331–338; 1964.

THE ORIGIN OF THE SUCTION PUMP. Sheldon Shapiro in Technology and Culture, Vol. 5, No. 4, pages 566–574; Fall, 1964.

4. The Origins of Feedback Control

AUTOMATIC CONTROL: A SCIENTIFIC AMERICAN BOOK. Simon and Schuster, 1955.

SELECTED PAPERS ON MATHEMATICAL TRENDS IN CONTROL THEORY. Edited by Richard Bellman and Robert Kalaba. Dover Publications, Inc., 1964.

THE ORIGINS OF FEEDBACK CONTROL. Otto Mayr. The M.I.T. Press, in press.

5. Bicycle Technology

RIDING HIGH: THE STORY OF THE BICYCLE. Arthur Judson Palmer. E. P. Dutton & Co., Inc., 1956.

HANDBOOK OF THE COLLECTION ILLUSTRATING CYCLES. C. F. Caunter. Her Majesty's Stationery Office, 1958.

WHEELS WITHIN WHEELS: THE STORY OF THE STARLEYS OF COVENTRY. Geoffrey Williamson. Geoffrey Bles, 1966.

II   THE RISE OF SCIENTIFIC TECHNOLOGY

6. The Origins of the Steam Engine

THE EARLY GROWTH OF STEAM POWER. A. E. Musson and E. Robinson in *Economic History Review*, Series 2, Vol. 2, No. 3, pages 418–439; April, 1959.

JAMES WATT AND THE STEAM ENGINE. H. W. Dickinson and Rhys Jenkins. Oxford University Press, 1927.

KINEMATICS OF MECHANISMS FROM THE TIME OF WATT. Eugene S. Ferguson in *U.S. National Museum Bulletin 228*, Paper 27, pages 185–230. Smithsonian Institution, 1962.

SCIENCE AND THE STEAM ENGINE. Milton Kerker in *Technology and Culture*. Wayne State University Press, 1961.

A SHORT HISTORY OF THE STEAM ENGINE. H. W. Dickinson. Cambridge University Press, 1938.

THOMAS NEWCOMEN: THE PRE-HISTORY OF THE STEAM ENGINE. L. T. C. Rolt. David and Charles: Dawlish MacDonald, 1963.

A TREATISE ON THE STEAM ENGINE. John Farey. Longman, Rees, Orme, Brown and Green, 1827.

7. From Faraday to the Dynamo

EXPERIMENTAL RESEARCHES IN ELECTRICITY. Michael Faraday. B. Quaritch, 1839–1855.

A HISTORY OF PHYSICS IN ITS ELEMENTARY BRANCHES, INCLUDING THE EVOLUTION OF PHYSICAL LABORATORIES. Florian Cajori. The Macmillan Company, 1929.

A HISTORY OF TECHNOLOGY, VOL. 5. Edited by Charles Singer *et al.* Oxford University Press, 1958.

MICHAEL FARADAY: HIS LIFE AND WORK. Silvanus P. Thompson. The Macmillan Company, 1898.

MODERN VIEWS OF ELECTRICITY. Oliver J. Lodge. Macmillan and Company, 1889.

A SHORT HISTORY OF SCIENTIFIC IDEAS TO 1900. Charles Singer. Oxford University Press, 1959.

8. Technology and Economic Development

ECONOMIC DEVELOPMENT: PRINCIPLES, PROBLEMS, AND POLICIES. Benjamin Higgins. W. W. Norton & Company, Inc., 1959.

THE PROCESS OF ECONOMIC GROWTH. W. W. Rostow. W. W. Norton & Company, Inc., 1952.

SCIENCE AND THE NEW NATIONS: THE PROCEEDINGS OF THE INTERNATIONAL CONFERENCE ON SCIENCE IN THE ADVANCEMENT OF NEW STATES AT REHOVOTH, ISRAEL, edited by Ruth Gruber. Basic Books, Inc., 1961.

SOCIOLOGICAL ASPECTS OF ECONOMIC GROWTH. Bert F. Hoselitz. The Free Press of Glencoe, 1960.

THE THEORY OF ECONOMIC GROWTH. W. Arthur Lewis. George Allen & Unwin Ltd., 1955.

9. The Origin of the Automobile Engine

GAS ENGINE. *Encyclopaedia Britannica*, Eleventh Edition, Vol. 11, pages 495–501.

NIKOLAUS AUGUST OTTO: CREATOR OF THE INTERNAL-COMBUSTION ENGINE. Gustav Goldbeck in *From Engines to Autos*, by Eugen Diesel, Gustav Goldbeck and Friedrich Schildberger. Henry Regnery Company, 1960.

THE ORIGIN OF THE FOUR-STROKE CYCLE. Lynwood Bryant in *Technology and Culture*, April, 1967.

THE SILENT OTTO. Lynwood Bryant in *Technology and Culture*, Vol. 7, No. 2, pages 184–200; Spring, 1966.

10. Rudolf Diesel and His Rational Engine

THEORY AND CONSTRUCTION OF A RATIONAL HEAT MOTOR. Rudolf Diesel. Translated from the German by Bryan Donkin. Spon & Chamberlain, 1894.

DIESEL'S RATIONAL HEAT MOTOR. *Progressive Age*, Vol. 15, pages 575–578, December 1, 1897; Vol. 15, pages 602–607, December 15, 1897; Vol. 16, pages 5–6, January 1, 1898; Vol. 16, pages 30–35, January 15, 1898.

DIESEL'S RATIONAL HEAT MOTOR. E. D. Meier in *Journal of the Franklin Institute of the State of Pennsylvania*, Vol. 146, No. 4, pages 241–264; October, 1898.

RUDOLF DIESEL. Eugen Diesel in *From Engines to Autos: Five Pioneers in Engine Development and Their Contributions to the Automotive Industry*, by Eugen Diesel, Gustav Goldbeck and Friedrich Schildberger. Henry Regnery Company, 1960.

11. The Invention of the Electric Light

EDISON. Matthew Josephson, McGraw-Hill Book Co., 1959.

12. The First Electron Tube

FIFTY YEARS OF ELECTRICITY: THE MEMORIES OF AN ELECTRICAL ENGINEER. J. A. Fleming. The Wireless Press, Ltd., 1921.

MEMORIES OF A SCIENTIFIC LIFE. Sir Ambrose Fleming. Marshall, Morgan & Scott Ltd., 1934.

THERMIONIC VALVES 1904–1954: THE FIRST FIFTY YEARS. The Institution of Electrical Engineers, 1955.

III THE TRIUMPH OF SCIENTIFIC TECHNOLOGY

13. Steam Turbines

STEAM TURBINES. Edwin F. Church, Jr. McGraw-Hill Book Company, Inc., 1950.

STEAM TURBINES AND THEIR CYCLES. J. Kenneth Salisbury. John Wiley & Sons, Incorporated, 1950.

STEAM TURBINE PERFORMANCE AND ECONOMICS. Robert L. Bartlett. McGraw-Hill Book Company, Inc., 1958.

14. The Fastest Computer

THE SOLOMON COMPUTER. D. L. Slotnick, W. C. Borck and R. C. McReynolds in *Joint Computer Conference AFIPS Proceedings*, Vol. 22, pages 97–107; Fall, 1962.

COMPUTER NETWORK DEVELOPMENT TO ACHIEVE RESOURCE SHARING. Lawrence G. Roberts and Barry D. Wessler in *Joint Computer Conference AFIPS Proceedings*, Vol. 36, pages 543–549; Spring, 1970.

AN INTRODUCTORY DESCRIPTION OF THE ILLIAC IV SYSTEM. S. A. Denenberg in *ILLIAC IV Document*, No. 225; July 15, 1970.

15. Missile Submarines and National Security

POLARIS & POSEIDON, FBM FACTS. Strategic Systems Project Office. Navy Department, 1970.

SIPRI YEARBOOK OF WORLD ARMAMENTS AND DISARMAMENT 1969/70. Stockholm International Peace Research Institute. Humanities Press, 1971.

BEYOND SALT ONE. Herbert Scoville, Jr., in *Foreign Affairs*, Vol. 50, No. 3, pages 488–500; April, 1972.

REPORT ON ULMS. Members of Congress for Peace through Law. Washington, D.C., April, 1972.

TESTIMONY OF THE FEDERATION OF AMERICAN SCIENTISTS BEFORE DEFENSE SUBCOMMITTEE, APPROPRIATIONS COMMITTEE, U.S. House of Representatives on the FY 1973 Defense Program and Budget. Presented by Herbert Scoville. April, 1972.

16. Multiple-Warhead Missiles

RACE TO OBLIVION: A PARTICIPANT'S VIEW OF THE ARMS RACE, Herbert York. Simon and Schuster, 1970.

MISSILE SUBMARINES AND NATIONAL SECURITY. Herbert Scoville, Jr., in *Scientific American*, Vol. 226, No. 6, pages 15–27; June, 1972.

WORLD ARMAMENTS AND DISARMAMENT: SIPRI YEARBOOK 1973. Stockholm International Peace Research Institute. Almqvist & Wiksell, 1973.

17. The Economics of Technological Change

INPUT-OUTPUT ECONOMICS. Wassily Leontief. Oxford University Press, 1966.

INTERINDUSTRY ECONOMICS. Hollis B. Chenery and Paul G. Clark. John Wiley & Sons, Inc., 1959.

THE INTERINDUSTRY STRUCTURE OF THE UNITED STATES: A REPORT ON THE 1958 INPUT-OUTPUT STUDY. Morris R. Goldman, Martin L. Marimont and Beatrice N. Vaccara in *Survey of Current Business*, Vol. 44, No. 11, pages 10–17; November, 1964.

STRUCTURAL INTERDEPENDENCE AND ECONOMIC DEVELOPMENT: PROCEEDINGS OF AN INTERNATIONAL CONFERENCE ON INPUT-OUTPUT TECHNIQUES, 1961. Edited by Tibor Barna. St Martin's Press, 1963.

THE STRUCTURE OF THE U.S. ECONOMY. Wassily W. Leontief in *Scientific American*, Vol. 212, No. 4, pages 25–45; April, 1965.

18. Information

COMPUTERS AND THE WORLD OF THE FUTURE. Edited by Martin Greenberger. The M.I.T. Press, 1962.

CYBERNETICS: OR CONTROL AND COMMUNICATION IN THE ANIMAL AND THE MACHINE. Norbert Wiener. The M.I.T. Press, 1961.

19. Systems Analysis of Urban Transportation

THE URBAN TRANSPORTATION PROBLEM. J. R. Meyer, J. F. Kain and M. Wohl. Harvard University Press, 1965.

SYSTEMS ANALYSIS OF URBAN TRANSPORTATION. General Research Corporation. U.S. Department of Housing and Urban Development, 1968.

TOMORROW'S TRANSPORTATION: NEW SYSTEMS FOR URBAN FUTURE. Office of Metropolitan Development. U.S. Department of Housing and Urban Development, 1968.

20. Communication

CYBERNETICS OR CONTROL AND COMMUNICATION IN THE ANIMAL AND THE MACHINE. Norbert Wiener. The Technology Press of Massachusetts Institute of Technology and John Wiley & Sons, Inc., 1948.

SYNTACTIC STRUCTURES. Noam Chomsky. Mouton & Co., 1957.

THE MATHEMATICAL THEORY OF COMMUNICATION. Claude E. Shannon and Warren Weaver. The University of Illinois Press, 1959.

SYMBOLS, SIGNALS AND NOISES: THE NATURE AND PROCESS OF COMMUNICATION. J. R. Pierce. Harper & Brothers, 1961.

SCIENCE, ART AND COMMUNICATION. J. R. Pierce. Clarkson N. Potter, Inc., 1968.

IV ENERGY: THE ULTIMATE RESOURCE

21. The Energy Resources of the Earth

MAN AND ENERGY. A. R. Ubbelohde. Hutchinson's Scientific and Technical Publications, 1954.

ENERGY FOR MAN: WINDMILLS TO NUCLEAR POWER. Hans Thirring. Indiana University Press, 1958.

ENERGY RESOURCES. M. King Hubbert. National Academy of Sciences—National Research Council, Publication 1000-D, 1962.

RESOURCES AND MAN: A STUDY AND RECOMMENDATIONS. Committee on Resources and Man. W. H. Freeman and Company, 1969.

ENVIRONMENT: RESOURCES, POLLUTION AND SOCIETY. Edited by William W. Murdoch. Sinauer Associates, 1971.

22. The Flow of Energy in an Industrial Society

ENERGY IN THE UNITED STATES: SOURCES, USES, AND POLICY ISSUES. Hans H. Landsberg and Sam H. Schurr. Random House, 1968.

AN ENERGY MODEL FOR THE UNITED STATES, FEATURING ENERGY BALANCES FOR THE YEARS 1947 TO 1965 AND PROJECTIONS AND FORECASTS TO THE YEARS 1980 AND 2000. Warren E. Morrison and Charles L. Readling. U:S. Department of the Interior, Bureau of Mines, No. 8384, 1968.

THE ECONOMY, ENERGY, AND THE ENVIRONMENT: A BACKGROUND STUDY PREPARED FOR THE USE OF THE JOINT ECONOMIC COMMITTEE, CONGRESS OF THE UNITED STATES. Environmental Policy Division, Legislative Reference Service, Library of Congress. U.S. Government Printing Office, 1970.

ENERGY CONSUMPTION AND GROSS NATIONAL PRODUCT IN THE UNITED STATES: AN EXAMINATION OF A RECENT CHANGE IN THE RELATIONSHIP. National Economic Research Associates, Inc., 1971.

23. The Arrival of Nuclear Power

CIVILIAN NUCLEAR POWER: THE 1967 SUPPLEMENT TO THE 1962 REPORT TO THE PRESIDENT. U.S. Atomic Energy Commission. U.S. Government Printing Office, February, 1967.

THE NUCLEAR INDUSTRY—1967. U.S. Atomic Energy Commission. U.S. Government Printing Office, November 6, 1967.

NUCLEAR POWER, U.S.A. Walter H. Zinn, Frank K. Pittman and John F. Hogerton. McGraw-Hill Book Company, 1964.

24. Fast Breeder Reactors

THE TECHNOLOGY OF NUCLEAR REACTOR SAFETY, VOL. I: REACTOR PHYSICS AND CONTROL. Edited by T. G. Thompson and J. G. Beckerley. The M.I.T. Press, 1964.

FAST REACTOR TECHNOLOGY: PLANT DESIGN. Edited by John G. Yevick. The M.I.T. Press, 1966.

AEC AUTHORIZING LEGISLATION, FISCAL YEAR 1971, HEARINGS BEFORE THE JOINT COMMITTEE ON ATOMIC ENERGY, PART 3. U.S. Government Printing Office, 1970.

25. The Prospects of Fusion Power

PROGRESS IN CONTROLLED THERMONUCLEAR RESEARCH. R. W. Gould, H. P. Furth, R. F. Post and F. L. Ribe in Presentation Made before the President's Science Advisory Committee, December 15, 1970, and AEC's General Advisory Committee, December 16, 1970.

WORLD SURVEY OF MAJOR FACILITIES IN CONTROLLED FUSION. Nuclear Fusion, Special Supplement 1970, STI/Pub/23. International Atomic Energy Agency, 1970.

WHY FUSION? William C. Gough in Proceedings of the Fusion Reactor Design Symposium, Held at Texas Tech University, Lubbock, Texas, on June 2–5, 1970. In press.

26. Human Energy Production as a Process in the Biosphere

NATIONAL RESOURCES FOR U.S. GROWTH: A LOOK AHEAD TO THE YEAR 2000. Hans H. Landsberg. Resources for the Future, Inc., 1964.

CLEANING OUR ENVIRONMENT: THE CHEMICAL BASIS FOR ACTION. Subcommittee on Environmental Improvement, Committee on Chemistry and Public Affairs. American Chemical Society, 1969.

RESOURCES AND MAN: A STUDY AND RECOMMENDATIONS. The Committee on Resources and Man. W. H. Freeman and Company, 1969.

GLOBAL EFFECTS OF ENVIRONMENTAL POLLUTION. Edited by S. F. Singer. D. Reidel Publishing Company, 1970.

POWER GENERATION AND ENVIRONMENTAL CHANGE: SYMPOSIUM OF THE COMMITTEE ON ENVIRONMENTAL AFFAIRS, AMERICAN ASSOCIATION FOR THE ADVANCEMENT OF SCIENCE, BOSTON, DECEMBER 28, 1969. Edited by David A. Berkowitz and Arthur M. Squires. The M.I.T. Press, in press.

V   HEAT: THE ULTIMATE WASTE

27. Perpetual Motion Machines

ORDER AND CHAOS: LAWS OF ENERGY AND ENTROPY. Stanley W. Angrist and Loren G. Hepler. Basic Books, Inc., 1967.

PERPETUUM MOBILE. Henry Dircks. Rogers & Hall Co., 1916.

THE SEVEN FOLLIES OF SCIENCE. John Phin. D. Van Nostrand Company, 1906.

28. The Conversion of Energy

EFFICIENCY OF THERMOELECTRIC DEVICES. Eric T. B. Gross in American Journal of Physics, Vol. 29, No. 1, pages 729–731; November, 1961.

ELECTRICAL ENERGY BY DIRECT CONVERSION. Claude M. Summers. Publication No. 147, The Office of Engineering Research, Oklahoma State University, March, 1966.

APPROACHES TO NONCONVENTIONAL ENERGY CONVERSION EDUCATION. Eric T. B. Gross in IEEE Transactions on Education, Vol. E-10, No. 2, pages 98–99; June, 1967.

29. Thermal Pollution and Aquatic Life

THE PHYSIOLOGY OF FISHES. Edited by Margaret E. Brown. Academic Press, Inc., 1957.

THE PHYSIOLOGY OF CRUSTACEA, VOL. I: METABOLISM AND GROWTH. Edited by Talbot H. Waterman. Academic Press, Inc., 1960.

FISH AND RIVER POLLUTION. J. R. Erichsen Jones. Butterworths, 1964.

A FIELD AND LABORATORY INVESTIGATION OF THE EFFECT OF HEATED EFFLUENTS ON FISH. J. S. Alabaster in Fishery Investigations, Ministry of Agriculture, Fisheries, and Food, Series I, Vol. 6, No. 4; 1966.

THERMAL POLLUTION—1968. Subcommittee on Air and Water Pollution of the Committee on Public Works. U.S. Government Printing Office, 1968.

30. Cooling Towers

COOLING TOWERS. A. R. Thompson in Chemical Engineering, Vol. 75, No. 22, pages 100–102; October 14, 1968.

COOLING TOWER FUNDAMENTALS AND APPLICATION PRINCIPLES. The Marley Company, 1969.

COOLING TOWERS FOR LARGE STEAM-ELECTRIC GENERATING UNITS: SYMPOSIUM ON THERMAL CONSIDERATIONS OF ELECTRIC POWER. Black and Veatch/Consulting Engineers, 1970.

31. The Climate of Cities

THE DONORA SMOG DISASTER—A PROBLEM IN ATMOSPHERIC POLLUTION. Robert D. Fletcher in Weatherwise, Vol. 2, No. 3, pages 56–60; June, 1949.

LOCAL CLIMATOLOGICAL STUDIES OF THE TEMPERATURE CONDITIONS IN AN URBAN AREA. A. Sundborg in Tellus, Vol. 2, No. 3, pages 221–231; August, 1950.

ON THE CAUSES OF INSTRUMENTALLY OBSERVED SECULAR TEMPERATURE TRENDS. J. Murray Mitchell, Jr., in Journal of Meteorology, Vol. 10, No. 4, pages 244–261; August, 1953.

INDEX